INTERNATIONAL UNION OF PURE AND APPLIED CHEMISTRY

ANALYTICAL CHEMISTRY DIVISION
COMMISSION ON SOLUBILITY DATA

SOLUBILITY DATA SERIES

Volume 39

CUMULATIVE INDEX
VOLUMES 20–38

SOLUBILITY DATA SERIES

Editor-in-Chief
J. W. LORIMER
University of Western Ontario
London, Ontario, Canada

H. L. Clever (USA)
Sub-editor
Gas/Liquid Systems

A. F. M. Barton (Australia)
Sub-editor
Liquid/Liquid Systems

M. Salomon (USA)
Sub-editor
Solid/Liquid Systems

C. L. Young (Australia)
Sub-editor
Indexes

CUMULATIVE EDITORIAL BOARD
Volumes 20–38

Managing Editor
P. D. GUJRAL
IUPAC Secretariat, Oxford, UK

INTERNATIONAL UNION OF PURE AND APPLIED CHEMISTRY

IUPAC Secretariat, Bank Court Chambers,
2-3 Pound Way, Cowley Centre, Oxford OX4 3YF, UK

SOLUBILITY DATA SERIES

Volume 39

CUMULATIVE INDEX
VOLUMES 20–38

Compiled and Edited by

COLIN L. YOUNG

Department of Chemistry, University of Melbourne,
Parkville, Victoria, Australia

PERGAMON PRESS

Member of Maxwell Macmillan Pergamon Publishing Corporation

OXFORD · NEW YORK · BEIJING · FRANKFURT
SÃO PAULO · SYDNEY · TOKYO · TORONTO

U.K.	Pergamon Press plc, Headington Hill Hall, Oxford OX3 0BW, England
U.S.A.	Pergamon Press, Inc., Maxwell House, Fairview Park, Elmsford, New York 10523, U.S.A.
PEOPLE'S REPUBLIC OF CHINA	Pergamon Press, Room 4037, Qianmen Hotel, Beijing, People's Republic of China
FEDERAL REPUBLIC OF GERMANY	Pergamon Press GmbH, Hammerweg 6, D-6242 Kronberg, Federal Republic of Germany
BRAZIL	Pergamon Editora Ltda, Rua Eça de Queiros, 346, CEP 04011, Paraiso, São Paulo, Brazil
AUSTRALIA	Pergamon Press Australia Pty Ltd., P.O. Box 544, Potts Point, N.S.W. 2011, Australia
JAPAN	Pergamon Press, 5th Floor, Matsuoka Central Building, 1-7-1 Nishishinjuku, Shinjuku-ku, Tokyo 160, Japan
CANADA	Pergamon Press Canada Ltd., Suite No. 271, 253 College Street, Toronto, Ontario, Canada M5T 1R5

First edition 1989

The Library of Congress has catalogued this serial title as follows:

Solubility data series.—Vol. 1—Oxford; New York; Pergamon, c 1979-
v.; 28 cm.
Separately catalogued and classified in LC before no. 18.
ISSN 0191-5622 = Solubility data series.
1. Solubility—Tables—Collected works.
QD543.S6629   541.3'42'05-dc19   85-641351
AACR 2   MARC-S

British Library Cataloguing in Publication Data
Young, Colin L.
Cumulative index: volumes 20–38.—(Solubility data series; v. 39).
1. Solubility. Tables. Bibliographies
I. Title   II. Series
016.5413'42'0212

ISBN 0–08–037205–8

A complete list of volumes published in the Solubility Data Series will be found on p. 371.
(For Cumulative Index for Volumes 1–18, see Volume 19)

Printed in Great Britain by BPCC Wheatons Ltd, Exeter

CONTENTS

Foreword vii

Preface ix

Cumulative System Index, Vols. 20-38 1

Cumulative Registry Number Index, Vols. 20-38 205

Cumulative Author Index, Vols. 20-38 257

Solubility Data Series: Published and Forthcoming Volumes 371

FOREWORD

*If the knowledge is
undigested or simply wrong,
more is not better.*

The Solubility Data Series is a project of Commission V.8 (Solubility Data) of the International Union of Pure and Applied Chemistry (IUPAC). The project had its origins in 1973, when the Analytical Chemistry Division of IUPAC set up a Subcommission on Solubility Data under the chairmanship of the late Prof. A.S. Kertes. When publication of the Solubility Data Series began in 1979, the Committee became a full commission of IUPAC, again under the chairmanship of Prof. Kertes, who also became *Editor-in-Chief* of the Series. The Series has as its goal the preparation of a comprehensive and critical compilation of data on solubilities in all physical systems, including gases, liquids and solids.

The motivation for the Series arose from the realization that, while solubility data are of importance in a wide range of fields in science and technology, the existing data had not been summarized in a form that was at the same time comprehensive and complete. Existing compilations of solubility data indeed existed, but they contained many errors, were in general uncritical, and were seriously out-of-date.

It was also realized that a new series of compilations of data gave educational opportunities, in that careful compilations of existing data could be used to demonstrate what constitutes data of high and lasting quality. As well, if the data were summarized in a sufficiently complete form, any individual could prepare his or her own evaluation, independently of the published evaluation. Thus, a special format was established for each volume, consisting of individual data sheets for each separate publication, and critical evaluations for each separate system, provided sufficient data from different sources were available for comparison. The compilations and, especially, the evaluations were to be prepared by active scientists who were either involved in producing new data, or were interested in using data of high quality. With minor modifications in format, this strategy has continued throughout the Series.

In the standard arrangement of each volume, the Critical Evaluation gives the following information:

(*i*) A text which discusses the numerical solubility information which has been abstracted from the primary sources in the form of compilation sheets. The text concerns primarily the quality of the data, after consideration of the purity of the materials and their characterization, the experimental method used, the uncertainties in the experimental values, the reproducibility, the agreement with accepted test values, and, finally, the fitting of the data to suitable functions, along with statistical tests of the fitted data.

(*ii*) A set of recommended data, whenever possible, including weighted averages and estimated standard deviations. If applicable, one or more smoothing equations which have been computed or verified by the evaluator are also given.

(*iii*) A graphical plot of the recommended data, in the form of phase diagrams where appropriate.

The Compilation part consists of data sheets which summarize the experimental data from the primary literature. Here much effort is put into obtaining complete coverage; many good data have appeared in publications from the late nineteenth and early twentieth centuries, or in obscure journals. Data of demonstrably low precision are not compiled, but are mentioned in the Critical Evaluation. Similarly, graphical data, given the uncertainty of accurate conversion to numerical values, are

compiled only where no better data are available. The documentation of
data of low precision can serve to alert researchers to areas where more
work is needed.

A typical data sheet contains the following information:

  (i) list of components: names, formulas, Chemical Abstracts
       Registry Numbers;
 (ii) primary source of the data;
(iii) experimental variables;
 (iv) compiler's name;
  (v) experimental values as they appear in the primary source, in
       modern units with explanations if appropriate;
 (vi) experimental methods used;
(vii) apparatus and procedure used;
(viii) source and purity of materials used;
 (ix) estimated error, either from the primary source or estimated
       by the compiler;
  (x) references relevant to the generation of the data cited in
       the primary source.

Each volume also contains a general introduction to the particular
type of system, such as solubility of gases, of solids in liquids, etc.,
which contains a discussion of the nomenclature used, the principles of
accurate determination of solubilities, and related thermodynamic principles. This general introduction is followed by a specific introduction
to the subject matter of the volume itself.

The Series embodies a new approach to the presentation of numerical
data, and the details continue to be influenced strongly by the perceived
needs of prospective users. The approach used will, it is hoped, en-
courage attention to the quality of new published work, as authors become
more aware that their work will attain permanence only if it meets the
standards set out in these volumes. If the Series succeeds in this
respect, even partially, the Solubility Data Commission will have jus-
tified the labour expended by many scientists throughout the world in its
production.

January, 1989   J.W. Lorimer,
                                        London, Canada

PREFACE

This CUMULATIVE INDEX covers volumes 20 to 38 of the Solubility Data Series. The first cumulative index, volume 19, published in 1985, covered the first eighteen volumes in the series. Both indexes have been produced in the hope that they will enable workers to locate data both for the system of interest and closely related systems without the necessity of consulting the indexes of individual volumes.

Further cumulative indexes will be issued, each covering volumes published in the preceding four or five years. Publication of a COMPREHENSIVE CUMULATIVE INDEX is planned at the end of the series of some 100 volumes. The editor would welcome comments on any aspect of the first and second cumulative indexes.

The index consists of three sections: a System Index, an Author Index and a Chemical Abstracts Registry Number Index. The system index is arranged in a similar manner to the system indexes in individual volumes. Page numbers preceded by E refer to evaluation text whereas those not preceded by E refer to compiled tables. In general, compounds are listed as in Chemical Abstracts: for example toluene is listed as Benzene, methyl-; and Dimethylsulfoxide is listed as Methane, sulfinylbis-. Gas-liquid systems are listed by solvent. Solid-liquid systems are listed under the solid but not usually under the solvent as well. Liquid-liquid systems appear under the non-aqueous component but not under water.

I am indebted to members of the Commission on Solubility Data and the editors of the individual volumes for suggestions and their help in checking parts of the indexes. The help of Lesley Flannigan and Georgia Christou in the production of the index is gratefully acknowledged.

Melbourne   Colin Young
March 1989

SYSTEM INDEX

Page numbers preceded by E refer to evaluation text whereas those not
preceded by E refer to compiled tables. In general compounds are
indexed as listed in Chemical Abstracts; for example toluene is
listed as Benzene, methyl- and dimethylsulfoxide is listed as Methane,
sulfinylbis-. Gas-liquid systems are listed by solvent. Solid-liquid
systems are listed under the solid but not usually under the solvent
as well. Liquid-liquid systems appear only in volumes 20, 37 and 38 and
are listed under the organic compound not under water.

A

A 306
      see benzenesulfonamide, 4-amino-N-2-pyrimidinyl-
A 307
      see benzenesulfonamide, 4-amino-N-[(aminoiminomethyl)-
A 310
      see benzenesulfonamide, 4-amino-N-(4-methyl-2-pyrimidinyl)-
A-319
      see acetamide, N-[4-(aminosulfonyl)phenyl]-
A 435
      see benzenesulfonamide, 4-amino-N-(aminocarbonyl)-
A-499
      see benzenesulfonamide, 4-amino-N-2-pyridinyl-
A-500
      see Acetamide, N-[(4-aminophenyl)sulfonyl]-
A 502
      see benzenesulfonamide, 4-amino-N-(4,6-dimethyl-2-pyrimidinyl)-
Abdominal muscle
      see rat abdominal muscle
Abiguanil
      see benzenesulfonamide, 4-amino-N-[(aminoiminomethyl)-
Acenaphthene
      + water   vol 38   E332 -E334,
                                    335 - 339
Acenaphthene
      + seawater   vol 38   340
Acetamide (aqueous)
      + bromic acid, potassium salt   vol 30   253
Acetamide, N-[[(4-acetylamino)phenyl]sulfonyl]-
      + water   vol 34   265
Acetamide, N-[[(4-acetylamino)phenyl]sulfonyl]- (aq)
      + calcium chloride   vol 34   269
      + ethanol   vol 34   270
      + magnesium chloride   vol 34   269
      + phosphoric acid, disodium salt
                      vol 34   265, 267, 268
      + phosphoric acid, monoammonium salt
                      vol 34   269
      + phosphoric acid, monopotassium salt
                      vol 34   266 - 268
      + potassium chloride   vol 34   269
      + sodium chloride   vol 34   269
      + urea   vol 34   269
Acetamide, N-[[4-(acetylamino)phenyl]sulfonyl]-N-(3,4-dimethyl-
5-isoxazolyl)-
      + water   vol 35   101
Acetamide, N-[[4-(acetylamino)phenyl]sulfonyl]-N-(3,4-dimethyl-
5-isoxazolyl)-, (aq)
      + phosphoric acid, disodium salt
                      vol 35   101
      + phosphoric acid,
        monopotassium salt   vol 35   101
Acetamide, N-[[4-(acetylamino)phenyl]sulfonyl]-N-(4,6-dimethyl-
2-pyrimidinyl)-
      + phosphoric acid, disodium salt
                      vol 36   363
      + phosphoric acid,
        monopotassium salt   vol 36   363
      + water   vol 36   363

```
Acetamide, N-[[4-(acetylamino)phenyl]sulfonyl]-N-(4-methoxy-
1,2,5-thiadiaxol-3-yl)-
                + water                           vol 35                          255
Acetamide, N-[[4-(acetylamino)phenyl]sulfonyl]-N-(4-methoxy-
1,2,5-thiadiaxol-3-yl)-, (aq)
                + phosphoric acid, disodium salt
                                                  vol 35                          255
                + phosphoric acid,
                  monopotassium salt              vol 35                          255
Acetamide, N-[4-[[(aminoiminomethyl)amino]sulfonyl]phenyl]-
                + ethanol                         vol 34                          321
                + water                           vol 34                 320,     321
Acetamide, N-[(4-aminophenyl)sulfonyl]-
                + cottonseed oil                  vol 34                 260,     261
                + petrolatum (white)              vol 34                 258,     259
                + poly(oxy-1,2-ethanediyl),
                  hydro- -hydroxy-                vol 34                          264
                + 2-propanone                     vol 34                          263
                + sorbitan, (Z)-9-octadecenoate
                                                  vol 34                 259,     261
                + trichloromethane                vol 34                          262
                + water                           vol 34        E204,     205  -  257
Acetamide, N-[(4-aminophenyl)sulfonyl]- (aq)
                + acetamide                       vol 34                          241
                + ammonium chloride              vol 34                          228
                + barium chloride                 vol 34                          231
                + butanamide                      vol 34                          244
                + calcium chloride                vol 34                 230,     238
                + N,N-diethyl-3-pyridinecarboxamide
                                                  vol 34                          249
                + N,N-dimethylacetamide           vol 34                          247
                + N,N-dimethylformamide           vol 34                          245
                + ethanethioamide                 vol 34                          242
                + ethanol                         vol 34                          239
                + formamide                       vol 34                          240
                + lithium chloride                vol 34                          215
                + magnesium chloride              vol 34                 229,     238
                + N-methylacetamide               vol 34                          246
                + phosphoric acid, disodium salt
                                                  vol 34        218,     233  -  237
                + phosphoric acid,
                  monoammonium salt               vol 34                          238
                + phosphoric acid,
                  monopotassium salt              vol 34        227,     233  -  236
                + phosphoric acid,
                  monosodium salt                 vol 34                          232
                + potassium bromide               vol 34                 221,     222
                + potassium chloride              vol 34                 219,     220,
                                                                         232,     238
                + potassium iodide                vol 34                 223,     224
                + poly(oxy-1,2-ethanediyl),
                  -hydro- -hydroxy-               vol 34                 250  -  253
                + propanamide                     vol 34                          243
                + 1,2,3-propanetricarboxylic acid,
                  2-hydroxy-                       vol 34                          237
                + 3-pyridinecarboxamide           vol 34                          248
                + sodium chloride                 vol 34        216,     217,     238
                + sorbitan monolaurate,
                  polyoxyethylene derivatives     vol 34                 254,     255
                + sorbitan monooleate,
                  polyoxyethylene derivatives     vol 34                          257
                + sorbitan monopalmitate,
                  polyoxyethylene derivatives     vol 34                          256
                + thiocyanic acid, potassium salt
                                                  vol 34                 225,     226
                + urea                            vol 34                          238
Acetamide, N-[4-[[[4-(aminophenyl)sulfonyl]amino]sulfonyl]-
phenyl]-
                + 2-propanone                     vol 36                           10
Acetamide, N-[(4-aminophenyl)sulfonyl]-N-(3,4-dimethyl-
5-isoxazolyl)-,(aq)
                + 1,2-benzenedicarboxylic acid,
                  monopotassium salt              vol 35                           95
                + phosphoric acid,
                  monopotassium salt              vol 35                           94
                + sodium hydroxide                vol 35                  94,       95
```

Acetamide, N-[(4-aminophenyl)sulfonyl]-N-(4-methoxy-
1,2,5-thiadiazol-3-yl)-, (aq)
         + phosphoric acid, disodium salt
                                        vol 35   253

         + phosphoric acid,
           monopotassium salt   vol 35   253
         + water   vol 35   253
Acetamide, N-[(4-aminophenyl)sulfonyl]-N-(5-methyl-
3-isoxazolyl)-,(aq)
         + phosphoric acid, disodium salt
                                          vol 35   62,   63

         + phosphoric acid,
           monopotassium salt   vol 35   62,   63
         + sodium chloride   vol 35   61
         + water   vol 35   61 - 63
Acetamide, N-[4-(aminosulfonyl)phenyl]-
         + 2-propanone   vol 34   192
         + water   vol 34   181 - 191
         + water   vol 36   450
Acetamide, N-[4-(aminosulfonyl)phenyl]- (aq)
         + benzenesulfonamide, 4-amino- vol 36   450
         + N,N-dimethylurea   vol 34   189
         + N,N´-dimethylurea   vol 34   188
         + methylurea   vol 34   187
         + phosphoric acid, disodium salt
                                          vol 34   184, 185

         + phosphoric acid,
           monpotassium salt   vol 34   183, 185
         + tetramethylurea   vol 34   190
         + thiourea   vol 34   191
         + urea   vol 34   186
Acetamide, N-[4-[[(5,6-dimethoxy-4-pyrimidinyl)amino]sulfonyl]-
phenyl]-
         + water   vol 36   435
Acetamide, N-[4-[[(5,6-dimethoxy-4-pyrimidinyl)amino]sulfonyl]-
phenyl]-, (aq)
         + phosphoric acid, disodium salt
                                          vol 36   435

         + phosphoric acid,
           monopotassium salt   vol 36   435
Acetamide, N,N-dimethyl-
         + hydrogen sulfide   vol 32   E180, E181,
                                                    E183, 299
Acetamide, N-[4-[[(3,4-dimethyl-5-isoxazolyl)-amino]sulfonyl]-
phenyl]-
         + trichloromethane   vol 35   100
         + water   vol 35   96 - 99
Acetamide, N-[4-[[(3,4-dimethyl-5-isoxazolyl)-amino]sulfonyl]-
phenyl]-,(aq)
         + calcium chloride   vol 35   99
         + 2-hydroxy-1,2,3-
           propanetricarboxylic acid   vol 35   98
         + magnesium chloride   vol 35   99
         + phosphoric acid,
           monoammonium salt   vol 35   99
         + phosphoric acid, disodium salt
                                        vol 35   96 - 98

         + phosphoric acid,
           monopotassium salt   vol 35   96,   97
         + potassium chloride   vol 35   99
         + sodium chloride   vol 35   99
         + urea   vol 35   99
Acetamide, N-[4-[[(2,6-dimethyl-4-pyrimidinyl)amino]sulfonyl]-
phenyl]-
         + phosphoric acid disodium salt
                                        vol 36   403

         + phosphoric acid
           monopotassium salt   vol 36   403
         + water   vol 36   403
Acetamide, N-[4-[[(4,6-dimethyl-2-pyrimidinyl)amino]sulfonyl]-
phenyl]-
         + water   vol 36   358, 362, 477,
                                                             480 - 482

Acetamide, N-[4-[[(4,6-dimethyl-2-pyrimidinyl)amino]sulfonyl]-
phenyl]- (aq)
            + calcium chloride   vol 36   481, 482
            + 2-hydroxy-1,2,3-
              propanetricarboxylic acid   vol 36   480
            + 2-hydroxy-1,2,3-propanetricarboxylic acid,
              disodium salt   vol 36   362
            + magnesium chloride   vol 36   481, 482
            + N-4-[[(4-methyl-2-pyrimidinyl)-
              amino]sulfonyl]phenyl]acetamide
                                            vol 36   477, 480 - 482
            + phosphoric acid, disodium salt
                                            vol 36   359 - 361,
                                                                477, 480
            + phosphoric acid,
              monoammonium salt   vol 36   481, 482
            + phosphoric acid,
              monopotassium salt   vol 36   359 - 361, 477
            + potassium chloride   vol 36   481, 482
            + N-[4-[(2-pyrimidinylamino)-
              sulfonyl]phenyl]acetamide   vol 36   477, 480 - 482
            + sodium chloride   vol 36   481, 482
            + sodium hydroxide   vol 36   358, 362
            + urea   vol 36   481, 482
Acetamide, N-[4-[[(5-ethyl-1,3,4-thiadiazol-2-yl)amino]sulfonyl]-
phenyl]-
            + water   vol 35   302 - 309
Acetamide, N-[4-[[(5-ethyl-1,3,4-thiadiazol-2-yl)amino]sulfonyl]-
phenyl]-, (aq)
            + calcium chloride   vol 35   309
            + magnesium chloride   vol 35   309
            + phosphoric acid, disodium salt
                                            vol 35   304, 306 - 308
            + phosphoric acid,
              monoammonium salt   vol 35   309
            + phosphoric acid,
              monopotassium salt   vol 35   305 - 308
            + potassium chloride   vol 35   309
            + sodium chloride   vol 35   309
            + urea   vol 35   309
Acetamide, N-[4-[[5-ethyl-1,3,4-thiadiazol-2-yl)amino]sulfonyl]-
phenyl]-, (aq)
            + 4-amino-N-(5-ethyl-1,3,4-thiadiazol-2-yl)-
              benzenesulfonamide   vol 36   452
            + water   vol 36   452
Acetamide, N-[4-[[(3-methoxypyrazinyl)amino]sulfonyl]phenyl]-
            + 2-hydroxy-1,2,3-
              propanetricarboxylic acid   vol 36   442
            + phosphoric acid, disodium salt
                                            vol 36   442
            + water   vol 36   442
Acetamide, N-[4-[[(6-methoxy-2-pyridazinyl)amino]sulfonyl]-
phenyl]-
            + water   vol 36   120, 121
Acetamide, N-[4-[[(6-methoxy-2-pyridazinyl)amino]sulfonyl]-
phenyl]-, (aq)
            + calcium chloride   vol 36   121
            + magnesium chloride   vol 36   121
            + phosphoric acid, disodium salt
                                            vol 36   120
            + phosphoric acid,
              monoammonium salt   vol 36   121
            + phosphoric acid,
              monopotassium salt   vol 36   120
            + potassium chloride   vol 36   121
            + sodium chloride   vol 36   121
            + urea   vol 36   121
Acetamide, N-[4-[[(4-methoxy-1,2,5-thiadiazol-3-yl)amino]sulfonyl]-
phenyl]-, (aq)
            + phosphoric acid, disodium salt
                                            vol 35   254
            + phosphoric acid,
              monopotassium salt   vol 35   254
            + water   vol 35   254

```
Acetamide,N-methyl-
          + bromic acid, potassium salt    vol 30                    256
          + bromic acid, sodium salt       vol 30                    218
Acetamide, N-[4-[[[4-[(methylamino)sulfonyl]phenyl]amino]sulfonyl]-
phenyl]-
          + water                          vol 36            15 -   17
Acetamide, N-[4-[[[4-[(methylamino)sulfonyl]phenyl]amino]sulfonyl]-
phenyl]-, (aq)
          + phosphoric acid, disodium salt
                                           vol 36            15,    17
          + phosphoric acid,
            monopotassium salt             vol 36            16,    17
Acetamide, N-[4-[[(5-methyl-3-isoxaxolyl)amino]sulfonyl]-
phenyl]-
          + water                          vol 35                    63
Acetamide, N-[4-[[(5-methyl-3-isoxaxolyl)amino]sulfonyl]-
phenyl]-, (aq)
          + phosphoric acid, disodium salt
                                           vol 35                    63
          + phosphoric acid,
            monopotassium salt             vol 35                    63
Acetamide, N-4-[[(4-methyl-2-pyrimidinyl)amino]sulfonyl]-
phenyl]-
          + water                          vol 36      E269,  270 - 278,
                                                       464,   465,   473,
                                                       480 - 482,   485
Acetamide, N-4-[[(4-methyl-2-pyrimidinyl)amino]sulfonyl]-
phenyl]-, (aq)
          + 4-amino-N-(4-methyl-2-pyrimidinyl)-
            benzenesulfonamide             vol 36                    485
          + 4-amino-N-2-pyrimidinyl-
            benzenesulfonamide             vol 36                    485
          + 4-amino-N-2-thiazolyl-
            benzenesulfonamide             vol 36                    485
          + calcium chloride               vol 36            481,   482
          + N-[4-[[(4,6-dimethyl-2-pyrimidinyl)-
            amino]sulfonyl]phenyl]acetamide
                                           vol 36      477,   480 - 482
          + 2-hydroxy-1,2,3-
            propanetricarboxylic acid      vol 36                    480
          + 2-hydroxy-1,2,3-
            propanetricarboxylic acid,
            disodium salt                  vol 36                    273
          + mannitol                       vol 36            277,   278
          + magnesium chloride             vol 36            481,   482
          + methylbenzene                  vol 36            277,   278
          + phosphoric acid, disodium salt
                                           vol 36      274 - 278,   464,
                                                       465,   473,   477,
                                                       480 - 482,   485
          + phosphoric acid,
            monoammonium salt              vol 36            481,   482
          + phosphoric acid,
            monopotassium salt             vol 36            274 - 278,
                                                             464,   465,
                                                       473,   477,   485
          + potassium chloride             vol 36            481,   482
          + N-[4-[(2-pyrimidinylamino)-
            sulfonyl]phenyl]acetamide      vol 36      465,   473,   477,
                                                       480 - 482,   485
          + sodium chloride                vol 36            277,   278,
                                                             481,   482
          + sodium hydroxide               vol 36                    273
          + N-[4-[(2-thiazolylamino)-
            sulfonyl]phenyl]acetamide      vol 36      464,   473,   485
          + trichloromethane               vol 36                    278
          + urea                           vol 36            481,   482
Acetamide, N-[4-[[(5-methyl-1,3,4-thiadiazol-2-yl)amino]sulfonyl]-phenyl]-
          + water                          vol 35      282 - 288
Acetamide, N-[4-[[(5-methyl-1,3,4-thiadiazol-2-yl)amino]sulfonyl]-
phenyl]-, (aq)
          + calcium chloride               vol 35                    288
```

```
Acetamide, N-[4-[[(5-methyl-1,3,4-thiadiazol-2-yl)amino]sulfonyl]-
phenyl]-, (aq)
                + magnesium chloride              vol 35                        288
                + phosphoric acid, disodium salt
                                                  vol 35           283,  285 - 287
                + phosphoric acid,
                  monoammonium salt               vol 35                        288
                + phosphoric acid,
                  monopotassium salt              vol 35                  284 - 287
                + potassium chloride              vol 35                        288
                + sodium chloride                 vol 35                        288
                + urea                            vol 35                        288
Acetamide, N-[4-[[(4-methyl)-2-thiazolylamino]sulfonyl]phenyl]-
                + water                           vol 35          E239,  240 - 245
Acetamide, N-[4-[[(4-methyl)-2-thiazolylamino]sulfonyl]phenyl]-,(aq)
                + phosphoric acid, disodium salt
                                                  vol 35                  243,  245
                + phosphoric acid,
                  monopotassium salt              vol 35                  244,  245
Acetamide, N-phenyl- (aqueous)
                + iodic acid                      vol 30                        489
Acetamide, N-[4-[(2-pyridinylamino)sulfonyl]phenyl]-
                + water                           vol 36          E63,   64 -  75
Acetamide, N-[4-[(2-pyridinylamino)sulfonyl]phenyl]-, (aq)
                + phosphoric acid, disodium salt
                                                  vol 36          71,    73 -  75
                + phosphoric acid,
                  monopotassium salt              vol 36                  72 -  75
                + sodium hydroxide                vol 36                         70
Acetamide, N-[4-[(2-pyrimidinylamino)sulfonyl]phenyl]-
                + water
                                                  vol 36          E228,  229 - 242,
                                                                 462,   463,  465,
                                                                 473,   480 - 482,
                                                                                485
Acetamide, N-[4-[(2-pyrimidinylamino)sulfonyl]phenyl]- (aq)
                + 4-amino-N-(4-methyl-2-pyrimidinyl)-
                  benzenesulfonamide              vol 36          481,   482,  485
                + 4-amino-N-2-pyridinyl-
                  benzenesulfonamide              vol 36          462,   475,  485
                + 4-amino-N-2-thiazolyl-
                  benzenesulfonamide              vol 36                        485
                + calcium chloride                vol 36                 481,   482
                + N-[4-[[(4,6-dimethyl-2-pyrimidinyl)-
                  amino]sulfonyl]phenyl]acetamide
                                                  vol 36          477,   480 - 482
                + 2-hydroxy-1,2,3-propane-
                  tricarboxylic acid              vol 36                        480
                + 2-hydroxy-1,2,3-propane-
                  tricarboxylic acid,
                  disodium salt                   vol 36                 233,   234
                + magnesium chloride              vol 36                 481,   482
                + N-[4-[[(4-methyl-2-pyrimidinyl)-
                  amino]sulfonyl]phenyl]acetamide
                                                  vol 36          465,   473,  477,
                                                                 480 - 482,    485
                + phosphoric acid, disodium salt
                                                  vol 36          236 - 242,    463,
                                                                 465,   473,  477,
                                                                        480,   485
                + phosphoric acid,
                  monoammonium salt               vol 36                 481,   482
                + phosphoric acid,
                  monopotassium salt              vol 36          235,   237 - 242,
                                                                 463,   465,  473,
                                                                        477,   485
                + potassium chloride              vol 36                 481,   482
                + sodium chloride                 vol 36                 481,   482
                + sodium hydroxide                vol 36                 232 - 234
                + N-[4-[(2-thiazolylamino)-
                  sulfonyl]phenyl]acetamide       vol 36          463,   473,  485
                + urea                            vol 36                 481,   482
Acetamide, N-[4-[(2-thiazolylamino)sulfonyl]phenyl]-
                + water                           vol 36                 451,   463,
                                                                        464,   473
```

```
Acetamide, N-[4-[(2-thiazolylamino)sulfonyl]phenyl]-, (aq)
          + 4-amino-N-(4-methyl-2-pyrimidinyl)-
            benzenesulfonamide              vol 36                            485
          + 4-amino-N-2-pyrimidinyl-
            benzenesulfonamide              vol 36                            485
          + 4-amino-N-2-thiazolyl-
            benzenesulfonamide              vol 36                   451,     485
          + N-[4-[[(4-methyl-2-pyrimidinyl)-
            amino]sulfonyl]phenyl]acetamide
                                            vol 36         464,     473,     485
          + phosphoric acid, disodium salt
                                            vol 35         223,     225 -    229
                                            vol 36         463,     464,     473
          + phosphoric acid, monopotassium salt
                                            vol 35                   224 -    229
                                            vol 36         463,     464,     473
          + N-[4-[(2-pyrimidinylamino)-
            sulfonyl]phenyl]acetamide       vol 36         463,     473,     485
          + sodium hydroxide               vol 35                            222
          + water                          vol 35                   219 -    229
2-(p-Acetamidobenzenesulfonamido)thiazole
          see acetamide, N-[4-[(2-thiazolylamino)sulfonyl]phenyl]-
p-Acetamidobenzenesulfonamide
          see acetamide, N-[4-(aminosulfonyl)phenyl]-
4-Acetamidobenzenesulfonamide
          see acetamide, N-[4-(aminosulfonyl)phenyl]-
N-(4-Acetamidobenzenesulfonyl)acetamide
          see acetamide, N-[[(4-acetylamino)phenyl]sulfonyl]-
6-Acetamido-4-hydroxy-3-[(p-sulfamoylphenyl)azo]-2,7-naphthalenedisulfonic
acid, disodium salt
          see 2,7-naphthalenedisulfonic acid, 6-acetylamino-3-[[4-
             (aminosulfonyl)phenyl]azo]-4-hydroxy-, disodium salt
2-[(p-Acetamidophenyl)sulfonamido]pyrimidine
          see acetamide, N-[4-[(2-pyrimidinylamino)sulfonyl]phenyl]-
Acetate,butyl
          see acetic acid,butyl ester
Acetate,ethyl
          see acetic acid,ethyl ester
Acetic acid
          + hydrogen sulfide               vol 32                   E176,    270
          + methane                        vol 27/28                         580
          + phosphine                      vol 21          E281,   E282,     290
          + propane                        vol 24          E375   -E378,     379
Acetic acid (aqueous)
          + bromic acid, potassium salt    vol 30                            248
          + selenious acid, copper (ll) salt
                                            vol 26                            379
Acetic acid (aqueous and multicomponent)
          + nitric acid, sodium salt       vol 26                   301,     302
          + sulfurous acid, calcium salt   vol 26                  E191,    E192,
                                                                    232,     233
          + sulfurous acid, lead salt      vol 26                   301,     302
Acetic acid (multi component)
          + mercury (II) oxide             vol 23                            339
Acetic acid, ammonium salt (aqueous)
          + sulfurous acid, lead salt      vol 26                   299,     300
Acetic acid, anhydride
          + butane                         vol 24          E375   -E378,     397
          + hydrogen sulfide               vol 32                   E176,    271
          + propane                        vol 24          E375   -E378,     383
Acetic acid, cadmium salt
          + acetic acid, caesium salt      vol 33                             25
          + acetic acid, potassium salt    vol 33           E26,     27 -     30
          + acetic acid, rubidium salt     vol 33                             34
          + acetic acid, sodium salt       vol 33           E31,     32,      33
Acetic acid, cesium salt
          + acetic acid, cadmium salt      vol 33                             25
          + acetic acid, lithium salt      vol 33           E40,     41,      42
          + acetic acid, potassium salt    vol 33           E35,     36 -     39
          + acetic acid, rubidium salt     vol 33           E46,     47,      48
          + acetic acid, sodium salt       vol 33           E43,     44,      45
          + acetic acid, zinc salt         vol 33                             49
```

```
Acetic acid, cesium salt
            + nitric acid, cesium salt        vol 33      E134,  134,   135
            + nitrous acid, caesium salt      vol 33                    133
Acetic acid, dichloro-
            + butane                          vol 24      E257, E258,   272
Acetic acid, 2,2-dichloro-
            + hydrogen sulfide                vol 32      E178, E179,   282
Acetic acid, ethyl ester
            + N,N-dimethylmethanamine         vol 21            E85,    199
            + hydrogen sulfide                vol 32                    199
Acetic acid, lead salt
            + acetic acid, potassium salt     vol 33                     72
            + acetic acid, zinc salt          vol 33                     95
Acetic acid, lithium salt
            + acetic acid, cesium salt        vol 33      E40,   41,    42
            + acetic acid, potassium salt     vol 33      E50,   51 -   55
            + acetic acid, rubidium salt      vol 33      E81,   82,    83
            + acetic acid, sodium salt        vol 33      E77,   78 -   80
            + acetic acid, zinc salt          vol 33                     84
            + formic acid, lithium salt       vol 33                    206
            + nitric acid, lithium salt       vol 33      E210,  211 - 213
            + thiocyanic acid, lithium salt
                                              vol 33                    209
Acetic acid, magnesium salt
            + acetic acid, potassium salt     vol 33                     56
            + acetic acid, sodium salt        vol 33      E85,   86,    87
Acetic acid, methyl ester
            + methane                         vol 27/28                662
Acetic acid, methyl ester (aqueous)
            + bromic acid, potassium salt     vol 30                    250
Acetic acid, methyl ester (multicomponent)
            + copper (I) oxide                vol 23                      9
            + copper (II) oxide               vol 23      E10 - E26,    82
Acetic acid, octyl ester
            + ammonia                         vol 21            E8,     60
Acetic acid, potassium salt
            + acetic acid, cadmium salt       vol 33      E26,   27 -   30
            + acetic acid, cesium salt        vol 33      E35,   36 -   39
            + acetic acid, lead salt          vol 33                     72
            + acetic acid, lithium salt       vol 33      E50,   51 -   55
            + acetic acid, magnesium salt     vol 33                     56
            + acetic acid, rubidium salt      vol 33      E73,   74,    75
            + acetic acid, sodium salt        vol 33            E57 - E60,
                                                                 61 -   71
            + acetic acid, zinc salt          vol 33                     76
            + butanoic acid, 2-methyl-,
              potassium salt                  vol 33            E161,   162
            + butanoic acid, potassium salt   vol 33      E154,  155,   156
            + formic acid, potassium salt     vol 33                    139
            + hexanoic acid, potassium salt   vol 33            E163,   164
            + nitric acid, potassium salt     vol 33      E175,  176 - 179
            + nitrous acid, potassium salt    vol 33      E172,  173,   174
            + pentanoic acid, potassium salt
                                              vol 33            E159,   160
            + potassium bromide               vol 33            137,   138
            + potassium chloride              vol 33            165,   166
            + potassium iodide                vol 33            170,   171
            + propanoic acid, 2-methyl-,
              potassium salt                  vol 33            E157,   158
            + propanoic acid, potassium salt
                                              vol 33                    153
            + thiocyanic acid, potassium salt
                                              vol 33      E167,  168,   169
Acetic acid, potassium salt (aqueous)
            + phosphoric acid,
              monopotassium salt              vol 31                    221
Acetic acid, rubidium salt
            + acetic acid, cadmium salt       vol 33                     34
            + acetic acid, cesium salt        vol 33      E46,   47,    48
            + acetic acid, lithium salt       vol 33      E81,   82,    83
            + acetic acid, potassium salt     vol 33      E73,   74,    75
            + acetic acid, sodium salt        vol 33      E88,   89,    90
            + acetic acid, zinc salt          vol 33                     96
            + nitric acid, rubidium salt      vol 33      E332,  333 - 335
```

```
Acetic acid, sodium salt
          + acetic acid, cadmium salt      vol 33        E31,    32,    33
          + acetic acid, caesium salt      vol 33        E43,    44,    45
          + acetic acid, lithium salt      vol 33        E77,    78  -  80
          + acetic acid, magnesium salt    vol 33        E85,    86,    87
          + acetic acid, potassium salt    vol 33                E57  - E60,
                                                                 61  -  71
          + acetic acid, rubidium salt     vol 33        E88,    89,    90
          + acetic acid, zinc salt         vol 33        E91,    92  -  94
          + benzoic acid, sodium salt      vol 33                       256
          + butanoic acid, 2-methyl-,
            sodium salt                    vol 33        E250,  251,   252
          + butanoic acid, sodium salt     vol 33        E243,  244  - 246
          + formic acid, sodium salt       vol 33                       221
          + hexanoic acid, sodium salt     vol 33        E253,  254,   255
          + nitric acid, sodium salt       vol 33        E267,  268  - 272
          + nitrous acid, sodium salt      vol 33               264  - 266
          + pentanoic acid, sodium salt    vol 33                E248,  249
          + propanoic acid, 2-methyl-,
            sodium salt
                                           vol 33                       247
          + propanoic acid, sodium salt    vol 33                       242
          + sodium bromide                 vol 33               219,   220
          + sodium chloride                vol 33               257,   258
          + sodium iodide                  vol 33               262,   263
          + thiocyanic acid, sodium salt   vol 33               259  - 261
Acetic acid, sodium salt (aqueous and multicomponent)
          + formic acid, sodium salt       vol 26        E191, E192,   234
          + nitric acid, sodium salt       vol 26               301,   302
          + phosphoric acid, sodium salt   vol 26        E191, E192,   234
          + sulfurous acid, calcium salt   vol 26        E191, E192,   234
          + sulfurous acid, lead salt      vol 26               301,   302
Acetic acid, sodium salt (multicomponent)
          + mercury (II) oxide             vol 23                       339
Acetic acid, trifluoro-
          + phosphine                      vol 21               E281, E282,
                                                                 290,   301
Acetic acid, zinc salt
          + acetic acid, cesium salt       vol 33                        49
          + acetic acid, lead salt         vol 33                        95
          + acetic acid, lithium salt      vol 33                        84
          + acetic acid, potassium salt    vol 33                        76
          + acetic acid, rubidium salt     vol 33                        96
          + acetic acid, sodium salt       vol 33        E91,    92  -  94
Acetocid
          see acetamide, N-[(4-aminophenyl)sulfonyl]-
Acetic anhydride
          see acetic acid, anhydride
Acetone
          see 2-propanone
Acetonitrile
          + arsine                         vol 21               E302,   303
          + N,N-dimethylmethanamine        vol 21               E86,    207
          + hydrogen sulfide               vol 32               E180 -E182,
                                                                 298,   299
Acetonitrile (aqueous)
          + copper (I) oxide               vol 23                         9
          + copper (II) oxide              vol 23        E10  - E26,     82
Acetophenone
          see ethanone, 1-phenyl-
Acetosulfamide
          see acetamide, N-[(4-aminophenyl)sulfonyl]-
Acetosulfamin
          see acetamide, N-[(4-aminophenyl)sulfonyl]-
Acetothioamide
          see ethanethioamide
6-Acetylamino-3-[[4-(aminosulfonyl)phenyl]azo]-4-hydroxy-
2,7-naphthalenedisulfonic acid, di-sodium salt
          see 2,7-naphthalenedisulfonic acid, 6-acetylamino-
               3-[[4-(aminosulfonyl)phenyl]azo]-4-hydroxy-, disodium salt
4-(Acetylamino)benzenesulfonamide
          see acetamide, N-[4-(aminosulfonyl)phenyl]-
4-Acetylaminobenzenesulfonamide
          see acetamide, N-[4-(aminosulfonyl)phenyl]-
```

N-[[(4-Acetylamino)phenyl]sulfonyl]acetamide
     see acetamide, N-[[(4-acetylamino)phenyl]sulfonyl]-
N-[[(4-Acetylamino)phenyl]sulfonyl]-3,4-dimethylbenzamide
     see benzamide, N-[[4-(acetylamino)phenyl]sulfonyl]-3,4-dimethyl-
N-[[4-(Acetylamino)phenyl]sulfonyl]-N-(3,4-dimethyl-5-isoxazolyl)acetamide
     see acetamide, N-[[4-(acetylamino)phenyl]sulfonyl]-
       N-(3,4-dimethyl-5-isoxazolyl)-
N-[[4-(Acetylamino)phenyl]sulfonyl]-N-(4,6-dimethyl-2-pyrimidinyl)-
acetamide
     see acetamide, N-[[4-(acetylamino)phenyl]sulfonyl]-N-
       (4,6-dimethyl-2-pyrimidinyl)-
N-[[4-(Acetylamino)phenyl]sulfonyl]-N-(4-methoxy-1,2,5-thiadiaxol-3-yl)
acetamide
     see acetamide, N-[[4-(acetylamino)phenyl]sulfonyl]-
       N-(4-methoxy-1,2,5-thiadiaxol-3-yl)-
N-[[4-(Acetylamino)phenyl]sulfonyl]-3-methyl-2-butenamide
     see 2-butenamide, N-[[4-(Acetylamino)phenyl]sulfonyl]-3-methyl-
Acetylcholine chloride
     see ethanaminium, 2-(acetyloxy)-N,N,N-trimethyl-, chloride
N1-Acetyl-N1-(3,4-dimethyl-5-isoxazolylsulfanilamide
     see acetamide, N-[(4-aminophenyl)sulfonyl]-
       N-(3,4-dimethyl-5-isoxazolyl)-
Acetyl disulfanilamide
     see acetamide, N-[4-[[[4-(aminophenyl)sulfonyl]amino]sulfonyl]-
       phenyl]-
Acetylene
     see ethyne
Acetylene tetrachloride
     see ethane, 1,1,2,2-tetrachloro-
Acetylgantrisin
     see acetamide, N-[(4-aminophenyl)sulfonyl]-
       N-(3,4-dimethyl-5-isoxazolyl)-
Acetyl irgafene
     see benzamide, N-[[4-(acetylamino)phenyl]sulfonyl]-3,4-dimethyl-
Acetyl irgamide
     see 2-Butenamide, N-[[4-(Acetylamino)phenyl]sulfonyl]-3-methy-
Acetyl-Neo-uliron
     see acetamide, N-[4-[[[4-[(Methylamino)sulfonyl]phenyl]amino]-
       sulfonyl]-phenyl]-
Acetylmidicel
     see acetamide, N-[4-[[(6-methoxy-2-pyridazinyl)amino]sulfonyl]-
       phenyl]-
Acetylsulfacetamide
     see acetamide, N-[[(4-acetylamino)phenyl]sulfonyl]-
Acetyl sulfadiazine
     see acetamide, N-[4-[(2-pyrimidinylamino)sulfonyl]phenyl]-
N4-Acetylsulfadiazine
     see acetamide, N-[4-[(2-pyrimidinylamino)sulfonyl]phenyl]-
N4-Acetylsulfadimidine
     see acetamide, N-[4-[[(4,6-dimethyl-2-pyrimidinyl)amino]-
       sulfonyl]-phenyl]-
N4-Acetylsulfadoxine
     see acetamide, N-[4-[[(5,6-dimethoxy-4-pyrimidinyl)amino]-
       sulfonyl]phenyl]-
Acetylsulfadimethylisoxazole
     see acetamide, N-[4-[[(3,4-dimethyl-5-isoxazolyl)-amino]sulfonyl]-
       phenyl]-
Acetyl sulfaethylthiadiazole
     see acetamide, N-[4-[[(5-ethyl-1,3,4-thiadiazol-2-yl)amino]-
       sulfonyl]-phenyl]-
N4-Acetylsulfafurazole
     see acetamide, N-[4-[[(3,4-dimethyl-5-isoxazolyl)-amino]sulfonyl]-
       phenyl]-
N4-Acetylsulfaguanidine
     see acetamide, N-[4-[[(aminoiminomethyl)amino]sulfonyl]phenyl]-
4N-Acetylsulfaisodimidine
     see acetamide, N-[4-[[(2,6-dimethyl-4-pyrimidinyl)amino]-
       sulfonyl]phenyl]-
Acetyl sulfamerazine
     see acetamide, N-4-[[(4-methyl-2-pyrimidinyl)amino]sulfonyl]-
       phenyl]-
N4-Acetylsulfamerazine
     see acetamide, N-4-[[(4-methyl-2-pyrimidinyl)amino]sulfonyl]-
       phenyl]-

```
Acetylsulfamethazine
         see acetamide, N-[4-[[(4,6-dimethyl-2-pyrimidinyl)amino]sulfonyl]-
             phenyl]-
N4-Acetylsulfamethizole
         see acetamide. N-[4-[[(5-methyl-1,3,4-thiadiazol-2-yl)amino]-
             sulfonyl]-phenyl]-
Acetylsulfamethoxazole
         see acetamide, N-[4-[[(5-methyl-3-isoxaxolyl)amino]sulfonyl]phenyl]
N4-Acetylsulfamethoxazole
         see acetamide, N-[4-[[(5-methyl-3-isoxaxolyl)amino]-
             sulfonyl]phenyl]-
Acetylsulfamethoxypyrazine
         see acetamide, N-[4-[[(3-methoxypyrazinyl)amino]sulfonyl]phenyl]-
Acetylsulfamethoxypyridazine
         see acetamide, N-[4-[[(6-methoxy-2-pyridazinyl)amino]sulfonyl]-
             phenyl]-
N1-Acetyl sulfametrole
         see acetamide, N-[(4-aminophenyl)sulfonyl]-N-(4-methoxy-
             1,2,5-thiadiazol-3-yl)-
N4-Acetyl sulfametrole
         see acetamide, N-[4-[[(4-methoxy-1,2,5-thiadiazol-3-yl)amino]-
             sulfonyl]-phenyl]-
4´-(Acetylsulfamoyl)acetanilide
         see acetamide, N-[[(4-acetylamino)phenyl]sulfonyl]-
4´-Acetyl-3-sulfa-5-methylisoxazole
         see acetamide, N-[4-[[(5-methyl-3-isoxaxolyl)amino]sulfonyl]phenyl]
Acetyl sulfamethylthiadiazole
         see acetamide, N-[4-[[(5-methyl-1,3,4-thiadiazol-2-yl)amino]-
             sulfonyl]-phenyl]-
Acetyl sulfamethylthiazole
         see acetamide, N-[4-[[(4-methyl)-2-thiazolylamino]sulfonyl]phenyl]-
Acetylsulfanilamide
         see acetamide, N-[4-(aminosulfonyl)phenyl]-
N1-Acetylsulfanilamide
         see acetamide, N-[(4-aminophenyl)sulfonyl]-
N´-Acetylsulfanilamide
         see acetamide, N-[(4-aminophenyl)sulfonyl]-
N4-Acetylsulfanilamide
         see acetamide, N-[4-(aminosulfonyl)phenyl]-
N-4-Acetylsulfanilamide
         see acetamide, N-[4-(aminosulfonyl)phenyl]-
3-(N1-Acetylsulfanilamido)-6-methoxypyridazine
         see acetamide, N-[4-[[(6-methoxy-2-pyridazinyl)amino]sulfonyl]-
             phenyl]-
Acetylsulfapyridine
         see acetamide, N-[4-[(2-pyridinylamino)sulfonyl]phenyl]-
N4-Acetylsulfapyridine
         see acetamide, N-[4-[(2-pyridinylamino)sulfonyl]phenyl]-
Acetyl sulfapyrimidine
         see acetamide, N-[4-[(2-pyrimidinylamino)sulfonyl]phenyl]-
Acetylsulfathiazole
         see acetamide, N-[4-[(2-thiazolylamino)sulfonyl]phenyl]-
N4-Acetylsulfismezole
         see acetamide, N-[4-[[(5-methyl-3-isoxaxolyl)amino]sulfonyl]phenyl]
N4-Acetylsulfisomidine
         see acetamide, N-[4-[[(2,6-dimethyl-4-pyrimidinyl)amino]sulfonyl]-
             phenyl]-
N1-Acetylsulfisoxazole
         see acetamide, N-[(4-aminophenyl)sulfonyl]-
             N-(3,4-dimethyl-5-isoxazolyl)-
N´-Acetylsulfisoxazole
         see acetamide, N-[(4-aminophenyl)sulfonyl]-N-(3,4-dimethyl-
             5-isoxazolyl)-
4-N-Acetyl-sulfaisoxazole
         see acetamide, N-[4-[[(3,4-dimethyl-5-isoxazolyl)-amino]sulfonyl]-
             phenyl]-
N-Acetylsulfisoxazole
         see acetamide, N-[4-[[(3,4-dimethyl-5-isoxazolyl)-amino]sulfonyl]-
             phenyl]-
N4-Acetylsulfisoxazole
         see acetamide, N-[4-[[(3,4-dimethyl-5-isoxazolyl)-amino]sulfonyl]-
             phenyl]-
N´-Acetylsulphanilamide
         see acetamide, N-[4-(aminosulfonyl)phenyl-
```

```
Acids
        see under individual acids
Acrylic acid
        see 2-propenoic acid
Active amyl alcohol
        see 1-propanol, 2,2-dimethyl-
Actinium
                + mercury                       vol 25                    E421
Adiazin
        see benzenesulfonamide, 4-amino-N-2-pyrimidinyl-
Adiazine
        see benzenesulfonamide, 4-amino-N-2-pyrimidinyl-
Adiplon
        see benzenesulfonamide, 4-amino-N-2-pyridinyl-
Adiponitrile
        see hexanedinitrile
ADP
        see adenosine, 5-(trihydrogen phosphate)
Adrenals
        see guinea pig adrenals
Aethazol
        see benzenesulfonamide, 4-amino-N-(5-methyl-
            1,3,4-thiadiazol-2-yl)-
Ag succinylsulfadiazine
        see butanoic acid, 4-oxo-4-[[[4-(2-pyrimidinylamino)sulfonyl]-
            phenyl]-amino]-, disilver(1+) salt
Ag sulfadiazine
        see silver, (4-amino-N-2-pyrimidinylbenzenesulfonamidato-NN-01)-
Ag sulfadiazine imidazole
        see silver, (4-amino-N-2-pyrimidinylbenzene-sulfonamidato-NN,O)-
            bis(1-imidazole-N3)-, (T-4)-
Ag sulfamethazine
        see benzenesulfonamide, 4-amino-N-(4,6-dimethyl-2-pyrimidinyl)-
            monosilver(1+) salt
Ag-sulfamethizole
        see benzenesulfonamide, 4-amino-N-(5-methyl-
            1,3,4-thiadiazol-2-yl)-
AHR 857
        see benzenesulfonamide, 4-amino-N-(5-methoxy-2-pyrimidinyl)-
Al sulfadiazine
        see aluminium, tris(4-amino-N-2-pyrimidinyl-
            benzenesulfonamidato-NN,O)-
L-Alanine (aqueous)
                + mercury                       vol 29                E49,  72
Albasil
        see benzenesulfonamide, 4-[[(4-aminophenyl)sulfonyl]amino]-
            N,N-dimethyl-
            or benzenesulfonamide, 1-amino-N-[5-(1,1´dimethylethyl)-
            1,3,4-thiadiazol-2-yl)-
Albamine
        see acetamide, N-[(4-aminophenyl)sulfonyl]-
Albexan
        see benzenesulfonamide, 4-amino-
Albosal
        see benzenesulfonamide, 4-amino-
Albucid
        see acetamide, N-[(4-aminophenyl)sulfonyl]-
Albumen
        see bovine serum albumen, egg albumen, human albumen, serum
            albumen
Alcohols
        see under individual alcohols
Alentin
        see benzenesulfonamide, 4-amino-N-[(butylamino)carbonyl]-
Alert-Pep
        see 1H-purine-2,6-dione, 3,7-dihydro-1,3,7-trimethyl-
Alesten
        see acetamide, N-[(4-aminophenyl)sulfonyl]-
Alkali halides
        see under individual halides
Allyl alcohol
        see 2-propen-1-ol
2-Allyl-1,3,5-trimethylbenzene
        see benzene, 1,3,5-trimethyl-3-(2-propenyl)-
```

```
Alphazole
         see benzenesulfonamide, 4-amino-N-(3,4-dimethyl-5-isoxazolyl)-
Altezol
         see benzenesulfonamide, 4-amino-N-(6-methoxy-3-pyridazinyl)-
Aluminium
            + mercury                          vol 25               E84,  E85,
                                                                     86 -  92
                                               vol 29               E214,  217
Aluminium chloride (molten)
            + mercury                          vol 29        E207, E208,  209
Aluminium nitrate
         see nitric acid, aluminium salt
Aluminium sulfate
         see sulfuric acid, aluminium salt
Aluminium, tris(4-amino-N-2-pyrimidinylbenzenesulfonamidato-NN,O)-
            + water                            vol 36                      208
Ambamide
         see benzenesulfonamide, 4-(aminomethyl)-
Ambeside
         see benzenesulfonamide, 4-amino-
4´-(Amidinosulfamoyl)acetanilide
         see acetamide, N-[4-[[(aminoiminomethyl)amino]sulfonyl]phenyl]-
Nl-Amidinosulfanilamide, monohydrate
         see benzenesulfonamide, 4-amino-N-[(aminoiminomethyl)-, monohydrate
Nl-Amidinosulfanilamide
         see benzenesulfonamide, 4-amino-N-[(aminoiminomethyl)-
Amidoxal
         see benzenesulfonamide, 4-amino-N-(3,4-dimethyl-5-isoxazolyl)-
Aminoacetic acid
         see glycine
7-Aminoacetoxycephalosporanic acid
         see 5-thia-1-azabicyclo[4,2,0]oct-2-ene-2-carboxylic acid,
            7-amino-3-(hydroxymethyl)-8-oxo, acetate ester
4-Amino-N-(aminocarbonyl)benzenesulfonamide
         see benzenesulfonamide, 4-amino-N-(aminocarbonyl)-
4-Amino-N-[(aminoiminomethyl)benzenesulfonamide
         see benzenesulfonamide, 4-amino-N-[(aminoiminomethyl)-
Amino-N-(aminoiminomethyl)-N-methylglycine
         see glycine, amino-N-(aminoiminomethyl)-N-methyl-
4-Amino-N-[(aminoiminomethyl)benzenesulfonamide, monohydrate
         see benzenesulfonamide, 4-amino-N-[(aminoiminomethyl)-, monohydrate
4-Amino-N-(6-amino-3-pyridinyl)benzenesulfonamide
         see benzenesulfonamide, 4-amino-N-(6-amino-3-pyridinyl)-
4-Amino-N-(5-amino-2-pyridinyl)benzenesulfonamide
         see benzenesulfonamide, 4-amino-N-(5-amino-2-pyridinyl)-
4-Amino-N-(2-amino-5-pyrimidinyl)benzenesulfonamide
         see benzenesulfonamide, 4-amino-N-(2-amino-5-pyrimidinyl)-
4-Amino-N-(4-amino-2-pyrimidinyl)benzenesulfonamide
         see sulfanilamide, Nl-(4-amino-2-pyrimidinyl)-
4-Amino-N-[4-(aminosulfonyl)phenyl]benzenesulfonamide
         see benzenesulfonamide, 4-amino-N-[4-(aminosulfonyl)phenyl]-
4-Amino-N-[4-(aminosulfonyl)phenyl]benzenesulfonamide monosodium salt
         see benzenesulfonamide, 4-amino-N-[4-(aminosulfonyl)phenyl]-
            monosodium salt
4-Amino-N-4-[(aminosulfonyl)phenyl]benzenesulfonamide monosodium salt
monohydrate
         see benzenesulfonamide, 4-amino-N-4-[(aminosulfonyl)phenyl]-,
            monosodium salt monohydrate
4-Amino-N-4-[(aminosulfonyl)phenyl]benzenesulfonamide monohydrochloride
         see benzenesulfonamide, 4-amino-N-4-[(aminosulfonyl)phenyl]-,
            monohydrochloride
4-Amino-N-[(aminothioxomethyl)benzenesulfonamide
         see benzenesulfonamide, 4-amino-N-[(aminothioxomethyl)-
2-(p-Aminobenzenesulfamido)-5-ethyl-1,3,4-thiadiazole
         see benzenesulfonamide, 4-amino-N-(5-methyl-
            1,3,4-thiadiazol-2-yl)-
p-Aminobenzenesulfoguanidide
         see benzenesulfonamide, 4-amino-N-[(aminoiminomethyl)-
p-Aminobenzenesulfonacetamide
         see acetamide, N-[(4-aminophenyl)sulfonyl]-
m-Aminobenzenesulfonamide
         see benzenesulfonamide, 3-amino-
2-Aminobenzenesulfonamide
         see benzenesulfonamide, 2-amino-
```

3-Aminobenzenesulfonamide
     see benzenesulfonamide, 3-amino-
4-Aminobenzenesulfonamide
     see benzenesulfonamide, 4-amino-
o-Aminobenzenesulfonamide
     see benzenesulfonamide, 2-amino-
p-Aminobenzenesulfonamide
     see benzenesulfonamide, 4-amino-
4-Aminobenzenesulfonamide monohydrate
     see benzenesulfonamide, 4-amino-, monohydrate
4-Aminobenzenesulfonamide monohydrochloride
     see benzenesulfonamide, 4-amino-, monohydrochloride
4-Aminobenzenesulfonamide monosodium salt
     see benzenesulfonamide, 4-amino-, monosodium salt
2-(p-Amino-N-benzenesulfonamide)thiazole sodium salt
     see benzenesulfonamide, 4-amino-N-2-thiazolyl-,
        monosodium salt
4-Aminobenzenesulfonamide zinc salt
     see benzenesulfonamide, 4-amino-, zinc salt (2:1)
4-(4´-Aminobenzenesulfonamido)benzenesulfonamide
     see benzenesulfonamide, 4-amino-N-[4-(aminosulfonyl)phenyl]-
5-(p-Aminobenzenesulfonamido)-3,4-dimethylisoxazole
     see benzenesulfonamide, 4-amino-N-(3,4-dimethyl-
        5-isoxazolyl)-
5-(4-Aminobenzenesulfonamido)-3,4-dimethylisoxazole
     see benzenesulfonamide, 4-amino-N-(3,4-dimethyl-
        5-isoxazolyl)-
6-(p-Aminobenzenesulfonamido)-2,4-dimethylpyrimidine
     see benzenesulfonamide, 4-amino-N-(2,6-dimethyl-4-pyrimidinyl)-
6-(4-Aminobenzenesulfonamido)-2,4-dimethylpyrimidine
     see benzenesulfonamide, 4-amino-N-(2,6-dimethyl-4-pyrimidinyl)-
2-(p-Aminobenzenesulfonamido)-4,6-dimethylpyrimidine
     see benzenesulfonamide, 4-amino-N-(4,6-dimethyl-2-pyrimidinyl)-
2-(4-Aminobenzenesulfonamido)-4,6-dimethylpyrimidine
     see benzenesulfonamide, 4-amino-N-(4,6-dimethyl-2-pyrimidinyl)-
2-(p-Aminobenzenesulfonamido)-5-ethylthiadiazole
     see benzenesulfonamide, 4-amino-N-(5-methyl-
        1,3,4-thiadiazol-2-yl)-
p-Aminobenzenesulfamidoisopropylthiadiazole
     see benzenesulfonamide, 4-amino-N-[5-(1-methylethyl)-
        1,3,4-thiadiazol-2-yl]-
3-(p-Aminobenzenesulfamido)-6-methoxypyridazine
     see benzenesulfonamide, 4-amino-N-(6-methoxy-3-pyridazinyl)-
2-(4-Aminobenzenesulfonamido)-5-methoxypyrimidine
     see benzenesulfonamide, 4-amino-N-(5-methoxy-2-pyrimidinyl)-
2-(4-Aminobenzenesulfonamido)-4-methylpyrimidine
     see benzenesulfonamide, 4-amino-N-(4-methyl-2-pyrimidinyl)-
2-(p-Aminobenzenesulfonamido)-4-methylthiazole
     see benzenesulfonamide, 4-amino-N-(4-methyl-
        2-thiazolyl)-
3-(p-Aminobenzenesulfonamido)-2-phenylpyrazole
     see benzenesulfonamide, 4-amino-N-(1-phenyl-
        1H-pyrazol-5-yl)-
2-(4-Aminobenzenesulfonamido)pyrimidine
     see benzenesulfonamide, 4-amino-N-2-pyrimidinyl-
2-(p-Aminobenzenesulfonamido)thiazole
     see benzenesulfonamide, 4-amino-N-2-thiazolyl-
6-[(p-Aminobenzenesulfonyl)amino]-2,4-dimethylpyrimidine
     see benzenesulfonamide, 4-amino-N-(2,6-dimethyl-4-pyrimidinyl)-
N-(p-Aminobenzenesulfonyl)benzamide
     see benzamide, N-[(4-aminophenyl)sulfonyl]-
N´-(4-Aminobenzenesulfonyl)-N-butylurea
     see benzenesulfonamide, 4-amino-N-[(butylamino)carbonyl]-
4-Aminobenzenesulfonylguanidine
     see benzenesulfonamide, 4-amino-N-[(aminoiminomethyl)-
p-Aminobenzenesulfonylguanidine
     see benzenesulfonamide, 4-amino-N-[(aminoiminomethyl)-
p-Aminobenzenesulfonylguanidine monohydrate
     see benzenesulfonamide, 4-amino-N-[(aminoiminomethyl)-, monohydrate
p-Aminobenzenesulfonylthiourea
     see benzenesulfonamide, 4-amino-N-[(aminothioxomethyl)-
p-Aminobenzenesulfonylurea
     see benzenesulfonamide, 4-amino-N-(aminocarbonyl)-

N-p-Aminobenzenesulphonylguanidine monohydrate
        see benzenesulfonamide, 4-amino-N-[(aminoiminomethyl)-, monohydrate
2-(p-Aminobenzosulfonamido)-4,5-dimethyloxazole
        see benzenesulfonamide, 4-amino-N-(4,5-dimethyl-2-oxazolyl)-
4-Amino-N-2-benzothiazolylbenzenesulfonamide
        see benzenesulfonamide, 4-amino-N-2-benzothiazolyl-
4-Amino-N-4-[1,1´-biphenyl]-4-yl-2-thiazolyl)]benzenesulfonamide
        see benzenesulfonamide, 4-amino-
              N-4-[1,1´-biphenyl]-4-yl-2-thiazolyl)]
4-Amino-N-4-[4-(4-biphenylyl)-2-thiazolyl)]benzenesulfonamide
        see benzenesulfonamide, 4-amino-
              N-4-[1,1´-biphenyl]-4-yl-2-thiazolyl)]-
4-Amino-N-(5-bromo-2-pyridinyl)benzenesulfonamide
        see benzenesulfonamide, 4-amino-N-(5-bromo-2-pyridinyl)-
4-Amino-N-(2-bromo-5-pyridinyl)benzenesulfonamide
        see benzenesulfonamide, 4-amino-N-(2-bromo-5-pyridinyl)-
4-Amino-N-[(butylamino)carbonyl]benzenesulfonamide
        see benzenesulfonamide, 4-amino-N-[(butylamino)carbonyl]-
4-Amino-N-[5-butyl-1,3,4-thiadiazol-2-yl)-benzenesulfonamide
        see benzenesulfonamide, 4-amino-N-[5-butyl-
              1,3,4-thiadiazol-2-yl)-
4-Amino-N-(6-chloro-3-pyridazinyl)benzenesulfonamide
        see benzenesulfonamide, 4-amino-N-(6-chloro-3-pyridazinyl)-
4-Amino-N-(2-chloro-5-pyridinyl)benzenesulfonamide
        see benzenesulfonamide, 4-amino-N-(2-chloro-5-pyridinyl)-
4-Amino-N-(5-chloro-2-pyrimidinyl)benzenesulfonamide
        see benzenesulfonamide, 4-amino-N-(5-chloro-2-pyrimidinyl)-
4-Amino-N-(2-chloro-5-pyrimidinyl)benzenesulfonamide
        see benzenesulfonamide, 4-amino-N-(2-chloro-5-pyrimidinyl)-
Aminocyclohexane
        see cyclohexanamine
7-Aminodeacetoxycephalosporanic acid
        see 5-thia-1-azabicyclo[4,2,0]oct-2-ene-2-carboxylic acid,
              7-amino-3-methyl-8-oxo-
4-Amino-N-[4-(diethylamino)-2-pyrimidinyl)benzenesulfonamide
        see benzenesulfonamide, 4-amino-N-[4-(diethylamino)-
              2-pyrimidinyl)-
4-Amino-N-(2,3-dihydro-2,5-dimethyl-3-isoxazolyl)benzenesulfonamide
        see benzenesulfonamide, 4-amino-N-(2,3-dihydro-2,5-dimethyl-
              3-isoxazolyl)-
4-Amino-N-(2,3-dihydro-3-methyl-2-thiazolyl)benzenesulfonamide
        see benzenesulfonamide, 4-amino-N-(2,3-dihydro-3-methyl-
              2-thiazolyl)-
4-Amino-N-(2,3-dihydro-3-methyl-2-thiazolyl)benzenesulfonamide
        see benzenesulfonamide, 4-amino-N-(2,3-dihydro-3-methyl-
              2-thiazolyl)-
4-Amino-N-(5,6-dimethoxy-4-pyrimidinyl)benzenesulfonamide
        see benzenesulfonamide, 4-amino-N-(5,6-dimethoxy-4-pyrimidinyl)-
4-Amino-N-(4,6-dimethoxy-2-pyrimidinyl)benzenesulfonamide
        see benzenesulfonamide, 4-amino-N-(4,6-dimethoxy-2-pyrimidinyl)-
4-Amino-N-(2,6-dimethoxy-4-pyrimidinyl)benzenesulfonamide
        see benzenesulfonamide, 4-amino-N-(1,2-dimethoxy-4-pyrimidinyl)-
4-Amino-N-(2,6-dimethoxy-4-pyrimidinyl)benzenesulfonamide, cobalt complex
        see cobalt, bis[4-amino-N-(2,6-dimethoxy-4-pyrimidinyl)-
              benzenesulfonamidato]-, hydrate
4-Amino-N-(2,6-dimethoxy-4-pyrimidinyl)benzenesulfonamide, copper complex
        see copper, bis[4-amino-N-(2,6-dimethoxy-4-pyrimidinyl)-
              benzenesulfonamidato]-, hydrate
4-Amino-N-(2,6-dimethoxy-4-pyrimidinyl)benzenesulfonamide, magnesium
complex
        see magnesium, bis-[4-amino-N-(2,6-dimethoxy-4-pyrimidinyl)-
              benzenesulfonamidato]-, hydrate, (T-4)-
4-Amino-N-(2,6-dimethoxy-4-pyrimidinyl)benzenesulfonamide, manganese
complex
        see manganese, bis-[4-amino-N-(2,6-dimethoxy-4-pyrimidinyl)-
              benzenesulfonamidato]-, hydrate, (T-4)-
4-Amino-N-(2,6-dimethoxy-4-pyrimidinyl)benzenesulfonamide, nickel complex
        see nickel, bis-[4-amino-N-(2,6-dimethoxy-4-pyrimidinyl)-
              benzenesulfonamidato]-, hydrate
4-Amino-N-[4-[(dimethylamino)sulfonyl]phenyl]benzenesulfonamide
        see benzenesulfonamide, 4-[[(4-aminophenyl)sulfonyl]amino]-
              N,N-dimethyl-

4-Amino-N,N-dimethylbenzenesulfonamide
      see benzenesulfonamide, 4-amino-N,N-dimethyl-
p-Amino-N,N-dimethylbenzenesulfonamide
      see benzenesulfonamide, 4-amino-N,N-dimethyl-
4-Amino-N-(3,4-dimethyl-5-isoxazolyl)benzenesulfonamide
      see benzenesulfonamide, 4-amino-N-(3,4-dimethyl-5-isoxazolyl)-
4-Amino-N-(4,5-dimethyl-2-oxazolyl)benzenesulfonamide
      see benzenesulfonamide, 4-amino-N-(4,5-dimethyl-2-oxazolyl)-
4-Amino-N-(2,6-dimethyl-4-pyrimidinyl)benzenesulfonamide
      see benzenesulfonamide, 4-amino-N-(2,6-dimethyl-4-pyrimidinyl)-
4-Amino-N-(2,6-dimethyl-4-pyrimidinyl)benzenesulfonamide zinc complex
      see zinc, bis[4-amino-N-(2,6-dimethyl-4-pyrimidinyl)-
         benzenesulfonamidato-,(T-4)-
4-Amino-N-(4,6-dimethyl-2-pyrimidinyl)benzenesulfonamide, manganese
complex
      see manganese, bis[4-amino-N-(4,6-dimethyl-2-pyrimidinyl)-
         benzenesulfonamidato-NN,Ol]-, hydrate
4-Amino-N-(4,5-dimethyl-2-pyrimidinyl)benzenesulfonamide
      see sulfanilamide, Nl-(4,5-dimethyl-2-pyrimidinyl)-
4-Amino-N-(4,6-dimethyl-2-pyrimidinyl)benzenesulfonamide
      see benzenesulfonamide, 4-amino-N-(4,6-dimethyl-2-pyrimidinyl)-
4-Amino-N-(4,6-dimethyl-2-pyrimidinyl)benzenesulfonamide, copper complex
      see copper,bis[4-amino-N-(4,6-dimethyl-2-pyrimidinyl)-
         benzenesulfonamidato-NN,Ol]-, hydrate
4-Amino-N-(4,6-dimethyl-2-pyrimidinyl)benzenesulfonamide, monosilver salt
      see benzenesulfonamide, 4-amino-N-(4,6-dimethyl-2-pyrimidinyl)-
         monosilver(1+) salt
4-Amino-N-(4,6-dimethyl-2-pyrimidinyl)benzenesulfonamide, nickel complex
      see nickel, bis[4-amino-N-(4,6-dimethyl-2-pyrimidinyl)-
         benzenesulfonamidato-NN,Ol]diaqua-
4-Amino-N-(4,6-dimethyl-2-pyrimidinyl)benzenesulfonamide, cobalt complex
      see cobalt, bis[4-amino-N-(4,6-dimethyl-2-pyrimidinyl)-
         benzenesulfonamidato-NN,Ol]-, hydrate
4-Amino-N-(4,6-dimethyl-2-pyrimidinyl)benzenesulfonamide, hemihydrate
      see benzenesulfonamide, 4-amino-N-(4,6-dimethyl-2-pyrimidinyl)-,
         hemihydrate
Aminoethane
      see ethanamine
2-Aminoethanol
      see ethanol, 2-amino-
4-Amino-N-(3-ethoxy-2-pyridinyl)benzenesulfonamide
      see benzenesulfonamide, 4-amino-N-(3-ethoxy-2-pyridinyl)-
4-Amino-N-(6-ethoxy-3-pyridinyl)benzenesulfonamide
      see benzenesulfonamide, 4-amino-N-(6-ethoxy-3-pyridinyl)-
4-Amino-N-(4-ethoxy-2-pyrimidinyl)benzenesulfonamide
      see benzenesulfonamide, 4-amino-N-(4-ethoxy-2-pyrimidinyl)-
4-Amino-N-[1-(2-hydroxyethyl)-1,2-dihydro-2-pyridinyl]-benzenesulfonamide
      see benzenesulfonamide, 4-amino-N-[1,2-dihydro-1-(2-hydroxyethyl)-
         2-pyridinyl]-
4-Amino-N-(2-hydroxy-5-pyridinyl)benzenesulfonamide
      see benzenesulfonamide, 4-amino-N-(2-hydroxy-5-pyridinyl)-
4-Amino-N-hydroxy-N-2-pyridinylbenzenesulfonamide, calcium salt
      see benzenesulfonamide, 4-amino-N-hydroxy-N-2-pyridinyl-,
         calcium salt
4-Amino-N-hydroxy-N-2-pyridinylbenzenesulfonamide, calcium salt (1:1),
dihydrate
      see benzenesulfonamide, 4-amino-N-hydroxy-N-2-pyridinyl-,
         calcium salt(1:1), dihydrate
4-Amino-N-1H-imidazol-2-ylbenzenesulfonamide
      see benzenesulfonamide, 4-amino-N-1H-imidazol-2-yl-
N-[4-[[(Aminoiminomethyl)amino]sulfonyl]phenyl]acetamide
      see acetamide, N-[4-[[(aminoiminomethyl)amino]sulfonyl]phenyl]-
N-(Aminoiminomethyl)-N-methylglycine
      see glycine, N-(aminoiminomethyl)-, N-methyl-
4-Amino-N-[imino(methylthio)methyl]benzenesulfonamide,
      see benzenesulfonamide, 4-Amino-N[imino(methylthio)methyl]-
4-Amino-N-(5-iodo-2-pyridinyl)benzenesulfonamide
      see benzenesulfonamide, 4-amino-N-(5-iodo-2-pyridinyl)-
4-Amino-N-(3-methoxypyrazinyl)benzenesulfonamide
      see benzenesulfonamide. 4-amino-N-(3-methoxypryrazinyl)-
4-Amino-N-(6-methoxy-3-pyridazinyl)benzenesulfonamide
      see benzenesulfonamide, 4-amino-N-(6-methoxy-3-pyridazinyl)-

4-Amino-N-(6-methoxy-3-pyridazinyl)benzenesulfonamide, magnesium complex
        see magnesium, bis-[4-amino-N-(6-methoxy-3-pyridazinyl)-
            benzenesulfon-amidato]-, hydrate, (T-4)-
4-Amino-N-(6-methoxy-3-pyridazinyl)benzenesulfonamide, manganese complex
        see manganese, bis-[4-amino-N-(6-methoxy-3-pyridazinyl)-
            benzenesulfon-amidato]-, hydrate
4-Amino-N-(6-methoxy-3-pyridazinyl)benzenesulfonamide, nickel complex
        see nickel, bis-[4-amino-N-(6-methoxy-3-pyridazinyl)-
            benzenesulfon-amidato]-, hydrate
4-Amino-N-(4-methoxy-2-pyrimidinyl)benzenesulfonamide
        see benzenesulfonamide, 4-amino-N-(4-methoxy-2-pyrimidinyl)-
4-Amino-N-(2-methoxy-5-pyrimidinyl)benzenesulfonamide
        see benzenesulfonamide, 4-amino-N-(2-methoxy-5-pyrimidinyl)-
4-Amino-N-(5-methoxy-2-pyrimidinyl)benzenesulfonamide
        see benzenesulfonamide, 4-amino-N-(5-methoxy-2-pyrimidinyl)-
4-Amino-N-(6-methoxy-4-pyrimidinyl)benzenesulfonamide
        see benzenesulfonamide, 4-amino-N-(6-methoxy-4-pyrimidinyl)-
4-Amino-N-(6-methoxy-4-pyrimidinyl)benzenesulfonamide-
1,4,7,10,13,16-hexaoxacyclooctadecane-complex (1:1)
        see benzenesulfonamide, 4-amino-N-(6-methoxy-4-pyrimidinyl)-,
            comp. with 1,4,7,10,13,16-hexaoxacyclooctadecane (1:1)
4-Amino-N-(4-methoxy-1,2,5-thiadiazol-3-yl)benzenesulfonamide
        see benzenesulfonamide, 4-amino-N-(4-methoxy-
            1,2,5-thiadiazol-3-yl)-
4-Amino-N-[4-[(methylamino)sulfonyl]phenyl]benzenesulfonamide
        see benzenesulfonamide, 4-amino-N-[4-[(methylamino)sulfonyl]-
            phenyl]-
4-Amino-N-methylbenzenesulfonamide
        see benzenesulfonamide, 4-amino-N-methyl-
4-(Aminomethyl)benzenesulfonamide
        see benzenesulfonamide, 4-(aminomethyl)-
4-(Aminomethyl)benzenesulfonamide, monosodium salt
        see benzenesulfonamide, 4-(aminomethyl)-, monosodium salt
4-Amino-N-[5-(3-methylbutyl)-1,3,4-thiadiazol-2-yl)benzenesulfonamide
        see benzenesulfonamide, 4-amino-N-[5-(3-methylbuty)l-
            1,3,4-thiadiazol-2-yl)-
4-Amino-N-[5-(1-methylethyl)-1,3,4-thiadiazol-2-yl]-
benzenesulfonamide
        see benzenesulfonamide, 4-amino-N-[5-(1-methylethyl)-
            1,3,4-thiadiazol-2-yl]-
4-Amino-N-(5-methyl-3-isoxazolyl)benzenesulfonamide
        see benzenesulfonamide, 4-amino-N-(5-methyl-3-isoxazolyl)-
4-Amino-N-(5-methyl-3-isoxazolyl)benzenesulfonamide
1,4,7,10,13,16-hexaoxacyclooctadecane complex (1:1)
        see benzenesulfonamide, 4-amino-N-(5-methyl-3-isoxazolyl)-, compd.
            with 1,4,7,10,13,16-hexaoxacyclooctadecane complex (1:1)
4-Amino-N-methyl-N-(5-methyl-3-isoxazolyl)benzenesulfonamide
        see benzenesulfonamide, 4-amino-N-methyl-N-
            (5-methyl-3isoxazolyl)-
4-Amino-N-(5-methyl-1,2,4-oxadiazol-3-yl)benzenesulfonamide
        see benzenesulfonamide, 4-amino-N-(5-methyl-1,2,4-oxadiazo-3-yl)-
4-Amino-N-(4-methyl-5-pentyl-2-pyrimidinyl)benzenesulfonamide
        see benzenesulfonamide, 4-amino-N-(4-methyl-5-pentyl-
            2-pyrimidinyl)-
4-Amino-N-(3-methyl-1-phenyl-1H-pyrazol-5-yl)benzenesulfonamide
        see benzenesulfonamide, 4-amino-N-(3-methyl-1-phenyl-
            1H-pyrazol-5-yl)-
1-Amino-N-[5-(2-methyl-2-propyl)-1,3,4-thiadiazol-2-yl)benzenesulfonamide
        see benzenesulfonamide, 1-amino-N-[5-
            (1,1´dimethylethyl)-1,3,4-thiadiazol-2-yl)-
4-Amino-N-methyl-N-2-pyridinylbenzenesulfonamide
        see benzenesulfonamide, 4-amino-N-methyl-N-2-pyridinyl-
4-Amino-N-(4-methyl-2-pyrimidinyl)benzenesulfonamide
        see benzenesulfonamide, 4-amino-N-(4-methyl-2-pyrimidinyl)-
4-Amino-N-(2-methyl-4-pyrimidinyl)benzenesulfonamide
        see benzenesulfonamide, 4-amino-N-(2-methyl-4-pyrimidinyl)-
4-Amino-N-(4-methyl-2-pyrimidinyl)benzenesulfonamide, zinc complex
        see zinc, bis[4-amino-N-(4-methyl-2-pyrimidinyl-
            benzenesulfonamidato-NN,O)-
4-Amino-N-(5-methyl-1,3,4-thiadiazol-2-yl)benzenesulfonamide
        see benzenesulfonamide, 4-amino-N-(5-methyl-1,3,4-thiadiazol-2-yl)-
4-Amino-N-(5-methyl-1,3,4-thiadiazol-2-yl)benzenesulfonamide, monosilver salt
        see benzenesulfonamide, 4-amino-N-(5-methyl-1,3,4-thiadiazol-2-yl)-
            monosilver salt

```
Amino-N-(4-methyl-2-thiazolyl)benzenesulfonamide
        see benzenesulfonamide, 4-amino-N-methyl-N-2-thiazolyl)-
4-Amino-N-methyl-N-thiazolybenzenesulfonamide
        see benzenesulfonamide, 4-amino-N-methyl-N-2-thiazolyl)-
4-Amino-N-[6-(methylthio)-3-pyridazinyl)benzenesulfonamide
        see benzenesulfonamide, 4-amino-N-[6-(methylthio)-3-pyridazinyl)-
4-Amino-N-(5-nitro-2-pyridinyl)-benzenesulfonamide
        see benzenesulfonamide, 4-amino-N-(5-nitro-2-pyridinyl)-
4-Amino-N-2-oxazolylbenzenesulfonamide
        see benzenesulfonamide, 4-amino-N-2-oxazolyl-
4-Amino-N-(5-pentyl-1,3,4-thiadiazol-2-yl)benzenesulfonamide
        see benzenesulfonamide, 4-amino-N-(5-pentyl-1,3,4-thiadiazol-2-yl)-
4-Amino-N-(1-phenyl-2,3-dimethyl-5-oxopyrazol-4-yl)benzenesulfonamide
        see benzenesulfonamide, 4-amino-N-(1-phenyl-2,3-dimethyl-5-oxo-
            pyrazol-4-yl)-
4-Amino-N-(1-phenyl-1H-pyrazol-5-yl)benzenesulfonamide
        see benzenesulfonamide, 4-amino-N-(1-phenyl-1H-pyrazol-5-yl)-
4-Aminophenylsulfonamide
        see benzenesulfonamide, 4-amino-
6-(p-Aminophenylsulfonamido)-2,4-dimethylpyrimidine
        see benzenesulfonamide, 4-amino-N-(2,6-dimethyl-4-pyrimidinyl)-
4-(4´-Aminophenylsulfonamido)phenylsulfondimethylamide
        see benzenesulfonamide, 4-[[(4-aminophenyl)sulfonyl]amino]-
            N,N-dimethyl-
4-(4´-Aminophenylsulfonamido)phenylsulfondimethylamide
        see benzenesulfonamide, 1-amino-N-[5-(1,1´dimethylethyl)-
            1,3,4-thiadiazol-2-yl]-
N-[(4-Aminophenyl)sulfonyl]acetamide
        see Acetamide, N-[(4-aminophenyl)sulfonyl]-
N-[(p-Aminophenyl)sulfonyl]acetamide
        see Acetamide, N-[(4-aminophenyl)sulfonyl]-
4-[[(4-Aminophenyl)sulfonyl]amino]-N,N-dimethylbenzenesulfonamide
        see benzenesulfonamide, 4-[[(4-aminophenyl)sulfonyl]amino]-
            N,N-dimethyl-
N-[4-[[[4-(Aminophenyl)sulfonyl]amino]sulfonyl]phenyl]acetamide
        see acetamide, N-[4-[[[4-(aminophenyl)sulfonyl]amino]sulfonyl]-
            phenyl]-
N-[(4-Amino)phenyl)sulfonyl]benzamide
        see benzamide, N-[(4-aminophenyl)sulfonyl]-
N´-(4-Aminophenylsulfonyl)-N-butylurea
        see benzenesulfonamide, 4-amino-N-[(butylamino)carbonyl]-
N-[(4-Aminophenyl)sulfonyl]-3,4-dimethylbenzamide
        see benzamide, N-[(4-aminophenyl)sulfonyl]-3,4-dimethyl-
N-[(4-Aminophenyl)sulfonyl]-N-(3,4-dimethyl-5-isoxazolyl)acetamide
        see acetamide, N-[(4-aminophenyl)sulfonyl]-N-(3,4-dimethyl-
            5-isoxazolyl)-
[(p-Aminophenyl)sulfonyl]guanidide
        see benzenesulfonamide, 4-amino-N-[(aminoiminomethyl)-
1-[(p-Aminophenyl)sulfonyl]guanidine
        see benzenesulfonamide, 4-amino-N-[(aminoiminomethyl)-
4-Aminophenylsulfonylguanidine
        see benzenesulfonamide, 4-amino-N-[(aminoiminomethyl)-
N-[(4-Aminophenyl)sulfonyl]-N-(4-methoxy-1,2,5-thiadiazol-3-yl)acetamide
        see acetamide, N-[(4-aminophenyl)sulfonyl]-N-
            (4-methoxy-1,2,5-thiadiazol-3-yl)-
N-[(4-Aminophenyl)sulfonyl]-3-methyl-2-butenamide
        see 2-butenamide, N-[(4-aminophenyl)sulfonyl]-3-methyl-
N-[(4-Aminophenyl)sulfonyl]-3-methyl-2-butenamide, monosodum salt
        see 2-butenamide, N-[(4-aminophenyl)sulfonyl]-3-methyl-,
            monosodium salt
N-[(4-Aminophenyl)sulfonyl]-N-(5-methyl-3-isoxazolyl)acetamide
        see acetamide, N-[(4-aminophenyl)sulfonyl]-N-(5-methyl-
            3-isoxazolyl)-
p-Aminophenylsulfonylthiourea
        see benzenesulfonamide, 4-amino-N-[(aminothioxomethyl)-
[(p-Aminophenyl)sulfonyl]urea
        see benzenesulfonamide, 4-amino-N-(aminocarbonyl)-
1-Aminopropane
        see 1-propanamine
2-Aminopropane
        see 2-propanamine
N,N´-bis-(3-Aminopropyl)-ethylenediamine
        see 1,3-propanediamine, N-(2-aminoethyl)-N-(3-aminopropyl)-
```

```
N-(3-Aminopropyl)-1,3-propanediamine
        see 1,3-Propanediamine, N-(3-aminopropyl)-
4-Amino-N-[5-(2-propyl)-1,3,4-thiadiazol-2-yl]benzenesulfonamide
        see benzenesulfonamide, 4-amino-N-[5-(1-methylethyl)-
            1,3,4-thiadiazol-2-yl)-
4-Amino-N-(5-propyl-1,3,4-thiadiazol-2-yl)benzenesulfonamide
        see benzenesulfonamide, 4-amino-N-(5-propyl-
            1,3,4-thiadiazol-2-yl)-
4-Amino-N-pyrazinylbenzenesulfonamide
        see benzenesulfonamide, 4-amino-N-pyrazinyl-
4-Amino-N-pyrazinylbenzenesulfonamide, monosodium salt
        see benzenesulfonamide, 4-amino-N-pyrazinyl-, monosodium salt
4-Amino-N-3-pyridazinylbenzenesulfonamide
        see benzenesulfonamide, 4-amino-N-3-pyridazinyl-
(4-Amino-N-2-pyridinylbenzenesulfonamidato)-hydroxy-, calcium salt, dihydrate
        see benzenesulfonamide, 4-amino-N-hydroxy-N-2-pyridinyl-,
            calcium salt(1:1), dihydrate
4-Amino-N-2-pyridinylbenzenesulfonamide
        see benzenesulfonamide, 4-amino-N-2-pyridinyl-
4-Amino-N-3-pyridinylbenzenesulfonamide
        see benzenesulfonamide, 4-amino-N-3-pyridinyl-
4-Amino-N-2-pyridinylbenzenesulfonamide monosodium salt
        see benzenesulfonamide, 4-amino-N-2-pyridinyl-, monosodium salt
4-Amino-N-2-pyridinylbenzenesulfonamide, zinc complex
        see zinc, bis-(4-amino-N-2-pyridinylbenzenesulfonamidato-NN,O)-,
            (T-4)-
N1-(5-Amino-2-pyridyl)sulfanilamide
        see benzenesulfonamide, 4-amino-N-(5-amino-2-pyridinyl)-
N1-(6-Amino-3-pyridyl)sulfanilamide
        see benzenesulfonamide, 4-amino-N-(6-amino-3-pyridinyl)-
4-Amino-N-4-pyrimidinylbenzenesulfonamide
        see benzenesulfonamide, 4-amino-N-4-pyrimidinyl-
4-Amino-N-5-pyrimidinylbenzenesulfonamide
        see benzenesulfonamide, 4-amino-N-5-pyrimidinyl-
4-Amino-N-2-pyrimidinylbenzenesulfonamide, aluminium complex
        see aluminium, tris(4-amino-N-2-pyrimidinyl-
            benzenesulfonamidato-NN,O)-
4-Amino-N-2-pyrimidinylbenzenesulfonamide, cobalt complex
        see cobalt, bis(4-amino-N-2-pyrimidinylbenzenesulfonamidato-NN,O)-
4-Amino-N-2-pyrimidinylbenzenesulfonamide, copper complex
        see copper, bis(4-amino-N-2-pyrimidinylbenzenesulfonamidato-NN,O)-
4-Amino-N-2-pyriminylbenzenesulfonamide, iron complex
        see iron, tris(4-amino-N-2-pyrimidinylbenzenesulfonamidato-NN,O)-
4-Amino-N-2-pyrimidinylbenzenesulfonamide, monosodium salt
        see benzenesulfonamide, 4-amino-N-2-pyrimidinyl-, monosodium salt
4-Amino-N-2-pyrimidinylbenzenesulfonamide, monosilver(1+) salt
        see silver, (4-amino-N-2-pyrimidinylbenzenesulfonamidato-NN-01)-
4-Amino-N-2-pyrimidinylbenzenesulfonamide, silver complex
        see silver, (4-amino-N-2-pyrimidinylbenzene-sulfonamidato-NN,O)-
            bis(1-imidazole-N3)-, (T-4)-
N1-(4-Amino-2-pyrimidinyl)sulfanilamide
        see sulfanilamide, N1-(4-amino-2-pyrimidinyl)-
(T-4)-(4-Amino-N-2-pyrimidinylbenzenesulfonamidato-NN,O)ccpper
        see copper, (4-Amino-N-2-pyrimidinylbenzenesulfonamidato-NN,O)-
(T-4)-(4-Amino-N-2-pyrimidinylbenzenesulfonamidato-NN,O)-
bis(1-imidazole-N3)silver
        see silver, (4-amino-N-2-pyrimidinylbenzene-sulfonamidato-NN,O)-
            bis(1-imidazole-N3)-,
4-Amino-N-2-pyrimidinylbenzenesulfonamide
        see benzenesulfonamide, 4-amino-N-2-pyrimidinyl-
4-Amino-N-2-pyrimidinylbenzenesulfonamide, cerium complex
        see cerium, tris(4-amino-N-2-pyrimidinyl-
            benzenesulfonamidato-NN,O)-
4-Amino-N-2-pyrimidinylbenzenesulfonamide, chromium complex
        see chromium, bis(4-amino-N-2-pyrimidinyl-
            benzenesulfonamidato-NN,O)-chromium
4-Amino-N-2-pyrimidinylbenzenesulfonamide, zinc salt
        see zinc, bis(4-amino-N-2-pyrimidinylbenzenesulfonamidato-NN,O)-,
            (T-4)-
2-(p-Amino-N-sodiobenzenesulfonamide)thiazole
        see benzenesulfonamide, 4-amino-N-2-thiazolyl-, monosodium salt
N-[(4-Aminosulfonyl)phenyl]acetamide
        see acetamide, N-[4-(aminosulfonyl)phenyl]-
```

```
[(4-Aminosulfonyl)phenyl]aminomethanesulfonate, sodium
        see methanesulfonic acid, [[4-(aminosulfonyl)phenyl]amino]-,
            monosodium salt
[4-[[[4-(Aminosulfonyl)phenyl]amino]sulfonyl]phenyl]acetamide
        see acetamide, N-[4-[[[4-(aminophenyl)sulfonyl]amino]sulfonyl]-
            phenyl]-
4-[[(4-Aminosulfonyl)phenyl]azo]-2-hydroxybenzoic acid, monopotassium salt
        see benzoic acid, 4-[[(4-Aminosulfonyl)phenyl]azo]-2-hydroxy-,
            monopotassium salt
N-[(4-Aminosulfonyl)phenyl]glycine
        see glycine, N-[(4-Aminosulfonyl)phenyl]-
N-[(4-Aminosulfonyl)phenyl]glycine, monosodium salt
        see glycine, N-[(4-Aminosulfonyl)phenyl]-, monosodium salt
Amino-p-toluenesulfonamide
        see benzenesulfonamide, 4-(aminomethyl)-
4-Amino-N-1,3,4-thiadiazol-2-yl-benzenesulfonamide
        see benzenesulfonamide, 4-amino-N-1,3,4-thiadiazol-2-yl-
N1-(5-Amino-1,3,4-thiadiazol-2-yl)sulfanilamide
        see benzenesulfonamide, 4-amino-N-(5-amino-1,3,4-thiadiazol-2-yl)-
4-Amino-N-2-thiazolylbenzenesulfonamide
        see benzenesulfonamide, 4-amino-N-2-thiazolyl-
4-Amino-N-2-thiazolylbenzenesulfonamide, cobalt complex
        see cobalt,bis(4-amino-N-2-thiazolyl-
            benzenesulfonamidato-NN,O1)-, hydrate
4-Amino-N-2-thiazolylbenzenesulfonamide, copper complex
        see copper, bis(4-amino-N-2-thiazolyl-
            benzenesulfonamido-NN,O1)-, hydrate
4-Amino-N-2-thiazolyl-benzenesulfonamide monohydrochloride
        see benzenesulfonamide, 4-amino-N-2-thiazolyl-, monohydrochloride
4-Amino-N-2-thiazolylbenzenesulfonamide, monosodium salt
        see benzenesulfonamide, 4-amino-N-2-thiazolyl-, monosodium salt
4-Amino-N-2-thiazolylbenzenesulfonamide, monosodium salt, hexahydrate
        see benzenesulfonamide, 4-amino-N-2-thiazolyl-,
            monosodium salt, hexahydrate
4-Amino-N-1H-1,2,4-triazol-3-ylbenzenesulfonamide
        see benzenesulfonamide, 4-amino-N-1H-1,2,4-triazol-3-yl-
4-Amino-N-1H-1,2,4-triazol-4-ylbenzenesulfonamide
        see benzenesulfonamide, 4-amino-N-1H-1,2,4-triazol-4-yl-
```

Ammonia			
+ ammonium chloride	vol 30		424
+ bromic acid, potassium salt	vol 30		257
+ chloric acid, potassium salt	vol 30		161
+ iodic acid, potassium salt	vol 30	420	- 426
+ mercury	vol 29		E191
+ methane	vol 27/28		736
+ potassium bromide	vol 30		425
+ potassium chloride	vol 30	421,	422
+ potassium iodide	vol 30		426
+ sodium chloride	vol 30		423

Ammonia (aqueous)				
+ bromic acid, potassium salt	vol 30			253
+ cadmium oxide	vol 23	E277,	E279,	294
+ hydrogen sulfide	vol 32		E54,	E55
			56 -	65
+ diphosphoric acid	vol 31		269,	270
+ phosphoric acid	vol 31		269,	270
+ phosphoric acid, lithium salt				
	vol 31		E3,	8
+ phosphoric acid, tripotassium salt				
	vol 31	E297,	300,	301
+ potassium oxide	vol 31		269,	270
+ selenious acid, copper (11) salt				
	vol 26			379
+ sulfur dioxide	vol 26		124,	125
+ sulfurous acid, diammonium salt				
	vol 26	E115,	121 -	123
				126
+ zinc hydroxide	vol 23	E178,	E179,	214

Ammonia (multicomponent)				
+ hydrogen sulfide	vol 32		66 -	78
+ zinc oxide	vol 23	E178,	269,	303

```
Ammonium acetate
        see acetic acid, ammonium salt
```

```
Ammonium bisulfite
        see sulfurous acid, monoammonium salt
Ammonium bromide (aqueous)
        + methane                          vol 27/28        E59,   90
        + propane                          vol 24           E58 - E72,
                                                            78,   79
Ammonium carbonate
        see carbonic acid, ammonium salt
Ammonium chloride (aqueous)
        + hydrogen sulfide                 vol 32                   39
        + methane                          vol 27/28        E59,   89
        + phenol, pentachloro-             vol 20           E6,    E7,
                                                            13,    14
        + phosphoric acid, monoammonium salt
                                           vol 31       90,  267,  268
        + phosphoric acid, monopotassium salt
                                           vol 31           262,  263,
                                                            267,  268
        + phosphoric acid, monosodium salt
                                           vol 31                   90
        + potassium chloride               vol 31           267,  268
        + sodium chloride                  vol 26      E74,   87,   88
        + sodium chloride                  vol 31       90,  119,  120
        + sulfurous acid, diammonium salt
                                           vol 26      E115,  132 - 135
Ammonium chloride (aqueous and multicomponent)
        + disulfurous acid, diammonium salt
                                           vol 26      E74,   87,   88
        + disulfurous acid, disodium salt
                                           vol 26      E74,   87,   88
        + sodium chloride                  vol 26      E5,    66 -  70
        + sulfurous acid, diammonium salt
                                           vol 26      E5,    66 -  70,
                                                            E115,  139
        + sulfurous acid, disodium salt
                                           vol 26      E5,    66 -  70
Ammonium chloride (in liquid ammonia)
        + iodic acid, potassium salt       vol 30                  424
Ammonium chloride (aqueous in ammonia solution)
        + hydrogen sulfide                 vol 32                   66
Ammonium chloride (multicomponent)
        + zinc hydroxide                   vol 23                  264
Ammonium dihydrogen phosphate
        see phosphoric acid, monoammonium salt
Ammonium fluoride (aq)
        + iodic acid, ammonium salt        vol 30           E462,  464
Ammonium hydroxide (aqueous)
        + copper (I) oxide                 vol 23       E1 - E3,    7
        + silver (I) oxide                 vol 23           E83,  E84,
                                                            E93,  109
        + zinc hydroxide                   vol 23          E178, E179,
                                                            254,  255
        + zinc oxide                       vol 23      E178, E179,  200
Ammonium hydroxide (multicomponent)
        + copper (II) hydroxide            vol 23           E10 - E26,
                                                            64 -  78
        + silver (I) oxide                 vol 23          142 - 144
Ammonium iodate
        see iodic acid, ammonium salt
Ammonium nitrate
        see nitric acid, ammonium salt
Ammonium pyroselenite
        see diselenious acid, diammonium salt
Ammonium pyrosulfite
        see disulfurous acid, diammonium salt
Ammonium selenite
        see selenious acid, ammonium salt
Ammonium sulfate
        see sulfuric acid, ammonium salt
Ammonium sulfite
        see sulfurous acid, diammonium salt
Ammonium tetrabutyl-, bromide
        see 1-butanaminium, N,N,N-tributyl-, bromide
```

```
Ammonium tetraethyl-, bromide
         see ethanaminium, N,N,N-triethyl-, bromide
Ammonium tetramethyl bromide
         see methanaminium, N,N,N-trimethyl-, bromide
Ammonium tetramethyl iodide
         see methanaminium, N,N,N-trimethyl-, iodide
Ammonium tetrapropyl bromide
         see 1-propanaminium, N,N,N-tripropyl-, bromide
Ammonium thiocyanate
         see thiocyanic acid, ammonium salt
Ammonium trihydrogen diselenite
         see selenious acid, diammonium salt
Amyl acetate
         see acetic acid, pentyl ester
iso-Amyl acetate
         see 1-butanol, 3-methyl-, acetate
Amyl alcohol
         see 1-pentanol
iso-Amyl alcohol
         see 1-butanol, 3-methyl-
tert-Amyl alcohol
         see 2-butanol, 2-methyl-
Anabion
         see benzenesulfonamide, 4-amino-N-[4-[(methylamino)sulfonyl]-
             phenyl]-
Anacardone
         see 3-pyridinecarboxamide, N,N-diethyl-
Anacordone
         see 3-pyridinecarboxamide, N,N-diethyl-
Anhydohexitol sesquioleate
         see sorbitan (z)-9-octadecenoate (2:3)
Aniline
         see benzenamine
Aniline-p-sulfonic amide
         see benzenesulfonamide, 4-amino-
p-Anilinesulfonamide
         see benzenesulfonamide, 4-amino-
Animal oil
         see oil, animal
Anisole
         see benzene, methoxy-
Anthracene
            + water                              vol 38                    E396 -E398,
                                                                           399 - 404
Anthracene (aqueous)
            + sodium chloride                    vol 38                    406,   407
Anthracene
            + seawater                           vol 38                         E405
Anthracene, 9,10-dimethyl-
            + water                              vol 38                            461
Anthracene, 2-methyl-
            + water                              vol 38         E433,   434 - 435
Anthracene, 2-methyl- (aqueous)
            + sodium chloride                    vol 38                            436
Anthracene, 9-methyl-
            + hydrogen sulfide                   vol 32         E170, E171,   237
            + water                              vol 38                            437
Antimony
            + mercury                            vol 25                    E173, E174,
                                                                           175 - 181
Antistrept
         see benzenesulfonamide, 4-amino-
L-Arginine, monohydrochloride (aqueous)
            + mercury                            vol 29         E50,   E51,    86
L-Arginine, monohydrochloride (in Hank´s balanced salt solution)
            + mercury                            vol 29         E50,   E51,  E87,
                                                                           88 -  94
Argon
            + mercury                            vol 29                    E190 -E192
                                                                           195,  196
Aristamid
         see benzenesulfonamide, 4-amino-N-(2,6-dimethyl-4-pyrimidinyl)-
```

```
Aristogyn
         see benzenesulfonamide, 4-amino-N-(2,6-dimethyl-4-pyrimidinyl)-
Arlacel 83
         see sorbitan (z)-9-octadecenoate (2:3)
Arlacel C
         see sorbitan (z)-9-octadecenoate (2:3)
Arsenic
                + mercury                           vol 25                          E172
Arsenic acid, monopotassium salt (aqueous)
                + phosphoric acid,
                  monopotassium salt                vol 31                   E208,    220
Arsenic (V) oxide (aqueous)
                + zinc oxide                        vol 23           E174,   262,    263
Arsenic (VI) oxide (aqueous)
                + cadmium oxide                     vol 23               E279,  E280,
                                                                         298,    299
Arsonium, tetraphenyl-, nitrate
                + tetraphenylarsonium ????
                  tetraphenylborate ????
                + nitric acid silver salt ????
                + acetonitrile ????                 vol 18                           191
Artificial seawater
         see seawater
AS 18908
         see benzenesulfonamide. 4-amino-N-(3-methoxypryrazinyl)-
L-Aspartic acid (aqueous)
                + mercury                           vol 29                    E49,     73
Aseptil
         see glycine, N-[(4-Aminosulfonyl)phenyl]-, monosodium salt
Aseptil 2
         see benzenesulfonamide, 4-amino-N-(4-methyl-2-thiazolyl)-
Astreptine
         see benzenesulfonamide, 4-amino-
Astrocar
         see 3-pyridinecarboxamide, N,N-diethyl-
Astrocid
         see benzenesulfonamide, 4-amino-
Aterian
         see benzenesulfonamide, 4-amino-N-[(aminoiminomethyl)-
ATP
         see adenosine, 5-(tetrahydrogen phosphate)
Ayerlucil
         see benzenesulfonamide, 4-amino-N-(5-methyl-
                  1,3,4-thiadiazol-2-yl)-
2-H-Azepin-2-one, hexahydro-1-methyl-
                + hydrogen sulfide                   vol 32                   E180,  E181,
                                                                             E183,    314
Azolmetazin
         see benzenesulfonamide, 4-amino-N-(4,6-dimethyl-2-pyrimidinyl)-
Azoquimiol
         see benzenesulfonamide, 4-amino-N-2-thiazolyl-
Azoseptale
         see benzenesulfonamide, 4-amino-N-2-thiazolyl-
Azosulfamide
         see 2,7-naphthalenedisulfonic acid, 6-acetylamino-
                  3-[[4-(Aminosulfonyl)phenyl]azo]-4-hydroxy-, disodium salt
```

```
B

BA 10,370
        see benzenesulfonamide, 4-amino-N-(6-chloro-3-pyridazinyl)-
Bacteramid
        see benzenesulfonamide, 4-amino-
Bactesid
        see benzenesulfonamide, 4-amino-
Bactolin
        see benzenesulfonamide, 4-amino-N-(6-chloro-3-pyridazinyl)-
Badional
        see benzenesulfonamide, 4-amino-N-[(aminothioxomethyl)-
Baldinol
        see benzenesulfonamide, 4-amino-N-[(aminothioxomethyl)-
Barium
                + mercury                       vol 25              E76, E77,
                                                                    78 - 82
Barium bromate
        see bromic acid, barium salt
Barium chlorate
        see chloric acid, barium salt
Barium chloride (aqueous)
                + butane                        vol 24       E58 - E72,   96
                + chloric acid, sodium salt     vol 30              E28,  E29
                + mercury                       vol 29              E47,   57
Barium hydroxide (aqueous)
                + copper (II) hyroxide          vol 23       E10 - E26,   65
                + silver (I) oxide              vol 23              E83,  E84,
                                                                    E88,  119
Barium iodate
        see iodic acid, barium salt
Barium nitrate
        see nitric acid, barium salt
Barium selenite
        see selenious acid, barium salt
Barium sulfite
        see sulfurous acid, barium salt
Barium tellurite
        see tellurous acid, barium salt
Barium thiocyanate
        see thiocyanic acid, barium salt
BAY 5400
        see benzenesulfonamide, 4-amino-N-(5-methoxy-2-pyrimidinyl)-
Benz [j] aceanthrylene, 1,2-dihydro-
                + water                         vol 38                     521
Benz [j] aceanthrylene, 1,2-dihydro-3-methyl-
                + water                         vol 38       E528,  529,  530
Benz [a] anthracene
                + water                         vol 3              E477, E478,
                                                                    479 - 482
Benz [a] anthracene (aqueous)
                + seawater                      vol 38                    E483
                + sodium chloride               vol 38              484,  485
Benz [a] anthracene, 10-butyl-
                + water                         vol 38                     541
Benz [a] anthracene, 7,12-dimethyl-
                + water                         vol 38                     522
Benz [a] anthracene, 9,10-dimethyl-
                + water                         vol 38                     523
Benz [a] anthracene, 10-ethyl-
                + water                         vol 38                     524
Benz [a] anthracene, 1-methyl-
                + water                         vol 38                     508
Benz [a] anthracene, 9-methyl-
                + water                         vol 38                     509
Benz [a] anthracene, 10-methyl-
                + water                         vol 38                     510
Benz [a] anthracene, 7-pentyl-
                + water                         vol 38                     542
Benzamide, N-[[4-(acetylamino)phenyl]sulfonyl]-3,4-dimethyl-
                + phosphoric acid, disodium salt
                                                vol 34                     282
```

```
Benzamide, N-[[4-(acetylamino)phenyl]sulfonyl]-3,4-dimethyl-
          + phosphoric acid,
            monopotassium salt          vol 34                           282
          + water                       vol 34                           282
Benzamide, N-[(4-aminophenyl)sulfonyl]-
          + phosphoric acid,
            monopotassium salt          vol 34                           276
          + sodium hydroxide            vol 34                           276
          + water                       vol 34                    275,   276
Benzamide, N-[(4-aminophenyl)sulfonyl]-
          + water                       vol 36         466 -  468,   484
Benzamide, N-[(4-aminophenyl)sulfonyl]- (aq)
          + 4-amino-N-2-pyrimidinyl-
            benzenesulfonamide          vol 36         466 -  468,   484
          + 4-amino-N-(4-methyl-2-pyrimidinyl-
            benzenesulfonamide          vol 36         466 -  468,   484
          + 4-amino-N-2-thiazolyl-
            benzenesulfonamide          vol 36                468,   484
          + phosphoric acid,
            monopotassium salt          vol 36                467,   484
          + sodium hydroxide            vol 36                467,   484
Benzamide, N-[(4-aminophenyl)sulfonyl]-3,4-dimethyl-
          + 2-propanone                 vol 34                           281
          + trichloromethane            vol 34                           279
          + water                       vol 34         277,   278,   280
Benzamide, N-[(4-aminophenyl)sulfonyl]-3,4-dimethyl- (aq)
          + phosphoric acid, disodium salt
                                        vol 34                277,   280
          + phosphoric acid,
            monopotassium salt          vol 34                277,   280
Benzamsulfonamide
        see benzenesulfonamide, 4-(aminomethyl)-
Benzenamine
          + ammonia                     vol 21         E10,   E11,    16,
                                                        74,    75,   E90
          + butane                      vol 24         E299  -E302,
                                                              322,   323
          + N,N-dimethylmethanamine     vol 21                E86,   213
          + hydrogen sulfide            vol 32         E180  -E182,
                                                              290 -  293
          + methanamine                 vol 21         E81,   E91,   119
          + methane                     vol 27/28             580,   709
          + N-methylmethanamine         vol 21                E83,   162
          + phosphine                   vol 21         E281,  E282,   286
          + propane                     vol 24         E299  -E302,   309
                                                              311,   313
Benzenamine (multicomponent)
          + copper (II) hydroxide       vol 23         E10 -  E26,    78
Benzenamine, N,N-diethyl-
          + ammonia                     vol 21                          16
          + N,N-dimethylmethanamine     vol 21                E86,   222
          + hydrogen sulfide            vol 32         E180  -E182,   295
          + methanamine                 vol 21         E81,   E90,   127
          + N-methylmethanamine         vol 21                E83,   170
Benzenamine, N,N-dimethyl-
          + butane                      vol 24         E299  -E302,   326
          + N,N-dimethylmethanamine     vol 21                E86,   221
          + hydrogen sulfide            vol 32         E180  -E182,
                                                              293,   294
          + methanamine                 vol 21         E81,   E90,   126
          + N-methylmethanamine         vol 21         E83,   E91,   168
          + propane                     vol 24         E299  -E302,   317
Benzenamine, N-ethyl-
          + N,N-dimethylmethanamine     vol 21                E86,   219
          + hydrogen sulfide            vol 32         E180  -E182,   295
          + methanamine                 vol 21         E81,   E90,   124
          + N-methylmethanamine         vol 21                E83,   167
Benzenamine, N-methyl-
          + butane                      vol 24         E299  -E302,   326
          + N,N-dimethylmethanamine     vol 21                          214
          + hydrogen sulfide            vol 32         E180  -E182,   293
          + methanamine                 vol 21         E81,   E90,   123
          + N-methylmethanamine         vol 21                E83,   165
          + propane                     vol 24         E299  -E302,   316
```

```
Benzene
            + ammonia                         vol 21        E3,     E4,     13,
                                                            28  -   32,    E90
          +⊠ N,N-dimethylmethanamine          vol 21       E84,    184,    185
            + hydrogen sulfide                vol 32               E170, E171,
                                                            224  -  226
            + mercury                         vol 29      E149,   154 - 156
            + methanamine                     vol 21       E80,   E81,    100
            + methane                         vol 27/28          E493, E494,
                                                            498  - 501,  E511,
                                                            512  - 521,   580

            + N-methylmethanamine             vol 21                E82,   140
            + 2-methylpropane                 vol 24      E218 -E221,   246
            + phosphine                       vol 21      E281, E282,   286
            + propane                         vol 24        15, E218 -E221,
                                                            225  - 229, E251,
                                                            252  - 254,   338,
                                                                   386,    415
            + water                           vol 37             E63 - E85,
                                                             86  - 183,
                                                            190  - 192
            + water-d2                        vol 37            184 - 187
Benzene (aqueous)
            + sodium chloride                 vol 37            190 - 192
Benzene + seawater                            vol 37           E188,    189
Benzene (ternary and multicomponent)
            + methane                         vol 27/28         551,    579
Benzene, 1,3-bis(trifluoromethyl)-
            + water                           vol 20                     237
Benzene, bromo-
            + ammonia                         vol 21             E10,     28
            + N,N-dimethylmethanamine         vol 21             E86,    228
            + hydrogen sulfide                vol 32      E178, E179,    224,
                                                                   286,    288
            + mercury                         vol 29            E178,    186
            + methanamine                     vol 21             E81,    131
            + N-methylmethanamine             vol 21       E83,   161,    175
            + 2-methylpropane                 vol 24      E257, E258,    281
            + water                           vol 20            E132 -E136,
                                                            137  - 148
            + water-d2                        vol 20                     149
Benzene, 1-bromo-2-chloro-
            + water                           vol 20             E49,     50
Benzene, 1-bromo-3-chloro-
            + water                           vol 20             E51,     52
Benzene, 1-bromo-4-chloro-
            + water                           vol 20             E53,     54
Benzene, 1-bromo-4-iodo-
            + water                           vol 20             E55,     56
Benzene, 1-bromo-3-methyl-
            + butane                          vol 24      E257, E258,    275
            + N,N-dimethylmethanamine         vol 21             E87,    230
            + methanamine                     vol 21             E81,    132
            + N-methylmethanamine             vol 21                     176
            + propane                         vol 24      E257, E258,    266
Benzene, 4-bromo-1-methyl-
            + water                           vol 20                     230
Benzene, butyl-
            + methane                         vol 27/28                  539
            + water                           vol 38       E277, E278,
                                                            279  - 284
            + seawater                        vol 38                     285
Benzene, butyl- (ternary)
            + methane                         vol 27/28         540,    542
Benzene, chloro-
            + ammonia                         vol 21              E9,    E10,
                                                             28,     70
            + butane                          vol 24      E257, E258,    273
            + N,N-dimethylmethanamine         vol 21                     E87
            + hydrogen sulfide                vol 32      E178, E179,    199,
                                                            224,   285 - 287
```

```
Benzene, chloro-
              + mercury                    vol 29              E179,   185
              + methane                    vol 27/28      E692,  693 - 695
              + N-methylmethanamine        vol 21              E83,   175
              + 2-methylpropane            vol 24         E257, E258,
                                                           280,   291
              + propane                    vol 24         E257, E258,
                                                           263,   283
              + water                      vol 20         E153 -E157,
                                                           158 - 181
              + water-d2                   vol 20                 182
Benzene, (1-chloroethyl)-
              + methane                    vol 27/28              696
              + 2-methylpropane            vol 24         E257, E258,  297
              + propane                    vol 24         E257, E258,  288
Benzene, 1-chloro-2-iodo-
              + water                      vol 20              E67,   68
Benzene, 1-chloro-3-iodo-
              + water                      vol 20              E69,   70
Benzene, 1-chloro-4-iodo-
              + water                      vol 20              E71,   72
Benzene, (chloromethyl-)
              + ammonia                    vol 21              E10,   28
Benzene, 1-chloro-4-methyl-
              + 2-methylpropane            vol 24         E257, E258,  292
              + propane                    vol 24         E257, E258,  284
Benzene, 1,2-dibromo-
              + water                      vol 20              E57,   58
Benzene, 1,4-dibromo-
              + water                      vol 20         E62,   63,   64
Benzene, 1,2-dichloro-
              + water                      vol 20              E73 - E77,
                                                            78 -  88
Benzene, 1,3-dichloro-
              + water                      vol 20              E89 - E91,
                                                            92 -  98
Benzene, 1,4-dichloro-
              + water                      vol 20         E99, E100,  101,
                                                           102, E103 -E105,
                                                           106 - 115
Benzene, 3,4-dichloro-1-methyl-
              + water                      vol 20                 225
Benzene, diethyl-
              + water                      vol 38                 264
Benzene, 1,2-difluoro-
              + water                      vol 20              E118,  119
Benzene, 1,3-difluoro-
              + water                      vol 20              E120,  121
Benzene, 1,4-difluoro-
              + water                      vol 20         E122,  123,  124
Benzene, 1,2-diiodo-
              + water                      vol 20              E125,  126
Benzene, 1,3-diiodo-
              + water                      vol 20              E127,  128
Benzene, 1,4-diiodo-
              + water                      vol 20         E129,  130,  131
Benzene, 1,2-dimethyl-
              + water                      vol 38              E9 - E12,
                                                            13 -  21
              + seawater                   vol 38                  22
Benzene, dimethyl-
              + butane                     vol 24                 240
              + methane                    vol 27/28              580
              + phosphine                  vol 21         E281, E282,  286
Benzene, 1,2-dimethyl-
              + hydrogen sulfide           vol 32         E170, E171,  230
              + mercury                    vol 29              E151,  160
              + methane                    vol 27/28      E495, E496,  504
Benzene, 1,2-dimethyl- (multicomponent)
              + methane                    vol 27/28              579
Benzene, 1,3-dimethyl-
              + butane                     vol 24         E218 -E221,  242
              + N,N-dimethylmethanamine    vol 21              E91,  185
```

```
Benzene, 1,3-dimethyl-
          + hydrogen sulfide          vol 32       E170, E171,   231
          + methanamine               vol 21                     100
          + methane                   vol 27/28    E495, E496,   505
                                                           506, E533
                                                           534 - 536
          + N-methylmethanamine       vol 21                     140
          + propane                   vol 24       E218 -E221
                                                           233, 234
          + water                     vol 38       E23 - E28
                                                           29 -  40
          + seawater                  vol 38                      41
Benzene 1,4-dimethyl-
          + methane                   vol 27/28    E495, E496,   507
          + water                     vol 38       E42 - E46
                                                            47 -  57
          + seawater                  vol 38                      58
Benzene, 1,4-dimethyl- (ternary)
          + methane                   vol 27/28                  579
Benzene, (1,1-dimethylethyl)-
          + water                     vol 38       E265, E266
                                                           267 - 269
          + seawater                  vol 38                     270
Benzene, (1,1-dimethylpropyl)-
          + water                     vol 38                     323
Benzene, diphenyl-
     see terphenyl
Benzene, ethoxy-
          + butane                    vol 24       E375 -E378,   400
          + N,N-dimethylmethanamine   vol 21              E85,   203
          + hydrogen sulfide          vol 32       E173, E174,   256
          + methanamine               vol 21              E81,   108
          + N-methylmethanamine       vol 21              E82,   155
          + propane                   vol 24       E218 -E221,   232
                                                   E375 -E378,   391
Benzene, 1,1'-(1,2-ethenediyl)bis-
          + water                     vol 38                     424
Benzene, ethenyl-
          + water                     vol 38       E1,   E2,    3-8
Benzene, ethyl-
          + hydrogen sulfide          vol 32       E170, E171,   233
          + propane                   vol 24       E218 -E221,   232
          + seawater                  vol 38               90,    91
          + water                     vol 38       E59 - E66
                                                            67 -  89
Benzene, (2-ethylcyclopentyl)-
          + water                     vol 38                     393
Benzene, 2-ethyl-1,3,5-trimethyl-
          + water                     vol 38                     322
Benzene, fluoro-
          + water                     vol 20       E202 -E204,
                                                           205 - 209
Benzene, hexachloro-
          + water                     vol 20       E1,            2
Benzene, hexafluoro-
          + methane                   vol 27/28                  683
          + phosphine                 vol 21       E281, E282,   286
Benzene, hexyl-
          + water                     vol 38                     375
Benzene, iodo-
          + butane                    vol 24       E257, E258,   274
          + N,N-dimethylmethanamine   vol 21              E87,   229
          + hydrogen sulfide          vol 32       E178, E179,   289
          + N-methylmethanamine       vol 21       E83,   155,   175
          + propane                   vol 24       E257, E258,   264
          + water                     vol 20       E210 -E213,
                                                           214 - 220
Benzene, methoxy-
          + methane                   vol 27/28    E643, E644,   654
Benzene, methyl-
          + ammonia                   vol 21              E4,    13,
                                                            28,    33
```

Benzene, methyl-
                  + butane   vol 24   E218 -E221,
                                                                        240,   241
                  + N,N-dimethylmethanamine   vol 21   185
                  + hydrogen sulfide   vol 32   E170, E171, 224,
                                                                        227 - 229
                  + mercury   vol 29   E150, 157 - 159
                  + methanamine   vol 21   E81,   100
                  + methane   vol 27/28   E494, 502,   503,
                                                                  E522, 523 - 532
                  + N-methylmethanamine   vol 21   E82,   140
                  + phosphine   vol 21   E281, E282, 287
                  + propane   vol 24   E218 -E221,
                                                                        230,   231
                  + water-d2   vol 37   423,   424
                  + water   vol 37   E369 -E379,
                                                                        380 - 422,
                                                                  427, 428,   430
Benzene, methyl- (aqueous)
                  + sodium chloride   vol 37   427, 428,   430
                  + seawater   vol 37   E425, 426,
                                                                        429,   431
Benzene, methyl- (ternary and multicomponent)
                  + methane   vol 27/28   553 - 556,   579
Benzene, 1,1´-methylenebis-
                  + hydrogen sulfide   vol 32   E170, E171, 235
                  + methane   vol 27/28   E496, 509,
                                                                        548,   549
                  + propane   vol 24   E218 -E221, 239
                  + water   vol 38   391
Benzene, (1-methylethyl)-
                  + hydrogen sulfide   vol 32   233
                  + mercury   vol 29   E151, E152, 161
                  + phosphine   vol 21   E281, E282, 286
                  + water   vol 38   E187 -E189,
                                                                        190 - 199
                  + seawater   vol 38   200
Benzene, 1-methyl-4-(1-methylethyl)-
                  + water   vol 38   E260, 261 - 263
Benzene, 1-methyl-2-nitro-
                  + butane   vol 24   E299 -E302, 324
                  + N,N-dimethylmethanamine   vol 21   E86, E91,
                                                                        216,   217
                  + methanamine   vol 21   121
                  + N-methylmethanamine   vol 21   164
                  + phosphine   vol 21   E281, E282, 287
Benzene, (1-methylpropyl)-
                  + water   vol 38   E271, 272 - 275
                  + seawater   vol 38   276
Benzene, (2-methylpropyl)-
                  + mercury   vol 29   E151, E153, 162
Benzene nitro-
                  + arsine   vol 21   E302, 304,   305
                  + butane   vol 24   E299 -E302, 324
                  + N,N-dimethylmethanamine   vol 21   E86, E91,
                                                                        211,   212
                  + germane   vol 21   E315, 321,   322
                  + hydrogen sulfide   vol 32   E180, E182, 277,
                                                                        278, 315,   316
                  + hydrogen selenide   vol 32   E330, 332,   333
                  + mercury   vol 29   E150, 157 - 159
                  + methanamine   vol 21   118
                  + methane   vol 27/28   580,   722
                  + N-methylmethanamine   vol 21   161
                  + 2-methylpropane   vol 24   E299 -E302,
                                                                        329,   330
                  + phosphine   vol 21   E281, E282, 292,
                                                                        298,   299
                  + propane   vol 24   E299 -E302,
                                                                        307,   308
                  + silane   vol 21   E315, 316,   317
                  + stannane   vol 21   E315, 325,   326
                  + stibine   vol 21   E310, 313,   314

```
Benzene, pentachloro-
         + water                              vol 20          E3,    4,     5
Benzene perfluoro-
         see benzene, hexafluoro-
Benzene, propyl-
         + hydrogen sulfide                   vol 32       E170, E171,  233
         + water                              vol 38             E201, E202,
                                                                   203 - 210
Benzene, 1,2,4,5-tetrabromo-
         + water                              vol 20                 E15,   16
Benzene, 1,2,3,4-tetrachloro-
         + water                              vol 20                 E17,   18
Benzene, 1,2,3,5-tetrachloro-
         + water                              vol 20          E19,   20,    21
Benzene, 1,2,4,5-tetrachloro-
         + water                              vol 20                 E22,   23
Benzene, 1,2,3,4-tetramethyl-
         + water                              vol 38                258,   259
Benzene, 1,2,4-tribromo-
         + water                              vol 20                 E26,   27
Benzene, 1,3,5-tribromo-
         + water                              vol 20                 E28,   29
Benzene, 1,2,3-trichloro-
         + water                              vol 20                 E33,   34
Benzene, 1,2,4-trichloro-
         + water                              vol 20          E35,   36 -   39
         + water-d2                           vol 20                        40
Benzene, 1,3,5-trichloro-
         + water                              vol 20                 E41,   42
Benzene, 1,2,3-trimethyl-
         + water                              vol 38             E164, E165,
                                                                   166 - 168
Benzene, 1,2,4-trimethyl-
         + water                              vol 38             E169, E170,
                                                                   171 - 176
Benzene, 1,3,5-trimethyl-
         + ammonia                            vol 21                        16
         + butane                             vol 24       E218 -E221,  247
         + N,N-dimethylmethanamine            vol 21                       185
         + hydrogen sulfide                   vol 32             E170, E171,
                                                                   232 - 234
         + methanamine                        vol 21                       100
         + methane                            vol 27/28          537 - 539
         + N-methylmethanamine                vol 21                       141
         + seawater                           vol 38                       186
         + water                              vol 38             E177, E178,
                                                                   179 - 185
Benzene 1,3,5-trimethyl (multicomponent)
         + methane                            vol 27/28                    579
Benzene, 1,3,5-trimethyl-3-(2-propenyl)-
         + water                              vol 38                       373
Benzene, 1,3,5-trimethyl-2-propyl-
         + water                              vol 38                       374
Benzene alcohol
         see benzenemethanol
1,2-Benzenedicarboxylic acid
         + stannane                           vol 21       E315,  327,  328
1,2-Benzenedicarboxylic acid, dibutyl ester
         + phosphine                          vol 21       E281, E282,  286
1,2-Benzenedicarboxylic acid, didecyl ester
         + arsine                             vol 21       E302,  306,  307
         + germane                            vol 21       E315,  323,  324
         + hydrogen sulfide                   vol 32             E176, E178,
                                                                   277,  278
         + hydrogen selenide                  vol 32             334,  335
         + phosphine                          vol 21             E281, E282,
                                                                   296,  297
         + silane                             vol 21       E315,  318,  319
         + stannane                           vol 21       E315,  327,  328
Benzeneethanamine
         + ammonia                            vol 21                        16
Benzenemethanamine
         + ammonia                            vol 21                        16
         + butane                             vol 24       E299 -E302,  325
```

```
Benzenemethanamine
        + N,N-dimethylmethanamine      vol 21              E86,    218
        + methanamine                  vol 21                      122
        + N-methylmethanamine          vol 21              E91,    166
        + propane                      vol 24        E299 -E302,   317
Benzenemethanol
        + ammonia                      vol 21                       16
        + butane                       vol 24        E331 -E334
                                                            361,    400
        + N,N-dimethylmethanamine      vol 21     E84,   E87,    E91
                                                            192 -   194
        + hydrogen sulfide             vol 32     E171 -E172,    245
        + methanamine                  vol 21              E90,    106
        + methane                      vol 27/28                   677
        + N-methylmethanamine          vol 21              E91,    153
        + 2-methylpropane              vol 24        E331 -E334,   373
        + propane                      vol 24        E331 -E335
                                                            344,    347
Benzenesulfonamide, 2-amino-
        + water                        vol 34                1 -    6
Benzenesulfonamide, 2-amino- (aq)
        + 1,2-benzenedicarboxylic acid,
          monopotassium salt           vol 34                       6
        + boric acid                   vol 34                       4
        + hydrochloric acid            vol 34                3,     6
        + phosphoric acid, disodium salt
                                       vol 34                2,     5
        + phosphoric acid,
          monopotassium salt           vol 34                       2
        + potassium chloride           vol 34                3,     4
        + 1,2,3-propanetricarboxylic acid,
          2-hydroxy-                   vol 34                       5
        + sodium chloride              vol 34                       4
Benzenesulfonamide, 3-amino-
        + water                        vol 34                7 -   12
Benzenesulfonamide, 3-amino- (aq)
        + 1,2-benzenedicarboxylic acid,
          monopotassium salt           vol 34                      12
        + boric acid                   vol 34                      10
        + hydrochloric acid            vol 34                9,    12
        + phosphoric acid, disodium salt
                                       vol 34                8,    11
        + phosphoric acid,
          monopotassium salt           vol 34                       8
        + potassium chloride           vol 34                9,    10
        + 1,2,3-propanetricarboxylic acid,
          2-hydroxy-                   vol 34                      11
        + sodium chloride              vol 34                      10
Benzenesulfonamide, 4-amino-
        + benzene                      vol 34                     158
        + 1-butanol                    vol 34                     151
        + 1,2-ethanediol               vol 34                     153
        + methylcyclohexane            vol 34                     162
        + methylcyclohexanone          vol 34                     163
        + 1,1´-oxybisbutane            vol 34                     166
        + 2,2´-oxybisethanol           vol 34                     157
        + 1-pentanol                   vol 34                     152
        + poly(oxy-1,2-ethanediyl),
          -hydro- -hydroxy-            vol 34              155,   156
        + 1,2-propanediol              vol 34                     154
        + 2-propanol                   vol 34                     150
        + 2-propanone                  vol 34                     161
        + sorbitan, monolaurate        vol 34                     164
        + sorbitan monooleate,
          polyoxyethylene derivatives  vol 34                     165
        + trichloromethane             vol 34              159,   160
        + water                        vol 34              E13 -  E15
                                                             16 -  149
Benzenesulfonamide, 4-amino- (aq)
        + acetic acid, sodium salt     vol 34               85,    86
        + 4-amino-2-hydroxybenzoic acid,
          monosodium salt              vol 34                     104
        + 2-aminopropanoic acid,
          monosodium salt              vol 34                      94
```

Benzenesulfonamide, 4-amino- (aq)
            + ammonium chloride   vol 34   68
            + barium chloride   vol 34   71
            + 1,2-benzenedicarboxylic acid,
              monopotassium salt   vol 34   83
            + benzoic acid, sodium salt   vol 34   98,   99
            + boric acid   vol 34   79
            + calcium chloride   vol 34   70
            + carbamic acid, ethyl ester   vol 34   126
            + carbonic acid, disodium salt   vol 34   57,   76
            + carbonic acid, monosodium salt
                                               vol 34   56,   76
            + N,N-diethyl-3-pyridinecarboxamide
                                               vol 34   119
            + 2,3-dihydroxybutanedioic acid,
              disodium salt   vol 34   95
            + N,N-dimethylurea   vol 34   122, 123
            + 1,2-ethanediol   vol 34   112
            + ethanol   vol 34   109 - 111,
                                                                            114, 127
            + 2,2´-oxybisethanol   vol 34   115
            + N-ethylethanamine   vol 34   84
            + ethylurea   vol 34   121
            + formic acid, sodium salt   vol 34   85
            + D-glucose   vol 34   149
            + glycine   vol 34   91
            + hexanoic acid, sodium salt   vol 34   89,   90
            + hydrochloric acid   vol 34   46,   77,
                                                                             72,   83
            + 2-hydroxybenzoic acid,
              monosodium salt   vol 34   E13 - E15,
                                                                            100 - 103
            + 3-hydroxybenzoic acid,
              monosodium salt   vol 34   106
            + 2-hydroxy-1,2,3-propane-
              tricarboxylic acid   vol 34   80 - 82,   96
            + 2-hydroxy-1,2,3-propane-
              tricarboxylic acid,
              trisodium salt   vol 34   97
            + 2-hydroxypropanoic acod,
              monosodium salt   vol 34   93
            + lithium chloride   vol 34   49
            + magnesium chloride   vol 34   69
            + 4-methylbenzenesulfonic acid,
              sodium salt   vol 34   105
            + methylurea   vol 34   120
            + nitric acid   vol 34   48
            + pectin   vol 34   107
            + pectinic acid, sodium salt   vol 34   108
            + pentanoic acid, sodium salt   vol 34   88,   92
            + perchloric acid   vol 34   47
            + phosphoric acid, disodium salt
                                               vol 34   58,   72 - 75,
                                                                             80 - 82
            + phosphoric acid,
              monopotassium salt   vol 34   67,   72 - 75
            + potassium bromide   vol 34   61,   62
            + potassium chloride   vol 34   59,   60,
                                                                             78,   79
            + potassium iodide   vol 34   63,   64
            + poly(oxy-1,2-ethanediyl),
              -hydro- -hydroxy-   vol 34   128 - 139
            + 1,2-propanediol   vol 34   113
            + 1,2,3-propanetriol   vol 34   114, 127
            + propanoic acid, sodium salt   vol 34   87
            + 3-pyridinecarboxamide   vol 34   118
            + sodium bromide   vol 34   54
            + sodium chloride   vol 34   50 - 53,
                                                                             77,   84
            + sodium hydroxide   vol 34   79
            + sodium iodide   vol 34   54,   55
            + sorbitan monolaurate,
              polyoxyethylene derivatives   vol 34   140 - 142

```
Benzenesulfonamide, 4-amino- (aq)
          + sorbitan monooleate,
            polyoxyethylene derivatives    vol 34              146 - 148
          + sorbitan monopalmitate,
            polyoxyethylene derivatives    vol 34                    143
          + sorbitan monostearate,
            polyoxyethylene derivatives    vol 34              144,   145
          + tetramethylurea               vol 34                    124
          + thiocyanic acid, potassium salt
                                          vol 34               65,    66
          + thiourea                      vol 34                    125
          + urea                          vol 34       116,   117,  127
Benzenesulfonamide, 4-amino-, (aq)
          + N-[(4-aminosulfonyl)phenyl]acetamide
                                          vol 36                    450
          + water                         vol 36                    450
Benzenesulfonamide, 4-amino-, monohydrate
          + 2-propanone                   vol 34                    177
          + water                         vol 34              167 - 176
Benzenesulfonamide, 4-amino-, monohydrate (aq)
          + aminoacetic acid              vol 34              174,   175
          + boric acid, trisodium salt    vol 34                    172
          + ethanol                       vol 34                    176
          + hydrochloric acid             vol 34       169,   171,  174
Benzenesulfonamide, 4-amino-, monohydrate (aq)
          + phosphoric acid, disodium salt
                                          vol 34                    171
          + phosphoric acid,
            monopotassium salt            vol 34                    171
          + 1,2,3-propanetricarboxylic acid,
            disodium salt                 vol 34                    173
          + sodium chloride               vol 34              174,   175
          + sodium hydroxide              vol 34              170,   175
Benzenesulfonamide 4-amino- compd wit benzenesulfonamide 4-amino-N-2-
thiazolyl
          + water                         vol 34                    178
Benzenesulfonamide, 4-amino- monohydrochloride
          + 2-propanone                   vol 34                    177
Benzenesulfonamide, 4-amino-, monosodium salt
          + 2-propanone                   vol 34                    179
Benzenesulfonamide, 4-amino-, zinc salt (2:1)
          + water                         vol 34                    180
Benzenesulfonamide, 4-amino-N-(aminocarbonyl)-
          + mannitol                      vol 34              283,   284
          + methylbenzene                 vol 34                    284
          + phosphoric acid, disodium salt
                                          vol 34              283,   284
          + phosphoric acid,
            monopotassium salt            vol 34              283,   284
          + sodium chloride               vol 34              283,   284
          + trichloromethane              vol 34                    283
Benzenesulfonamide, 4-amino-N-[(aminoiminomethyl)-
          + 2-propanol                    vol 34                    316
          + 2-propanone                   vol 34                    317
          + water                         vol 34       E302,   303 - 315
Benzenesulfonamide, 4-amino-N-[(aminoiminomethyl)- (aq)
          + ethanol                       vol 34              311 - 313
          + D-glucose                     vol 34                    314
          + pectin                        vol 34                    309
          + pectinic acid, sodium salt    vol 34                    310
          + 1,2,3-propanetriol            vol 34              312,   313
          + sodium chloride               vol 34              307,   308
          + sorbitan monooleate,
            polyoxyethylene derivatives   vol 34                    315
          + urea                          vol 34                    313
Benzenesulfonamide, 4-amino-N-[(aminoiminomethyl)-, monohydrate
          + ethanol                       vol 34              318,   319
          + 1,2,3-propanetriol            vol 34              318,   319
          + urea                          vol 34                    319
          + water                         vol 34              318,   319
Benzenesulfonamide, 4-amino-N-(2-amino-5-pyridinyl)-
          + water                         vol 36                     86
Benzenesulfonamide, 4-amino-N-(5-amino-2-pyridinyl)-
          + water                         vol 36                     79
```

```
Benzenesulfonamide, 4-amino-N-(4-amino-2-pyrimidinyl)-
              + water                        vol 36                          299
Benzenesulfonamide, 4-amino-N-(2-amino-5-pyrimidinyl)-
              + water                        vol 36                          300
Benzenesulfonamide, 4-amino-N-[4-(aminosulfonyl)phenyl]-
              + 2-propanone                  vol 36                            6
              + water                        vol 36              1  -         5
Benzenesulfonamide, 4-amino-N-[4-(aminosulfonyl)phenyl]- (aq)
              + D-glucose                    vol 36                            5
              + pectin                       vol 36                            3
              + pectinic acid, sodium salt   vol 36                            4
              + sodium chloride              vol 36                            2
Benzenesulfonamide, 4-amino-N-[4-(aminosulfonyl)phenyl]-, monohydrochloride
              + 2-propanone                  vol 36                            7
Benzenesulfonamide, 4-amino-N-[4-(aminosulfonyl)phenyl]-, monosodium salt
              + 2-propanone                  vol 36                            8
Benzenesulfonamide, 4-amino-N-[4-(aminosulfonyl)phenyl]-, monosodium salt,
monohydrate
              + water                        vol 36                            9
Benzenesulfonamide, 4-amino-N-(5-amino-1,3,4-thiadiazol-2-yl)-
              + water                        vol 35                          316
Benzenesulfonamide, 4-amino-N-[(aminothioxomethyl)-
              + ethanol                      vol 34              293 -      296
              + 1,2,3-propanetriol           vol 34              295,       296
              + urea                         vol 34                          296
              + water                        vol 34              293 -      296
Benzenesulfonamide, 4-amino-N-2-benzothiazolyl-
              + water                        vol 35                          250
Benzenesulfonamide, 4-amino-N-4-[1,1´-biphenyl]-4-yl-2-thiazolyl)]-
              + water                        vol 35                          251
Benzenesulfonamide, 4-amino-N-(2-bromo-5-pyridinyl)-
              + water                        vol 36                           84
Benzenesulfonamide, 4-amino-N-(5-bromo-2-pyridinyl)-
              + water                        vol 36                           76
Benzenesulfonamide, 4-amino-N-[(butylamino)carbonyl]-
              + water                        vol 34        E285,  286 -     292
Benzenesulfonamide, 4-amino-N-[(butylamino)carbonyl]- (aq)
              + acetic acid                  vol 34                          288
              + acetic acid, sodium salt     vol 34                          288
              + hydrochloric acid            vol 34                    286,  287
              + 2-hydroxy-1,2,3-propane-
                tricarboxylic acid           vol 34                    291,  292
              + potassium chloride           vol 34        287,    288,     289
                                                                   290,     292
              + phosphoric acid, disodium salt
                                             vol 34        E285,   289,     291
              + phosphoric acid,
                monopotassium salt           vol 34               E285,     289
              + sodium hydroxide             vol 34                          290
Benzenesulfonamide, 4-amino-N-[5-butyl-1,3,4-thiadiazol-2-yl]-
              + water                        vol 35                          312
Benzenesulfonamide, 4-amino-N-[5-butyl-1,3,4-thiadiazol-2-yl]- (aq)
              + 2-hydroxy-1,2,3-propane-
                tricarboxylic acid           vol 35                          312
              + phosphoric acid, disodium salt
                                             vol 35                          312
Benzenesulfonamide, 4-amino-N-(1-carboxymethyl-1,2-dihydro-2-pyridinyl)-
        see 1(2H)-pyridineacetic acid, 2-[[(4-aminophenyl)sulfonyl]amino]-
Benzenesulfonamide, 4-amino-N-(6-chloro-3-pyridazinyl)-
              + trichloromethane            vol 36                           99
              + water                        vol 36                           98
Benzenesulfonamide, 4-amino-N-(6-chloro-3-pyridazinyl)-, (aq)
              + phosphoric acid, disodium salt
                                             vol 36                           98
              + phosphoric acid,
                monopotassium salt           vol 36                           98
Benzenesulfonamide, 4-amino-N-(2-chloro-5-pyridinyl)-
              + water                        vol 36                           83
Benzenesulfonamide, 4-amino-N-(2-chloro-5-pyrimidinyl)-
              + water                        vol 36                          298
Benzenesulfonamide, 4-amino-N-(5-chloro-2-pyrimidinyl)-
              + water                        vol 36                          297
Benzenesulfonamide, 4-amino-N-(4,6-diamino-1,3,5-triazin-2-yl)-
              + water                        vol 36                          443
```

```
Benzenesulfonamide, 4-amino-N-[4-(diethylamino)-2-pyrimidinyl)-
               + water                     vol 36                          301
Benzenesulfonamide, 4-amino-N-(2,3-dihydro-2,5-dimethyl-3-isoxazolyl)-
               + trichloromethane          vol 35                          103
               + water                     vol 35                          102
Benzenesulfonamide, 4-amino-N-[1,2-dihydro-1-(2-hydroxyethyl)-2-pyridinyl]-
               + water                     vol 36                           81
Benzenesulfonamide, 4-amino-N-(1,2-dihydro-1-methyl-2-pyridinyl)-
               + trichloromethane          vol 36                           95
               + water                     vol 36          E92,    93,    94
Benzenesulfonamide, 4-amino-N-(2,3-dihydro-3-methyl-2-thiazolyl)-,
               + trichloromethane          vol 35                          248
               + water                     vol 35                  246,   247
Benzenesulfonamide, 4-amino-N-(2,6-dimethoxy-4-pyrimidinyl)-
               + 1-butanol                 vol 36                  418,   419
               + 1-decanol                 vol 36                  424,   425
               + ethanol                   vol 36                  414,   415
               + methanol                  vol 36                  412,   413
               + 1-octanol                 vol 36                  422,   423
               + 1-pentanol                vol 36                  420,   421
               + 1-propanol                vol 36                  416,   417
               + trichloromethane          vol 36                  426,   427
               + water                     vol 36          E404,  405 -  411
Benzenesulfonamide, 4-amino-N-(2,6-dimethoxy-4-pyrimidinyl)-, (aq)
               + hydrochloric acid         vol 36          E404,  408,   409
               + phosphoric acid, disodium salt
                                           vol 36                  410,   411
               + phosphoric acid,
                 monopotassium salt        vol 36                  410,   411
Benzenesulfonamide, 4-amino-N-(4,6-dimethoxy-2-pyrimidinyl)-
               + water                     vol 36            365,  474
Benzenesulfonamide, 4-amino-N-(4,6-dimethoxy-2-pyrimidinyl)-, (aq)
               + 4-amino-N-(4-methyl-2-pyrimidinyl)-
                 benzenesulfonamide        vol 36                          474
               + 4-amino-N-2-pyrimidinyl-
                 benzenesulfonamide        vol 36                          474
               + 2-hydroxy-1,2,3-propanetri-
                 carboxylic acid           vol 36                          365
               + phosphoric acid, disodium salt
                                           vol 36                  365,   474
               + phosphoric acid,
                 monopotassium salt        vol 36                          474
Benzenesulfonamide, 4-amino-N-(5,6-dimethoxy-4-pyrimidinyl)-
               + water                     vol 36                          434
Benzenesulfonamide, 4-amino-N-(5,6-dimethoxy-4-pyrimidinyl)-, (aq)
               + phosphoric acid, disodium salt
                                           vol 36                          434
               + phosphoric acid,
                 monopotassium salt        vol 36                          434
Benzenesulfonamide, 4-amino-N,N-dimethyl-
               + trichloromethane          vol 34                          203
               + water                     vol 34                          202
Benzenesulfonamide, 4-amino-N-[4-[(dimethylamino)sulfonyl]phenyl]-
               + 2-propanone               vol 36                           18
Benzenesulfonamide, 1-amino-N-[5-(1,1´-dimethylethyl)-
1,3,4-thiadiazol-2-yl)-, (aq)
               + 2-hydroxy-1,2,3-propane-
                 tricarboxylic acid        vol 35                          313
               + phosphoric acid, disodium salt
                                           vol 35                          313
Benzenesulfonamide, 4-amino-N-(3,4-dimethyl-5-isoxazolyl)-
               + acetic acid, ethyl ester  vol 35                           91
               + 1-butanol                 vol 35                           85
               + 1-decanol                 vol 35                           88
               + ethanol                   vol 35          E64,    82,    83
               + 1-ethenyl-2-pyrrolidinone polymer
                                           vol 35                           90
               + 2-ethoxyethanol           vol 35                           89
               + methanol                  vol 35                           81
               + 1-octanol                 vol 35                           87
               + 1-pentanol                vol 35                           86
               + 1-propanol                vol 35                           84
               + trichloromethane          vol 35                   92,    93
               + water                     vol 35          E64,    65 -   80
```

Benzenesulfonamide, 4-amino-N-(3,4-dimethyl-5-isoxazolyl)- (aq)
            + calcium chloride   vol 35   76
            + carbonic acid, disodium salt   vol 35   70,   71
            + carbonic acid, monosodium salt
                                               vol 35   69,   71
            + hydrochloric acid   vol 35   68
Benzenesulfonamide, 4-amino-N-(3,4-dimethyl-5-isoxazolyl)- (aq)
            + 2-hydroxy-1,2,3-propanetri-
              carboxylic acid   vol 35   77 -   79
            + magnesium chloride   vol 35   76
            + phosphoric acid, disodium salt
                                               vol 35   72 -   75,
                                                                        78,   79
            + phosphoric acid,
              monoammonium salt   vol 35   76
            + phosphoric acid,
              monopotassium salt   vol 35   72 -   75
            + potassium chloride   vol 35   76
            + sodium chloride   vol 35   76
            + sorbitan, monododecanoate
              poly (oxy-1,2-ethanediyl) derivs.
                                               vol 35   80
            + urea   vol 35   76
Benzenesulfonamide, 4-amino-N-(4,5-dimethyl-2-oxazolyl)-
            + water   vol 35   105 - 107
Benzenesulfonamide, 4-amino-N-(4,5-dimethyl-2-oxazolyl)-, (aq)
            + hydrochloric acid   vol 35   105,   107
            + 2-hydroxy-1,2,3-propanetri-
              carboxylic acid,
              disodium salt   vol 35   107
            + phosphoric acid, disodium salt
                                               vol 35   106
            + phosphoric acid,
              monpotassium salt   vol 35   106
            + sodium chloride   vol 35   105
Benzenesulfonamide, 4-amino-N-(2,6-dimethyl-4-pyrimidinyl)-
            + 1-butanol   vol 36   390,   391
            + 1-decanol   vol 36   396,   397
            + ethanol   vol 36   384 - 386,   401
            + ethoxyethanol   vol 36   387
            + hexane   vol 36   398,   401
            + methanol   vol 36   382,   383
            + octanol   vol 36   394,   395
            + 1-pentanol   vol 36   392,   393
            + 1-propanol   vol 36   388,   389
            + trichloromethane   vol 36   399,   400
            + water   vol 36   E367,   368 - 381
Benzenesulfonamide, 4-amino-N-(2,6-dimethyl-4-pyrimidinyl)-, (aq)
            + carbonic acid, disodium salt   vol 36   373,   374
            + carbonic acid, monosodium salt
                                               vol 36   372,   374
            + ethanol   vol 36   380,   381
            + formic acid   vol 36   381
            + formic acid, sodium salt   vol 36   381
            + 2-hydroxy-1,2,3-propanetri-
              carboxylic acid   vol 36   378,   379
            + hydrochloric acid   vol 36   371
            + phosphoric acid, disodium salt
                                               vol 36   375 - 378
            + phosphoric acid,
              monopotassium salt   vol 36   375 - 377
Benzenesulfonamide, 4-amino-N-(4,5-dimethyl-2-pyrimidinyl)-
            + water   vol 36   302
Benzenesulfonamide, 4-amino-N-(4,6-dimethyl-2-pyrimidinyl)-
            + water   vol 36   E303,   304 - 341,
                                                              476,   478,   479
Benzenesulfonamide, 4-amino-N-(4,6-dimethyl-2-pyrimidinyl)-, (aq)
            + 4-amino-N-(4-methyl-
              2-pyrimidinyl)-
              benzenesulfonamide   vol 36   474,   476,
                                                                       478,   479
            + 4-amino-N-(2-pyrimidinyl)-
              benzenesulfonamide   vol 36   474,   476,
                                                                       478,   479

Benzenesulfonamide, 4-amino-N-(4,6-dimethyl-2-pyrimidinyl)-, (aq)
          + ammonium chloride   vol 36   326
          + barium chloride   vol 36   325
          + benzoic acid   vol 36   337,   341
          + calcium chloride   vol 36   327, 478,   479
          + 2-ethoxyethanol   vol 36   342
Benzenesulfonamide, 4-amino-N-(4,6-dimethyl-2-pyrimidinyl)-, (aq)
          + ethanol   vol 36   338
          + 1-ethenyl-2-pyrrolidinone vol 36   340,   341
          + hydrochloric acid   vol 36   328
          + -hydro- -hydroxypoly-
            (oxy-1,2-ethanediyl)   vol 36   345
          + 2-hydroxy-1,2,3-propanetri-
            carboxylic acid   vol 36   332,   476
          + 2-hydroxy-1,2,3-propanetri-
            carboxylic acid,
            disodium salt   vol 36   335,   336
          + lithium chloride   vol 36   313
          + magnesium chloride   vol 36   324, 478,   479
          + phosphoric acid, disodium salt
                                      vol 36   329 - 334,   336
                                                       474,   476
          + phosphoric acid,
            monoammonium salt   vol 36   478,   479
          + phosphoric acid,
            monopotassium salt   vol 36   329 -   331
                                                       334,   474
          + phosphoric acid, monosodium salt
                                      vol 36   333
          + potassium bromide   vol 36   318,   319
          + potassium chloride   vol 36   315 -   317
                                                       478,   479
          + potassium iodide   vol 36   320,   321
          + 2-propanone   vol 36   343
          + sodium chloride   vol 36   314,   328
                                                       478,   479
          + sodium hydroxide   vol 36   335
          + thiocyanic acid, potassium salt
                                      vol 36   322,   323
          + trichloromethane   vol 36   344
          + urea   vol 36   478,   479
          + sorbitan monooleate,
            polyoxyethylene deriv.   vol 36   339
Benzenesulfonamide, 4-amino-N-(4,6-dimethyl-2-pyrimidinyl)-, hemihydrate
          + sodium hydroxide   vol 36   347
          + water   vol 36   347
Benzenesulfonamide, 4-amino-N-(4,6-dimethyl-2-pyrimidinyl)-, monosilver(1+)
salt
          + 4-morpholinepropanesulfonic acid
                                      vol 36   356
          + 4-morpholinepropanesulfonic acid,
            sodium salt   vol 36   356
          + nitric acid   vol 36   352 -   355
          + nitric acid, potassium salt vol 36   352 -   356
          + water   vol 36   352 -   356
Benzenesulfonamide, 4-amino-N-(2-ethoxy-5-pyridinyl)-
          + water   vol 36   87
Benzenesulfonamide, 4-amino-N-(3-ethoxy-2-pyridinyl)-
          + water   vol 36   80
Benzenesulfonamide, 4-amino-N-(4-ethoxy-2-pyrimidinyl)-
          + water   vol 36   296
Benzenesulfonamide, 4-amino-N-(5-ethyl-1,3,4-thiadiazol-2-yl)-
          + trichloromethane   vol 35   301
          + water   vol 35   E289, 290 - 300
Benzenesulfonamide, 4-amino-N-(5-ethyl-1,3,4-thiadiazol-2-yl)-, (aq)
          + aqueous phosphate buffers vol 35   E289
          + calcium chloride   vol 35   300
          + magnesium chloride   vol 35   300
          + phosphoric acid, disodium salt
                                     vol 35   293, 295 - 299
          + phosphoric acid,
            monoammonium salt   vol 35   300
          + phosphoric acid,
            monopotassium salt   vol 35   294 -   299

```
Benzenesulfonamide, 4-amino-N-(5-ethyl-1,3,4-thiadiazol-2-yl)-, (aq)
            + potassium chloride        vol 35                    300
            + sodium chloride           vol 35              292,  300
            + urea                      vol 35                    300
Benzenesulfonamide, 4-amino-N-(5-ethyl-1,3,4-thiadiazol-2-yl), (aq)
            + N-[4-[[(5-ethyl-1,3,4-thiadiazol-
              2-yl)amino]sulfonyl]phenyl]-
              acetamide                 vol 36                    452
            + water                     vol 36                    452
Benzenesulfonamide, 4-amino-N-2-[3-(2-hydroxyethyl)-2,3-dihydro-
2-thiazolyl]-
            + water                     vol 35                    249
Benzenesulfonamide, 4-amino-N-(2-hydroxy-5-pyridinyl)-
            + water                     vol 36                     85
Benzenesulfonamide, 4-amino-N-hydroxy-N-2-pyrimidinyl-, calcium salt (1:1)
            + 2-propanone               vol 36                     55
Benzenesulfonamide, 4-amino-N-hydroxy-N-2-pyrimidinyl-, calcium salt (1:1)
dihydrate
            + 2-propanone               vol 36                     56
Benzenesulfonamide, 4-amino-N-1H-imidazol-2-yl-
            + water                     vol 35                      1
Benzenesulfonamide, 4-amino-N-[imino(methylthio)methyl]-
            + D-glucose                 vol 34                    301
            + pectin                    vol 34                    300
            + pectinic acid, sodium salt  vol 34                  299
            + sodium chloride           vol 34                    298
            + water                     vol 34                    297
Benzenesulfonamide, 4-amino-N-(5-iodo-2-pyridinyl)-
            + water                     vol 36                     77
Benzenesulfonamide, 4-amino-N-(4-methoxy-1,2,5-oxadiazol-3-yl)-
            + water                     vol 35                    109
Benzenesulfonamide, 4-amino-N-(3-methoxypyrazinyl)-
            + 2-hydroxy-1,2,3-propanetri-
              carboxylic acid           vol 36                    441
            + phosphoric acid, disodium salt
                                        vol 36                    441
            + water                     vol 36                    441
Benzenesulfonamide, 4-amino-N-(6-methoxy-3-pyridazinyl)-
            + trichloromethane          vol 36              113,  114
            + water                     vol 36              102 - 114
Benzenesulfonamide, 4-amino-N-(6-methoxy-3-pyridazinyl)-
            + calcium chloride          vol 36                    108
            + ethanol                   vol 36              110,  111
            + hydrochloric acid         vol 36                    103
            + -hydro- -hydroxy-
              poly(oxy-1,2-ethanediyl)  vol 36              111,  112
            + 2-hydroxy-1,2,3-propanetri-
              carboxylic acid           vol 36                    104
            + 2,2´-iminodiethanol       vol 36                    109
            + magnesium chloride        vol 36                    108
            + phosphoric acid, disodium salt
                                        vol 36              104 - 107
            + phosphoric acid,
              monoammonium salt         vol 36                    108
            + phosphoric acid,
              monopotassium salt        vol 36              105 - 107
            + piperazine                vol 36              110 - 112
            + potassium chloride        vol 36                    108
Benzenesulfonamide, 4-amino-N-(6-methoxy-3-pyridazinyl)-
            + 1,3-propanediol           vol 36              109,  112
            + sodium chloride           vol 36                    108
            + -[(tetrahydro-2-furanyl)methyl]- -hydroxy-
              (glylcofurol)poly(oxy-1,2-ethanediyl)
                                        vol 36                    110
            + urea                      vol 36                    108
Benzenesulfonamide, 4-amino-N-(6-methoxy-3-pyridazinyl)-, cobalt complex
        see cobalt, bis[4-amino-N-(6-methoxy-3-pyridazinyl)benzenesulfon-
            amidato]diaqua
Benzenesulfonamide 4-amino-N-(6-methoxy-3-pyridazinyl)-
monosodium salt (aq)
            + acetic acid, cobalt(2+)   vol 36                    116
```

```
Benzenesulfonamide 4-amino-N-(6-methoxy-3-pyridazinyl)-
monosodium salt (aq)
          + cobalt, bis[4-amino-N-(6-methoxy-
            3-pyridazinyl)benzenesulfon-
            amidato]benzenesulfonamide, diaqua
                                      vol 36                      116
          + water                     vol 36              116
Benzenesulfonamide, 4-amino-N-(2-methoxy-5-pyrimidinyl)-
          + water                     vol 36                      284
Benzenesulfonamide, 4-amino-N-(5-methoxy-2-pyrimidinyl)-
          + 2-ethoxyethanol           vol 36                      282
          + water                     vol 36           280,  281
Benzenesulfonamide, 4-amino-N-(5-methoxy-2-pyrimidinyl)-, (aq)
          + 1,4,7,10,13,16-hexaoxacyclooctadecane
                                      vol 36                      283
          + hydrochloric acid         vol 36                      283
          + -hydro- -hydroxy-
            poly(oxy-1,2-ethanediyl)  vol 36           280,  281
Benzenesulfonamide, 4-amino-N-(6-methoxy-4-pyrimidinyl)-
          + benzene                   vol 36    290,  292,  293
          + 1,4,7,10,13,16-hexaoxacyclooctane
                                      vol 36           292,  293
          + trichloromethane          vol 36                      291
          + water                     vol 36           285 - 289
Benzenesulfonamide, 4-amino-N-(6-methoxy-4-pyrimidinyl)-, aq
          + hydrochloric acid         vol 36                      287
          + phosphoric acid, disodium salt
                                      vol 36           288,  289
          + phosphoric acid, monosodium salt
                                      vol 36           288,  289
Benzenesulfonamide, 4-amino-N-(6-methoxy-4-pyrimidinyl)-, comp. with
1,4,7,10,13,16-hexaoxacyclooctadecane (1:1)
          + water                     vol 36                      294
Benzenesulfonamide, 4-amino-N-(6-methoxy-4-pyrimidinyl)-, comp. with
1,4,7,10,13,16-hexaoxacyclooctadecane (1:1), (aq)
          + hydrochloric acid         vol 36                      295
          + phosphoric acid, disodium salt
                                      vol 36                      294
          + phosphoric acid, monosodium salt
                                      vol 36                      294
Benzenesulfonamide, 4-amino-N-(4-methoxy-2-pyrimidinyl)-
          + water                     vol 36                      279
Benzenesulfonamide, 4-amino-N-(4-methoxy-1,2,5-thiadiazol-3-yl)-
          + water                     vol 35                      252
Benzenesulfonamide, 4-amino-N-(4-methoxy-1,2,5-thiadiazol-3-yl)- (aq)
          + phosphoric acid, disodium salt
                                      vol 35                      252
          + phosphoric acid,
            monopotassium salt        vol 35                      252
Benzenesulfonamide, 4-amino-N-methyl-
          + trichloromethane          vol 34                      201
          + water                     vol 34                      200
Benzenesulfonamide, 4-(aminomethyl)-
          + 2-propanol                vol 34                      322
Benzenesulfonamide, 4-(aminomethyl)-, monosodium salt
          + 2-propanol                vol 34                      323
Benzenesulfonamide, 4-amino-N-[4-[(methylamino)sulfonyl]phenyl]-
          + 2-propanone               vol 36                       14
          + water                     vol 36              11 -  13
Benzenesulfonamide, 4-amino-N-[4-[(methylamino)sulfonyl]phenyl]- (aq)
          + phosphoric acid, disodium salt
                                      vol 36               12,   13
          + phosphoric acid,
            monopotassium salt        vol 36               11,   13
Benzenesulfonamide, 4-amino-N-[5-(3-methylbuty)1-
1,3,4-thiadiazol-2-yl)-, (aq)
          + 2-hydroxy-1,2,3-propanetri-
            carboxylic acid           vol 35                      315
          + phosphoric acid, disodium salt
                                      vol 35                      315
Benzenesulfonamide, 4-amino-N-[5-(1-methylethyl)-
1,3,4-thiadiazol-2-yl]-, (aq)
          + 2-hydroxy-1,2,3-propanetri-
            carboxylic acid           vol 35                      311
```

Benzenesulfonamide, 4-amino-N-[5-(1-methylethyl)-
1,3,4-thiadiazol-2-yl]-, (aq)
       + phosphoric acid, disodium salt
                               vol 35   311
Benzenesulfonamide, 4-amino-N-(5-methyl-3-isoxazolyl)-
       + benzene   vol 35   50 -   53
       + ethanol   vol 35   47
       + ethoxyethanol   vol 35   46
       + 1,4,7,10, 13, 16-hexaoxacyclo-
         octadecane (in benzene)   vol 35   52,   53
       + methanol   vol 35   E14, E15,   43,
                                             44,   47
       + 1,1´-oxybisethane   vol 35   48
       + petroleum ether   vol 35   49
       + 2-propanol   vol 35   45
       + trichloromethane   vol 35   54 -   57
       + water   vol 35   E14, E15,
                                           16 -   42
Benzenesulfonamide, 4-amino-N-(5-methyl-3-isoxazolyl)-, (aq)
       + aminoacetic acid   vol 35   31
       + bovine serum albumin   vol 35   35
       + ethanol   vol 35   E14, E15,
                                         33,   34
       + -D-fructofuranosyl- -D-glucopyranoside
                                vol 35   41
       + galactose   vol 35   40
       + D-glucitol   vol 35   37
       + 4-O- -D-glucopyranosyl- -D-glucose
                                vol 35   42
       + glucose   vol 35   39
Benzenesulfonamide, 4-amino-N-(5-methyl-3-isoxazolyl)-, (aq)
       + 1,4,7,10, 13, 16-hexaoxacyclo-
         octadecane   vol 35   36
       + hydrochloric acid   vol 35   E14, E15,   22,
                                      23,   31,   36
       + mannitol   vol 35   38
       + phosphoric acid, disodium salt
                                vol 35   27 -   30,
                                         32,   35
       + phosphoric acid,
         monopotassium salt   vol 35   27 -   30,
                                       32,   35
       + sodium chloride   vol 35   26,   31,
                                       32,   35
       + sodium hydroxide   vol 35   E14, E15,
                                       24,   25
Benzenesulfonamide, 4-amino-N-(5-methyl-3-isoxazolyl)-, compd. with
1,4,7,10,13,16-hexaoxacyclooctadecane complex (1:1)
          + hydrochloric acid   vol 35   58
          + water   vol 35   58
Benzenesulfonamide, 4-amino-N-methyl-N-(5-methyl-3-isoxazolyl)-
          + trichloromethane   vol 35   60
          + water   vol 35   59
Benzenesulfonamide, 4-amino-N-(4-methyl-1,2,5-oxadiazol-3-yl)-
          + water   vol 35   109
Benzenesulfonamide, 4-amino-N-(5-methyl-1,2,4-oxadiazol-3-yl)-
          + water   vol 35   108
Benzenesulfonamide, 4-amino-N-(3-methyl-1-phenyl-1H-pyrazol-5-yl)-
          + trichloromethane   vol 35   10
          + water   vol 35   9
Benzenesulfonamide, 4-amino-N-(3-methyl-1-phenyl-1H-pyrazol-5-yl)-, (aq)
       + phosphoric acid, disodium salt
                                vol 35   9
       + phosphoric acid,
         monopotassium salt   vol 35   9
Benzenesulfonamide, 4-amino-N-(4-methyl-5-pentyl-2-pyrimidinyl)-
          + water   vol 36   364
Benzenesulfonamide, 4-amino-N-(6-methyl-3-pyridazinyl)-
          + trichloromethane   vol 36   101
          + water   vol 36   100
Benzenesulfonamide, 4-amino-N-methyl-N-2-pyridinyl-
          + trichloromethane   vol 36   91
          + water   vol 36   E88,   89,   90

```
Benzenesulfonamide, 4-amino-N-(4-methyl-2-pyrimidinyl)-
          + ethanol                      vol 36             262,   267
          + 2-ethoxyethanol              vol 36                    264
          + l-ethynyl-2-pyrrolidinone polymer
                                         vol 36                    267
          + 2-propanol                   vol 36                    263
          + trichloromethane             vol 36             265,   266
          + water                        vol 36      E243,  244 - 261,
                                                     474,   483 - 485
Benzenesulfonamide, 4-amino-N-(4-methyl-2-pyrimidinyl)- (aq)
          + N-[(4-aminophenyl)sulfonyl]benzamide
                                         vol 36      466,   468,   484
          + 4-amino-N-(4,6-dimethyl-2-pyrimidinyl)-
            benzenesulfonamide           vol 36             474,   476,
                                                            478,   479
          + 4-amino-N-2-pyrimidinyl-
            benzenesulfonamide           vol 36      459,   460,   461,
                                                     466,   468 - 472,
                                                     474,   476,   478,
                                                     479,   483 - 485
Benzenesulfonamide, 4-amino-N-(4-methyl-2-pyrimidinyl)- (aq)
          + 4-amino-N-2-thiazolyl-
            benzenesulfonamide           vol 36      458,   468 - 472,
                                                            483 - 485
          + calcium chloride            vol 36             478,   479
          + ethanol                     vol 36             260,   261
          + 2-hydroxy-1,2,3-propanetri-
            carboxylic acid             vol 36                     476
          + 2-hydroxy-1,2,3-propanetri-
            carboxylic acid, disodium salt
                                        vol 36                     257
          + magnesium chloride         vol 36             478,   479
          + mannitol                   vol 36             258,   259
          + methylbenzene              vol 36                     258
          + N-[4[[(4-methyl-2-pyrimidinyl)-
            amino]sulfonyl]phenyl]acetamide
                                        vol 36                     485
          + phosphoric acid, disodium salt
                                        vol 36      251 - 256,   258,
                                                    259,   458,   461,
                                                    472,   474,   476,
                                                                  485
          + phosphoric acid,
            monoammonium salt           vol 36             478,   479
          + phosphoric acid,
            monopotassium salt          vol 36      251 - 256,   258,
                                                    259,   458,   461,
                                                    467,   471,   472,
                                                    474,   484,   485
          + potassium chloride          vol 36             478,   479
          + 1,2,3-propanetriol          vol 36             260,   261
          + N-[4-[(2-pyrimidinylamino)-
            sulfonyl]phenyl]acetamide   vol 36                     485
          + sodium chloride             vol 36             258,   259
                                                           478,   479
          + sodium hydroxide            vol 36      250,   257,   459,
                                                    460,   469 - 471,
                                                                  484
          + N-[4-[(2-thiazoylyamino)-
            sulfonyl]phenyl]acetamide   vol 36                     485
          + trichloromethane            vol 36                     259
          + urea                        vol 36      261,   478,   479
Benzenesulfonamide, 4-amino-N-(2-methyl-4-pyrimidinyl)-
          + water                       vol 36                     366
Benzenesulfonamide, 4-amino-N-(5-methyl-1,3,4-thiadiazol-2-yl)-
          + ethanol                     vol 35             275,   276
          + l-ethenyl-2-pyrrolidinone polymer
                                        vol 35                     276
          + water                       vol 35      E257,  258 - 274
Benzenesulfonamide, 4-amino-N-(5-methyl-1,3,4-thiadiazol-2-yl)- (aq)
          + calcium chloride            vol 35                     273
          + hydrochloric acid           vol 35             263,   264
          + 2-hydroxy-1,2,3-propane-
            carboxylic acid             vol 35             271,   272
```

```
Benzenesulfonamide, 4-amino-N-(5-methyl-1,3,4-thiadiazol-2-yl)- (aq)
            + magnesium chloride           vol 35                      273
            + phosphoric acid              vol 35                      270
            + phosphoric acid, disodium salt
                                           vol 35          265 - 269,
                                                           271,  272
            + phosphoric acid,
              monoammonium salt            vol 35                      273
            + phosphoric acid,
              monopotassium salt           vol 35          266 - 269
            + phosphoric acid,
              monosodium salt              vol 35                      270
            + potassium chloride           vol 35                      273
            + sodium chloride              vol 35          264,  270,
                                                           271,  273
            + sulfuric acid,
              monododecyl ester,
              sodium salt                  vol 35                      274
            + urea                         vol 35                      273
Benzenesulfonamide, 4-amino-N-(5-methyl-1,3,4-thiadiazol-2-yl)-,
monosilver salt
            + water                        vol 35          277 - 281
Benzenesulfonamide, 4-amino-N-(5-methyl-1,3,4-thiadiazol-2-yl)-,
monosilver salt, (aq)
            + 4-morpholinepropanesulfonic acid
                                           vol 35                      277
            + 4-morpholinepropanesulfonic acid,
              sodium salt                  vol 35                      277
            + nitric acid                  vol 35          278 - 281
            + nitric acid, potassium salt  vol 35          277 - 281
Benzenesulfonamide, 4-amino-N-(4-methyl-N-2-thiazolyl)-
            + 2-propanone                  vol 35                      238
            + water                        vol 35     E231,  232 - 237
Benzenesulfonamide, 4-amino-N-(4-methyl-N-2-thiazolyl)- (aq)
            + phosphoric acid, disodium salt
                                           vol 35          235,  237
            + phosphoric acid,
              monopotassium salt           vol 35          236,  237
Benzenesulfonamide, 4-amino-N-(methyl-N-2-thiazolyl)-
            + trichloromethane             vol 35                      218
            + water                        vol 35          216,  217
Benzenesulfonamide, 4-amino-N-methyl-N-2-thiazolyl)-
            + trichloromethane             vol 35                      218
            + water                        vol 35          216,  217
Benzenesulfonamide, 4-amino-N-[6-(methylthio)-3-pyridazinyl)-
            + trichloromethane             vol 36                      123
            + water                        vol 36                      122
Benzenesulfonamide, 4-amino-N-[6-(methylthio)-3-pyridazinyl)- (aq)
            + phosphoric acid, disodium salt
                                           vol 36                      122
            + phosphoric acid,
              monopotassium salt           vol 36                      122
Benzenesulfonamide, 4-amino-N-(5-nitro-2-pyridinyl)-
            + water                        vol 36                       78
Benzenesulfonamide, 4-amino-N-2-oxazolyl-
            + water                        vol 35                      104
Benzenesulfonamide, 4-amino-N-(5-pentyl-1,3,4-thiadiazol-2-yl)- (aq)
            + 2-hydroxy-1,2,3-propanetri-
              carboxylic acid              vol 35                      314
            + phosphoric acid, disodium salt
                                           vol 35                      314
Benzenesulfonamide, 4-amino-N-(1-phenyl-2,3-dimethyl-5-oxopyrazol-4-yl)-
            + water                        vol 35                       11
Benzenesulfonamide, 4-amino-N-(1-phenyl-1H-pyrazol-5-yl)-
            + trichloromethane             vol 35                7,    8
            + water                        vol 35                2 -   6
Benzenesulfonamide, 4-amino-N-(1-phenyl-1H-pyrazol-5-yl)-
            + hydrochloric acid            vol 35                       3
            + 2-hyroxy-1,2,3-propanetri-
              carboxylic acid              vol 35                       6
            + phosphoric acid, disodium salt
                                           vol 35                4 -   6
```

```
Benzenesulfonamide, 4-amino-N-(1-phenyl-1H-pyrazol-5-yl)-
          + phosphoric acid, monopotassium
                                    vol 35                    4,    5
Benzenesulfonamide, 4-amino-N-(5-propyl-1,3,4-thiadiazol-2-yl)-
          + 2-hydroxy-1,2,3-propanetri-
            carboxylic acid        vol 35                         310
          + phosphoric acid, disodium salt
                                    vol 35                         310
Benzenesulfonamide, 4-amino-N-pyrazinyl-
          + 2-propanol             vol 36                         439
Benzenesulfonamide, 4-amino-N-pyrazinyl-, monosodium salt
          + 2-propanol             vol 36                         440
Benzenesulfonamide, 4-amino-N-3-pyridazinyl-
          + water                  vol 36                          97
Benzenesulfonamide, 4-amino-N-3-pyridinyl-
          + water                  vol 36                          54
Benzenesulfonamide, 4-amino-N-2-pyridinyl-
          + methylcyclohexanone    vol 36               52,    53
          + 2-propanol             vol 36                          49
          + 2-propanone            vol 36               50,    51
          + trichloromethane       vol 36               47,    48
          + water                  vol 36              E19,   E20,
                                                        21 -   46
Benzenesulfonamide, 4-amino-N-2-pyridinyl-, (aq)
          + ethanol                vol 36         42,   44,    45
          + calcium chloride       vol 36                          41
          + phosphoric acid, disodium salt
                                    vol 36         34,   36 -   40
          + phosphoric acid,
            monopotassium salt     vol 36               35 -   40
          + potassium chloride     vol 36                          41
          + 1,2,3-propanetriol     vol 36               44,    45
          + 1H-purine-2,6-dione,
            3,7-dihydro-1,3,7-trimethyl-  vol 36                   46
          + N-[4[(2-pyridinylamino)-
            sulfonyl]phenyl]acetamide   vol 36                    462
          + sodium chloride        vol 36               33,    41
          + sodium hydroxide       vol 36                          32
          + urea                   vol 36               43,    45
Benzenesulfonamide, 4-amino-N-2-pyridinyl-, monosilver salt
          + water                  vol 36               59 -   62
Benzenesulfonamide, 4-amino-N-2-pyridinyl-, monosilver salt, (aq)
          + 4-morpholinepropanesulfonic acid
                                    vol 36               60,    61
          + 4-morpholinepropanesulfonic acid,
            sodium salt            vol 36               60,    61
          + nitric acid            vol 36               59,    62
          + nitric acid, potassium salt   vol 36        59 -   62
Benzenesulfonamide, 4-amino-N-2-pyridinyl-, monosodium salt
          + water                  vol 36                          57
Benzenesulfonamide, 4-amino-N-2-pyrimidinyl-
          + butanol                vol 36              193,   194
          + 1-decanol              vol 36              199,   200
          + N,N-dimethylacetamide  vol 36                         201
          + N,N-dimethylformamide  vol 36                         202
          + ethanol                vol 36              188,   189
          + hexane                 vol 36                         203
          + methanol               vol 36              186,   187
          + 1-pentanol             vol 36              195,   196
          + 1-propanol             vol 36              190,   191
          + 2-propanol             vol 36                         192
          + 1-octanol              vol 36              197,   198
          + trichloromethane       vol 36              204,   205
          + water                  vol 36       E124,  125 -  185,
                                                 453,  466,   467,
                                                 469 -  472,   476,
                                                        478,   479,
                                                        483 -  485
Benzenesulfonamide, 4-amino-N-2-pyrimidinyl- (aq)
          + acetic acid            vol 36                         163
          + acetic acid, sodium salt   vol 36                     163
```

Benzenesulfonamide, 4-amino-N-2-pyrimidinyl- (aq)
+ 4-amino-N-(4,6-dimethyl-
  2-pyrimidinyl)benzenesulfonamide
                                   vol 36   474 - 476,
                                                       478, 479
+ 4-amino-N-(4-methyl-
  2-pyrimidinyl)benzenesulfonamide
                                   vol 36   459, 460, 461,
                                                 466 - 472, 474,
                                                 476, 478, 479,
                                                 483 - 485
+ N-[(aminophenyl)sulfonyl]benzamide
                                   vol 36   466 - 468, 484
+ 4-amino-N-2-thiazolyl-
  benzenesulfonamide   vol 36   453 - 457,
                                                       468 - 472,
                                                       483 - 485
+ boric acid, disodium salt   vol 36   162
+ carbonic acid, disodium salt   vol 36   143, 151
+ carbonic acid, monosodium salt
                                   vol 36   144, 151
+ calcium chloride   vol 36   161, 478, 479
+ dextrin   vol 36   181
+ N,N-dimethylformamide   vol 36   176 - 178
+ ethanesulfonic acid,
  2-[[(3α, 5β, 7α, 12α)-3,7,12-trihydroxy-
  24-trihydroxy-24-oxocholan-24-yl]amino]-,
  monosodium salt   vol 36   169
+ ethanesulfonic acid,
  2-[[[[(3α, 5β, 7α, 12α)-3,7,12-trihydroxy-
  24-trihydroxy-24-oxocholan-24-yl]-
  amino]acetyl]amino]-,
  sodium salt   vol 36   170
+ ethanol   vol 36   172, 173
+α-D-glucopyranoside,β-D-fructofuranosyl-
                                   vol 36   185
+ hydrochloric acid   vol 36   137, 138, 150
+ 2,2´-iminobisethanol   vol 36   174
+ 2,2´,2´´-nitrilotrisethanol   vol 36   175
+ magnesium chloride   vol 36   478, 479
+ N-[4-[[(4-methyl-2-pyrimidinyl)-
  amino]sulfonyl]phenyl]acetamide
                                   vol 36   485
+ phosphoric acid   vol 36   162, 163
+ phosphoric acid, disodium salt vol 36   149, 152 - 160,
                                                 167 - 170, 455,
                                                 460, 461, 472,
                                                 474, 476, 485
+ phosphoric acid,
  monoammonium salt   vol 36   478, 479
+ phosphoric acid,
  monopotassium salt   vol 36   148, 152 - 160,
                                                 455, 460, 461,
                                                 467, 471, 472,
                                                 474, 484, 485
+ phosphoric acid, monosodium salt
                                   vol 36   162, 163,
                                                      169, 170
+ phosphoric acid, trisodium salt
                                   vol 36   150
+ poly(oxy-1,2-ethanediyl), α-hexadecylω-hydroxy-, mixt with
  -octadecyl- -hydroxypoly(oxy-1,2-ethanediyl
                                   vol 36   182
+ poly(oxy-1,2-ethanediyl),
  α-hydro-ω-hydroxy-   vol 36   183
+ potassium hydroxide   vol 36   142
+ potassium chloride   vol 36   161, 478, 479
+ 1,2,3-propanetricarboxylic acid,
  2-hydroxy-   vol 36   166 - 168, 476
+ 1,2,3-propanetricarboxylic acid,
  2-hydroxy-, disodium salt   vol 36   164, 165
+ 1,2,3-propanetriol   vol 36   172, 173
+ 1H-purine-2,6-dione,
  3,7-dihydro-1,3,7-trimethyl-   vol 36   184

Benzenesulfonamide, 4-amino-N-2-pyrimidinyl- (aq)
          + N-[4-[(2-pyridinylamino)-
            sulfonyl]phenyl]acetamide   vol 36   462,   475,   485
          + sodium chloride   vol 36   138,   145 - 147
                                                               161,   163
                                                               478,   479
          + sodium hydroxide   vol 36   139 - 141,   164
                                                               165,   453,   454
                                                               456,   457,   459
                                                                      460,   467
                                                               469 - 471,   484
          + sulfuric acid, monododecyl ester,
            sodium salt   vol 36   171
          + urea   vol 36   173,   179,   180
                                                               478,   479
Benzenesulfonamide, 4-amino-N-4-pyrimidinyl-
          + water   vol 36   206
Benzenesulfonamide, 4-amino-N-5-pyrimidinyl-
          + water   vol 36   207
Benzenesulfonamide, 4-amino-N-2-pyrimidinyl-, monosilver salt
          + water   vol 36   215 - 222
Benzenesulfonamide, 4-amino-N-2-pyrimidinyl-, monosilver salt (aq)
          + 4-morpholineethanesulfonic acid
                                               vol 36   221,   222
          + 4-morpholineethanesulfonic acid,
            sodium salt   vol 36   221,   222
          + nitric acid   vol 36   218 - 220
          + nitric acid, potassium salt   vol 36   218 - 222
Benzenesulfonamide, 4-amino-N-2-pyrimidinyl-, monosodium salt
          + 2-propanol   vol 36   226
          + water   vol 36   225
Benzenesulfonamide, 4-amino-N-2-pyrimidinyl-, zinc salt
          + water   vol 36   227
Benzenesulfonamide, 4-amino-N-2-quinoxalinyl-
          + hydrochloric acid   vol 36   444
          + nitric acid   vol 36   446
          + perchloric acid   vol 36   445
          + water   vol 36   444 - 446
Benzenesulfonamide, 4-amino-N-(1,2,3,4-tetrahydro-2,4-dioxo-5-pyrimidinyl)-
          + water   vol 36   438
Benzenesulfonamide, 4-amino-N-[5,6,7,8-tetrahydro-5-methyl-8-(2-propyl)-
quinozalinyl]-
          + water   vol 36   447
Benzenesulfonamide, 4-amino-N-[5,6,7,8-tetrahydro-2-quinozalinyl]-
          + water   vol 36   448
Benzenesulfonamide 4-amino-N-[5,6,7,8-tetrahydro-8,9,9-trimethyl-
5,8-methanoquinozalin-2-yl]-
          + water   vol 36   449
Benzenesulfonamide, 4-amino-N-1,3,4-thiadiazol-2-yl-
          + water   vol 35   256
Benzenesulfonamide, 4-amino-N-2-thiazolyl-
          + 2-butanol   vol 35   190
          + cottonseed oil   vol 35   200,   202
          + ethanol   vol 35   186,   190
          + 2-ethoxyethanol   vol 35   189
          + -hydro- -hydroxypoly-
            (oxy-1,2-ethanediyl)-   vol 35   197 - 199
          + methylcyclohexanone   vol 35   195,   196
          + petrolatum (white)   vol 35   201
          + 2-propanol   vol 35   187,   188
          + 2-propanone   vol 35   194
          + sorbitan, (z)-2-octadecenoate (2:3)
                                               vol 35   201,   202
          + trichloromethane   vol 35   191 - 193
          + water   vol 35   E110 -E112,
                                                               113 - 185,
                                                               203 - 206
Benzenesulfonamide, 4-amino-N-2-thiazolyl-, (aq)
          + 4-amino-N-2-thiazolylbenzenesulfonamide-
            1-ethenyl-2-pyrrolidinone complex
                                               vol 35   206
          + aqueous phosphate buffer   vol 35   E110 -E112
          + calcium chloride   vol 35   149
          + carbamic acid, ethyl ester   vol 35   174

Benzenesulfonamide, 4-amino-N-2-thiazolyl-, (aq)
        + 3,7-dihydro-1,3,7-trimethyl-
          1H-purine-2,6-dione   vol 35   175,   176
        + ethanol   vol 35   168 - 172
        + 1-ethenyl-2-pyrrolidinone
          polymer   vol 35   177,   206
        + Ext D and C   vol 35   166
        + F.D and C   vol 35   167
        + D-glucose   vol 35   178
        + hydrochloric acid   vol 35   160,   203 - 205
        + -hydro- -hydroxypoly-
          (oxy-1,2-ethanediyl)   vol 35   180,   181
        + 2-hydroxy-1,2,3-propanetri-
          carboxylic acid   vol 35   161
        + 2-hydroxy-1,2,3-propane-
          tricarboxylic acid, sodium salt
                                         vol 35   160
        + Klucel MF   vol 35   204
        + Methocel 65 HG   vol 35   205
        + pectin   vol 35   179
        + pectinic acid, sodium salt   vol 35   164
        + phosphoric acid, disodium salt
                                         vol 35   147,   150 - 159,
                                                             161 - 163
        + phosphoric acid,
          monopotassium salt   vol 35   148,   150 - 159
        + phosphoric acid, monosodium salt
                                         vol 35   162,   163
        + potassium chloride   vol 35   146,   149
        + 1,2,3-propanetriol   vol 35   171,   172
        + sodium chloride   vol 35   144,   145,   149,
                                                             203 - 205
        + sodium hydroxide   vol 35   142,   143
        + sorbitan, monododecanoate,
          poly(oxy-1,2-ethanediyl)- derivs.
                                         vol 35   182,   183
        + sorbitan, monohexadecanoate,
          poly(oxy-1,2-ethanediyl)- derivs.
                                         vol 35   184
        + sorbitan, monooctadecanoate,
          poly(oxy-1,2-ethanediyl)- derivs.
                                         vol 35   185
        + (3 ,5 ,7 ,12)-3,7,12-
          trihydroxy-24-oxocholan-24-oic acid,
          monosodium salt   vol 35   165,   166,   167
        + 2-[[[[(3 ,5 ,7 ,12)-3,7,12-
          trihydroxy-24-oxocholan-24-yl]-
          amino]acetyl]amino]ethane-
          sulfonic acid, sodium salt   vol 35   163
        + 2-[[(3 ,5 ,7 ,12)-3,7,12-
          trihydroxy-24-oxocholan-24-yl]-
          aminoethanesulfonic acid,
          monosodium salt   vol 35   162
        + urea   vol 35   172,   173
Benzenesulfonamide, 4-amino-N-2-thiazolyl-
        + water
                                         vol 36   451,   453 - 458,
                                                             468 - 472,
                                                             483 - 485
Benzenesulfonamide, 4-amino-N-2-thiazolyl-, (aq)
        + N-[(4-aminophenyl)sulfonyl]benzamide
                                         vol 36   468,   484
        + 4-amino-N-(4-methyl-2-pyrimidinyl)-
          benzenesulfonamide   vol 36   458 - 472,
                                                             483 - 485
        + 4-amino-N-2-pyrimidinyl-
          benzenesulfonamide   vol 36   453 - 457,
                                                             468 - 472,
                                                             483 - 485
        + N-[4-[[(4-methyl-2-pyrimidinyl)-
          amino]sulfonyl]phenyl]acetamide
                                         vol 36   485
        + phosphoric acid, disodium salt
                                         vol 36   458,   472,   485

```
Benzenesulfonamide, 4-amino-N-2-thiazolyl-, (aq)
          + phosphoric acid,
            monopotassium salt            vol 36          458,  471,  472,
                                                                484,  485
          + sodium hydroxide              vol 36          469 - 472,  484
          + N-[4-[(2-thiazolylamino)-
            sulfonyl]phenyl]acetamide      vol 36                451,  485
Benzenesulfonamide, 4-amino-N-2-thiazolyl- 2-pyrrolidinone,
1-ethenyl- homopolymer complex
          + 4-amino-N-2-thiazolyl-
            benzenesulfonamide             vol 35                        206
          + 1-ethenyl-2-pyrrolidinone homopolymer
                                           vol 35                        206
Benzenesulfonamide, 4-amino-N-2-thiazolyl-, monohydrochloride
          + water                          vol 35                        207
Benzenesulfonamide, 4-amino-N-2-thiazolyl-, monosodium salt
          + water                          vol 35                        213
Benzenesulfonamide, 4-amino-N-2-thiazolyl-, monosodium salt, hexahydrate
          + water                          vol 35                        214
Benzenesulfonamide, 4-amino-N-1H-1,2,4-triazol-3-yl-
          + water                          vol 35                         12
Benzenesulfonamide, 4-amino-N-1H-1,2,4-triazol-4-yl-
          + water                          vol 35                         13
Benzenesulfonamide, 4-[(2,4-diaminophenyl)azo]-
          + 2-propanone                    vol 34                        198
Benzenesulfonamide, 4-(galactosylamino)-
          + 2-propanone                    vol 34                        196
Benzenesulfonamide, 4-(4,5-dihydro-3-methyl-5-oxo-1H-pyrazol-1-yl)-
          + water                          vol 35                        317
Benzenesulfonamide, N,N´-(2,5-pyridinediyl)bis[4-amino-
          + water                          vol 36                         96
Benzenesulfonamide, N,N´-(2,5-pyrimidinediyl)bis[4-amino-
          + water                          vol 36                        437
Benzo [a] fluorene
          + water                          vol 38                        475
Benzo [b] fluorene
          + water                          vol 38                        476
Benzo [g,h,i] perylene
          + water                          vol 38                        531
          + salt water                     vol 38                        533
Benzo [a] pyrene
          + water                          vol 38          E512,  513,  514
          + salt water                     vol 38                        515
Benzo [e] pyrene
          + water                          vol 38                        516
Benzo [e] pyrene (aqueous)
          + salt water                     vol 38                        518
          + seawater                       vol 38                       E517
          + sodium chloride                vol 38                        519
Benzo [a] pyrene, 5-methyl-
          + water                          vol 38                        527
Benzo [b] triphenylene
          + water                          vol 38                        532
Benzoic acid, 4-[[(4-aminosulfonyl)phenyl]azo]-2-hydroxy-,
monopotassium salt
          + 2-propanone                    vol 34                        197
Benzoic acid, 5-[[4-[[2,6-dimethoxy-4-pyrimidinyl)-
amino]sulfonyl]phenyl]azo]-2-hydroxy-
          + water                          vol 36                        436
Benzoic acid, ethyl ester
          + N,N-dimethylmethanamine        vol 21                E85,  E91,
                                                                204,  205
          + methanamine                    vol 21                E90,  110
          + N-methylmethanamine            vol 21                E82,  156
Benzoic acid, 2-hydroxy-, methyl ester
          + butane                         vol 24          E375 -E378,  401
          + hydrogen sulfide               vol 32          E176, E178,  273
          + propane                        vol 24          E375 -E378,  390
Benzoic acid, sodium salt
          + acetic acid, sodium salt       vol 33                        256
          + butanoic acid, sodium salt     vol 33                        282
          + butanoic acid, 2-methyl-,
            sodium salt                     vol 33                E313,  314
          + hexanoic acid, sodium salt     vol 33                E324,  325
```

```
Benzoic acid, sodium salt
          + octadecanoic acid, sodium salt
                                           vol 33                    E330,   331
          + propanoic acid, 2-methyl-,
            sodium salt                    vol 33                            296
Benzonitrile
          + butane                         vol 24        E299 -E302,         325
          + N,N-dimethylmethanamine        vol 21                    E86,    215
          + hydrogen sulfide               vol 32        E180 -E182,         300
          + methanamine                    vol 21                            120
          + N-methylmethanamine            vol 21                    E91,    163
          + propane                        vol 24        E299 -E302,         315
Nl-2-Benzothiazolylsulfanilamide
          see benzenesulfonamide, 4-amino-N-2-benzothiazolyl-
Benzotrifluoride
          see Benzene, (trifluoromethyl)-
Nl-Benzoylsulfanilamide
          see benzamide, N-[(4-aminophenyl)sulfonyl]-
Benzyl acetate
          see Acetic acid, phenylmethyl ester
Benzyl alcohol
          see Benzenemethanol
Benzylamine
          see Benzenemethanamine
Benzyl benzoate
          see Benzoic acid, phenylmethyl ester
Benzyl chloride
          see benzene,(chloromethyl)-
Benzyl cyanide
          see Benzeneacetonitrile
Benzyl ether
          see Benzene, 1,1-[oxybis(methylene)]bis-
Berlicid
          see benzenesulfonamide, 4-amino-N-(5-methoxy-2-pyrimidinyl)-
Beryllium
          + mercury                        vol 25                    E55,    56
Beryllium nitrate
          see nitric acid, beryllium salt
Beryllium sulfate
          see sulfurous acid, beryllium salt
Beryllium selenite
          see selenious acid, beryllium salt
1,1´-Biphenyl
          + water                          vol 38                    E341 -E343,
                                                                      344 - 350
1,1´-Biphenyl (aqueous)
          + sodium chloride                vol 38                            353
1,1´-Biphenyl
          + seawater                       vol 38                    E352,   354
1,1´-Biphenyl
          + water-d2                       vol 38                            351
1,1-Bicyclohexyl
          + ammonia                        vol 21        E2,    E3,    24
                                                               25,  E90
          + butane                         vol 24        E218 -E221,  245
          + hydrogen sulfide               vol 32        E166 -E169,  223
          + methane                        vol 27/28            E453,  464
          + 2-methylpropane                vol 24        E218 -E221,  249
          + propane                        vol 24        E218 -E221,  238
1,1-Bicyclohexyl (multicomponent)
          + methane                        vol 27/28                   487
Bicyclo(3,1,1)-hept-2-ene, 2,6,6-trimethyl-
          + methane                        vol 27/28            E491,  492
          + phosphine                      vol 21        E281, E282,  285
1,1´-Biphenyl (multicomponent)
          + methane                        vol 27/28                   579
Bis[4-amino-N-(2,6-dimethoxy-4-pyrimidinyl)benzenesulfonamidato]-cobalt,
hydrate
          see cobalt, bis[4-amino-N-(2,6-dimethoxy-4-pyrimidinyl)-
            benzenesulfonamidato]-, hydrate
Bis[4-amino-N-(2,6-dimethoxy-4-pyrimidinyl)benzenesulfonamidato]-copper,
hydrate
          see copper, bis[4-amino-N-(2,6-dimethoxy-4-pyrimidinyl)-
            benzenesulfonamidato]-, hydrate
```

(T-4)-Bis-[4-amino-N-(2,6-dimethoxy-4-pyrimidinyl)-benzenesulfonamidato]-
magnesium, hydrate
          see magnesium, bis-[4-amino-N-(2,6-dimethoxy-4-pyrimidinyl)-
              benzenesulfonamidato]-, hydrate, (T-4)-
Bis[4-amino-N-(2,6-dimethoxy-4-pyrimidinyl)benzenesulfonamidato]-manganese,
hydrate
          see manganese. bis-[4-amino-N-(2,6-dimethoxy-4-pyrimidinyl)
              benzenesulfonamidato]-, hydrate, (T-4)-
Bis[4-amino-N-(2,6-dimethoxy-4-pyrimidinyl)benzenesulfonamidato]-nickel,
hydrate
          see nickel, bis-[4-amino-N-(2,6-dimethoxy-4-pyrimidinyl)-
              benzenesulfonamidato]-, hydrate (T-4)
Bis[4-amino-N-(2,6-dimethyl-4-pyrimidinyl)benzenesulfonamidato-zinc
          see zinc, bis[4-amino-N-(2,6-dimethyl-4-pyrimidinyl)-
              benzenesulfonamidato-,(T-4)-
Bis[4-amino-N-(4,6-dimethyl-2-pyrimidinyl)benzenesulfonamide-NN,0l]-cobalt,
hydrate
          see cobalt, bis[4-amino-N-(4,6-dimethyl-2-pyrimidinyl)-
              benzenesulfonamidato-NN,0l]-, hydrate
Bis[4-amino-N-(4,6-dimethyl-2-pyrimidinyl)benzenesulfonamide-NN,0l]-copper,
hydrate
          see copper, bis[4-amino-N-(4,6-dimethyl-2-pyrimidinyl)-
              benzenesulfonamidato-NN,0l]-, hydrate
(T-4)-Bis[4-amino-N-(4,6-dimethyl-2-pyrimidinyl)benzenesulfonamide-NN,0l]-
manganese, hydrate
          see manganese, bis[4-amino-N-(4,6-dimethyl-2-pyrimidinyl)-
              benzenesulfonamidato-NN,0l]-, hydrate
Bis[4-amino-N-(4,6-dimethyl-2-pyrimidinyl)benzenesulfonamide-NN,0l]-nickel,
hydrate
          see nickel, bis[4-amino-N-(4,6-dimethyl-2-pyrimidinyl)-
              benzenesulfonamidato-NN,0l]diaqua-
(T-4)-Bis[4-amino-N-(4,6-dimethyl-2-pyrimidinyl)-benzenesulfonamide-NN,0)-
zinc
          see zinc, bis[4-amino-N-(4,6-dimethyl-2-pyrimidinyl)-
              benzenesulfonamidato-NN,0]-
Bis[4-amino-N-(6-methoxy-3-pyridazinyl)benzenesulfonamidato]cobalt, diaqua
          see cobalt, bis[4-amino-N-(6-methoxy-3-pyridazinyl)benzenesulfon-
              amidato]diaqua
Bis-[4-amino-N-(6-methoxy-3-pyridazinyl)benzenesulfonamidato]-manganese,
hydrate
          see manganese, bis-[4-amino-N-(6-methoxy-3-pyridazinyl)-
              benzenesulfon-amidato]-, hydrate
Bis-[4-amino-N-(6-methoxy-3-pyridazinyl)benzenesulfonamidato]-magnesium,
hydrate
          see magnesium, bis-[4-amino-N-(6-methoxy-3-pyridazinyl)-
              benzenesulfon-amidato]-, hydrate, (T-4)-
Bis-[4-amino-N-(6-methoxy-3-pyridazinyl)benzenesulfonamidato-]nickel,
hydrate
          see nickel, bis-[4-amino-N-(6-methoxy-3-pyridazinyl)benzenesulfon-
              amidato]-, hydrate
Bis[4-amino-N-(4-methyl-2-pyrimidinylbenzenesulfonamidato-NN,0)-zinc
          see zinc, bis[4-amino-N-(4-methyl-2-pyrimidinylbenzenesulfon-
              amidato-NN,0)-
2,5-Bis[[(4-aminophenyl)-sulfonyl]amino-pyridine
          see benzenesulfonamide, N,N´-2,5-pyridinedylbis[4-amino-
2,5-Bis[[(4-aminophenyl)sulfonyl]amino]pyrimidine
          see benzenesulfonamide, N,N´-2,5-pyridinedylbis[4-amino-
(T-4)-Bis-(4-amino-N-2-pyridinyl-benzenesulfonamidato-NN,0) zinc
          see zinc, bis-(4-amino-N-2-pyridinylbenzenesulfonamidato-NN,0)-,
              (T-4)-
Bis(4-amino-N-2-pyrimidinylbenzenesulfonamidato-NN,0)-chromium
          see chromium, bis(4-amino-N-2-pyrimidinyl-
              benzenesulfonamidato-NN,0)-
Bis(4-amino-N-2-pyrimidinylbenzenesulfonamidato-NN,0)cobalt
          see cobalt, bis(4-amino-N-2-pyrimidinylbenzenesulfonamidato-NN,0)-
Bis(4-amino-N-2-pyrimidinylbenzenesulfonamidato-NN,0)copper
          see copper, bis(4-amino-N-2-pyrimidinylbenzenesulfonamidato-NN,0)-
Bis(Nl-2-pyrimidinylsulfanilamido)copper
          see copper, bis(4-amino-N-2-pyrimidinylbenzenesulfonamidato-NN,0)-
Bis(4-amino-N-2-thiazolylbenzenesulfonamidato-NN,0l)-cobalt, hydrate
          see cobalt, bis(4-amino-N-2-thiazolylbenzenesulfonamidato-NN,0l)-,
              hydrate

Bis(4-amino-N-2-thiazolylbenzenesulfonamido-NN,01)-copper, hydrate
        see copper, bis(4-amino-N-2-thiazolylbenzenesulfonamido-NN,01)-,
            hydrate
Bis-(4-amino-N-2-thiazolylbensenesulfonamidato-NN,O)-magnesium, hydrate
        see magnesium, bis(4-amino-N-2-thiazolylbensenesulfonamidato-
            NN,O)-, hydrate, (T-4)-
Bis(4-amino-N-2-thiazolylbenzenesulfonamidato-NN,O)-manganese, hydrate
        see manganese, bis-4-amino-N-2-thiazolylbensenesulfonamidato-NN,O)
            hydrate
Bis(4-amino-N-2-thiazolylbenzenesulfonamidato-NN,O)-nickel, hydrate
        see nickel, bis(4-amino-N-2-thiazolyl-benzenesulfonamidato-NN,O)-,
            hydrate
(T-4)-Bis(4-amino-N-2-thiazolylbenzenesulfamidato-NN,O)-zinc
        see zinc, bis(4-amino-N-2-thiazolylbenzenesulfamidato-NN,O)-,
            (T-4)-
Bis-2-chloroethyl ether
        see ethane, 1,1´-oxybis(2-chloro)-
Bis(1-methylethyl)benzene
        see benzene, bis(1-methylethyl)-
Bismuth
        + mercury   vol 25   E182, E183,
                                                            198 - 193
Bis(Nl-2-thiazolylsulfanilamidato)zinc
        see zinc, bis(4-amino-N-2-thiazolylbenzenesulfamidato-NN,O)-,
1,3-Bis(trifluoromethyl)benzene
        see benzene, 1,3-bis(trifluoromethyl)-
Bitumen, athabasca
        + methane   vol 27/28   576,   577
Blood
        see also bovine blood, dog blood, human blood and
            rabbit blood
Blood, human
        + mercury   vol 29   E51,   97
Blood cells
        see human red blood cells
Bone marrow
        see Guinea pig bone marrow
Boric acid
        + phosphoric acid, disodium salt
                                        vol 31   E98,   118
Boric acid, potassium salt (aqueous)
        + phosphoric acid,
          monopotassium salt   vol 31   E208,   257,   258
        + phosphoric acid,
          tripotassium salt   vol 31   E297, 304
Boric acid, sodium salt
        + water   vol 30   37,   40
        + water   vol 31   73,   74
Boric acid, sodium salt (aqueous)
        + chloric acid, sodium salt   vol 30   E30,   37,   40
        + phosphoric acid, monosodium salt
                                        vol 31   E47,   73,   74
        + phosphoric acid, trisodium salt
                                        vol 31   E127,   147
Boron
        + mercury   vol 25   E83
Boron (III) oxide (aqueous)
        + silver (I) oxide   vol 23   E83,   E84,   E92,
                                                            130 - 133
Bovine serum albumen (aqueous)
        + butane   vol 22   E414,   421
Brain
        see guinea pig brain, homogenized brain, human fetal brain,
            rabbit brain
Brain matter
        see dog brain grey matter, dog brain white matter, human brain
            white matter
Brine
        + methane   vol 27/28   144,   145
Bromic acid, barium salt (aqueous)
        + bromic acid, potassium salt   vol 30   239
Bromic acid, cesium salt
        + water   vol 30   E263, E264,
                                                            265 - 267

```
Bromic acid, cesium salt (aqueous)
        + bromic acid, potassium salt    vol 30                   238
        + bromic acid, rubidium salt     vol 30                   262
Bromic acid, lithium salt
        + 2-propanone                    vol 30                   194
        + water                          vol 30        E184 -E187,
                                                       188 - 191
Bromic acid, lithium salt (aqueous)
        + bromic acid, lithium salt      vol 30                   192
        + bromic acid, sodium salt       vol 30                   192
        + sulfuric acid, lithium salt    vol 30                   193
Bromic acid, potassium salt
        + ammonia                        vol 30                   257
        + 2-furan carboxaldehyde         vol 30                   255
        + N-methylacetamide              vol 30                   256
        + water                          vol 30        E220 -E223,
                                                       224 - 253
        + water-d2                       vol 30                   227
Bromic acid, potassium salt (aqueous)
        + acetamide                      vol 30                   253
        + acetic acid                    vol 30                   248
        + acetic acid, methyl ester      vol 30                   250
        + ammonia                        vol 30                   253
        + bromic acid, barium salt       vol 30                   239
        + bromic acid, cesium salt       vol 30                   238
        + bromic acid, rubidium salt     vol 30                   237
        + bromic acid, samarium salt     vol 30                   241
        + bromic acid, silver salt       vol 30                   242
        + bromic acid, yttrium salt      vol 30                   240
        + carbamic acid, ethyl ester     vol 30                   253
        + chloric acid, potassium salt   vol 30        E109, 142, 143
        + dimethylpyrone                 vol 30                   252
        + 1,2-ethanediol                 vol 30                   243
        + ethanol                        vol 30                   243
        + N-ethylethanamine              vol 30                   251
        + formaldehyde                   vol 30                   246
        + formamide                      vol 30                   253
        + D-glucose                      vol 30                   245
        + glycine                        vol 30                   248
        + D-mannitol                     vol 30                   245
        + methanol                       vol 30                   243
        + 2-methyl-2-butanol             vol 30                   240
        + nitric acid, potassium salt    vol 30                   230
        + nitric acid, sodium salt       vol 30                   228
        + 1,1´-oxybisethane              vol 30                   244
        + phenol                         vol 30                   249
        + piperidine                     vol 30                   251
        + potassium bromide              vol 30             233 - 235
        + potassium chloride             vol 30                   232
        + potassium iodide               vol 30                   236
        + 1,2,3-propanetriol             vol 30                   243
        + 1-propanol                     vol 30                   243
        + 2-propanone                    vol 30                   247
        + pyridine                       vol 30                   251
        + sodium chloride                vol 30                   229
        + sulfuric acid, potassium salt
                                         vol 30                   231
        + urea                           vol 30                   253
Bromic acid, potassium salt (in ethanol)
        + lithium chloride               vol 30                   254
Bromic acid, rubidium salt
        + water                          vol 30        E258, E259,
                                                       260 - 262
Bromic acid, rubidium salt (aqueous)
        + bromic acid, cesium salt       vol 30                   262
        + bromic acid, potassium salt    vol 30                   237
Bromic acid, samarium salt (aqueous)
        + bromic acid, potassium salt    vol 30                   241
Bromic acid, silver salt (aqueous)
        + bromic acid, potassium salt    vol 30                   242
        + bromic acid, sodium salt       vol 30             214, 215
Bromic acid, sodium salt
        + hydrazine                      vol 30                   219
        + N-methylacetamide              vol 30                   218
```

```
Bromic acid, sodium salt
          + water                             vol 30               E195 -E197
                                                                   198 - 217
Bromic acid, sodium salt (aqueous)
          + bromic acid, lithium salt   vol 30                          192
          + bromic acid, silver salt    vol 30                  214,   215
          + carbonic acid, disodium salt  vol 30               201,   216
          + carbonic acid, monosodium salt
                                        vol 30                  202,   217
          + chloric acid, sodium salt   vol 30          E29,     62,    63
          + disodium (I-4)-tetroxomolybdate (2-)
                                        vol 30                          200
          + nitric acid, sodium salt    vol 30                          203
          + sodium bromide              vol 30          E196,   210 - 212
                                                                216,   217
          + sodium chloride             vol 30                  208,   209
          + sodium iodide               vol 30                          213
          + sulfuric acid, sodium salt  vol 30                  204 - 207
          + water-d2                    vol 30                          199
Bromic acid, yttrium salt (aqueous)
          + bromic acid, potassium salt vol 30                          240
Bromides
       see under individual bromides
Bromide, cetyl trimethyl ammonium
        see 1-hexadecanaminium, N,N,N-trimethyl-, bromide
Bromide, tetrabutyl ammonium
        see 1-butanaminium, N,N,N-tributyl-, bromide
Bromide, tetraethyl ammonium
        see ethanaminium, N,N,N-triethyl-, bromide
Bromide, tetramethyl ammonium
        see methanaminium, N,N,N-trimethyl-, bromide
Bromide, tetrapropyl ammonium
        see 1-propanaminium, N,N,N-tripropyl-, bromide
Bromine (multicomponent)
          + copper (I) oxide            vol 23                            9
          + copper (II) oxide           vol 23          E10 - E26,       82
Bromobenzene
        see benzene, bromo-
1-Bromobutane
        see butane, 1-bromo-
Bromoethane
        see ethane, bromo-
Bromoform
        see methane, tribromo-
1-Bromoheptane
        see Heptane, 1-bromo-
1-Bromo-2-chlorobenzene
        see benzene, 1-bromo-2-chloro-
1-Bromo-3-chlorobenzene
        see benzene, 1-bromo-3-chloro-
1-Bromo-4-chlorobenzene
        see benzene, 1-bromo-4-chloro-
1-Bromo-4-iodobenzene
        see benzene, 1-bromo-4-iodo-
1-Bromo-3-methylbenzene
        see benzene, 1-bromo-3-methyl-
1-Bromo-naphthalene
        see naphthalene, 1-bromo-
1-Bromooctane
        see octane, 1-bromo-
1-Bromo-1,1,2,2,3,3,4,4,5,6,6,6-dodecafluoro-5-(trifluoromethyl)hexane
        see hexane, 1-bromo-1,1,2,2,3,3,4,4,5,6,6,6-dodecafluoro
               -5-(trifluoromethyl)
1-Bromo-1,1,2,2,3,3,4,4,5,5,6,6,7,7,8,8,8-heptadecafluoro-octane
        see octane, 1-bromo-1,1,2,2,3,3,4,4,5,5,6,6,7,7,8,8,8-
               hepadecafluoro-
4-Bromophenol
        see phenol, 4-bromo-
4-Bromo-1-methylbenzene
        see benzene, 4-bromo-1-methyl-
N1-(5-Bromo-2-pyridyl)sulfanilamide
        see benzenesulfonamide, 4-amino-N-(5-bromo-2-pyridinyl)-
N1-(5-Bromo-3-pyridyl)sulfanilamide
        see benzenesulfonamide, 4-amino-N-(2-bromo-5-pyridinyl)-
```

```
m-Bromotoluene
        see benzene, 1-bromo-3-methyl-
Bucarban
        see benzenesulfonamide, 4-amino-N-[(butylamino)carbonyl]-
Bucrol
        see benzenesulfonamide, 4-amino-N-[(butylamino)carbonyl]-
Bukarban
        see benzenesulfonamide, 4-amino-N-[(butylamino)carbonyl]-
1,3-Butadiene, 2-methyl-
        + water                         vol 37                 E7,    E8,
                                                                9,    10
1-Butanamine
        + 2-aminoethanol                vol 21                        99
Butanaminium, 1,1,2,2,3,3,4,4,4-nonafluoro-N,N-bis(nonafluorobutyl)-
        + methane                       vol 27/28                    721
Butanaminium, N,N,N-tributyl- bromide (aqueous)
        + butane                        vol 24         E58 - E72,     93
        + methane                       vol 27/28        E61,   E62,
                                                          96 -   98
        + propane                       vol 24         E58 - E72,     92
Butane
        + ethanamine                    vol 21         E252,  253,   254
        + ethanamine -d2                vol 21         E252,  257,   258
        + hydrogen sulfide              vol 32           E166 -E169,
                                                         192,   193
        + mercury                       vol 29           E191, E192
                                                         201,   202
        + methanamine                   vol 21           E80,   E90,
                                                          92,    93
        + methane                       vol 27/28      E281,  282 - 296
        + propane                       vol 24           E196, E197,
                                                         198 - 200
Butane (aqueous)
        + methane                       vol 27/28        196 - 199
Butane (ternary)
        + methane                       vol 27/28        387 - 389,
                                                         407 - 409,
                                                         420 - 422,
                                                         436 - 440,
                                                         447 - 449
isoButane
        see Propane, 2-methyl-
Butane, 1-bromo-
        + hydrogen sulfide              vol 32         E166 -E169,  192,
                                                         193,   283
Butane, 1,4-dicyclopentyl-
        + water                         vol 38                      425
1,3-Butanediol
        + 1-propanamine                 vol 21         E252,   267
Butane, 2,2-dimethyl-
        + butane                        vol 24         E130 -E134,  170
        + mercury                       vol 29           E107 -E109,
                                                         120,   121
        + methane                       vol 27/28      E212, E213,  223
        + 2-methylpropane               vol 24         E130 -E134,  191
        + propane                       vol 24         E130 -E134,
                                                         143,   144
        + water                         vol 37         E283, E284,
                                                         285 - 288
                                                       E289 -E291,
                                                         292 - 295
Butane, 2,3-dimethyl-
        + mercury                       vol 29         E107 -E109,  122
        + propane                       vol 24         E130 -E134,  145
Butane, 2-methyl-
        + mercury                       vol 29         E107 -E109,  112
        + methane                       vol 27/28        317 - 318
        + water                         vol 37           E31 - E34,
                                                          35 -   41
Butane, 2-methyl- (ternary)
        + methane                       vol 27/28                   411
Butane, 1,1´-oxybis-
        + mercury                       vol 29         E172, E173,  175
```

```
Butane, 2,2,3-trimethyl-
          + methane                       vol 27/28            551,   552
          + water                         vol 37                      464
Butanenitrile
          + arsine                        vol 21            E302,   303
Butanoic acid, lithium salt
          + butanoic acid, sodium salt    vol 33      E113,  114,   115
          + nitric acid, lithium salt     vol 33                     217
          + thiocyanic acid, lithium salt
                                          vol 33                     216
Butanoic acid, magnesium salt
          + butanoic acid, potassium salt
                                          vol 33      E105,  106,   107
          + butanoic acid, sodium salt    vol 33      E116,  117,   118
Butanoic acid, 2-methyl-, potassium salt
          + acetic acid, potassium salt   vol 33            E161,   162
          + butanoic acid, 2-methyl-,
            sodium salt                   vol 33      E125,  126,   127
          + nitric acid, potassium salt   vol 33      E199,  200,   201
          + nitrous acid, potassium salt  vol 33            E197,   198
Butanoic acid, 2-methyl-, sodium salt
          + acetic acid, sodium salt      vol 33      E250,  251,   252
          + benzoic acid, sodium salt     vol 33            E131,   314
          + butanoic acid, 2-methyl-,
            potassium salt                vol 33      E125,  126,   127
          + formic acid, sodium salt      vol 33            E226,   227
          + hexanoic acid, sodium salt    vol 33            E311,   312
          + nitric acid, sodium salt      vol 33      E321,  322,   323
          + nitrous acid, sodium salt     vol 33            E319,   320
          + octadecanoic acid, sodium salt
                                          vol 33            E315,   316
          + propanoic acid, 2-methyl-
            sodium salt                   vol 33            E292,   293
          + thiocyanic acid, sodium salt  vol 33            E317,   318
Butanoic acid,3-oxo-, ethyl ester
          + phosphine                     vol 21      E275,  276,   284
Butanoic acid, 4-oxo-4-[[[4-(2-pyrimidinylamino)sulfonyl]phenyl]amino]-,
disilver(1+) salt
          + water                         vol 36                     224
Butanoic acid, 4-oxo-4-[[4-[(2-thiazolylamino)sulfonyl]phenyl]amino]-
          + 2-propanol                    vol 35                     230
Butanoic acid, potassium salt
          + acetic acid, potassium salt   vol 33      E154,  155,   18?
          + butanoic acid, magnesium salt vol 33      E105,  106,   107
          + butanoic acid, sodium salt    vol 33            E108, E109,
                                                            110 -  112
          + nitrous acid, potassium salt  vol 33            E185,   186
          + nitric acid, potassium salt   vol 33            E187,   188
          + thiocyanic acid, potassium salt
                                          vol 33            E183,   184
Butanoic acid, sodium salt
          + acetic acid, sodium salt      vol 33      E243,  244 -  246
          + benzoic acid, sodium salt     vol 33                     282
          + butanoic acid, lithium salt   vol 33      E113,  114,   115
          + butanoic acid, magnesium salt
                                          vol 33      E116,  117,   118
          + butanoic acid, potassium salt
                                          vol 33            E108, E109,
                                                            110 -  112
          + formic acid, sodium salt      vol 33            E223,   224
          + hexanoic acid, sodium salt    vol 33            E280,   281
          + nitric acid, sodium salt      vol 33      E289,  290,   291
          + nitrous acid, sodium salt     vol 33            E287,   288
          + octadecanoic acid, sodium salt
                                          vol 33            E283,   284
          + propanoic acid, 2-methyl-,
            sodium salt                   vol 33      E277, E278,   279
          + thiocyanic acid, sodium salt  vol 33            E285,   286
1-Butanol
          + ammonia                       vol 21              E5,    E6
                                                             45 -    47
          + butane                        vol 24            E331 -E335,
                                                            356,   358
          + N,N-dimethylmethanamine       vol 21             E91,   189
```

```
1-Butanol
          + hydrogen sulfide            vol 32       E171, E172,   243
          + methanamine                vol 21               E90,  105
          + methane                    vol 27/28    E582 -E582, E589,
                                                     E590,  611 - 613
          + N-methylmethanamine        vol 21               E82,  148
          + 2-methylpropane            vol 24       E331- E334,   369
          + propane                    vol 24       E331- E334,   342
2-Butanol
          + ammonia                    vol 21                E6,   46
iso-Butanol
       see 1-propanol, 2-methyl-
sec-Butanol
       see 2-butanol
tert-Butanol
       see 2-propanol, 2-methyl-
2-Butanol, 2-methyl- (aqueous)
          + bromic acid, potassium salt  vol 30                   243
1-Butanol, 3-methyl-
          + ammonia                    vol 21                E6,   45
          + butane                     vol 24       E331 -E334,   359
          + N,N-dimethylmethanamine    vol 21                     190
          + methane                    vol 27/28     580, E591,   619
          + propane                    vol 24       E331 -E334,   343
2-Butanone
          + N,N-dimethylmethanamine    vol 21               181,  182
2-Butenamide, N-[[4-(acetylamino)phenyl]sulfonyl]-3-methyl-
          + phosphoric acid, disodium salt
                                       vol 34                     274
          + phosphoric acid,
            monopotassium salt         vol 34                     274
          + water                      vol 34                     274
2-Butenamide, N-[(4-aminophenyl)sulfonyl]-3-methyl-
          + 2-propanone                vol 34                     272
          + water                      vol 34                     271
2-Butenamide, N-[(4-aminophenyl)sulfonyl]-3-methyl- (aq)
          + phosphoric acid, disodium salt
                                       vol 34                     271
          + phosphoric acid, monopotassium salt
                                       vol 34                     271
2-Butenamide, N-[(4-aminophenyl)sulfonyl]-3-methyl-, monosodium salt
          + 2-propanone                vol 34                     273
1-Butene, 2,3-dimethyl-
          + water                      vol 37                     272
2-Butene, 2-methyl-
          + water                      vol 37               E22,  E23,
                                                              24 -  26
1-Butene, 3-methyl-
          + water                      vol 37                      27
Butisulfina
       see benzenesulfonamide, 4-amino-N-[(butylamino)carbonyl]-
2-Butoxyethanol
       see Ethanol, 2-butoxy-
Butoxy triethylene gylcol acetate
       see ethanol, 2-[2-(2-butoxyethoxy)-ethoxy]-, acetate
Butrolactone
       see 2(3H)-furanone, dihydro-
Butter oil
       see oil, butter
Butylacetate
       see acetic acid, butyl ester
iso-Butyl acetate
       see acetic acid, 2-methylpropyl ester
Butyl alcohol
       see 2-propanol,2-methyl-
sec-Butyl alcohol
       see 2-butanol
iso-Butylamine
       see 1-propanamine, 2-methyl-
Butylamine
       see 1-butanamine
10-Butylbenz[a]anthracene
       see benz [a] anthracene, 10-butyl-
```

```
Butylbenzene
        see benzene,butyl-
sec-Butylbenzene
        see benzene (1-methylpropyl)-
tert-Butylbenzene
        see benzene, (2-methylpropyl)-
Butyl carbinol
        see 1-pentanol
iso-Butyl carbinol
        see 1-butanol, 3-methyl-
sec-Butyl carbinol
        see 1-butanol, 2-methyl-
Butyl cellosolve
        see ethanol,2-butoxy-
1,3-Butylene glycol
        see 1,3-butanediol
Butyl ether
        see butane, 1,1´-oxybis-
Butylmethyl carbinol
        see 2-hexanol
iso-Butylmethyl carbinol
        see 2-pentanol, 4-methyl-
sec-Butylmethyl carbinol
        see 2-pentanol, 3-methyl-
Butyl propionate
        see propanoic acid, butyl ester
Butyronitrile
        see butanenitrile
1-Butyl-3-sulfanilylurea
        see benzenesulfonamide, 4-amino-N-[(butylamino)carbonyl]-
N1-[5-Butyl-1,3,4-thiadiazol-2-yl)sulfanilamide
        see benzenesulfonamide, 4-amino-N-[5-butyl-1,3,4-thiadiazol-2-yl)-
BZ 55
        see benzenesulfonamide, 4-amino-N-[(butylamino)carbonyl]-
```

C

Cadmium
   + mercury   vol 25   E402- E404,
                            405 - 420
Cadmium hydroxide
   + water   vol 23   E270 -E282,
                           284, 285
Cadmium hydroxide (aqueous)
   + hydrogen peroxide   vol 23   291, 292
   + perchloric acid, sodium salt vol 23   290
   + sodium hydroxide   vol 23   E270 -E277, 286,
                           287, 288
Cadmium hydroxide (multicomponent)
   + sodium sulfide   vol 23   293
   + sulfuric acid, cadmium salt vol 23   289
   + lithium hydroxide   vol 23   300
   + perchloric acid   vol 23   301
   + perchloric acid, sodium salt vol 23   303
   + potassium hydroxide vol 23   300
   + sodium hydroxide   vol 23   300, 301
Cadmium nitrate
   see nitric acid, cadmium salt
Cadmium oxide
   + water   vol 23   E270 -E282, 284
Cadmium oxide (aqueous)
   + ammonia   vol 23   E277, E279, 294
   + arsenic (VI) oxide   vol 23   E279, E280,
                           298, 299
   + ethanol, 2-amino-   vol 23   E279, 304
   + ethanol, 2,2´-imino- vol 23   E279, 304
   + sodium hydroxide   vol 23   E270 -E277, 286,
                        287, 296, 297
   + sulfuric acid, ammonium salt vol 23   295
Cadmium oxide (multicomponent)
   + carbon dioxide   vol 23   E280, 305, 306
   + sodium chloride   vol 23   E280, 305, 306
Cadmium perchlorate
   see perchloric acid, cadmium salt
Cadmium selenite
   see selenious acid, cadmium salt
Cadmium sulfate
   see sulfuric acid, cadmium salt
Cadmium sulfite
   see sulfurous acid, cadmium salt
Cafeina
   see 1H-purine-2,6-dione, 3,7-dihydro-1,3,7-trimethyl-
Caffeine
   see 1H-purine-2,6-dione, 3,7-dihydro-1,3,7-trimethyl-
Cafipel
   see 1H-purine-2,6-dione, 3,7-dihydro-1,3,7-trimethyl-
Calcium
   + mercury   vol 25   E66, E67, 68-70
Calcium bisulfite
   see sulfurous acid, calcium salt
Calcium chlorate
   see chloric acid, calcium salt
Calcium chloride (aqueous)
   + hydrogen sulfide   vol 32   E25, E26,
                           40,   44
   + mercury   vol 29   E87, 88 - 94
   + methane   vol 27/28   E66 - E68,
                         107 - 113
   + phosphoric acid, calcium salt
               vol 31   271
   + phosphoric acid, monopotassium salt
               vol 31   271
   + potassium chloride   vol 31   271
Calcium chloride (ternary and multicomponent)
   + methane   vol 27/28   E68, E75, E80,
                         E81, 113,
                         146, 147
Calcium dihydrogen phosphate
   see phosphoric acid, calcium salt

```
Calcium hydrogen phosphate
        see phosphoric acid, calcium salt (1:1)
Calcium hydroxide phosphate (aqueous and multicomponent)
              + sulfurous acid, calcium salt    vol 26              E191, E192
                                                                     229,  230
Calcium iodate
        see iodic acid, calcium salt
Calcium nitrate
        see nitric acid, calcium salt
Calcium selenite
        see selenious acid, calcium salt
Calcium sulfite
        see sulfurous acid, calcium salt
Calcium thiocyanate
        see thiocyanic acid, calcium salt
Camphor
        see bicyclo(2,2,1)heptan-2-one,1,7,7-trimethyl-
Caproic acid
        see hexanoic acid
Caprylic alcohol
        see 2-octanol
Carbamidal
        see 3-pyridinecarboxamide, N,N-diethyl-
Carbamic acid, ethyl ester (aqueous)
              + bromic acid, potassium salt    vol 30                      253
Carbon
              + mercury                        vol 25                      E134
Carbon dioxide
              + methane                        vol 27/28        E737,  738 - 748
Carbon dioxide (aqueous)
              + hydrogen sulfide               vol 32           E20,   21,  E27
                                                                 48,   49,   50
Carbon dioxide (aqueous and multicomponent)
              + sulfurous acid, manganese salt
                                               vol 26                  E248,  251
Carbon dioxide (multicomponent)
              + cadmium oxide                  vol 23           E280,  305,  306
              + copper (I) oxide               vol 23           E1 -  E3,    8
              + copper (II) oxide              vol 23           E10 - E26,   80
              + hydrogen sulfide               vol 32                  68 -   79
                                                                E79 - E83
                                                                118 - 165
              + mercury (II) oxide             vol 23           E316,  340,  341
              + sodium chloride                vol 23           E181,  268
Carbon dioxide (ternary and multicomponent)
              + methane                        vol 27/28              200,  201
                                                                      436 - 466
                                                                      663 - 665
                                                                      667 - 671
                                                                      757 - 760
Carbon disulfide
              + hydrogen sulfide               vol 32                  E185,  318
              + methane                        vol 27/28              580, E702
                                                                      703,  704
              + phosphine                      vol 21                 E281, E282
                                                                      295,  300
              + propane                        vol 24           E218 -E221,  222
Carbon monoxide (multicomponent)
              + hydrogen sulfide               vol 32                   75 -   77
Carbon tetrachloride
        see methane, tetrachloro-
Carbonic acid, ammonium salt (multicomponent)
              + copper (II) hydroxide          vol 23           E10 - E26,   74
Carbonic acid, dilithium salt (aqueous)
              + iodic acid, lithium salt       vol 30                 E272,  293
Carbonic acid, dipotassium salt (aqueous)
              + hydrogen sulfide               vol 32           E26,  E27,   46
                                                                 47,   49,   50
              + methane                        vol 27/28        E83,  E84,  170
              + phosphoric acid, monopotassium salt
                                               vol 31                 E212,  222
Carbonic acid, dipotassium salt (aqueous and multicomponent)
              + sulfuric acid, dipotassium salt
                                               vol 26           E92,  106,  107
```

```
Carbonic acid, dipotassium salt (ternary)
          + methane                    vol 27/28                        166
Carbonic acid, disodium salt
          + water                      vol 30            139,   157,   158
Carbonic acid, disodium salt (aqueous)
          + bromic acid, sodium salt   vol 30                   201,   216
          + chloric acid, sodium salt  vol 30            E31,   E32,    82
                                                                85,   E338
          + chlorous acid, sodium salt vol 30            E31,   E32,    85
          + hydrogen sulfide           vol 32                   E26,   E27
                                                                 45,    48
          + iodic acid, sodium salt    vol 30                  E337,  E338
                                                                344,   345
          + methane                    vol 27/28         E78,   E79,   158
          + phosphoric acid, trisodium salt
                                       vol 31           E127,  E129,   139
                                                         152,   157,   158
          + selenious acid, sodium salt vol 26                 314,   315
          + sodium bromide             vol 30                          216
          + sodium chloride            vol 30            E31,   E32,    82
          + sodium hydroxide           vol 31                   157,   158
          + tellurous acid, sodium salt vol 26          E403,   411 -  413
Carbonic acid, disodium salt (aqueous and multicomponent)
          + methane                    vol 27/28         E78,   E79,   159
          + sulfurous acid, barium salt vol 26          E240,  E241,   247
Carbonic acid, monopotassium salt (aqueous)
          + methane                    vol 27/28                E83,   169
Carbonic acid, monosodium salt (aqueous)
          + bromic acid, sodium salt   vol 30                   202,   217
          + hydrogen sulfide           vol 32                   E27,    48
          + mercury                    vol 29            E87,    88 -   94
          + methane                    vol 27/28         E77,   E78,   157
          + sodium bromide             vol 30                          217
Carbonic acid, monopotassium salt (aqueous)
          + methane                    vol 27/28                E83,   169
Carbonic acid, sodium salt (multicomponent)
          + mercury (II) oxide         vol 23                          342
Carbon tetrachloride
          see methane, tetrachloro-
N4-(Carboxymethyl)sulfanilamide
          see glycine, N-[(4-Aminosulfonyl)phenyl]-
Carbutamide
          see benzenesulfonamide, 4-amino-N-[(butylamino)carbonyl]-
Carbutamid
          see benzenesulfonamide, 4-amino-N-[(butylamino)carbonyl]-
Cardamine
          see 3-pyridinecarboxamide, N,N-diethyl-
Cardiamide
          see 3-pyridinecarboxamide, N,N-diethyl-
Cardiamid
          see 3-pyridinecarboxamide, N,N-diethyl-
Cardiamine
          see 3-pyridinecarboxamide, N,N-diethyl-
Cardimon
          see 3-pyridinecarboxamide, N,N-diethyl-
Carbonyl sulfide
          + methane                    vol 27/28                       749
Caroxin-D
          see butane,1,1,2,2,3,3,4,4-octafluoro-1,4-bis-
               (1,2,2,2-tetraflouro-1-(trifluoroethyl)ethoxy)-
Caroxin-F
          see hexane,1,1,1,2,2,3,3,4,4,5,5,6,6-tridecafluoro-
               6-(1,2,2,2,-tetrafluoromethyl)ethoxy)-
Ce sulfadiazine
          see cerium, tris(4-amino-N-2-pyrimidinylbenzenesulfonamidato-NN,O)-
Cefsulodin sodium
          see Pyridinium,4-(aminocarbonyl)-1[[2-carboxy-
               8-oxo-7-(phenyl-sulfoacetyl)-amino]-5-thia-1-
               azabicyclo[4,2,0]oct-2-en-3-yl]-methyl-hydroxide, inner salt,
               monosodium salt
Cellosolve
          see ethanol,2-ethoxy-
Cerelose
          see D-glucose
```

```
Cerium
            + mercury                       vol 25              E214, E215
                                                                216 - 220
Cerium ammonium nitrate
        see nitric acid, cerium ammonium salt
Cerium bromide
            + furan, tetrahydro-            vol 22                    109
            + pyridine                      vol 22              110 - 111
Cerium chloride
            + 1,3-butadiene, hexachloro-    vol 22                     98
            + 1,3-dioxolane                 vol 22                    103
            + ethane, 1,1'-oxybis-          vol 22                    101
            + ethanol (aqueous)             vol 22                     99
            + ethanol, 2-methoxy-           vol 22                    100
            + furan, tetrahydro-            vol 22                    104
            + heptane, 1-methoxy-           vol 22                    102
            + hydrazine                     vol 22                    107
            + nonane, 1-methoxy-            vol 22                    102
            + octane, 1-methoxy-            vol 23                    102
            + phosphoric triamide, hexamethyl-
                                            vol 22                    106
            + pyridine                      vol 22                    105
            + water (ternary)               vol 22                     99
Cerium (IV) chloride
            + phosphoryl chloride (ternary)
                                            vol 22                    108
            + stannane, tetrachloro- (ternary)
                                            vol 22                    108
Cerium cobalt nitrate
        see nitric acid, cerium cobalt salt
Cerium fluoride
            + acidic nitrosyl fluoride (nitrosyl
               fluoride compound with hydrofluoric acid)
                                            vol 22                     97
Cerium iodide
            + 1-butanol (aqueous)           vol 22                    113
            + ethanol (aqueous)             vol 22                    112
            + furan, tetrahydro-            vol 22                    114
            + water (ternary)               vol 22              112,   113
Cerium manganese nitrate
        see nitric acid, cerium manganese salt
Cerium nickel nitrate
        see nitric acid, cerium nickel
Cerium nitrate monohydrate
        see nitric acid, cerium salt, monohydrate
Cerium nitrate
        see nitric acid, cerium salt
Cerium, tris(4-amino-N-2-pyrimidinylbenzenesulfonamidato-NN,O)-
            + water                         vol 36                    209
Cesium
            + mercury                       vol 25         E50, E51, 52-54
Cesium bromate
        see bromic acid, cesium salt
Cesium chlorate
        see chloric acid, cesium salt
Cesium chloride
            + water                         vol 31                    327
Cesium chloride (aqueous)
            + chloric acid, cesium salt     vol 30         E164,  174,  175
                                                           E177, E178,  182
                                                                        183
            + chloric acid, rubidium salt   vol 30         E164,  174,  175,
                                                                        E178
            + chloric acid, sodium salt     vol 30                    E178
            + methane                       vol 27/28           E84,  171
            + phosphoric acid, monocesium salt
                                            vol 31          E318,      327
            + rubidium chloride             vol 30         E164,  174,  175,
                                                                        E178
            + sodium chloride               vol 30                    E178
Cesium dideuterium phosphate
        see phosphoric acid-d3, cesium salt (2:1)
Cesium dihydrogen phosphate
        see phosphoric acid, monocesium salt
```

```
Cesium iodate
        see iodic acid, cesium salt
Cesium nitrate
        see nitric acid, cesium salt
Cesium oxide (aqueous)
        + phosphoric acid, monocesium salt
                                        vol 31              E320, E322,
                                                            329 - 331
        + phosphoric acid-d3, monocesium salt
                                        vol 31              332,  333
Cesium sulfate
        see sulfuric acid, cesium salt
Cesium tellurite
        see tellurous acid, cesium salt
Cetamacrogol
        see poly(oxy-1,2-ethanediyl), a-hexadecyl- -hydroxy-, mixture with
            a-octadecyl- -hydroxy-poly(oxy-1,2-ethanediyl)
Cetyl trimethyl ammonium bromide
        see 1-hexadecanaminium,N,N,N-trimethyl-, bromide
Chemosept
        see benzenesulfonamide, 4-amino-N-2-thiazolyl-
Chemouag
        see benzenesulfonamide, 4-amino-N-(3,4-dimethyl-5-isoxazolyl)
Chloroform
        see methane, trichloro-
Chloral hydrate
        see 1,1-ethanediol,2,2,2-trichloro-
Chloric acid, barium salt (aqueous)
        + chloric acid, sodium salt    vol 30              E29,    73
Chloric acid, calcium salt (aqueous)
        + chloric acid, cesium salt    vol 30       E177, E178,  183
        + chloric acid, potassium salt vol 30              E108,  150
Chloric acid, cesium salt
        + water                        vol 30              E176 -E178
                                                           179 - 183
Chloric acid, cesium salt (aqueous)
        + cesium chloride              vol 30       E31, E164,   174,
                                                    175, E177, E178,
                                                    182,  183
        + chloric acid, calcium salt   vol 30       E177, E178,  183
        + chloric acid, potassium salt vol 30              E108,  149,
                                                           E177, E178
        + chloric acid, rubidium salt  vol 30       E164,  173 - 175,
                                                           E177, E178
        + chloric acid, sodium salt    vol 30       E29,   E31,   72,
                                                           E177, E178
        + rubidium chloride            vol 30       E164,  174,  175,
                                                                 E178
        + sodium chloride              vol 30              E31, E178
Chloric acid, lithium salt
        + 2-propanone                  vol 30                      23
        + water                        vol 30              E1 - E10
                                                           11 - 22
Chloric acid, lithium salt (aqueous)
        + iodic acid, calcium salt + nitric acid, lithium salt????????
                                        vol 14                      93
        + lithium chloride             vol 30       E4,    20 -  22
Chloric acid, potassium salt
        + 2-aminoethanol               vol 30                     158
        + ammonia                      vol 30                     161
        + dimethylformamide            vol 30                     160
        + 1,2-ethanediamine            vol 30                     159
        + 1,2-ethanediol               vol 30                     157
        + water                        vol 30              E105 -E112,
                                                           113 - 156
Chloric acid, potassium salt (aqueous)
        + 1,2,3-propanetriol           vol 30              E108,  153
        + 2-propanone                  vol 30       E108,  154 - 156
        + bromic acid, potassium salt  vol 30       E109,  142,  143
        + chloric acid, calcium salt   vol 30              E108,  150
        + chloric acid, cesium salt    vol 30       E108,  149, E177,
                                                                 E178
        + chloric acid, rubidium salt  vol 30       E108, E110,  139,
                                                    140,  148, E164
```

Chloric acid, potassium salt (aqueous)
        + chloric acid, sodium salt   vol 30   E29,   E31
                                                                       68 -   70
        + ethanol   vol 30   E108,   151,   152
        + glycine   vol 30   E108
        + iodic acid, potassium salt   vol 30   145,   146,   E380
        + nitric acid, potassium salt   vol 30   122
        + perchloric acid, potassium salt
                                             vol 30   E109, E110
                                                                     137,   138
        + potassium bromide   vol 30   E108, E109,   141
        + potassium chloride   vol 30   E31, E108, E110
                                                             125 - 140, E164
        + potassium hydroxide   vol 30   E108,   147
        + potassium iodide   vol 30   E108, E109,   144
        + rubidium chloride   vol 30   E164
        + rubidium chloride   vol 30   E110,   139,   140
        + sodium chloride   vol 30   E31, E108, E109
                                                                              121
        + sulfuric acid, potassium salt
                                             vol 30   123,   124
        + water-d2   vol 30   119,   120
Chloric acid, rubidium salt
        + water   vol 30   E162 -E165
                                                                    166 - 175
Chloric acid, rubidium salt (aqueous)
        + cesium chloride   vol 30   E164,   174
                                                                    175,   E178
        + chloric acid, cesium salt   vol 30   E164,   173 - 175
                                                                    E177, E178
        + chloric acid, potassium salt   vol 30   E108, E110,   139
                                                             140,   148, E164
        + chloric acid, sodium salt   vol 30   E29,   E31
                                                                     71,   E164
        + nitric acid, rubidium salt   vol 30   E164,   169,   170
        + potassium chloride   vol 30   E110, E164
        + rubidium chloride   vol 30   E31, E110,   139
                                                             140, E164,   171
                                                             172,   174,   175
                                                                            E178
        + sodium chloride   vol 30   E31, E164
Chloric acid, silver salt
        + water   vol 30   79,   80
Chloric acid, silver salt (aqueous)
        + chloric acid, sodium salt   vol 30   79,   80
Chloric acid, sodium salt
        + 2-aminoethanol   vol 30   99
        + 1,2-ethanediamine   vol 30   101
        + 1,2-ethanediol   vol 30   98
        + dimethylformamide   vol 30   103
        + hydrazine   vol 30   104
        + 2-propanone   vol 30   100
        + tetrahydrothiophene, 1,1-dioxide
                                             vol 30   103
        + water   vol 30   E24 - E33
                                                                      34 -   97
Chloric acid, sodium salt (aqueous)
        + barium chloride   vol 30   E28,   E29
        + boric acid, sodium salt   vol 30   E30,   37 -   40
        + bromic acid, sodium salt   vol 30   E29,   62,   63
        + carbonic acid, disodium salt   vol 30   E30,   E31,   E32
                                                                     82,   85
        + cesium chloride   vol 30   E31,   E178
        + chloric acid, barium salt   vol 30   E29,   73
        + chloric acid, cesium salt   vol 30   E29,   E31,   72
                                                                    E177, E178
        + chloric acid, potassium salt   vol 30   E29,   E31
                                                                     68 -   70
        + chloric acid, rubidium salt   vol 30   E29,   E31
                                                                     71,   E164
        + chloric acid, silver salt   vol 30   79,   80
        + chlorous acid, sodium salt   vol 30   E29,   E31,   E32
                                                                     83 -   89
        + chromic acid, disodium salt   vol 30   E30,   74,   75

Chloric acid, sodium salt (aqueous)
```
         + disodium salt (I-4)-teroxomolybdate (2-)
                                      vol 30        E30,   76,    77
         + iodic acid, sodium salt    vol 30               65,    66
         + nitric acid, calcium salt  vol 30                      78
         + nitric acid, sodium salt   vol 30        E30,          41
         + potassium chloride         vol 30  E28,  E29,   E31
         + rubidium chloride          vol 30        E31,   E164
         + sodium bromide             vol 30        E28,          61
         + sodium chloride            vol 30  E28,  E31,   E32
                                                    51  -   58
                                                    82  -   84
                                                    86  -   89
                                              E164,  E178
         + sodium fluoride            vol 30        E28,          50
         + sodium hydroxide           vol 30        E30,          67
         + sodium iodide              vol 30                      E28
         + sulfuric acid, sodium salt vol 30  E30,  42  -   49
         + urea                       vol 30                      81
```
Chloride
         see under individual chlorides
Chloroacetic acid
         see acetic acid, chloro-
Chlorobenzene
         see benzene, chloro-
1-Chlorobutane
         see butane, 1-chloro-
2-Chlorobutane
         see butane, 2-chloro-
Chlorodifluoromethane
         see methane, chlorodifluoro-
Chloroethane
         see ethane, chloro-
Chloroform
         see methane, trichloro-
1-Chloroheptane
         see heptane, 1-chloro-
1-Chlorohexane
         see hexane, 1-chloro-
Chloromethane
         see methane, chloro-
(Chloromethyl)benzene
         see benzene, (chloromethyl)-
1-Chloronaphthalene
         see naphtahalene, 1-chloro-
1-Chlorooctane
         see octane, 1-chloro-
1-Chloropentane
         see pentane, 1-chloro-
2-Chlorophenol
         see phenol, 2-chloro-
1-Chloropropane
         see propane, 1-chloro-
2-Chloropropane
         see propane, 2-chloro-
N1-(6-Chloro-3-pyridazinyl)sulfanilamide
         see benzenesulfonamide, 4-amino-N-(6-chloro-3-pyridazinyl)-
N1-(6-Chloro-3-pyridyl)sulfanilamide
         see benzenesulfonamide, 4-amino-N-(2-chloro-5-pyridinyl)-
N1-(2-Chloro-5-pyrimidinyl)sulfanilamide
         see benzenesulfonamide, 4-amino-N-(2-chloro-5-pyrimidinyl)-
3-Chloro-6-sulfanilamidopyridazine
         see benzenesulfonamide, 4-amino-N-(6-chloro-3-pyridazinyl)-
2-Chlorotoluene
         see benzene, 1-chloro-2-methyl-
3-Chlorotoluene
         see benzene, 1-chloro-3-methyl-
4-Chlorotoluene
         see benzene, 1-chloro-4-methyl-
Chlorotrifluoromethane
         see methane, chlorotrifluoro-
4-Chloro-2,5-dimethylphenol
         see phenol,4-chloro-2,5-dimethyl-

4-Chloro-2,6-dimethylphenol
          see phenol,4-chloro-2,6-dimethyl-
4-Chloro-3,5-dimethylphenol
          see phenol,4-chloro-3,5-dimethyl-
4-Chloro-3,5-dimethyl-2-(phenylmethyl)phenol
          see phenol,4-Chloro-3,5-dimethyl-2-(phenylmethyl)-
1-Chloro-1,1,2,2,3,3,4,4,5,6,6,6-dodecafluoro-5-(trifluoromethyl)hexane,
          see hexane, 1-chloro-1,1,2,2,3,3,4,4,5,6,6,6-dodecafluoro
                5-(trifluoromethyl)-
1-Chloro-2-iodobenzene
          see benzene, 1-chloro-2-iodo-
1-Chloro-3-iodobenzene
          see benzene, 1-chloro-3-iodo-
1-Chloro-4-iodobenzene
          see benzene, 1-chloro-4-iodo-
4-Chloro-3-methyl-5-ethylphenol
          see phenol, 4-chloro-3-methyl-5-ethyl-
1-Chloro-2-methylbenzene
          see benzene, 1-chloro-2-methyl-
1-Chloro-3-methylbenzene
          see benzene,1-chloro-3-methyl-
4-Chloro-2-methylphenol
          see phenol, 4-chloro-2-methyl-
4-Chloro-3-methylphenol
          see phenol, 4-chloro-3-methyl-
2-Chloro-6-methylphenol
          see phenol, 2-chloro-6-methyl-
1-Chloro-2-methylpropane
          see propane, 1-chloro-2-methyl-
2-Chloro-2-methylpropane
          see propane, 2-chloro-2-methyl-
1-Chloronaphthalene
          see naphthalene, 1-chloro-
1-Chloro-2-nitrobenzene
          see benzene,1-chloro-2-nitro-
1-Chloro-3-nitrobenzene
          see benzene,1-chloro-3-nitro
1-Chloro-4-nitrobenzene
          see benzene,1-chloro-4-nitro-
2-Chlorooctane
          see octane, 2-chloro-
2-Chlorophenol
          see phenol, 2-chloro-
3-Chlorophenol
          see phenol, 3-chloro-
4-Chlorophenol
          see phenol, 4-chloro-
2-Chloro-6-(trichloromethyl)pyridine
          see pyridine,2-chloro-6-(trichloromethyl)-
5-Chloro-2-(trichloromethyl)pyridine
          see pyridine,5-chloro-2-(trichloromethyl)-
Chloro(trifluoromethyl)benzene
          see benzene,chloro(trifluoromethyl)-
Chlorous acid, sodium salt (aqueous)
          + carbonic acid, disodium salt   vol 30   E1,   E32,   85
          + chloric acid, sodium salt   vol 30   E29, E31,   E32,
                                                                83 -   89
          + sodium chloride   vol 30   E31, E32,   59,
                                                                60,   83,   84,
                                                                86 -   89
Cholanthrene
          see benz [j] aceanthrylene, 1,2-dihydro-
Cholanthrene, 3-methyl-
          see benz [j] aceanthrylene, 1,2-dihydro-3-methyl-
N-Choloyltaurine, monosodium salt
          see ethanesulfonic acid,2-[[(3a,5B,7a,12a)-3,7,12-trihydroxy-,
                24-oxocholan-24-yl]amino]-, monosodium salt
Chrysene
          + water   vol 38   E486, E487,
                                                                488 - 491
Chrysene (aqueous)
          + sodium chloride   vol 38   492
Chrysene (multicomponent)
          + methane   vol 27/28   579

```
Chrysene, 5-methyl-
        + water                         vol 38                        511
Chrysoidine, 4´-aminosulfonyl)-
        see benzenesulfonamide, 4-[(2,4-diaminophenyl)azo]-
Cholesterol
        see Cholest-5-en-3 -ol
Choline chloride
        see ethanaminium, 2-hydroxy-N,N,N-trimethyl-, chloride
Chorex
        see ehane, 1,1´-oxybis (2-chloro-
Choroid layer
        see rbbit choroid layer
Chromic acid, disodium salt (aqueous)
        + chloric acid, sodium salt    vol 30            E30,  74,   75
        + iodic acid, magnesium salt   vol 30                 E270, E271
Chromic sulfate
        see Sulfuric acid, chromium salt
Chromium
        + mercury                      vol 25        E277,  278 - 281
        + mercury                      vol 29        E214,  217
Chromium, bis(4-amino-N-2-pyrimidinylbenzenesulfonamidato-NN,O)-
        + water                        vol 36                        210
Chromium (VI) oxide (aqueous)
        + silver (I) oxide             vol 23              E83,  E84
                                                           E92,  123
        + zinc oxide                   vol 23              E174 -E176
                                                           194,   230
Chromium sulfate
        see sulfuric acid, chromium salt
Chrysene (multicomponent)
        + methane                      vol 27/28                     579
Ciba 18,605-Ba
        see benzenesulfonamide, 4-amino-N-(3-methyl-1-phenyl-
           1H-pyrazol-5-yl)-
Ciba 3753
        see benzenesulfonamide, 4-amino-N-(4-methyl-2-thiazolyl)-
Cibazol
        see benzenesulfonamide, 4-amino-N-2-thiazolyl-
Cinnabar (aqueous)
        + mercury                      vol 29              E47,   56
Citric acid (aqueous and multicomponent)
        + sulfurous acid, calcium salt vol 26        E191, E192,  233
Clay slurry (multicomponent)
        + methane                      vol 27/28            192,  193
Co(II) sulfadiazine
        see cobalt, bis(4-amino-N-2-pyrimidinylbenzenesulfonamidato-NN,O)-
Coal creosote (lignite)
        + methane                      vol 27/28                     565
Coal liquid-distillate from Exxon donor solvent process
        + methane                      vol 27/28            561,  562
Coal liquid-distillate from solvent refined coal process 2
        + methane                      vol 27/28            563,  564
Cobalt
        + mercury                      vol 25            E310, E311
                                                         338 - 356
        + mercury                      vol 29            E214,  217
Cobalt acetate
        see acetic acid, cobalt (2+) salt
Cobalt, bis(4-amino-N-2-pyrimidinylbenzenesulfonamidato-NN,O)-
        + water                        vol 36                        213
Cobalt, bis[4-amino-N-(2,6-dimethoxy-4-pyrimidinyl)-benzenesulfonamidato]-,
hydrate
        + hydrochloric acid            vol 36                        429
        + water                        vol 36                        429
Cobalt, bis[4-amino-N-(4,6-dimethyl-2-pyrimidinyl)-
benzenesulfonamidato-NN,01]-, hydrate
        + hydrochloric acid            vol 36                        348
        + water                        vol 36                        348
Cobalt, bis[4-amino-N-(6-methoxy-3-pyridazinyl)benzenesulfonamidato]-diaqua
        + water                        vol 36               115,  116
```

Cobalt, bis[4-amino-N-(6-methoxy-3-pyridazinyl)benzenesulfonamidato]-
diaqua,(aq)
            + acetic acid, cobalt(2+) salt   vol 36   116
            + 4-amino-N-(6-methoxy-3-pyridazinyl)benzenesulfonamide,
              monosodium salt   vol 36   116
            + hydrochloric acid   vol 36   115
Cobalt, bis(4-amino-N-2-thiazolylbenzenesulfonamidato-NN,O1)-, hydrate
            + hydrochloric acid   vol 35   208
            + water   vol 35   208
Cobalt nitrate
        see nitric acid, cobalt salt
Cobalt (II) selenite
        see selenious acid, cobalt (11) salt
Cobalt sulfate
        see sulfuric acid, cobalt salt
Cobalt sulfite
        see sulfurous acid, cobalt (11) salt
Cobalt tellurite
        see tellurous acid, cobalt salt
Cobaltous sulfate
        see sulfuric acid, cobalt salt
Coccoclase
        see benzenesulfonamide, 4-amino-N-2-pyridinyl-
Coco-Diazine
        see benzenesulfonamide, 4-amino-N-2-pyrimidinyl-
Coffein
        see 1H-purine-2,6-dione, 3,7-dihydro-1,3,7-trimethyl-
Colistatin
        see butanoic acid, 4-oxo-4-[[4-[(2-thiaxolylamino)sulfonyl]-
            phenyl]amino]-
Collomide
        see benzenesulfonamide, 4-amino-
Colsulanyde
        see benzenesulfonamide, 4-amino-
Consulid
        see benzenesulfonamide, 4-amino-N-(6-chloro-3-pyridazinyl)-
Copper
            + mercury   vol 25   E335 -E337,
                                                                  338 - 356
            + water   vol 29   E228, 229
Copper acetate
        see Acetic acid, copper (2+) salt
Copper, (4-amino-N-2-pyrimidinylbenzenesulfonamidato-NN,O)-
            + water   vol 36   211
Copper, bis(4-amino-N-2-pyrimidinylbenzenesulfonamidato-NN,O)-
            + water   vol 36   212
Copper, bis[4-amino-N-(2,6-dimethoxy-4-pyrimidinyl)-benzenesulfonamidato]-,
hydrate
            + hydrochloric acid   vol 36   428,   430
            + water   vol 36   428,   430
Copper, bis[4-amino-N-(4,6-dimethyl-2-pyrimidinyl)benzenesulfonamidato-
NN,O1]-, hydrate
            + hydrochloric acid   vol 36   349
            + water   vol 36   349
Copper, bis(4-amino-N-2-thiazolylbenzenesulfonamido-NN,O1)-, hydrate
            + hydrochloric acid   vol 35   209
            + water   vol 35   209
Copper (II) hydroxide
            + water   vol 23   E10 - E26,   31
Copper (II) hydroxide (aqueous)
            + ammonium hydroxide   vol 23   E10 - E26,
                                                                  49,   51
            + hydrogen peroxide   vol 23   E10 - E26,
                                                                  54,   55
            + perchloric acid, sodium salt   vol 23   E10 - E26,   57
            + selenium oxide   vol 23   E10 - E26,   62
            + sodium hydroxide   vol 23   E10 - E26,
                                                                  34 -   39,
                                                                  52,   61
            + sulfuric acid   vol 23   E10 - E26,
                                                                  43 -   44
            + sulfuric acid, copper salt   vol 23   E10 - E26,
                                                                  45,   50

Copper (II) hydroxide (multicomponent)
          + ammonium hydroxide   vol 23   E10 - E26,
                                                                     64 -   78
          + barium hydroxide   vol 23   E10 - E26,   65
          + benzenamine   vol 23   E10 - E26,   78
          + carbonic acid, ammonium salt   vol 23   E10 - E26,   74
          + ethanol   vol 23   E10 - E26,   69
          + glucose   vol 23   E10 - E26,   71
          + lithium chloride   vol 23   E10 - E26,   78
          + lithium hydroxide   vol 23   E10 - E26,   79
          + methanol   vol 23   E10 - E26,   68
          + 1-propanol   vol 23   E10 - E26,   70
          + 2-propanone   vol 23   E10 - E26,   67
          + pyridine   vol 23   E10 - E26,   77
          + sodium hydroxide   vol 23   E10 - E26,
                                                                    65,   66
          + sucrose   vol 23   E10 - E26,   73
          + sulfuric acid, ammonium salt   vol 23   E10 - E26,
                                                                     64,   75
          + sulfuric acid, sodium salt   vol 23   E10 - E26,   64
          + thiocyanic acid, ammonium salt
                                             vol 23   E10 - E26,   76
          + urea   vol 23   E10 - E26,   73
Copper nitrate
      see nitric acid, copper salt
Copper (I) oxide
          + seawater   vol 23   E1 -   E3,
                                                                     4,   5
          + water   vol 23   E1 -   E3,
                                                                     6 -   8
Copper (I) oxide (aqueous)
          + ammonium hydroxide   vol 23   E1 -   E3,   7
Copper (I) oxide (multicomponent)
          + acetic acid, methyl ester   vol 23   9
          + acetonitrile   vol 23   9
          + bromine   vol 23   9
          + carbon dioxide   vol 23   E1 -   E3,   8
          + potassium chloride   vol 23   E1 -   E3,   8
          + sodium chloride   vol 23   E1 -   E3,   8
          + water   vol 23   E1 -   E3,   8
Copper (II) oxide
          + seawater   vol 23   E10 - E26,   28
          + water   vol 23   E10 - E26,   27,
                                                                    29,   30,
                                                                    32,   33
Copper (II) oxide (aqueous)
          + ammonium hydroxide   vol 23   E10 - E26,   59
          + nitric acid   vol 23   E10 - E26,   53
          + perchloric acid, sodium salt   vol 23   E10 - E26,   56
          + potassium hydroxide   vol 23   E10 - E26,
                                                                    47,   48
          + sodium hydroxide   vol 23   E10 - E26,   46,
                                                                    58,   60
          + sulfuric acid   vol 23   E10 - E26,
                                                                    40 -   42
Copper (II) oxide (multicomponent)
          + acetic acid, methyl ester   vol 23   E10 - E26,   82
          + acetonitrile   vol 23   E10 - E26,   82
          + bromine   vol 23   E10 - E26,   82
          + carbon dioxide   vol 23   E10 - E26,   80
          + hydrofluoric acid   vol 23   E10 - E26,   63
          + perchloric acid, copper (II) salt
                                             vol 23   E10 - E26,   81
          + potassium chloride   vol 23   E10 - E26,   80
          + potassium fluoride   vol 23   E10 - E26,   63
          + sodium chloride   vol 23   E10 - E26,   80
          + sodium hydroxide   vol 23   E10 - E26,   81
Copper perchlorate
      see perchloric acid, copper salt
Copper phosphate
      see phosphoric acid, copper(II) salt
Copper (II) selenite
      see selenious acid, copper (II) salt

```
Copper sulfate
        see sulfuric acid, copper (II) salt
Copper (II) sulfate
        see sulfuric acid, copper (II) salt
Copper (I) sulfite
        see sulfurous acid, copper (I) salt
Copper (I, II) sulfite
        see sulfurous acid, copper (I, II) salt
Copper tellurite
        see tellurous acid, copper salt
Copticide
        see benzenesulfonamide, 4-amino-
Coracon
        see 3-pyridinecarboxamide, N,N-diethyl-
Coramine
        see 3-pyridinecarboxamide, N,N-diethyl-
Cordiamin
        see 3-pyridinecarboxamide, N,N-diethyl-
Corediol
        see 3-pyridinecarboxamide, N,N-diethyl-
Cormed
        see 3-pyridinecarboxamide, N,N-diethyl-
Cormid
        see 3-pyridinecarboxamide, N,N-diethyl-
Cornotone
        see 3-pyridinecarboxamide, N,N-diethyl-
Coronene
            + water                         vol 38                    543
Corvin
        see 3-pyridinecarboxamide, N,N-diethyl-
Corvitol
        see 3-pyridinecarboxamide, N,N-diethyl-
Corvotone
        see 3-pyridinecarboxamide, N,N-diethyl-
Cordycepic acid
        see D-mannitol
Cotton seed oil
        see oil, cotton seed
Cr(II) sulfadiazine
        see chromium, bis(4-amino-N-2-pyrimidinyl-
            benzenesulfonamidato-NN,O)-chromium
Cremodiazine
        see benzenesulfonamide, 4-amino-N-2-pyrimidinyl-
Cremomerazine
        see benzenesulfonamide, 4-amino-N-(4-methyl-2-pyrimidinyl)-
Cremomethazine
        see benzenesulfonamide, 4-amino-N-(4,6-dimethyl-2-pyrimidinyl)-
Cremosuxidine
        see butanoic acid, 4-oxo-4-[[4-[(2-thiaxolylamino)sulfonyl]-
            phenyl]amino]-
Creatine
        see Glycine, amino-N-(aminoiminomethyl)-N-methyl-
Creosote oil
        see coal creosote
1,2-Cresol
        see phenol, 2-methyl-
1,3-Cresol
        see phenol, 3-methyl-
o-Cresol
        see phenol, 2-methyl-
m-Cresol
        see phenol, 3-methyl-
Crill 16
        see sorbitan (z)-9-octadecenoate (2:3)
Crill K 16
        see sorbitan (z)-9-octadecenoate (2:3)
Crotoamide, 3-methyl-N-[(4-aminophenyl)sulfonyl]-
        see 2-butenamide, N-[(4-aminophenyl)sulfonyl]-3-methyl-
12-Crown-4
        see cyclododecane, 1,4,7,10-tetraoxa-
15-Crown-5
        see cyclopentadecane, 1,4,7,10,13,-pentaoxa-
18-Crown-6
        see 1,4,7,10,13,16-hexaoxacylcooctadecane
```

```
CTAB
          see 1-Hexadecanaminium, N,N,N-trimethyl-, bromide
Cumene
          see benzene, (1-methylethyl)-
p-Cumeme
          see benzene, 1-methyl-4-(1-methylethyl)-
Cu(I) sulfadiazine
          see copper, (4-amino-N-2-pyrimidinylbenzenesulfonamidato-NN,O)-
Cu(II) sulfadiazine
          see copper, bis(4-amino-N-2-pyrimidinylbenzenesulfonamidato-NN,O)-
Cupric sulfadiazine
          see copper, bis(4-amino-N-2-pyrimidinylbenzenesulfonamidato-NN,O)-
Cuprous sulfadiazine
          see copper, (4-amino-N-2-pyrimidinylbenzenesulfonamidato-NN,O)-
Cupric bromide
          see copper bromide
Cupric chloride
          see copper chloride
Cuprous nitrate
          see nitric acid, copper salt
Cuprous sulfate
          see sulfuric acid, copper (2+) salt (1,1)
Cyanate, sodium
          see hydrocyanic acid, sodium salt
Cyanate, potassium
          see hydrocyanic acid, potassium salt
Cyanide, potassium
          see potassium cyanide
Cyanide, silver
          see silver cyanide
Cyanide, sodium
          see sodium cyanide
Cyclododecane, 1,4,7,10-tetraoxa-
          + hydrogen sulfide          vol 32          E173, E176,  265
Cycloheptane
          + water                     vol 37                       454
1,3,5-Cycloheptatriene
          + water                     vol 37          E433, E434,
                                                      435 - 437
Cycloheptene
          + water                     vol 37                       440
1,4-Cyclohexadiene
          + water                     vol 37          E193, E194,
                                                      195,  196
Cyclohexanamine
          + methane                   vol 27/28                    708
          + propane                   vol 24          E299 -E302,  310
Cyclohexane
          + ammonia                   vol 21          E2,   20 -  23,
                                                                  E90
          + hydrogen sulfide          vol 32          E166 -E169,
                                                      190,  216
          + mercury                   vol 29          E133, E134,
                                                      138 - 141
          + methane                   vol 27/28       E450,  455,  456,
                                                      463, E465,
                                                      466 - 474
          + phosphine                 vol 21          E281, E282,  285
          + propane                   vol 24          E218 -E221,
                                                      223,  224
          + water                     vol 37          E221 -E229,
                                                      230 - 269
          + water-d2                  vol 37                 270,  271
Cyclohexane, 1,2-dimethyl-, cis-
          + mercury                   vol 29          E134 -E137,  143
          + methane                   vol 27/28       E451, E452,  458
          + water                     vol 38                       100
Cyclohexane, 1,2-dimethyl-, trans-
          + mercury                   vol 29          E134 -E137,  144
          + methane                   vol 27/28       E451, E452,  459
Cyclohexane, 1,3-dimethyl-, cis-
          + methane                   vol 27/28       E451. E452,  461
Cyclohexane, 1,3-dimethyl-, trans-
          + methane                   vol 27/28       E451, E452,  461
```

Cyclohexane, 1,4-dimethyl-
        + water   vol 38   101 - 103
Cyclohexane, 1,4-dimethyl-, cis-
        + mercury   vol 29   E134 -E137,   145
        + methane   vol 27/28   E451, E452,   460
Cyclohexane, 1,4-dimethyl-, trans-
        + mercury   vol 29   E134 -E137,   146
        + methane   vol 27/28   E451, E452,   460
Cyclohexane, ethyl-
        + hydrogen sulfide   vol 32   E166 -E169,   218
        + water   vol 38   E104 -E106
                                                                  107 - 109
Cyclohexane, perfluoromethyl-
        see cyclohexane, undecafluoro(trifluoromethyl)-
Cyclohexane, methyl-
        + ammonia   vol 21   E2,   23
        + hydrogen sulfide   vol 32   E166 -E169,   217
        + mercury   vol 29   E134 -E137,   142
        + methane   vol 27/28   E450, E451,   457
        + water   vol 37   E445, E447
                                                                  448 - 453
Cyclohexane, (1-methylethyl)-
        + hydrogen sulfide   vol 32   E166 -E169,   221
Cyclohexane, 1-methyl-4-(1-methylethenyl)-(R)-
        + water   vol 38   286
Cyclohexane, propyl-
        + hydrogen sulfide   vol 32   E166 -E169
                                                                  219,   220
Cyclohexane, 1,1,3-trimethyl-
        + water   vol 38   214,   215
Cyclohexane, undecafluoro(trifluoromethyl)-
        + phosphine   vol 21   E281, E282,   285
Cyclohexanol
        + ammonia   vol 21   E6,   49
        + methane   vol 27/28   E637, 638,   639
        + phosphine   vol 21   E281, E282,   288
        + silane   vol 21   E315,   320
Cyclohexanone
        + ammonia   vol 21   E7,   E8
        + methane   vol 27/28   661
        + phosphine   vol 21   181,   182
Cyclohexene
        + ammonia   vol 21   E3,   27
        + mercury   vol 29   E147,   148
        + methane   vol 27/28   463, E491
        + phosphine   vol 21   E281, E282,   285
        + water   vol 37   E197 -E199
                                                                  200 - 208
Cyclohexene (ternary)
        + ammonia   vol 21   E3,   31,   32
Cyclohexene, 1-methyl-
        + water   vol 37   439
1-Cyclohexene, 4-vinyl-
        + water   vol 38   92
Cyclohexylamine
        see cyclohexanamine
Cyclooctane
        + methane   vol 27/28   E452, E453,   462
        + water   vol 38   110

Cyclopentadecane, 1,4,7,10,13-pentaoxa-
        + hydrogen sulfide   vol 32   E173, E176,   265
1,3-Cyclopentadiene
        + water   vol 37   1
Cyclopentane
        + water   vol 37   E13 - E15
                                                                   16 - 21
Cyclopentane, butyl-
        + water   vol 38   213
Cyclopentane, ethyl-
        + water   vol 37   443,   444
Cyclopentane, hexyl-
        + water   vol 38   325

```
Cyclopentane, methyl-
          + water                           vol 37              E212, E213,
                                                                214 - 220
Cyclopentane, methyl- (aqueous)
          + sodium chloride                 vol 37                       220
Cyclopentane, methylethyl-
          + water                           vol 38                        97
Cyclopentane, pentyl-
          + water                           vol 38                       291
Cyclopentane, 1-propenyl-
          + water                           vol 38                        93
Cyclopentane, propyl-
          + water                           vol 38               98,    99
Cyclopentane, 1,1,3-trimethyl-
          + water                           vol 38               95,    96
Cyclopentanone
          + N,N-dimethylmethanamine         vol 21              179,    180
          + methane                         vol 27/28                    660
Cyclopentene
          + water                           vol 37                E2,    E3,
                                                                  4 -    6
2-Cyclopentyloctane
      see octane, 2-cyclopentyl-
Cyclotetrasiloxane, octamethyl-
          + butane                          vol 24      E218 -E221,   243
          + methane                         vol 27/28   E724,  725,   726
          + 2-methylpropane                 vol 24      E218 -E221,   250
          + propane                         vol 24      E218 -E221,   235
L-Cysteine monohydrochloride (aqueous)
          + mercury                         vol 29             E50,    82
L-Cysteine monohydrochloride (in Hanks balanced salt solution)
          + mercury                         vol 29      E50,  E87,    91
```

D

Daimeton
          see benzenesulfonamide, 4-amino-N-(6-methoxy-4-pyrimidinyl)-
Davozil
          see benzenesulfonamide, 4-amino-N-(6-methoxy-3-pyridazinyl)-
DB 32
          see benzenesulfonamide, 4-amino-N-[4-(aminosulfonyl)phenyl]-
DB 87
          see benzenesulfonamide, 4-amino-N-[4-[(methylamino)sulfonyl]-
             phenyl]-
DB 90
          see benzenesulfonamide, 4-[[(4-aminophenyl)sulfonyl]amino]-
             N,N-dimethyl-
          or
          see benzenesulfonamide, 1-amino-N-[5-(1,1´dimethylethyl)-
             1,3,4-thiadiazol-2-yl)-
Debenal
          see benzenesulfonamide, 4-amino-N-2-pyrimidinyl-
Debenal-M
          see benzenesulfonamide, 4-amino-N-(4-methyl-2-pyrimidinyl)-
Decahydronaphthalene
          see Naphthalene decahydro-
Decalin
          see Naphthalene decahydro-
Decanaminium, N,N,N-trimethyl-, bromide (aqueous)
          + methane   vol 27/28   101
Decane
          + ammonia   vol 21   16
          + butane   vol 24   E130 -E134,
                                                           174 - 176, E196,
                                                           E197, 216, 217
          + N,N-dimethylmethanamine   vol 21   E84, 183
          + hydrogen sulfide   vol 32   E166 -E169, 188,
                                                                 204 - 207
          + mercury   vol 29   E102 -E104,
                                                           E106, 131, 132
          + methanamine   vol 21   E80, E89, E90,
                                                                  98,   99
          + methane   vol 27/28   E202 -E204, E208,
                                                           230, 231, E360,
                                                                 361 - 369,
                                                                 400 - 406
          + N-methylmethanamine   vol 21   E82, E90, 139
          + 2-methylpropane   vol 24   E130 -E134,
                                                                 193, 194
          + phosphine   vol 21   E281, E282, 273
          + propane   vol 24   E130 -E134,
                                                           149 - 151, E196,
                                                           E197, 208, 209
          + seawater   vol 38   E308, 309, 310
          + water   vol 38   E294 -E296,
                                                                 297 - 307
Decane (multicomponent)
          + methane   vol 27/28   418, 419, 579
Decanedioic acid, bis(2-ethylhexyl) ester
          + 2-methylpropane   vol 24   E374 -E377, 409
          + propane   vol 24   E374 -E377, 393
1-Decanol
          + methane   vol 27/28   E582 -E584, E593,
                                                           E594, 626, 627
1-Decene
          + water   vol 38   292
Deltazina
          see benzenesulfonamide, 4-amino-N-2-pyrimidinyl-
Depocid
          see benzenesulfonamide, 4-amino-N-(1-phenyl-1H-pyrazol-5-yl)-
Depotsulfonamide
          see benzenesulfonamide, 4-amino-N-(1-phenyl-1H-pyrazol-5-yl)-
Dermazin
          see silver, (4-amino-N-2-pyrimidinylbenzenesulfonamidato-NN-01)-
Deseptyl
          see benzenesulfonamide, 4-amino-

```
Deuterium phosphate
        see phosphoric acid-d3
Dextrose
        see D-glucose
Diaboral
        see benzenesulfonamide, 4-amino-N-[(butylamino)carbonyl]-
Nl,N4-Diacetyl-Nl-(3,4-dimethyl-5-isoxazolyl)sulfanilamide
        see acetamide, N-[[4-(acetylamino)phenyl]sulfonyl]-
           N-(3,4-dimethyl-5-isoxazolyl)-
Nl,N4-Diacetylsulfafurazole
        see acetamide, N-[[4-(acetylamino)phenyl]sulfonyl]-
           N-(3,4-dimethyl-5-isoxazolyl)-
Nl,N4-Diacetylsulfametrole
        see acetamide, N-[[4-(acetylamino)phenyl]sulfonyl]-
           N-(4-methoxy-1,2,5-thiadiaxol-3-yl)-
Nl,N4-Diacetylsulfanilamide
        see acetamide, N-[[(4-acetylamino)phenyl]sulfonyl]-
Nl,N4-Diacetylsulfisoxazole
        see acetamide, N-[[4-(acetylamino)phenyl]sulfonyl]-
           N-(3,4-dimethyl-5-isoxazolyl)-
1,2-Diaminoethane
        see 1,2-ethanediamine
1,6-Diaminohexane
        see 1,6-hexanediamine
Nl-(Diaminomethylene)sulfanilamide
        see benzenesulfonamide, 4-amino-N-[(aminoiminomethyl)-
4-[(2,4-Diaminophenyl)azo]benzenesulfonamide
        see benzenesulfonamide, 4-[(2,4-diaminophenyl)azo]-
p-[(2,4-Diaminophenyl)azo]benzenesulfonamide
        see benzenesulfonamide, 4-[(2,4-diaminophenyl)azo]-
1,2-Diaminopropane
        see 1,2-propanediamine
1,3-Diaminopropane
        see 1,3-propanediamine
Diammonium hydrogen phosphate
        see phosphoric acid, diammonium salt
Diazil
        see benzenesulfonamide, 4-amino-N-(4,6-dimethyl-2-pyrimidinyl)-
Diazin
        see benzenesulfonamide, 4-amino-N-2-pyrimidinyl-
Diazolone
        see benzenesulfonamide, 4-amino-N-2-pyrimidinyl-
Di-Azo-Mul
        see benzenesulfonamide, 4-amino-N-2-pyrimidinyl-
Diazovit
        see benzenesulfonamide, 4-amino-N-2-pyrimidinyl-
Diazyl
        see benzenesulfonamide, 4-amino-N-(4,6-dimethyl-2-pyrimidinyl)-
        or
        see benzenesulfonamide, 4-amino-N-2-pyrimidinyl-
Dibenz [a,h] anthracene
        + water                              vol 38      E534,  535,  536
        + salt water                         vol 38                   537
Dibenz [a,j] anthracene
        + water                              vol 38                   538
        + salt water                         vol 38                   539
Dibenzo-18-crown-6
        see [1,4,7,10,13,16] hexaoxocyclooctadecin,
           6,7,10,17,18,20,21-octahydrodibenzo [b,k]
Dibenzofuran
        see furan, dibenzo-
Dibenzyl ether
        see benzene, 1,1´-[oxybis(methylene)]bis-
1,2-Dibromobenzene
        see benzene, 1,2-dibromo-
1,3-Dibromobenzene
        see benzene, 1,3-dibromo-
1,4-Dibromobenzene
        see benzene, 1,4-dibromo-
1,2-Dibromoethane
        see ethane, 1,2-dibromo-
1,2-Dibromo-1,1,2,3,3,3-hexafluoropropane
        see propane, 1,2-dibromo-1,1,2,3,3,3-hexafluoro-
```

2,4-Dibromophenol
        see phenol, 2,4-dibromo-
Dibutylamine
        see 1-butanamine, N-butyl-
Dibutylether
        see butane, 1,1´-oxybis-
Diisobutyl ketone
        see 3-pentanone, 2,2,4,4-tetramethyl-
Dibutylphthalate
        see 1,2-benzene dicarboxylic acid, dibutyl ester
Dichloroacetic acid
        see acetic acid, dichloro-
2,2-Dichloroacetic acid
        see acetic acid, 2,2-dichloro-
1,2-Dichlorobenzene
        see benzene, 1,2-dichloro-
1,3-Dichlorobenzene
        see benzene,1,3-dichloro-
1,4-Dichlorobenzene
        see benzene,1,4-dichloro-
o-Dichlorobenzene
        see benzene, 1,2-dichloro-
p-Dichlorobenzene
        see benzene, 1,4-dichloro-
1,4-Dichlorobutane
        see butane, 1,4-dichloro-
2,4-Dichloro-6-butylphenol
        see phenol,2,4-dichloro-6-butyl-
3,5-Dichloro-2-(dichloromethyl)pyridine
        see pyridine, 3,5-dichloro-2-(dichloromenthyl)-
Dichlorodifluoromethane
        see methane, dichlorodifluoro-
Dichlorodioxochromium
        see chromium, dichlorodioxo
1,1-Dichloroethane
        see ethane, 1,1-dichloro-
1,2-Dichloroethane
        see ethane, 1,2-dichloro-
1,1-Dichloroethene
        see ethene, 1,1-dichloro-
cis-1,2-Dichloroethene
        see ethene, 1,2-dichloro-,cis-
trans-1,2-Dichloroethene
        see ethene, 1,2-dichloro-,trans-
Dichloroethylene
        see ethane, 1,2-dichloro-
1,1-Dichloroethylene
        see ethene, 1,1-dichloro-
cis-1,2-Dichloroethylene
        see ethene, 1,2-dichloro-,cis-
trans-1,2-Dichloroethylene
        see ethene, 1,2-dichloro-,trans-
2,4-Dichloro-6-ethylphenol
        see pPhenol, 2,4-dichloro-6-ethyl-
3,4-Dichloro-4-methylphenol
        see phenol, 3,4-dichloro-4-methyl-
2,6-Dichloro-4-methylphenol
        see phenol, 2,6-dichloro-4-methyl-
1,2-Dichloro-4-nitrobenzene
        see benzene, 1,2-dichloro-4-nitro-
1,4-Dichloro-2-nitrobenzene
        see benzene, 1,4-dichloro-2-nitro-
Dichloromethane
        see methane, dichloro-
(Dichloromethyl)benzene
        see benzene,(dichloromethyl)-
2,4-Dichlorophenol
        see phenol,2,4-dichloro-
2,4-Dichloro-6-(phenylmethyl)phenol
        see phenol, 2,4-dichloro-6-(phenylmethyl)-
1,2-Dichloropropane
        see propane, 1,2-dichloro-
2,4-Dichloro-6-propylphenol
        see phenol, 2,4-dichloro-6-propyl-

1,2-Dichloro-1,1,2,2-tetrafluoroethane
        see ethane, 1,2-dichloro-1,1,2,2-tetrafluoro-
Dichloro(trifluoromethyl)benzene
        see benzene, dichloro(trifloromethyl)-
3,5-Dichloro-2-(trichloromenthyl)pyridine
        see pyridine, 3,5-dichloro-2-(trichloromethyl)-
Dicumyl methane
        see benzene, 1,1´-methylenebis (1-methyl ethyl)-
Dicyclohexyl-18-crown-6
        see dibenzo(b,k)-(1,4,7,10,13,16)hexaoxacyclooctadecin,
            eicosahydro-
1,4-Dicyclopentylbutane
        see butane, 1,4-dicyclopentyl-
Dieptal A
        see benzenesulfonamide, 4-[[(4-aminophenyl)sulfonyl]amino]-
            N,N-dimethyl-
Diesel
            + ammonia   vol 21   19
Diethanolamine
        see ethanol, 2,2´-iminobis-
Diethylamine
        see ethanamine, N-ethyl-,
Diethylamine nitrate
        see ethanamine, N-ethyl-, nitrate
N,N-Diethylaniline
        see benzenamine, N,N-diethyl-
N,N-Diethylbenzenamine
        see benzenamine, N,N-diethyl-
Diethylene glycol
        see ethanol, 2,2´-oxybis-
Diethylene glycol dibutyl ether
        see butane, 1,1´-oxybis(2,1-ethanediyloxy)bis-
Diethylene glycol dimethyl ether
        see ethane, 1,1´-oxybis[2-methoxy-
Diethylene glycol monobutyl ether
        see , 3,6,9,12-tetraoxahexadecan-1-ol
Diethylene glycol monomethyl ether
        see ethanol, 2-(2-methoxyethoxy)-
Diethyl ether
        see ethane, 1,1´-oxybis-
Di-2-ethylhexyl sebacate
        see decanedioic acid, bis(2-ethylhexyl) ester
Diethylketone
        see 3-pentanone
Diethylmethyl carbinol
        see 3-pentanol, 3-methyl-
N,N-Diethylnicotinamide
        see 3-pyridinecarboxamide, N,N-diethyl-
2,2-Diethyl-1-pentanol
        see 1-pentanol, 2,2-diethyl-
Diethyl phenylamine
        see benzenamine, N,N-diethyl-
1,2-Difluorobenzene
        see benzene,1,2-difluoro-
1,3-Difluorobenzene
        see benzene, 1,3-difluoro-
1,4-Difluorobenzene
        see benzene, 1,4-difluoro-
Difluorodioxouranium
        see uranium, difluorodioxo-
Diglycolamine
        see ethanol, 2-(2-aminoethoxy)-
Diglyme
        see ethane, 1,1´-oxybis(2-methoxy)-
9,10-Dihydrophenanthrene
        see phenanthrene, 9,10-dihydro-
2,3-Dihydropyran
        see pyran, 2,3-dihydro-
Dihydroxyethane
        see 1,2-ethanediol
4-(4,5-Dihydro-3-methyl-5-oxo-1H-pyrazol-1-yl)benzenesulfonamide
        see benzenesulfonamide, 4-(4,5-dihydro-3-methyl-5-oxo-
            1H-pyrazol-1-yl)-

1,2-Diiodobenzene
        see benzene, 1,2-diiodo-
1,3-Diiodobenzene
        see benzene, 1,3-diiodo-
1,4-Diiodobenzene
        see benzene, 1,4-diiodo-
Diisopropanolamine
        see 2-propanol, 1,1´-iminobis-
Diisopropyl benzene
        see benzene, bis(1-methylethyl)-
Diisopropyl carbinol
        see 3-pentanol, 2,4-dimethyl-
Diisopropyl ether
        see propane, 1,1´-oxybis-
Diisopropyl phenylmethane
        see benzene, 1,1-methylenebis(1-methylethyl)-
Dilithium carbonate
        see carbonic acid, lithium salt
1,3-Dimethoxybenzene
        see benzene, 1,3-dimethoxy-
N-[4-[[(5,6-Dimethoxy-4-pyrimidinyl)amino]sulfonyl]phenyl]acetamide
        see acetamide, N-[4-[[(5,6-dimethoxy-4-pyrimidinyl)amino]-
              sulfonyl]phenyl]-
5-[[4-[[2,6-Dimethoxy-4-pyrimidinyl)amino]sulfonyl]-phenyl]azo]-
2-hydroxybenzoic acid
        see benzoic acid, 5-[[4-[[2,6-dimethoxy-4-pyrimidinyl)amino]-
              sulfonyl]-phenyl]azo]-2-hydroxy-
4´-[(5,6-Dimethoxy-4-pyrimidinyl)sulfamoyl]acetanilide
        see acetamide, N-[4-[[(5,6-dimethoxy-4-pyrimidinyl)amino]-
              sulfonyl]phenyl]-
N1-(5,6-Dimethoxy-4-pyrimidinyl)sulfanilamide
        see benzenesulfonamide, 4-amino-N-(5,6-dimethoxy-4-pyrimidinyl)-
N1-5,6-Dimethoxy-4-pyrimidinylsulfanilamide
        see benzenesulfonamide, 4-amino-N-(5,6-dimethoxy-4-pyrimidinyl)-
4,5-Dimethoxy-4-pyrimidinylsulfanilamide
        see benzenesulfonamide, 4-amino-N-(5,6-dimethoxy-4-pyrimidinyl)-
Dimethylacetamide
        see acetamide, N,N-dimethyl-
N1-(3,3-Dimethylacryloyl)sulfanilamide
        see 2-butenamide, N-[(4-aminophenyl)sulfonyl]-3-methyl-
N,N-Dimethylacetamide
        see acetamide, N,N-dimethyl-
Dimethylamine nitrate
        see methanamine, N-methyl-, nitrate
Dimethylaminocyclohexamine
        see cyclohexamine, N,N-dimethyl-
1-Dimethylamino-3-propylamine
        see 1,3-propanediamine, N,N-dimethyl-
N,N-Dimethylaniline
        see benzenamine, N,N-dimethyl-
N,N-Dimethylaniline
        see benzenamine, N,N-dimethyl-
9,10-Dimethylanthracene
        see anthracene, 9,10-dimethyl-
9,10-Dimethylbenz [a] anthracene
        see benz [a] anthracene, 9,10-dimethyl-
7,12-Dimethylbenz [a] anthracene
        see benz [a] anthracene, 7,12-dimethyl-
N,N-Dimethylbenzenamine
        see benzenamine, N,N-dimethyl-
Dimethylbenzene
        see benzene, dimethyl-
1,2-Dimethylbenzene
        see benzene, 1,2-dimethyl-
1,3-Dimethylbenzene
        see benzene, 1,3-dimethyl-
1,4-Dimethylbenzene
        see benzene, 1,4-dimethyl-
4-[(3,4-Dimethylbenzoyl)sulfamoylacetanilide
        see benzamide, N-[[4-(acetylamino)phenyl]sulfonyl]-3,4-dimethyl-
4´-(3,4-Dimethylbenzoylsulfamoyl)acetanilide
        see benzamide, N-[[4-(acetylamino)phenyl]sulfonyl]-3,4-dimethyl-
2,2-Dimethylbutane
        see butane, 2,2-dimethyl-

2,3-Dimethylbutane
          see butane, 2,3-dimethyl-
2,2-Dimethyl-1-butanol
          see 1-butanol, 2,2-dimethyl-
2,3-Dimethyl-2-butanol
          see 2-butanol, 2,3-dimethyl-
3,3-Dimethyl-2-butanol
          see 2-butanol, 3,3-dimethyl-
3,3-Dimethyl-2-butanone
          see 2-butanone, 3,3-dimethyl-
2,3-Dimethyl-1-butene
          see 1-butene, 2,3-dimethyl-
Dimethylcyclohexane
          see cyclohexane, dimethyl-
1,2-Dimethylcyclohexane
          see cyclohexane, 1,2-dimethyl-
1,3-Dimethylcyclohexane
          see cylcohexane, 1,3-dimethyl-
1,4-Dimethylcyclohexane
          see cyclohexane, 1,4-dimethyl-
cis-1,2-Dimethylcyclohexane
          see cyclohexane, cis-1,2-dimethyl-
1,1-Dimethylethylbenzene
          see benzene, 1,1-dimethylethyl-
Dimethylformamide
          see formamide, N,N-dimethyl-
Dimethyl glutarate
          see pentanedioic acid, dimethyl ester
2,6-Dimethyl-4-heptanol
          see 4-heptanol, 2,6-dimethyl-
3,5-Dimethyl-4-heptanol
          see 4-heptanol, 3,5-dimethyl-
2,3-Dimethylhexane
          see hexane, 2,3-dimethyl-
2,4-Dimethylhexane
          see hexane, 2,4-dimethyl-
1,1-Dimethylhydrazine
          see hydrazine, 1,1-dimethyl-
1,2-Dimethylhydrazine
          see hydrazine, 1,2-dimethyl-
4´-[(3,4-Dimethyl-5-isoazolyl)sulfamoyl]acetanilide
          see acetamide, N-[4-[[(3,4-dimethyl-5-isoxazolyl)-amino]-
               sulfonyl]phenyl]-
N-[4-[[(3,4-Dimethyl-5-isoxazolyl)-amino]sulfonyl]phenyl]acetamide
          see acetamide, N-[4-[[(3,4-dimethyl-5-isoxazolyl)-amino]sulfonyl]-
               phenyl]-
N1-(3,4-Dimethyl-5-isoxazolyl)sulfanilamide
          see benzenesulfonamide, 4-amino-N-(3,4-dimethyl-5-isoxazolyl)-
Dimethyl ketone
          see 2-propanone
Dimethyl maleate
          see (z)-2-butenedioic acid, dimethyl ester
1,3-Dimethylnaphthalene
          see naphthalene, 1,3-dimethyl-
2,7-Dimethyloctane
          see octane, 2,7-dimethyl-
N1-(4,5-Dimethyl-2-oxazolyl)sulfanilamide
          see benzenesulfonamide, 4-amino-N-(4,5-dimethyl-2-oxazolyl)-
N´-(4,5-Dimethyl-2-oxazolyl)sulfanilamide
          see benzenesulfonamide, 4-amino-N-(4,5-dimethyl-2-oxazolyl)-
2,2-Dimethylpentane
          see pentane, 2,2-dimethyl-
2,3-Dimethylpentane
          see pentane, 2,3-dimethyl-
2,4-Dimethylpentane
          see pentane, 2,4-dimethyl-
3,3-Dimethylpentane
          see pentane, 3,3-dimethyl-
2,2-Dimethyl-1-pentanol
          see 1-pentanol, 2,2-dimethyl-
2,4-Dimethyl-1-pentanol
          see 1-pentanol, 2,4-dimethyl-
4,4-Dimethyl-1-pentanol
          see 1-pentanol, 4,4-dimethyl-

2,3-Dimethyl-2-pentanol
        see 2-pentanol, 2,3-dimethyl-
2,4-Dimethyl-2-pentanol
        see 2-pentanol, 2,4-dimethyl-
2,2-Dimethyl-3-pentanol
        see 3-pentanol, 2,2-dimethyl-
2,3-Dimethyl-3-pentanol
        see 3-pentanol, 2,3-dimethyl-
2,4-Dimethyl-3-pentanol
        see 3-pentanol, 2,4-dimethyl-
N,N-Dimethyl-N-phenylamine
        see benzenamine, N,N-dimethyl-
Dimethyl phthalate
        see 1,2-benzenecarboxylic acid, dimethyl ester
2,2-Dimethylpropane
        see propane, 2,2-dimethyl-
Dimethylpropyl carbinol
        see 2-pentanol, 2-methyl-
Dimethylisopropyl carbinol
        see 2-butanol, 2,3-dimethyl-
2,2-Dimethylpropane
        see propane, 2,2-dimethyl-
N,N-Dimethyl-1,3-propanediamine
        see 1,3-propanediamine, N,N-dimethyl-
2,2-Dimethyl-1-propanol
        see 1-propanol, 2,2-dimethyl-
Dimethylphenylamine
        see benzenamine, N,N-dimethyl-
2,4-Dimethyl-4H-pyran-4-one
        see pyrone, dimethyl-
2,5-Dimethyl-4H-pyran-4-one
        see pyrone, dimethyl-
3,4-Dimethyl-2H-pyran-2-one
        see pyrone, dimethyl-
3,5-Dimethyl-2H-pyran-2-one
        see pyrone, dimethyl-
3,5-Dimethyl-4H-pyran-4-one
        see pyrone, dimethyl-
3,6-Dimethyl-2H-pyran-2-one
        see pyrone, dimethyl-
4,5-Dimethyl-2H-pyran-2-one
        see pyrone, dimethyl-
4,6-Dimethyl-2H-pyran-2-one
        see pyrone, dimethyl-
1,2-Dimethyl-1H-pyrazole
        see 1H-pyrazole, 1,2-dimethyl-
N1-(4,6-Dimethyl-2-pyridinyl)sulfanilamide
        see benzenesulfonamide, 4-amino-N-(4,6-dimethyl-2-pyrimidinyl)-
N-[4-[[(4,6-Dimethyl-2-pyrimidinyl)amino]sulfonyl]phenyl]acetamide
        see acetamide, N-[4-[[(4,6-dimethyl-2-pyrimidinyl)amino]-
            sulfonyl]-phenyl]-
N-[4-[[(2,6-Dimethyl-4-pyrimidinyl)amino]sulfonyl]phenyl]acetamide
        see acetamide, N-[4-[[(2,6-dimethyl-4-pyrimidinyl)amino]-
            sulfonyl]phenyl]-
4´-[(4,6-Dimethyl-2-pyrimidinyl)sulfamoyl]acetanilide
        see acetamide, N-[4-[[(4,6-dimethyl-2-pyrimidinyl)amino]-
            sulfonyl]phenyl]-
4´-[(2,6-Dimethyl-4-pyrimidinyl)sulfamoyl]acetanilide
        see acetamide, N-[4-[[(2,6-dimethyl-4-pyrimidinyl)amino]-
            sulfonyl]phenyl]-
N1-(2,6-Dimethyl-4-pyrimidinyl)sulfanilamide
        see benzenesulfonamide, 4-amino-N-(2,6-dimethyl-4-pyrimidinyl)-
N1-(4,6-Dimethyl-2-pyrimidinyl)sulfanilamide
        see benzenesulfonamide, 4-amino-N-(4,6-dimethyl-2-pyrimidinyl)-
N1-(4,5-Dimethyl-2-pyrimidinyl)sulfanilamide
        see sulfanilamide, N1-(4,5-dimethyl-2-pyrimidinyl)-
N-(4,6-Dimethyl-2-pyrimidyl)sulfanilamide
        see benzenesulfonamide, 4-amino-N-(4,6-dimethyl-2-pyrimidinyl)-
1,5-Dimethyl-2-pyrrolidinone
        see 2-Pyrrolidinone, 1,5-dimethyl-
N1,N1-Dimethyl-N4-sulanilylsulfanilamide
        see benzenesulfonamide, 1-amino-N-[5-(1,1´dimethylethyl)-
            1,3,4-thiadiazol-2-yl)-

p-(Dimethylsulfamoyl)aniline
        see benzenesulfonamide, 4-amino-N,N-dimethyl-
Nl,Nl-Dimethylsulfanilamide
        see benzenesulfonamide, 4-amino-N,N-dimethyl-
N,N-Dimethylsulfanilamide
        see benzenesulfonamide, 4-amino-N,N-dimethyl-
3,4-Dimethyl-5-sulfanilamidoisoxazole
        see benzenesulfonamide, 4-amino-N-(3,4-dimethyl-5-isoxazolyl)-
4,6-Dimethyl-2-sulfanilamidopyrimidine
        see benzenesulfonamide, 4-amino-N-(4,6-dimethyl-2-pyrimidinyl)-
2,6-Dimethyl-4-sulfanilamidopyrimidine
        see benzenesulfonamide, 4-amino-N-(2,6-dimethyl-4-pyrimidinyl)-
2,4-Dimethyl-6-sulfanilamidopyrimidine
        see benzenesulfonamide, 4-amino-N-(2,6-dimethyl-4-pyrimidinyl)-
Nl,Nl-Dimethyl-N4-sulfanilylsulfanilamide
        see benzenesulfonamide, 4-[[(4-aminophenyl)sulfonyl]amino]-
                N,N-dimethyl-
4´-(Dimethylsulfamoyl)sulfanilanilide
        see benzenesulfonamide, l-amino-N-[5-(1,1´dimethylethyl)-
3,4-Dimethyl-5-sulfonamidoisoxazole
        see benzenesulfonamide, 4-amino-N-(3,4-dimethyl-5-isoxazolyl)
Dimethyl sulfate
        see sulfuric acid, dimethyl ester
Dimethylsulfoxide
        see methane, 1,1-sulfinylbis-
7,8-Dimethyltetradecane
        see tetradecane, 7,8-dimethyl-
Dimezathine
        see benzenesulfonamide, 4-amino-N-(4,6-dimethyl-2-pyrimidinyl)-
Dinitrogen tetroxide
        see nitrogen oxide
Dinonylamine
        see l-nonanamine,N-nonyl-
Dioctylether
        see octane, 1,1´-oxybis
1,4-Dioxane
        + ammonia   vol 21   E8,   20
        + butane   vol 24   E374 -E377,   396
        + N,N-dimethylmethanamine   vol 21   E85,   198
        + hydrogen sulfide   vol 32   El73, El74,   264
        + iodic acid, potassium salt   vol 30   E378,   412
        + methanamine   vol 21   E81,   104
        + methane   vol 27/28   E642, E643,
                                                          650 - 653
        + N-methylmethanamine   vol 21   E82,   152
        + propane   vol 24   E374 -E377,   384
1,4-Dioxane (aqueous)
        + methane   vol 27/28   176
p-Dioxane
        see 1,4-dioxane
1,3-Dioxolane-2-one, 4-methyl-
        + butane   vol 24   E374 -E377,   406
        + hydrogen sulfide   vol 32   El73, El76, El77,
                                                    259, 266 - 269
        + methane   vol 27/28   666, 672 - 675
        + propane   vol 24   E374 -E377,   385
1,3-Dioxolan-2-one, 4-methyl- (multicomponent)
        + methane   vol 27/28   663 - 665,   670
Dioxosulfatouranium
        see uranium, dioxosulfato-
Dipentyl ether
        see pentane, 1,1´oxybis-
Diphenylbenzene
        see terphenyl
Diphenylmethane
        see benzene, 1,1´-methylenebis-
Diphosphoric acid (aqueous)
        + ammonia   vol 31   269,   270
        + phosphoric acid   vol 31   269,   270
        + potassium oxide   vol 31   269,   270
Dipotassium carbonate
        see carbonic acid, potassium salt
Dipotassium hydrogen phosphate
        see phosphoric acid, dipotassium salt

```
Dipotassium silicate
        see silicic acid, dipotassium salt
Dipropylene glycol
        see propanol, oxybis-
Dipropyl ether
        see propane, 1,1´-oxybis-
Diselenious acid, diammmonium salt
            + water                             vol 26          E331,  344,  345
Diselenious acid, dipotassium salt
            + water                             vol 26          E331,  338,  339
Diselenious acid, disodium salt
            + water                             vol 26          E331,  333 - 335
Diseptal A
        see benzenesulfonamide, 1-amino-N-[5-(1,1´dimethylethyl)-
            1,3,4-thiadiazol-2-yl)-
Diseptal B
        see benzenesulfonamide, 4-amino-N-[4-[(methylamino)sulfonyl]-
            phenyl]-
Diseptal C
        see benzenesulfonamide, 4-amino-N-[4-(aminosulfonyl)phenyl]-
Disodium carbonate
        see carbonic acid, sodium salt
Disodium ethylenediaminetetraacetate
        see glycine, N,N-1,2-ethanediylbis[N-(carboxymethyl)-,
            disodium salt
Disodium glutamate
        see DL-glutamic acid, disodium salt
Disodium hydrogen phosphate
        see phosphoric acid, disodium salt
Disodium Neoprontosil
        see 2,7-naphthalenedisulfonic acid, 6-acetylamino-
            3-[[4-(aminosulfonyl)phenyl]azo]-4-hydroxy-, disodium salt
Disodium silicate
        see silicic acid, disodium salt
Disodium sulfate
        see sulfuric acid, disodium salt
Disodium tetraborate
        see boric acid, sodium salt
Disodium (I-4)-tetroxomolybdate (2-)
            + bromic acid, sodium salt        vol 30                          200
            + chloric acid, sodium salt       vol 30          E30,   76,    77
            + iodic acid, potassium salt      vol 30                 E380,  403
            + iodic acid, sodium salt         vol 30          E337, E338,  343
Disodium tungstanate
        see tungstic acid, disodium salt
Disulfan-HCl
        see benzenesulfonamide, 4-amino-N-4-[(aminosulfonyl)phenyl]-,
            monohydrochloride
Disulfan
        see benzenesulfonamide, 4-amino-N-[4-(aminosulfonyl)phenyl]-
Disulfurous acid, diammonium salt
            + water                             vol 26                 E72 - E75,
                                                                        76 -  84,
                                                                      E144, E145,
                                                                       146 - 151
Disulfurous acid, diammonium salt (aqueous)
            + sulfuric acid, diammonium salt
                                                vol 26                E144, E145,
                                                                       147 - 151
            + sulfurous acid, diammonium salt
                                                vol 26          E115,  127,  128
Disulfurous acid, diammonium salt (aqueous and multicomponent)
            + ammonium chloride               vol 26          E74,   87,   88
            + disulfurous acid, disodium salt
                                                vol 26          E74,   87,   88
            + sodium chloride                 vol 26          E74,   87,   88

Disulfurous acid, dipotassium salt
            + water                             vol 26                E108, E109,
                                                                       110,  111
Disulfurous acid, disodium salt
            + water                             vol 26                 E72 - E75,
                                                                        76 -  89
```

Disulfurous acid, disodium salt (aqueous)
                    + ethanol   vol 26   84
                    + sodium chloride   vol 26   E74,   86
                    + sulfuric acid, disodium salt   vol 26   E74,   78 - 83
                    + sulfurous acid, disodium salt
                                                       vol 26   E5,   39 - 45

Disulfurous acid, disocium salt (aqueous and multicomponent)
                    + ammonium chloride   vol 26   E74,   87,   88
                    + 2,2´-iminobisethanol   vol 26   E74,   89
                    + disulfurous acid, diammonium salt
                                                       vol 26   E74,   87,   88
                    + sodium chloride   vol 26   E74,   87,   88
                    + sulfuric acid, disodium salt   vol 26   E5,   63 - 65
Disulon
          see benzenesulfonamide, 4-amino-N-[4-(aminosulfonyl)phenyl]-
Dithionous acid, sodium salt (aqueous)
                    + sulfurous acid, moncsodium salt
                                                       vol 26   85
DJ 1550
          see benzenesulfonamide, 4-amino-N-(6-methoxy-4-pyrimidinyl)-
DMSO
          see methane, sulfinylbis-
Docosane
                    + butane   vol 24   E130 -E134,   184
                    + methane   vol 27/28   E202 -E204
                                                                           E211,   246
                    + propane   vol 24   E130 -E134,   164
1-Dodecanamine, hydrochloride (aqueous)
                    + propane   vol 24   E58 - E72,   80
Dodecane
                    + ammonia   vol 21   E2,   13,   16
                    + butane   vol 24   E130 -E134,
                                                                             177,   178
                    + hydrogen sulfide   vol 32   E166 -E169,
                                                                     189,   190,   208
                    + methane   vol 27/28   E202 -E204,
                                                                     E208,   232,   233
                    + phosphine   vol 21   E281, E282,   283
                    + propane   vol 24   E130 -E134,   152
                    + salt water   vol 38   384
                    + seawater   vol 38   E381,   382,   383
                    + water   vol 38   E376, E377,
                                                                             378 - 380
1-Dodecanol
                    + methane   vol 27/28   E582 -E584,
                                                                             E594,   628
Dodecylamine hydrochloride
          see 1-dodecanamine hydrochloride
Dog blood
          see Blood, dog
Dog bone and skeletal muscle
                    + methane   vol 27/28   732,   733
Domian
          see benzenesulfonamide, 4-amino-N-(2,6-dimethyl-4-pyrimidinyl)-
Dorsulfan Warthausen
          see benzenesulfonamide, 4-amino-N-(3,4-dimethyl-5-isoxazolyl)
Dotriacontane
                    + butane   vol 24   E130 -E134,   187
                    + methane   vol 27/28   377
Dotricontane (ternary)
                    + methane   vol 27/28   418,   419,   540
Dow Corning silicone oil
          see Silicone oil
Drometil
          see 2,7-naphthalenedisulfonic acid, 6-acetylamino-
              3-[[4-(aminosulfonyl)phenyl]azo]-4-hydroxy-, disodium salt
DS 36
          see benzenesulfonamide, 4-amino-N-(6-methoxy-4-pyrimidinyl)-
Duatok
          see benzenesulfonamide, 4-amino-N-2-thiazolyl-
Dulana
          see benzenesulfonamide, 4-amino-N-2-thiazolyl-

Durasulf
        see benzenesulfonamide, 4-amino-N-(6-chloro-3-pyridazinyl)-
Durenat
        see benzenesulfonamide, 4-amino-N-(5-methoxy-2-pyrimidinyl)-
Dynacoryl
        see 3-pyridinecarboxamide, N,N-diethyl-
Dysprosium
        + mercury   vol 25   E245,   246
Dysprosium bromide
        + 1,4-dioxane   vol 22   302
        + ethane, 1,2-diethyoxy-   vol 22   300
        + furan, tetrahydro-   vol 22   301
        + 2-propanamine   vol 22   303
Dysprosium chloride
        + 1-butanamine   vol 22   297
        + 2-butanamine   vol 22   297
        + 1,4-dioxane   vol 22   294
        + ethane, 1,2-diethoxy-   vol 22   293
        + ethane, 1-ethoxy-2-methoxy- vol 22   294
        + ethane, 1,1,-oxybis-   vol 22   292
        + ethanol (aqueous)   vol 22   290
        + ethanol, 2-ethoxy-   vol 22   291
        + ethanol, 2-methoxy-   vol 22   291
        + furan, tetrahydro-   vol 22   295
        + pentane, 1-methoxy-   vol 22   294
        + phosphoric acid, tributyl ester
                                    vol 22   296
        + phosphoric triamide, hexamethyl-
                                    vol 22   298
        + phosphoryl chloride (ternary)
                                    vol 22   299
        + 2-propanamine   vol 22   297
        + 2-propen-1-amine   vol 22   297
        + stannane tetrachloro-(ternary)
                                    vol 22   299
        + water   vol 22   290
Dysprosium fluoride
        + butane, 1-(chloromethoxy)-   vol 22   286
        + decane, 1-methoxy-   vol 22   286
        + ethanol   vol 22   285
        + methane, 1,1´-sulfinylbis-   vol 22   288
        + phosphoric acid, tribuyl ester
                                    vol 22   287
        + pyridine   vol 22   289
Dysprosium iodide
        + 1-butanol (aqueous)   vol 22   305
        + ethanol (aqueous)   vol 22   304
        + furan, tetrahydro-   vol 22   306
        + water (ternary)   vol 22   304,   305
Dysprosium nitrate
        see nitric acid, dysprosium salt

E

EDTA
        see glycine, N,N´-1,2-ethanediylbis(N-carboxymethyl)-
Eftolon
        see benzenesulfonamide, 4-amino-N-(1-phenyl-1H-pyrazol-5-yl)-
Eicosane

+ butane	vol 24	E130 -E134,	183, 184
+ methane	vol 27/28	E202 -E204, 244, 245,	E210 376
+ methypropane	vol 24	E130 -E134,	195
+ propane	vol 24	E130 -E134,	162, 163
+ seawater	vol 38		526
+ water	vol 38		525

1,2-Epoxyethane
        see ethane, 1,2-epoxy-
Elcosine
        see benzenesulfonamide, 4-amino-N-(2,6-dimethyl-4-pyrimidinyl)-
Electrolytes
        see under individual electrolyte
Eleudron
        see benzenesulfonamide, 4-amino-N-2-thiazolyl-
Elkosil
        see benzenesulfonamide, 4-amino-N-(2,6-dimethyl-4-pyrimidinyl)-
Elkosin
        see benzenesulfonamide, 4-amino-N-(2,6-dimethyl-4-pyrimidinyl)-
Elkosine
        see benzenesulfonamide, 4-amino-N-(2,6-dimethyl-4-pyrimidinyl)-
1,2-Epoxyethylene
        see Oxirane
Emasol 41S
        see sorbitan (z)-9-octadecenoate (2:3)
Emedan
        see benzenesulfonamide, 4-amino-N-[(butylamino)carbonyl]-
Emerin
        see benzenesulfonamide, 4-amino-N-[(aminoiminomethyl)-
Emilene
        see benzenesulfonamide, 4-(aminomethyl)-
Emsorb 2502
        see sorbitan (z)-9-octadecenoate (2:3)
Emulgator 8972
        see sorbitan (z)-9-octadecenoate (2:3)
Entusul
        see benzenesulfonamide, 4-amino-N-(3,4-dimethyl-5-isoxazolyl)
Erbium

+ mercury	vol 25	E249,	250

Erbium bromide

+ 1-butanamine	vol 22		347
+ 2-butanamine	vol 22		347
+ 2-butanamine, N-(1-methylpropyl)-	vol 22		347
+ 1,4-dioxane	vol 22		346
+ ethane, 1,2-diethoxy-	vol 22		344
+ furan, tetrahydro-	vol 22		345
+ 1-propanamine	vol 22		347
+ 2-propanamine	vol 22		347

Erbium chloride

+ 1-butanamine	vol 22		341
+ 2-butanamine	vol 22		341
+ 2-butanamine, N-(1-methylpropyl)-	vol 22		341
+ butane, 1-ethoxy-	vol 22	E335, 337,	338
+ 1,4-dioxane	vol 22		338
+ ethanamine, N-ethyl-	vol 22		341
+ ethane, 1,2-diethoxy-	vol 22		333
+ ethane, 1-ethoxy-2-methoxy-	vol 22		338
+ ethanol	vol 22	328,	330
+ ethanol (aqueous)	vol 22		331
+ ethanol, 2-ethoxy-	vol 22		329
+ ethanol, 2-methoxy-	vol 22		329
+ furan, tetrahydro-	vol 22		339
+ methanol	vol 22		327

```
Erbium chloride
         + pentane, 1-methoxy-                vol 22        E336,  337,  338
         + phosphoric acid, tributyl ester
                                              vol 22                     340
         + phosphoric triamide, hexamethyl-
                                              vol 22                     342
         + phosphoryl chloride               vol 22                     343
         + 2-propanamine                     vol 22                     341
         + propane, 1,1´-oxybis-             vol 22        E334,  337,  338
         + 2-propanol                        vol 22                     332
         + 2-propen-1-amine                  vol 22                     341
         + stannane, tetrachloro-            vol 22                     343
         + water (ternary)                   vol 22                     331
Erbium fluoride
         + butane, 1-(chloromethoxy)-        vol 22                     323
         + decane, 1-methoxy-                vol 22                     323
         + ethanol                           vol 22                     322
         + methane, 1,1´-sulfinylbis-        vol 22                     325
         + methanol                          vol 22                     322
         + phosphoric acid, tributyl ester
                                              vol 22                     324
         + pyridine                          vol 22                     326
Erbium nitrate
         see nitric acid, erbium salt
Ergaseptine
         see benzenesulfonamide, 4-amino-
Erysipan
         see benzenesulfonamide, 4-amino-
Erythritol
         see (R,R)-1,2,3,4-Butanetetrol
Erythrocytes
         see Rabbit erythrocytes and Ox erythrocytes
Erytrin
         see acetamide, N-[4-(aminosulfonyl)phenyl]-
Eskadiazine
         see benzenesulfonamide, 4-amino-N-2-pyrimidinyl-
Estafilol
         see benzenesulfonamide, 4-amino-N-2-thiazolyl-
Estreptocida
         see benzenesulfonamide, 4-amino-
Etazole
         see benzenesulfonamide, 4-amino-N-(5-methyl-
             1,3,4-thiadiazol-2-yl)-
Etazol
         see benzenesulfonamide, 4-amino-N-(5-methyl-
             1,3,4-thiadiazol-2-yl)-
Ethanamine, N,N-diethyl-
         + ammonia                           vol 21                      75
Ethanamine, N,N-diethyl-, phosphate
         + water                             vol 31                     273
Ethanamine, N,N-diethyl-, phosphate (aqueous)
         + phosphoric acid, disodium salt
                                              vol 31        E101,        116
         + phosphoric acid, monopotassium salt
                                              vol 31                     273
Ethanamine, N-ethyl- (aqueous)
         + bromic acid, potassium salt    vol 30                        251
Ethanaminium, 2-hydroxy-, N,N,N-tris(2-hydroxyethyl)-, bromide (aqueous)
         + methane                           vol 27/28     E62,   E63,   99
Ethanaminium, N,N,N-triethyl-, bromide (aqueous)
         + butane                            vol 24        E58 - E72,    87,
                                                                         88
         + mercury                           vol 29               E47,   55
         + methane                           vol 27/28     E60,   92 -   94
         + propane                           vol 24        E58 - E72,    86
Ethane
         + methane                           vol 27/28           E248, E249,
                                                                  250 - 265
Ethane (multicomponent)
         + methane                           vol 27/28           194,  195,
                                                                  378 - 386,
                                                                  553 - 555
Ethane, bromo-
         + hydrogen sulfide                  vol 32        E178, E179,  280
```

1,2-Ethanediamine
          + chloric acid, potassium salt   vol 30   159
          + chloric acid, sodium salt   vol 30   101
1,2-Ethanediamine, N,N-bis(2-aminoethyl)-
          + 1-butanamine   vol 21   E252,   275
Ethane, 1,2-dibromo-
          + hydrogen sulfide   vol 32   E178, E179,
                                                                     281,   283
Ethane, 1,2-dichloro-
          + ammonia   vol 21   E9,   67
          + arsine   vol 21   E302,   309
          + hydrogen sulfide   vol 32   E178, E179,   281
          + phosphine   vol 21   E281, E282,   294
Ethane, 1,2-epoxy-
          + methane   vol 27/28   648,   649
1,2-Ethanediol
          + ammonia   vol 21   E6,   E7,   50
          + 1-butanamine   vol 21   E252,   275
          + 2-butanamine   vol 21   E252,   276
          + chloric acid, potassium salt   vol 30   157
          + chloric acid, sodium salt   vol 30   98
          + N,N-dimethylmethanamine   vol 21   E85, E86,   E89,
                                                                     E91,   195
          + N-ethyl-ethanamine   vol 21   E252,   278
          + methanamine   vol 21   E80,   101,   102
          + N-methylmethanamine   vol 21   E82,   149
          + 2-methyl-1-propanamine   vol 21   E252,   277
          + phosphine   vol 21   E281, E282,   290
          + 1-propanamine   vol 21   E252,   267
          + 2-propanamine   vol 21   E252,   274
1,2-Ethanediol (aqueous)
          + bromic acid, potassium salt   vol 30   243
2,2"-(1,2-ethanediylbis(oxy))bis ethanol
        see ethanol,2,2"-(1,2-ethanediylbis(oxy))bis-
Ethane, 1,1´-oxybis-
          + N,N-dimethylmethanamine   vol 21   E85,   200
          + hydrogen sulfide   vol 32   E173, E174,   254
          + methane   vol 27/28   580, E640, E641,
                                                                     645,   646
Ethane, 1,1´-oxybis - (aqueous)
          + bromic acid, potassium salt   vol 30   244
Ethane, 1,1-oxybis-(2-chloro)-
          + arsine   vol 21   E302,   304,   305
          + germane   vol 21   E315,   321,   322
          + hydrogen sulfide   vol 32   E178, E179,
                                                                     277,   278
          + hydrogen selenide   vol 32   E330,   332,   333
          + phosphine   vol 21   E281, E282,
                                                                     298,   299
          + silane   vol 21   E315,   316,   317
          + stanane   vol 21   E315,   325,   326
          + stibine   vol 21   E310,   313,   314
Ethane, 1,1´-oxybis(2-methoxy)-
          + hydrogen sulfide   vol 32   E173, E174,   257
Ethane, pentachloro-
          + hydrogen sulfide   vol 32   E178, E179,   280
Ethane, 1,1,2,2-tetrabromo-
          + hydrogen sulfide   vol 32   E178, E179,   280
Ethane, 1,1,2,2-tetrachloro-
          + arsine   vol 21   E302,   309
          + hydrogen sulfide   vol 32   E178, E179,   280
Ethane, 1,1,2-trichloro-1,2,2-trifluoro-
          + methane   vol 27/28   684
1,2-Ethanediol
          + hydrogen sulfide   vol 32   E171, E172,
                                                                 246 - 249
Ethanoic acid
        see acetic acid
Ethanolamine
        see Ethanol, 2-amino-
Ethanol
          + ammonia   vol 21   E5,   38 -   42,
                                                                              E90
          + bromic acid, potassium salt   vol 30   254

```
Ethanol
            + N,N-dimethylmethanamine      vol 21              E84,   187
            + hydrogen sulfide             vol 32        E171, E172,  242
            + mercury                      vol 29                    E165
            + methane                      vol 27/28     580, E582 -E582
                                                         601 - 605,   635
            + N-methylmethanamine          vol 21         E82,  144,  145
Ethanol (aqueous)
            + bromic acid, potassium salt  vol 30                     243
            + chloric acid, potassium salt vol 30        E108,  151,  152
            + disulfurous acid, disodium salt
                                           vol 26              E74,   84
            + hydrogen sulfide             vol 32              E20,   23
            + iodic acid, lithium salt     vol 30             E270,   328
            + iodic acid, potassium salt   vol 30                     411
            + methane                      vol 27/28                 174
            + selenious acid, dipotassium salt
                                           vol 26                     328
            + selenious acid, disodium salt
                                           vol 26                     310
            + silver (I) oxide             vol 23         E83,  E84,  E94,
                                                                      111
            + sulfurous acid, dipotassium salt
                                           vol 26              E92,   95
            + sulfurous acid, disodium salt
                                           vol 26          E5,   54,  55
Ethanol (multicomponent)
            + copper (II) hydroxide        vol 23        E10 - E26,   69
Ethanol, 2-amino- (aqueous)
            + cadmium oxide                vol 23             E279,   304
            + hydrogen sulfide             vol 32         E79,   84 -  94,
                                                                      313
            + methane                      vol 27/28                 184
            + silver (I) oxide             vol 23         E83,  E84,  E93,
                                                                      124
Ethanol, 2-amino-
            + ammonia                      vol 21               E6,   75
            + arsine                       vol 21             E302,   308
            + chloric acid, potassium salt vol 30                     158
            + chloric acid, sodium salt    vol 30                     99
Ethanol, 2-amino- (multicomponent)
            + hydrogen sulfide             vol 32              E81,  E82,
                                                              118 - 139
Ethanol, 2-amino (ternary)
            + methane                      vol 27/28                 716
Ethanol, 2-(2-aminoethoxy)- (aqueous)
            + hydrogen sulfide             vol 32              E80,  108
Ethanol, 2-(2-aminoethoxy)- (multicomponent)
            + hydrogen sulfide             vol 32         E82,  155 - 159,
                                                                      308
Ethanol, 2-(2-(2-butoxyethoxy)-ethoxy)-,acetate
            + methane                      vol 27/28                 671
Ethanol, 2-chloro-
            + ammonia                      vol 21               E7,   54
Ethanol, 2,2-dichloro-
            + ammonia                      vol 21               E7,   54
Ethanol, 2,2-(1,2-ethanediybis(oxy))bis-
            + 1-butanamine                 vol 21             E252,   275
            + hydrogen sulfide             vol 32             E173, E174,
                                                              249,   251
            + methane                      vol 27/28                 669
            + 1-propanamine                vol 21             E252,   267
            + 2-propanamine                vol 21             E252,   274
Ethanol, 2-ethoxy-
            + arsine                       vol 21        E302,  304,  305
            + germane                      vol 21        E315,  321,  322
            + hydrogen sulfide             vol 32             E171, E172,
                                                              277,   278
            + hydrogen selenide            vol 32        E330,  332,  333
            + phosphine                    vol 21        E281, E282,
                                                              298,   299
            + silane                       vol 21        E315,  316,  317
            + stannane                     vol 21        E315,  325,  326
            + stibine                      vol 21        E310,  313,  314
```

Ethanol, 2,2´-iminobis- (aqueous)
            + hydrogen sulfide   vol 32   E79, E80
                                                                  95 - 107
            + methane   vol 27/28   185
Ethanol, 2,2´-iminobis-(multicomponent)
            + hydrogen sulfide   vol 32   E82, 140 - 154
Ethanol, 2,2´-iminobis- (aqueous and multicomponent)
            + cadmium oxide   vol 23   E279,   304
            + disulfurous acid, disodium salt
                                            vol 26   E74,   89
            + silver (I) oxide   vol 23   E83, E84, E93
                                                                    125
            + sodium chloride   vol 26   E74,   89
Ethanol, 2,2-iminobis- (ternary)
            + methane   vol 27/28   717
Ethanol, 2-(2-methoxyethoxy-)
            + hydrogen sulfide   vol 32   E173 -E175,   257
Ethanol, 2-(2-(methoxyethoxy)-ethoxy)-, acetate
            + methane   vol 27/28   668
Ethanol, 2,2´-(methylimino)bis- (aqueous)
            + hydrogen sulfide   vol 32   E80, 109 - 111
Ethanol, 2,2´,2"nitrilotris-
            + arsine   vol 21   E302,   308
Ethanol, 2,2´,2"-nitrilotris- (aqueous)
            + hydrogen sulfide   vol 32   E80, E81
                                                                122 - 115
            + silver (I) oxide   vol 23   E83, E84
                                                                E93, 126
            + zinc oxide   vol 23   E179,   259
Ethanol, 2,2´-oxybis-
            + ammonia   vol 21   E7,   51, E90
            + 1-butanamine   vol 21   E252,   269
            + methanamine   vol 21   E80,   102
            + 1-propanamine   vol 21   E252,   267
            + 2-propanamine   vol 21   E252,   274
Ethanol, 2,2"-(oxybis(2,1-ethanediyloxy))bis
            + 1-butanamine   vol 21   E252,   275
Ethanol, 2,2,2-trichloro-
            + ammonia   vol 21   E7,   54
            + N,N-dimethylmethanamine   vol 21   E85, E91,   227
            + hydrogen sulfide   vol 32   E178, E179,   282
            + methanamine   vol 21   E90,   134
            + N-methylmethanamine   vol 21   E91,   174
Ethanolamine,
        see ethanol, 2-amino- and ethanol,2,2-iminobis-
Ethanone, 1-phenyl-
            + N,N-dimethylmethanamine   vol 21   E85, 201,   202
            + methanamine   vol 21   107
            + N-methylmethanamine   vol 21   E82,   154
Ethazole
        see benzenesulfonamide, 4-amino-N-(5-methyl-
            1,3,4-thiadiazol-2-yl)-
Ethene
            + methane   vol 27/28   E477, 478 - 482
Ethene (ternary)
            + methane   vol 27/28   483 - 490,
Ethene, 1,2-dichloro-
            + arsine   vol 21   E302,   309
Ethene, tetrachloro-
            + arsine   vol 21   E302,   309
            + hydrogen sulfide   vol 32   E178, E179,   281
Ethene, trichloro-
            + arsine   vol 21   E302,   309
Ethene, 1,1,2-trichloro-
            + hydrogen sulfide   vol 32   281
Ethoxybenzene
        see benzene, ethoxy-
2-Ethoxy-1,1,1,3,3,4,4,5,5,5-decafluoro-2-(trifluoromethyl)pentane
        see Pentane, 2-ethoxy-1,1,1,3,3,4,4,5,5,5-decafluoro-2-
            (trifluoromethyl)-
2-Ethoxyethanol
        see Ethanol, 2-ethoxy-
N1-(4-Ethoxy-2-pyrimidinyl)sulfanilamide
        see benzenesulfonamide, 4-amino-N-(4-ethoxy-2-pyrimidinyl)-

```
Ethyl acetate
        see acetic acid, ethyl ester
Ethyl acetoacetate
        see 3-oxobutanoic acid, ethyl ester
N-Ethylaniline
        see benzenamine, N-ethyl-
10-Ethylbenz [a] anthracene
        see benz [a] anthracene, 10-ethyl-
N-Ethylbenzenamine
        see benzenamine, N-ethyl-
Ethylbenzene
        see benzene,ethyl-
Ethylbenzoate
        see benzoic acid, ethylester
Ethyl carbonate
        see carbamic acid, ethyl ester
Ethyl cellosolve
        see ethanol, 2-ethoxy-
Ethylcyclohexane
        see cyclohexane,ethyl-
(2-Ethylcyclopentyl)benzene
        see benzene, (2-ethylcyclopentyl)-
Ethylcyclopentane
        see cyclopentane, ethyl-
Ethylene
        see ethene
Ethylene bromide
        see ethene, 1,2-dibromo-
Ethylene chloride
        see ethene, 1,2-dichloro-
Ethylene oxide
        see ethane, 1,2-epoxy-
Ethyl ethanoate
        see acetic acid, ethyl ester
Ethyl ether
        see Ethane, 1,1´-oxybis-
Ethylenediamine
        see 1,2-ethanediamine
Ethylenediamine dinitrate
        see 1,2-ethanediamine, dinitrate
Ethylene glycol
        see 1,2-ethanediol
Ethylene glycol, dimethyl ether
        see ethane, dimethoxy-
Ethylene glycol, monoethyl monoacetate
        see ethanol, 2-ethoxy-, acetate
Ethylene oxide
        see ethane, 1,2-epoxy-
Ethyl formate
        see formic acid, ethyl ester
2-Ethyl hexanol
        see 1-hexanol, 2-ethyl-
Ethyl laurate
        see dodecanoic acid, ethyl ester
Ethyl malonate
        see diethyl propanedioate
Ethyl methyl carbinol
        see 2-butanol, 2-methyl-
1-Ethylnaphthalene
        see naphthalene, 1-ethyl-
2-Ethylnaphthalene
        see naphthalene, 2-ethyl-
Ethyl palmitate
        see hexadecanoic acid, ethyl ester
3-Ethyl-3-pentanol
        see 3-pentanol, 3-ethyl-
Ethylpropylketone
        see 3-hexanone
Ethyl phenyl ether
        see benzene, ethoxy-
N-[4-[[(5-Ethyl-1,3,4-thiadiazol-2-yl)amino]sulfonyl]-phenyl]-
acetamide
        see acetamide, N-[4-[[(5-ethyl-1,3,4-thiadiazol-2-yl)amino]-
            sulfonyl]-phenyl]-
```

4´-[(5-Ethyl-1,3,4-thiadiazol-2-yl)sulfamoyl]acetanilide
     see acetamide, N-[4-[[(5-ethyl-1,3,4-thiadiazol-2-yl)-
       amino]sulfonyl]-phenyl]-
2-Ethyl-1,3,5-trimethylbenzene
     see benzene, 2-ethyl-1,3,5-trimethyl-
Eubasin
     see benzenesulfonamide, 4-amino-N-2-pyridinyl-
Eubasinum
     see benzenesulfonamide, 4-amino-N-2-pyridinyl-
Eucoran
     see 3-pyridinecarboxamide, N,N-diethyl-
Euvernil
     see benzenesulfonamide, 4-amino-N-(aminocarbonyl)-
Europium

+ mercury	vol 25	E234, 235 - 238	

Europium (11) bromide

+ butane, 1-methoxy-	vol 22	242
+ decane, 2-methoxy-	vol 22	242
+ heptane, 2-methoxy-	vol 22	242
+ nonane, 2-methoxy-	vol 22	242
+ octane, 2-methoxy-	vol 22	242
+ pentane, 2-methoxy-	vol 22	242

Europium bromide

+ 2-butanamine	vol 22	244
+ furan, tetrahydro-	vol 22	243
+ 1-propanamine	vol 22	244
+ 2-propanamine	vol 22	244

Europium chloride

+ 1-butanamine	vol 22	239
+ decane, 2-methoxy-	vol 22	236
+ 1,4-dioxane	vol 22	237
+ 1,3-dioxolane	vol 22	237
+ ethane, 1,2-diethoxy-	vol 22	235
+ ethane, 1-ethoxy-2-methoxy-	vol 22	237
+ ethanol (aqueous)	vol 22	234
+ ethanol, 2-ethoxy-	vol 22	233
+ ethanol, 2-methoxy-	vol 22	233
+ heptane, 2-methoxy-	vol 22	236
+ nonane, 2-methoxy-	vol 22	236
+ octane, 2-methoxy-	vol 22	236
+ pentane, 2-methoxy-	vol 22	236
+ phosphoric acid, tributyl ester		
	vol 22	238
+ phosphoric triamide, hexamethyl-		
	vol 22	240
+ phosphoryl chloride (ternary)		
	vol 22	241
+ 2-propanamine	vol 22	239
+ 1-propanol	vol 22	233
+ 2-propen-1-amine	vol 22	239
+ stannane tetrachloro-(ternary)		
	vol 22	241
+ water (ternary)	vol 22	234

Europium fluoride

+ butane, 2-(chloromethoxy)-	vol 22	230
+ decane, 1-methoxy-	vol 22	230
+ dimethylsulfoxide	vol 22	231
+ pyridine	vol 22	232

Europium nitrate
    see nitric acid, europium salt

```
F

F 1162
          see benzenesulfonamide, 4-amino-
F.I. 5978
          see benzenesulfonamide. 4-amino-N-(3-methoxypryrazinyl)-
Fanasil
          see benzenesulfonamide, 4-amino-N-(5,6-dimethoxy-4-pyrimidinyl)-
Fanasulf
          see benzenesulfonamide, 4-amino-N-(5,6-dimethoxy-4-pyrimidinyl)-
Fat
          see Butter fat, Dog fat, Human fat, Rat fat, Guinea pig omental
             fat, Guinea pig subcutaneous fat
FC-47
          see 1-Butanamine,1,1,2,2,3,3,4,4,4-nonafluoro-N,N-bis(nonfluoro-)
FC-80
          see furan, hexafluorotetrahydro(nonafluorobutyl)-
Fe(III) sulfadiazine
          see iron, tris(4-amino-N-2-pyrimidinylbenzenesulfonamidato-NN,O)-
Ferric chloride
          see iron chloride
Ferric hydroxide
          see iron hydroxide
Ferric oxide
          see iron (3+) oxide
Ferric sulfadiazine
          see iron, tris(4-amino-N-2-pyrimidinylbenzenesulfonamidato-NN,O)-
Ferric sulfate
          see sulfuric acid, iron (3+) salt
Ferrocyanide, potassium
          see ferrate (4-), hexakis (cyano-C-)-, tetrapotassium salt
Ferrous ammonium sulfate
          see sulfuric acid, iron (2+) ammonium salt
Ferrous bromide
          see iron bromide
Ferrous chloride
          see iron(II) chloride
Ferrous iodide
          see iron iodide
Ferrous nitrate
          see nitric acid, iron (2+) salt
Ferrous selenate
          see selenic acid, iron (2+) salt
Ferrous sulfate
          see sulfuric acid, iron (2+) salt
Firmazolo
          see benzenesulfonamide, 4-amino-N-(1-phenyl-1H-pyrazol-5-yl)-
Flamazine
          see silver, (4-amino-N-2-pyrimidinylbenzenesulfonamidato-NN-01)-
Fluoranthene (multicomponent)
          + methane                              vol 27/28                 579
Fluoranthrene
          + water                                vol 38         E440,  441 - 444
          + sodium chloride                      vol 38                    445
9-H-Fluorene
          + water                                vol 38         E385, E386,
                                                                 387 - 389
Fluorene (aqueous)
          + sodium chloride                      vol 38                    390
9-H-Fluorene (multicomponent)
          + methane                              vol 27/28                 579
Fluoride
          see lithium fluoride, potassium fluoride, sodium fluoride
Fluorobenzene
          see benzene, fluoro-
Fluorocarbon fluid FC-47
          see 1-Butanamine,1,1,2,2,3,3,4,4,4-nonafluoro-N,N-bis(nonfluoro-)
Fluorocarbon fluid FC-80
          see furan, hexafluorotetrahydro(nonafluorobutyl)-
Fontamide
          see benzenesulfonamide, 4-amino-N-[(aminothioxomethyl)-
Formaldehyde (aqueous)
          + bromic acid, potassium salt     vol 30                    246
          + methane                         vol 27/28                 172
```

Formamide (aqueous)
          + bromic acid, potassium salt   vol 30   253
          + phosphoric acid, monopotassium salt
                             vol 31   E212, 272,
                                            276, 277
Formamide, N,N-dimethyl-
          + ammonia   vol 21   16
          + butane   vol 24   E299 -E302, 319
          + chloric acid, potassium salt   vol 30   160, 414, 415
          + chloric acid, sodium salt   vol 30   102
          + N,N-dimethylmethanamine   vol 21   E86, E91, 209
          + hydrogen sulfide   vol 32   E180, E182,
                                              E183, 299,
                                              302 - 304
          + iodic acid, cesium salt   vol 30   458, 459
          + iodic acid, rubidium salt   vol 30   444, 445
          + methanamine   vol 21   E81, E91,
                                              113, 114
          + methane   vol 27/28   720
          + N-methylmethanamine   vol 21   159
          + propane   vol 24   E125, 127, 129,
                                            E299 -E302,
                                            303 - 305
Formamide, N-methyl-
          + methane   vol 27/28   719
Formic acid, barium salt
          + formic acid, potassium salt   vol 33   3
          + formic acid, sodium salt   vol 33   4
          + formic acid, thallium salt   vol 33   5
Formic acid, calcium salt
          + formic acid, potassium salt   vol 33   6
          + formic acid, sodium salt   vol 33   7
          + formic acid, thallium salt   vol 33   8
Formic acid, lithium salt
          + acetic acid, lithium salt   vol 33   206
          + formic acid, potassium salt   vol 33   E9, 10, 11
          + formic acid, sodium salt   vol 33   19, 20
          + nitric acid, lithium salt   vol 33   208
          + thiocyanic acid, lithium salt
                             vol 33   207
Formic acid, magnesium salt
          + formic acid, potassium salt   vol 33   E12, 13, 14
           + formic acid, sodium salt   vol 33   21
          + formic acid, thallium salt   vol 33   22
Formic acid, potassium salt
          + acetic acid, potassium salt   vol 33   139
          + butanoic acid, potassium salt
                             vol 33   141
          + formic acid, barium salt   vol 33   3
          + formic acid, calcium salt   vol 33   6
          + formic acid, lithium salt   vol 33   E9, 10, 11
          + formic acid, magnesium salt   vol 33   E12, 13, 14
          + formic acid, sodium salt   vol 33   E15, 16, 17
          + formic acid, strontium salt   vol 33   18
          + nitric acid, potassium salt   vol 33   E150, 151, 152
          + nitrous acid, potassium salt   vol 33   149
          + potassium bromide   vol 33   136
          + potassium chloride   vol 33   144
          + potassium iodide   vol 33   148
          + propanoic acid, 2-methyl-, potassium salt
                             vol 33   E142, 143
          + propanoic acid, potassium salt
                             vol 33   140
          + thiocyanic acid, potassium salt
                             vol 33   E145, 146, 147
Formic acid, sodium salt
          + acetic acid, sodium salt   vol 33   221
          + butanoic acid, sodium salt   vol 33   E223, 224
          + butanoic acid, 2-methyl-, sodium salt
                             vol 33   E226, 227
          + formic acid, barium salt   vol 33   4
          + formic acid, calcium salt   vol 33   7
          + formic acid, lithium salt   vol 33   19, 20
          + formic acid, magnesium salt   vol 33   21

Formic acid, sodium salt
            + formic acid, potassium salt   vol 33   E15,   16,   17
            + formic acid, strontium salt   vol 33   23
            + nitric acid, sodium salt   vol 33   E236, E237
                                                                  238 - 241
            + nitrous acid, sodium salt   vol 33   235
            + propanoic acid, sodium salt   vol 33   222
            + propanoic acid, 2-methyl-, sodium salt
                                               vol 33   225
            + sodium bromide   vol 33   218
            + sodium chloride   vol 33   228
            + thiocyanic acid, sodium salt   vol 33   E229, 230 - 233
Formic acid, sodium salt (aqueous and multicomponent)
            + acetic acid, sodium salt   vol 26   E191, E192,   234
            + phosphoric acid, sodium salt   vol 26   E191, E192,   234
            + sulfurous acid, calcium salt   vol 26   E191, E192,   234
Formic acid, strontium salt
            + formic acid, potassium salt   vol 33   18
            + formic acid, sodium salt   vol 33   23
            + formic acid, thallium salt   vol 33   24
Formic acid, thallium salt
            + formic acid, barium salt   vol 33   5
            + formic acid, calcium salt   vol 33   8
            + formic acid, magnesium salt   vol 33   22
            + formic acid, strontium salt   vol 33   24

Formosulfacetamide
        see acetamide, N-[(4-aminophenyl)sulfonyl]-
Fourneau 1162
        see benzenesulfonamide, 4-amino-
Francium
            + mercury   vol 25   E421
Freon-11
        see methane, trichlorofluoro-
Freon-12
        see methane, dichlorodifluoro-
Freon-13
        see methane, chlorotrifluoro-
Freon-22
        see methane, chlorodifluoro-
Freon-113
        see ethane, 1,1,2-trichloro-1,2,2-trifluoro-
Freon-114
        see ethane, 1,2-dichloro-1,1,2,2-tetrafluoro-
β-D-Fructofuranosyl-α-D-glucopyranoside (aqueous)
            + methane   vol 27/28   180,   181
Fuel hydrogenated
        see hydrogenated fuel
2-Furancarboxaldehyde
            + arsine   vol 21   E302,   306,   307
            + bromic acid, potassium salt   vol 30   255
            + germane   vol 21   E315,   323,   324
            + hydrogen sulfide   vol 32   E176, E178
                                                                  277,   278
            + hydrogen selenide   vol 32   E330,   334,   335
            + 2-methylpropane   vol 24   E374 -E377,   406
            + phosphine   vol 21   E281, E282
                                                                  296,   297
            + propane   vol 24   E374 -E377,   386
            + silane   vol 21   E315,   318,   319
            + stannane   vol 21   E315,   327,   328
            + stibine   vol 21   E310,   311,   312
Furan, dibenzo-
            + hydrogen sulfide   vol 32   E176, E178,   276
Furan, tetrahydro-
            + hydrogen sulfide   vol 32   E173, E174,   263
            + methane   vol 27/28   E643,   653
            + phosphine   vol 21   E281, E282,   291
Furfural alcohol
        see 2-furanmethanol
Furfuryl
        see 2-furancarboxaldehyde

```
G

G 867
          see benzamide, N-[(4-aminophenyl)sulfonyl]-3,4-dimethyl-
Gadolinium
          + mercury                          vol 25          E239,  240,   241
Gadolinium bromide
          + 1-butanamine                     vol 22                         265
          + 2-butanamine                     vol 22                         265
          + 2-butanamine, N-(1-methylpropyl)-
                                             vol 22                         265
          + 1,4-dioxane                      vol 22                         264
          + ethane, 1,2-diethoxy-            vol 22                         262
          + furan, tetrahydro-               vol 22                         263
          + 2-propanamine                    vol 22                         265
Gadolinium chloride
          + 1-butanamine                     vol 22                         259
          + 2-butanamine                     vol 22                         259
          + 1,4-dioxane                      vol 22                         256
          + ethane, 1,2-diethyoxy-           vol 22                         255
          + ethane, 1-ethoxy-2-methoxy-      vol 22                         256
          + ethane, 1,1-oxybis-              vol 22                         254
          + ethanol                          vol 22          247 -  249
          + ethanol (aqueous)                vol 22                         250
          + ethanol, 2-ethoxy-               vol 22          249,   253
          + ethanol, 2-methoxy-              vol 22          249,   252
          + furan, tetrahydro-               vol 22                         257
          + methanol                         vol 22                         246
          + pentane, 1-methoxy-              vol 22                         256
          + phosphoric acid, tributyl ester
                                             vol 22                         258
          + phosphoric triamide, hexamethyl-
                                             vol 22                         260
          + phosphoryl chloride              vol 22                         261
          + 1-propanamine                    vol 22                         259
          + 2-propanamine                    vol 22                         259
          + 2-propanol                       vol 22                         251
          + 2-propen-1-amine                 vol 22                         259
          + 2-propen-1-ol                    vol 22                         249
          + stannane, tetrachloro-           vol 22                         261
          + water (ternary)                  vol 22                         250
Gadolinium cobalt nitrate
          see nitric acid, cobalt gadolinium salt
Gadolinium fluoride
          + methane, 1,1´-sulfinylbis-       vol 22                         245
Gadolinium iodide
          + 1-butanamine                     vol 22                         268
          + 2-butanamine                     vol 22                         268
          + 1,4-dioxane                      vol 22                         267
          + formamide, N,N-dimethyl-         vol 22                         269
          + furan, tetrahydro-               vol 22                         266
          + 1-propanamine                    vol 22                         268
          + 2-propanamine                    vol 22                         268
Gadolinium nickel nitrate
          see nitric acid, gadolinium nickel salt
Gadolinium nitrate
          see nitric acid, gadolinium salt
Gadolinium zinc nitrate
          see nitric acid, gadolinium zinc salt
4-(Galactosylamino)benzenesulfonamide
          see benzenesulfonamide, 4-(galactosylamino)-
Gallium
          + mercury                          vol 25               E93 -  E95,
                                                                   96 -  102
                                             vol 29          E213 -E214,   216
Ganidan
          see benzenesulfonamide, 4-amino-N-[(aminoiminomethyl)-
Gantanol
          see benzenesulfonamide, 4-amino-N-(5-methyl-3-isoxazolyl)-
Gantrisin
          see benzenesulfonamide, 4-amino-N-(3,4-dimethyl-5-isoxazolyl)
```

```
Gantrisin acetyl
        see acetamide, N-[(4-aminophenyl)sulfonyl]-N-
            (3,4-dimethyl-5-isoxazolyl)-
Gantrisona
        see benzenesulfonamide, 4-amino-N-(3,4-dimethyl-5-isoxazolyl)
Gantrosan
        see benzenesulfonamide, 4-amino-N-(3,4-dimethyl-5-isoxazolyl)
Gas oil
        see oil gas
Gasoline
        see also Kerosene
Gasoline
            + methane                          vol 27/28                     571
Geigy 867
        see benzamide, N-[(4-aminophenyl)sulfonyl]-3,4-dimethyl-
Geothermal brine
        see Brine
Gerison
        see benzenesulfonamide, 4-amino-
Germanium
            + mercury                          vol 25        E135,  136 - 138
Globucid
        see benzenesulfonamide, 4-amino-N-(5-methyl-1,3,4-thiadiazol-2-yl)
Globucin
        see benzenesulfonamide, 4-amino-N-(5-methyl-1,3,4-thiadiazol-2-yl)
Globulin, bovine
        see bovine globulin
Glucidoral
        see benzenesulfonamide, 4-amino-N-[(butylamino)carbonyl]-
Glucono lactone
        see D-gluconic acid
D-Glucose
            + mercury                          vol 29        E87,   88 -  94
D-Glucose (aqueous)
            + bromic acid, potassium salt      vol 30                       245
Glucose (aqueous)
            + methane                          vol 27/28                    183
Glucose (aqueous and multicomponent)
            + sulfurous acid, calcium salt     vol 26        E191, E192,  220,
                                                                   221,   235
Glucose (multicomponent)
            + copper (II) hydroxide            vol 23        E10 - E26,    71
 -glucose-1-phosphate
        see  -D-glucopyranose, 1-(dihydrogen phosphate)-
L-Glutamine (aqueous)
            + mercury                          vol 29                E50,   76
Glybutamide
        see benzenesulfonamide, 4-amino-N-[(butylamino)carbonyl]-
Glycerin
        see 1,2,3-propanetriol
Glycerol
        see 1,2,3-propanetriol
Glycerol hexanoate
        see hexanoic acid, 1,2,3-propanetriyl ester
Glycerol octanoate
        see octanoic acid, 1,2,3-propanetriyl ester
Glycerol triacetate
        see 1,2,3-propanetriol, triacetate
Glycerol tributurate
        see butanoic acid, 1,2,3-propanetriyl ester
Glyceryl triacetate
        see 1,2,3-propanetriol triacetate
Glycine (aqueous)
            + bromic acid, sodium salt         vol 30                       248
            + chloric acid, potassium salt     vol 30              E108,    413
            + zinc oxide                       vol 23              E179,    257
Glycine, N-[(4-aminosulfonyl)phenyl]-
            + 2-propanone                      vol 34                       193
Glycine, N-[(4-aminosulfonyl)phenyl]-, monosodium salt
            + 2-propanone                      vol 34                       194
Glycol
        see 1,2-ethanediol
```

Glyprothiazole
     see benzenesulfonamide, 4-amino-N-[5-(1-methylethyl)-
        1,3,4-thiadiazol-2-yl]-
Gold
        + mercury   vol 25   E139 -E141, 142
                                           vol 29   E213, E214,
                                                              226, 227
Gold (III) hydroxide
        + water   vol 23   E148 -E150
Gold (III) hydroxide (aqueous)
        + nitric acid   vol 23   E148, 155
        + sodium hydroxide   vol 23   E148 -E150,
                                                              153, 154
        + sulfuric acid   vol 23   E148, E180,
                                                              151, 152
Gombardol
     see benzenesulfonamide, 4-amino-
Guaiacol
     see phenol,2-methoxy-
Guamide
     see benzenesulfonamide, 4-amino-N-[(aminoiminomethyl)-
Guanicil
     see benzenesulfonamide, 4-amino-N-[(aminoiminomethyl)-
Guanidan
     see benzenesulfonamide, 4-amino-N-[(aminoiminomethyl)-
Guanidine hydrochloride (aqueous)
        + methane   vol 27/28   E64, 103
Guanidine monohydrochloride (aqueous)
        + butane   vol 24   E58 - E72,   82
        + methane   vol 27/28   E64, 103
        + 2-methylpropane   vol 24   E58 - E72,   83
        + propane   vol 24   E58 - E72,   81
Guanidinium chloride
     see Guanidine monohydrochloride
p-(Guanidinosulfonyl)acetanilide
     see acetamide, N-[4-[[(aminoiminomethyl)amino]sulfonyl]phenyl]-
N1-Guanidylsulfanilamide
     see benzenesulfonamide, 4-amino-N-[(aminoiminomethyl)-
N-Guanylsulfanilamide
     see benzenesulfonamide, 4-amino-N-[(aminoiminomethyl)-
Guaranine
     see 1H-purine-2,6-dione, 3,7-dihydro-1,3,7-trimethyl-

H

Hafnium
              + mercury   vol 25   E267
Hank´s balanced salt solution
              + mercury   vol 29   E87,   88
Haptocil
        see benzenesulfonamide, 4-amino-N-2-pyridinyl-
1,4,7,10,13,16-Hexaoxacyclooctadecane, comp. with 4-amino-N-(6-methoxy-
4-pyrimidinyl)benzenesulfonamide
        see benzenesulfonamide, 4-amino-N-(6-methoxy-4-pyrimidinyl)-, comp
           with 1,4,7,10,13,16-hexaoxacyclooctadecane (1:1)
Heavy water
        see water-d2
Helium
              + mercury   vol 29   E190, E192,   193
Hemoglobin
        see also bovine hemoglobin, dog hemoglobin, whale hemoglobin
Hemoglobin
              + mercury   vol 29   E51,   99
Hemoglobin, human
              + butane   vol 24   E414,   421
Heptadecane
              + propane   vol 24   E130 -E134,   158
1,6-Heptadiene
              + water   vol 37   441
1,6-Heptadiyne
              + water   vol 37   438
2,2,4,4,6,8,8-Heptamethylnonane
        see nonane, 2,2,4,4,6,8,8-heptamethyl-
Heptane
              + butane   vol 24   E130 -E134,   171
                                                                     172, E196, E197
                                                                           210 - 212
              + N,N-dimethylmethanamine   vol 21   E84, 181, 182
              + hydrogen sulfide   vol 32   E166 -E169,   188
                                                                           200,   201
              + mercury   vol 29   E102 -E106,
                                                                           123,   124
              + methane   vol 27/28   E202 -E206,   219
                                                                     224 - 226, E338
                                                                           339 - 348
                                                                           394 - 396
              + 2-methylpropane   vol 24   E130 -E134,   192
              + phosphine   vol 21   E281, E282,   287
              + propane   vol 24   E130 -E134
                                                                           146,   147
              + seawater   vol 37   E516,   517
              + water   vol 37   E485 -E491
                                                                           492 - 514
              + water-d2   vol 37   515
Heptane (ternary)
              + methane   vol 27/28   385,   386
Heptane, 2,6-dimethyl-
              + phosphine   vol 21   E281, E282,   284
Heptane, 2,6-dimethyl-
              + water   vol 38   220
Heptane, hexadecafluoro-
              + methane   vol 27/28   682
              + propane   vol 24   E257, E258,   267
Heptane, 3-methyl-
              + water   vol 38   133,   134
Heptanoic acid, potassium salt
              + nitrous acid, potassium salt   vol 33   203
1-Heptanol
              + methane   vol 27/28   E582 -E584, E591
                                                                           622,   623
              + propane   vol 24   E331 -E334,   344
1-Heptene
              + water   vol 37   E455, E456
                                                                           457 - 459
2-Heptene
              + water   vol 37   E460, 461 - 463

```
2-Heptene, 2,6,6-trimethylbicyclo (3,1.1)-
            + methane                    vol 27/28           E491,  492
Heptene, 2,4,6-trimethyl-2-phenyl-
            + water                      vol 38                     462
1-Heptyne
            + water                      vol 37                     442
Heptyl alcohol
        see 1-heptanol
Hexachlorobenzene
        see benzene, hexachloro-
1,1,2,3,4,4-Hexachloro-1,3-butadiene
        see 1,3-butadiene,1,1,2,3,4,4-hexachloro-
Hexacosane
            + water                      vol 38               544,  545
            + seawater                   vol 38                     545
Hexadecafluoroheptane
        see heptane, hexadecafluoro-
1-Hexadecanaminium,N,N,N-trimethyl bromide (aqueous)
            + methane                    vol 27/28                  100
Hexadecane
            + ammonia                    vol 21         E2,   13,   16
                                                             17,  E90
            + butane                     vol 24       E130 -E134,  179
                                                            180,  182
            + hydrogen sulfide           vol 32       E166 - E69,  189
                                                       190,  210 - 212
            + methane                    vol 27/28         E202 -E204
                                                           E208 -E210
                                                            235 - 241
                                                    E370,  371 - 375
            + propane                    vol 24            E130 -E134
                                                       154 - 159,  166
            + seawater                   vol 38                     474
            + water                      vol 38            E464, E465
                                                            466 - 473
Hexadecane (multicomponent)
            + methane                    vol 27/28                  486
1,5-Hexadiene + water                    vol 37               209,  210
Hexafluorobenzene
        see Benzene, hexafluoro-
Hexafluorobis(trifluoromethyl)cyclobutane
        see cyclobutane, hexafluorobis(trifluoromethyl)-
Hexamethylenediamine
        see 1,6-Hexanediamine
Hexamethylphosphoric triamide
        see Phosphoric triamide, hexamethyl-
Hexane
            + ammonia                    vol 21         E1,  E88 - E90
                                                             13 -  15
            + butane                     vol 24            E130 -E134
                                                            167 - 169
            + N,N-diethyl-ethanamine     vol 21            E252,  280
            + N,N-dimethylmethanamine    vol 21       E83,  E84,  E88
                                                            E89,  E91
                                                            178 - 180
            + ethanamine                 vol 21       E252,  255,  256
            + ethanamine-d2              vol 21       E234,  242,  243
            + hydrogen sulfide           vol 32       E166 -E169,  188
                                                       190,  198,  199
            + mercury                    vol 29            E102 -E106
                                                       113 - 118, E165
            + methanamine-d, N-methyl-   vol 21       E234,  244,  245
            + methanamine                vol 21       E80,  E88 - E90
                                                             94,   95
            + methanamine-d2             vol 21       E234,  238,  239
            + methanamine-d5             vol 21       E234,  240,  241
            + methan-d3-amine            vol 21       E234,  235,  237
            + methane                                      E202 -E205
                                                            216 - 222
                                                    E230,  321 - 335
            + N-methylmethanamine        vol 21       E82,  E88,  E89
                                                       E91,  135 - 138
            + N-(methyl-d3)-methan-d3-amine
                                         vol 21       E234,  246,  247
```

Hexane
        + N-(methyl-d3)-methan-d3 amine-d
                                        vol 21   E234,   248,   249
              + 2-methylpropane   vol 24   E130 -E134,
                                                          189,   190
              + phosphine   vol 21   E281, E282,   283
              + 1-propanamine   vol 21   E252,   259 - 263
              + 2-propanamine   vol 21   E252,   268 - 270
              + 1-propanamine-d   vol 21   E234,   264 - 266
              + 2-propanamine-d   vol 21   E234,   271 - 273
              + propane   vol 24   E130 -E134,   135,
                                                          136 - 140
              + seawater   vol 37   E366,   367
              + water   vol 37   E314 -E323,
                                                          324 - 365,   368
Hexane (aqueous)
              + sodium chloride   vol 37   368
Hexane (ternary)
              + methane   vol 27/28   414 - 417,
                                                          427 - 435,
                                                          441 - 443
neo-Hexane
        see Butane, 2,2´-dimethyl-
Hexane, 1-chloro-
              + methane   vol 27/28   691
Hexane, 2,3-dimethyl-
              + water   vol 38   131
Hexane, 2,4-dimethyl-
              + water   vol 38   132
Hexane, 2-methyl-
              + water   vol 37   478 - 480
Hexane, 3-methyl-
              + water   vol 37   E481,   482 - 484
Hexane, tetradecafluoro-
              + propane   vol 24   E257, E258,   265
Hexane, 2,2,5-trimethyl-
              + water   vol 38   E216, E217,
                                                          218,   219
1,6-Hexanediamine
              + 1-butanamine   vol 21   E252,   275
Hexanedinitrile
              + ammonia   vol 21   E11,   76 -   78
Hexanoic acid
              + butane   vol 24   E374 -E377,   399
              + hydrogen sulfide   vol 32   E176,   272
              + 2-methylpropane   vol 24   E374 -E377,   408
              + propane   vol 24   E374 -E377,   389
Hexanoic acid, potassium salt
              + acetic acid, potassium salt   vol 33   E163,   164
              + hexanoic acid, sodium salt   vol 33   E128,   129
              + nitrous acid, potassium salt vol 33   202
Hexanoic acid, sodium salt
              + acetic acid, sodium salt   vol 33   E253,   254,   255
              + benzoic acid, sodium salt   vol 33   E324,   325
              + butanoic acid, sodium salt   vol 33   E280,   281
              + butanoic acid , 2-methyl-, sodium salt
                                        vol 33   E311,   312
              + hexanoic acid, potassium salt
                                        vol 33   E128,   129
              + nitric acid, sodium salt   vol 33   329
              + octadecanoic acid, sodium salt
                                        vol 33   E326,   327
              + propanoic acid, 2-methyl-, sodium salt
                                        vol 33   E294,   295
              + thiocyanic acid, sodium salt vol 33   328
1-Hexanol
              + methane   vol 27/28   E582- E584, E591,
                                                          E592,   618,
                                                          620,   621
1,4,7,10,13,16-Hexaoxacyclooctadecane compd. with 4-amino-N-(5-methyl-
3-isoxazolyl)benzenesulfonamide (1:1)
        see benzenesulfonamide, 4-amino-N-(5-methyl-3-isoxazolyl)-, compd.
            with 1,4,7,10,13,16-hexaoxacyclooctadecane

```
[1,4,7,10,13,16] hexaoxocyclooctadecin, 6,7,10,18,20,21-octahydro-
dibenzo[b,k]
                 + iodic acid, potassium salt     vol 30                      418
                 + iodic acid, rubidium salt      vol 30                      447
                 + iodic acid, sodium salt        vol 30                      373
                 + methanol                       vol 30         373,  418,  447
2,5,8,11,14,17-Hexaoxanonadecane,18-methyl-
                 + hydrogen sulfide               vol 32                      261
Hexatriacontane
                 + butane                         vol 24       E130 -E134,  188
                 + methane                        vol 27/28      E202 -E204,
                                                                 E212,  241
                 + propane                        vol 24       E130 -E134,  166
                 + water                          vol 38             546,  547
1-Hexene
                 + water                          vol 37         E275, E276,
                                                                 277 - 281
2-Hexene
                 + water                          vol 37                      282
Hexylcyclopentane
          see cyclopentane, hexyl-
1-Hexyne
                 + water                          vol 37                      211
Holmium
                 + mercury                        vol 25         E247,  248
Holmium bromide
                 + furan, tetrahydro-             vol 22                      321
Holmium chloride
                 + 2-butanamine, N-(1-methylpropyl)-
                                                  vol 22                      318
                 + butane, 1-ethoxy-              vol 22                      315
                 + decane, 2-methoxy-             vol 22                      314
                 + 1,4-dioxane                    vol 22                      315
                 + ethane, 1-ethoxy-2-methoxy-    vol 22                      315
                 + ethane, 1,2-diethoxy-          vol 22                      313
                 + ethanol (aqueous)              vol 22                      311
                 + ethanol, 1-ethoxy-             vol 22                      312
                 + ethanol, 1-methoxy-            vol 22                      312
                 + furan, tetrahydro-             vol 22                      316
                 + heptane, 2-methoxy-            vol 22                      314
                 + nonane, 2-methoxy-             vol 22                      314
                 + octane, 2-methoxy-             vol 22                      314
                 + pentane, 2-methoxy-            vol 22              314,  315
                 + phosphoric acid, tributyl ester
                                                  vol 22                      317
                 + phosphoric triamide, hexamethyl-
                                                  vol 22                      319
                 + phosphoryl chloride (ternary)
                                                  vol 22                      320
                 + 1-propanamine                  vol 22                      318
                 + propane, 1,1´-oxybis-          vol 22                      315
                 + stannane tetrachloro- (ternary)
                                                  vol 22                      320
                 + water (ternary)                vol 22                      311
Holmium fluoride
                 + butane, 1-(chloromethoxy)-     vol 22                      308
                 + decane, 1-methoxy-             vol 22                      308
                 + ethanol                        vol 22                      307
                 + methane, 1,1´-sulfinylbis-     vol 22                      310
                 + methanol                       vol 22                      307
                 + phosphoric acid, tributyl ester
                                                  vol 22                      309
Holmium nitrate
          see Nitric acid, holmium salt
Homonal
          see benzenesulfonamide, 4-(aminomethyl)-
Homosul
          see benzenesulfonamide, 4-(aminomethyl)-
Homosulfanilamide
          see benzenesulfonamide, 4-(aminomethyl)-
Honey Diazine
          see benzenesulfonamide, 4-amino-N-2-pyrimidinyl-
Human blood
          see blood, human
```

```
Humus (aqueous)
          + mercury                        vol 29         E51,    95,    96
Hydrated aluminium oxide
        see Aluminium oxide (aqueous)
Hydrated ferric oxide
        see Ferric oxide, hydrated
Hydrated iron oxide
        see Iron oxide (aqueous)
Hydrazine
          + ammonia                        vol 21         E10,    71,    E90
          + bromic acid, sodium salt       vol 30                        219
          + chloric acid, sodium salt      vol 30                        104
          + iodic acid, potassium salt     vol 30                        427
Hydrazine 1,1-dimethyl-
          + ammonia                        vol 21         E10,    73,    E90
Hydrazine, methyl-
          + ammonia                        vol 21         E10,    72,    E90
Hydrocarbon blend
        see Gas oil, Heavy naphtha
Hydrocarbon blend (gas oil)
        see oil, gas
Hydrocarbon blend (heavy naphtha)
          + methane                        vol 27/28                     559
Hydrocarbon oil
          + butane                         vol 24         E421,   427,   429
          + methane                        vol 27/28              557,   558
          + propane                        vol 24         E421,   425,   427
Hydrochloric acid
          + butane                         vol 24         E58 - E72,     74
          + mercury                        vol 29                 E47,   52
          + mercury (II) oxide             vol 23                 328,   329
          + methane                        vol 27/28      E57,    E58,   86
          + propane                        vol 24         E58 - E72,     73
Hydrochloric acid (aqueous)
          + mercury (II) oxide             vol 23                 328,   329
          + selenious acid, calcium salt   vol 26                        357
          + selenious acid, copper (II) salt
                                           vol 26                        379
          + selenious acid, lead (II) salt
                                           vol 26                        401
          + selenious acid, magnesium salt
                                           vol 26         E348,   E349,  355
          + selenious acid, strontium salt
                                           vol 26         E348,   E349,  359
          + tellurous acid, barium salt    vol 26                 E404,  420
          + tellurous acid, copper salt    vol 26                 E404,  423
          + tellurous acid, nickel salt    vol 26                 E404,  422
Hydrochloric acid (aqueous and multicomponent)
          + sulfurous acid, calcium salt   vol 26                 E191,  E192
                                                                  224,   225
Hydrochloric acid (multicomponent)
          + zinc hydroxide                 vol 23                 266,   267
Hydrocortisone sodium succinate
        see pregn-4-ene-3,20-dione,21(3-carboxy-1-oxopropoxy)
             -11,17-dihydroxy-monosodium salt, (11-beta)-
Hydrocyanic acid
        see Hydrogen cyanide
          + arsine                         vol 21                 E302,  303
Hydrofluoric acid (aqueous)
          + iodic acid                     vol 30                 E470,  478
          + mercury (II) oxide             vol 23         E313,   319,   333
Hydrofluoric acid (multicomponent)
          + copper (II) oxide              vol 23         E10 - E26,     63
Hydroformer still bottoms (multicomponent)
          + methane                        vol 27/28                     490
 -Hydro- -hydroxy-poly(oxy-1,2-ethandiyl)-
          + hydrogen sulfide               vol 32         E173 -E175,    252
                                                                         253
Hydrogen
          + mercury                        vol 29         E190,   E191,  198
Hydrogen (multicomponent)
          + hydrogen sulfide               vol 32                 75  -  77
                                                                  240,   276
          + methane                        vol 27/28              663 -  665
```

Hydrogen chloride
          + hydrogen sulfide   vol 32   E25,   E26,   29,
                                                          30,   34 - 37
Hydrogen iodide
          + hydrogen sulfide   vol 32   E26,   31,   339
Hydrogen peroxide (aqueous)
          + cadmium hydroxide   vol 23   291,   292
          + phosphoric acid, disodium salt
                                           vol 31   E98,   107,   117
          + phosphoric acid, monopotassium salt
                                           vol 31   E212, 225
          + phosphoric acid, trisodium salt
                                           vol 31   E127,   141,   142
          + zinc oxide   vol 23   E180,   237 - 241
Hydrogen selenide
          + 1,2-benzenedicarboxylic acid, didecyl ester
                                           vol 32   E330,   334,   335
          + 2-ethoxyethanol   vol 32   E330,   332,   333
          + 2-furancarboxaldehyde   vol 32   E330,   334,   335
          + 1,1´-oxybis[2-chloroethane]   vol 32   E330,   332,   333
          + nitrobenzene   vol 32   E330,   332,   333
          + paraffin (liquid)   vol 32   E330, 332
          + triethoxysilane   vol 32   E330,   334,   335
          + silicic acid, tetraethyl ester
                                           vol 32   E330,   334,   335
Hydrogen sulfide
          + methane   vol 27/28   E750,   751 - 756
Hydrogen sulfide (ternary and multicomponent)
          + methane   vol 27/28   200,   201,
                                                                 447 - 449,
                                                                 757 - 760

Hydrogenated cotton seed oil
          see oil, hydrogenated cotton seed
Hydroquinone, dimethyl ether
          see benzene, 1,4-dimethoxy-
Hydroterpin
          see Hydroterpinol
Hydroxide
          see ammonium hydroxide, potassium hydroxide, sodium hydroxide
2-Hydroxybenzoic acid, methyl ester
          see benzoic acid, 2-hydroxy-, methylester
2-Hydroxy-N,N,N-tris(2-Hydroxyethyl)-ethanaminium bromide)
          see ethanamine, 2-hydroxy-, N,N,N-tris (2-hydroxyethyl)-,
                bromide
2-Hydroxy ethyl ether
          see ethanol, 2,2´-oxybis-
4-Hydroxyproline
          see proline, 4-hydroxy-
L-Hystidine, monochloride (aqueous)
          + mercury   vol 29   E50,   E51,   83
L-Hystidine, monochloride (in Hank´s balanced salt solution)
          + mercury   vol 29   E50,   E87,   90

```
I

ICI 32525
        see benzenesulfonamide, 4-amino-N-(6-methoxy-4-pyrimidinyl)-
ICI 3435
        see benzenesulfonamide, 4-amino-N-(4,6-dimethoxy-2-pyrimidinyl)-
1H-Imidazole, silver complex
        see silver, (4-amino-N-2-pyrimidinylbenzene-sulfonamidato-NN,O)-
            bis(1-imidazole-N3)-,(T-4)-
N1-Imidazol-2-ylsulfanilamide
        see benzenesulfonamide, 4-amino-N-1H-imidazol-2-yl-
2,2-Iminobisethanol
        see ethanol, 2,2-iminobis-
1,1´-Iminobis-2-propanol
        see 2-propanol, 1,1´-iminobis-
Inamil
        see benzenesulfonamide, 4-amino-N-(1-phenyl-1H-pyrazol-5-yl)-
Inbuton
        see benzenesulfonamide, 4-amino-N-[(butylamino)carbonyl]-
Indan
            + water                          vol 38        E161,   162,   163
Indium
            + mercury                        vol 25              E103 - 105,
                                                                 106 - 118
Inuline
        see aconitane-7,8-diol,4-[[(2-aminobenzoyl)oxy]-methyl]-20-
            ethyl-1,6,14,16-tetramethoxy-, (1b,6b,14a,16b)-
Invenol
        see benzenesulfonamide, 4-amino-N-[(butylamino)carbonyl]-
Iodic acid (aqueous)
            + hydrofluoric acid              vol 30              E470,   478
            + iodic acid, aluminium salt     vol 30        E272,   323,   324,
                                                           467, E470, E471,
                                                                          483
            + iodic acid, ammonium salt      vol 30        E462,   465,   466,
                                                                          E467
            + iodic acid, barium salt        vol 14              E246,   296
            + iodic acid, barium salt        vol 30                      E471
            + iodic acid, cadmium salt       vol 30              E471,   481
            + iodic acid, cesium salt        vol 30        E448, E449,   457,
                                                                          E467
            + iodic acid, lanthanium salt    vol 30              E471,   486
            + iodic acid, lithium salt       vol 30        E269, E270, E272,
                                                           279,   314 - 319,
                                                           321 - 324, E380,
                                                                 E467, E470
            + iodic acid, magnesium salt     vol 30                      E471
            + iodic acid, neodynium salt     vol 30        E471,   487,   488
            + iodic acid, potassium salt     vol 30        E272,   321,   322,
                                                           E378, E380,   407,
                                                           408, E467, E470
            + iodic acid, rubidium salt      vol 30        E428, E430, E467
            + iodic acid, scandium salt      vol 30              E471,   484
            + iodic acid, sodium salt        vol 30              E335,   371,
                                                                 372, E467
            + iodic acid, strontium salt     vol 30                      E471
            + iodic acid, zinc salt          vol 30              E471,   482
            + nitric acid                    vol 30        E470,   475 - 477
            + N-phenylacetamide              vol 30                      489
            + water                          vol 30              E467 -E473,
                                                                 474 - 489
Iodic acid, aluminium salt
            + water                          vol 30              E461, E462,
                                                                 463 - 466
Iodic acid, aluminium salt (aqueous)
            + iodic acid                     vol 30        E272,   323,   324,
                                                           E470, E471,   483
            + iodic acid, cesium salt        vol 30              E448,   455
            + iodic acid, lithium salt       vol 30        E271 -E273,   298,
                                                           299,   323, E470
            + iodic acid, rubidium salt      vol 30              E430,   440
            + iodic acid, sodium salt        vol 30              E335,   369
            + iodic acid, potassium salt     vol 30              E379,   398
```

```
Iodic acid, ammonium salt (aqueous)
          + ammonium fluoride             vol 30            E462,   464
          + iodic acid                    vol 30                    E462
          + iodic acid, lithium salt      vol 30      E271,   280, E462
          + iodic acid, magnesium salt    vol 30                    E462
Iodic acid, barium salt (aqueous)
          + iodic acid                    vol 30                    E471
          + iodic acid, potassium salt    vol 30            E378,   397
Iodic acid, cadmium salt (aqueous)
          + iodic acid                    vol 30            E471,   481
Iodic acid, calcium salt
          + water                         vol 30            E448 -E450,
                                                            451 - 460
Iodic acid, calcium salt (aqueous)
          + iodic acid, lithium salt      vol 30            E270, E271
          + iodic acid, sodium salt       vol 30                    E335
Iodic acid, cesium salt (aqueous)
          + N,N-dimethylformamide         vol 30            458,   459
          + iodic acid                    vol 30            E448, E449,
                                                            457, E467
          + iodic acid, aluminium salt    vol 30            E448,   455
          + iodic acid, hafnium salt      vol 30                    456
          + iodic acid, lithium salt      vol 30            E270,   291
                                                            292, E448
          + iodic acid, potassium salt    vol 30      E378,   396, E448
          + iodic acid, rubidium salt     vol 30            E430,   438
                                                            E448, E449
          + iodic acid, sodium salt       vol 30      E335,   367, E448
          + nitric acid, cesium salt      vol 30      E448, E449,   454
          + 1,1´sulfinylbismethane        vol 30                    460
Iodic acid, cobalt salt (aqueous)
          + iodic acid, potassium salt    vol 30            E379,   400
Iodic acid, copper salt (aqueous)
          + iodic acid, potassium salt    vol 30                    E379
Iodic acid, gallium salt (aqueous)
          + iodic acid, lithium salt      vol 30      E271,   300,   301
Iodic acid, hafnium salt (aqueous)
          + iodic acid, cesium salt       vol 30                    456
          + iodic acid, lithium salt      vol 30            E271, E272
                                                            308,   309
          + iodic acid, rubidium salt     vol 30                    441
          + iodic acid, sodium salt       vol 30            335,   370
Iodic acid, indium salt (aqueous)
          + iodic acid, lithium salt      vol 30            E271, E272
                                                            302,   303
Iodic acid, lanthanum salt (aqueous)
          + iodic acid                    vol 30            E471,   486
Iodic acid, lithium salt
          + 2-propanone                   vol 30                    329
          + water                         vol 30            E268 -E275,
                                                            276 - 328
Iodic acid, lithium salt (aqueous)
          + carbonic acid, dilithium salt vol 30            E272,   293
          + ethanol                       vol 30            E270,   328
          + iodic acid                    vol 30      E269, E270, E272
                                                      279,   314 - 319
                                                      321 - 324, E380
                                                            E467, E470
          + iodic acid, aluminium salt    vol 30      E271 -E273,   298
                                                            299,   323
                                                            324, E470
          + iodic acid, ammonium salt     vol 30      E271, E272,   280
          + iodic acid, calcium salt      vol 30            E270, E271
          + iodic acid, cesium salt       vol 30            E270,   291
                                                            292, E448
          + iodic acid, gallium salt      vol 30      E271,   300,   301
          + iodic acid, hafnium salt      vol 30            E271, E272
                                                            308,   309
          + iodic acid, indium salt       vol 30            E271, E272
                                                            302,   303
          + iodic acid, magnesium salt    vol 30            E270, E271
          + iodic acid, neodymium salt    vol 30            E271,   311
```

```
Iodic acid, lithium salt (aqueous)
          + iodic acid, potassium salt    vol 30      E270, E272, E273,
                                                       286,  287,   321,
                                                       322,  325 - 327,
                                                       E378, E380, E470
          + iodic acid, rubidium salt     vol 30      E270,  288 - 290,
                                                       E428, E430
          + iodic acid, samarium salt     vol 30      E271,   312
          + iodic acid, silver salt       vol 30      E271, E272,   297
          + iodic acid, sodium salt       vol 30      E270,  285, E335
          + iodic acid, thallium salt     vol 30      E271,   304
          + iodic acid, titanium salt     vol 30      E271, E272,
                                                       305,   306
          + iodic acid, zirconium salt    vol 30      E271, E272,   307
          + lithium bromide               vol 30      E272,   283
          + lithium chloride              vol 30      E272,   282
          + lithium hydroxide             vol 30      E270, E272,   313
                                                       325 - 327
          + lithium iodide                vol 30      E272,   284
          + lithium  (I-4)-tetraoxorhenate (1-)
                                          vol 30      E272,   310
          + nitric acid, lithium salt     vol 30      E272,  294,   295
          + phosphoric acid               vol 30              320
          + phosphoric acid, monolithium salt
                                          vol 30              320
          + phosphoric acid, trilithium salt
                                          vol 30      E272,   296
          + potassium hydroxide           vol 30      E273,  325 - 327,
                                                       E380
          + sulfuric acid, lithium salt   vol 30      E272,   281
Iodic acid, magnesium salt (aqueous)
          + iodic acid                    vol 30              E471
          + iodic acid, lithium salt      vol 30      E270, E271
          + iodic acid, sodium salt       vol 30              E335
          + iodic acid, potassium salt    vol 30              E378
Iodic acid, manganese salt (aqueous)
          + iodic acid, potassium salt    vol 30      E379,   399
Iodic acid, neodymium salt (aqueous)
          + iodic acid                    vol 30      E271,  311, E471,
                                                       487,   488
          + iodic acid, lithium salt      vol 30      E271,   311
          + iodic acid, rubidium salt     vol 30      E430,   442
          + iodic acid, potassium salt    vol 30      E379,   404
Iodic acid, nickel salt (aqueous)
          + iodic acid, potassium salt    vol 30      E379,   401
Iodic acid, potassium salt
          + ammonia                       vol 30       420 - 426
          + ammonium chloride             vol 30              424
          + hydrazine                     vol 30              427
          + methanol                      vol 30       417,   418
          + 6,7,9,10,17,18,20,21-octahydrodibenzo[b,k]-
            [1,4,7,10,13,16]hexaoxacyclooctadecin
                                          vol 30              418
          + potassium bromide             vol 30              425
          + potassium chloride            vol 30       421,   422
          + potassium iodide              vol 30              426
          + sodium chloride               vol 30              423
          + 1,1-sulfinylbismethane        vol 30              419
          + water                         vol 30      E374 -E382,
                                                       383 - 416
          + water-d2                      vol 30              387
Iodic acid, potassium salt (aqueous)
          + chloric acid, potassium salt  vol 30       145,  146, E380
          + N,N-dimethylformamide         vol 30       414,   415
          + 1,4-dioxane                   vol 30      E378,   412
          + disodium (I-4), tetraoxomolybdate (2-)
                                          vol 30      E380,   430
          + ethanol                       vol 30              411
          + glycine                       vol 30              413
          + iodic acid                    vol 30      E272,  321,   322,
                                                       E378, E380,   407,
                                                       408, E467, E470
          + iodic acid, aluminium salt    vol 30      E378,  398, E470
          + iodic acid, barium salt       vol 30      E378, E397
```

```
Iodic acid, potassium salt (aqueous)
          + iodic acid, cesium salt       vol 30      E378,   396, E448
          + iodic acid, cobalt salt       vol 30              E379,   400
          + iodic acid, copper salt       vol 30                      E379
          + iodic acid, lithium salt      vol 30      E270, E272,   286
                                                            287,   321,   322
                                                            E378, E380
          + iodic acid, magnesium salt    vol 30                      E378
          + iodic acid, manganese salt    vol 30              E379,   399
          + iodic acid, neodymium salt    vol 30              E379,   404
          + iodic acid, nickel salt       vol 30              E379,   401
          + iodic acid, rubidium salt     vol 30      E378,   395, E430
          + iodic acid, sodium salt       vol 30      E335,   366, E378
          + iodic acid, zinc salt         vol 30              E378,   402
          + lithium hydroxide             vol 30      E273,   325 - 327
                                                                    E380
          + nitric acid, potassium salt   vol 30      E380,   388,   389
          + potassium bromide             vol 30      E379,   392,   393
          + potassium chloride            vol 30              E379,   391
          + potassium hydoxide            vol 30      E273,   325 - 327
                                                      E380,   405,   406
                                                            409,   410
          + potassium iodide              vol 30      E379, E380,   394
                                                            409,   410
          + 1,1´sulfinylbismethane        vol 30                      416
          + sulfuric acid, potassium salt vol 30              E380,   390
Iodic acid, rubidium salt
          + methanol                      vol 30                      447
          + 6,7,10,17,18,20,21-octahydrodibenzo[b,k]-
            [1,4,7,10,13,16]hexaoxacyclooctadecin
                                          vol 30                      447
          + water                         vol 30              E428 -E431
                                                            432 - 447
Iodic acid, rubidium salt (aqueous)
          + N,N-dimethylformamide         vol 30              444,   445
          + iodic acid                    vol 30              E428, E430
                                                            443, E467
          + iodic acid, aluminium salt    vol 30              E430,   440
          + iodic acid, cersium salt      vol 30      E430,   438, E448
          + iodic acid, hafnium salt      vol 30                      441
          + iodic acid, lithium salt      vol 30      E270,   288 - 290
                                                            E428, E430
          + iodic acid, magnesium salt    vol 30                      E430
          + iodic acid, neodymium salt    vol 30              E430,   442
          + iodic acid, potassium salt    vol 30      E378,   395, E430
          + iodic acid, sodium salt       vol 30      E335,   367, E430
          + iodic acid, zinc salt         vol 30              E430,   439
          + nitric acid                   vol 30              E428,   435
          + nitric acid, potassium salt   vol 30              E428,   436
          + rubidium hydroxide            vol 30      E428, E430,   437
          + 1,1´-sulfinylbismethane       vol 30                      446
Iodic acid, samarium salt (aqueous)
          + iodic acid, lithium salt      vol 30              E271,   312
Iodic acid, scandium salt (aqueous)
          + iodic acid                    vol 30      E471,   484,   485
Iodic acid, silver salt (aqueous)
          + iodic acid, lithium salt      vol 30      E271, E272,   297
Iodic acid, sodium salt
          + methanol                      vol 30                      373
          + 6,7,10,17,18,20,21-octahydrodibenzo[b,k]-
            [1,4,7,10,13,16]hexaoxacyclooctadecin
                                          vol 30                      373
          + water                         vol 30              E330 -E339
                                                            340 - 372
Iodic acid, sodium salt (aqueous)
          + carbonic acid, disodium salt
                                          vol 30              E337, E338
                                                            344,   345
          + chloric acid, sodium salt     vol 30               65,    66
                                                            E337, E338
          + disodium (I-4) tetraoxomolybdate (2-)
                                          vol 30              E337,   343
          + iodic acid                    vol 30              E335,   371
                                                            372, E467
```

Iodic acid, sodium salt (aqueous)
+ iodic acid, aluminium salt	vol 30	E335, 369
+ iodic acid, calcium salt	vol 30	E335
+ iodic acid, cesium salt	vol 30	E335, 368, E448
+ iodic acid, hafnium salt	vol 30	E335, 370
+ iodic acid, lithium salt	vol 30	E270, 285, E335
+ iodic acid, magnesium salt	vol 30	E335, E430
+ iodic acid, potassium salt	vol 30	E335, 366, E378
+ iodic acid, rubidium salt	vol 30	E335, 367, E430
+ nitric acid, sodium salt	vol 30	E336, E337 346 - 353
+ sodium bromide	vol 30	E335, E336 359 - 363
+ sodium chloride	vol 30	E335, E336 357, 358
+ sodium iodide	vol 30	E335, E336 364, 365
+ sulfuric acid, sodium salt	vol 30	E337, 354 - 356

Iodic acid, strontium salt (aqueous)
+ iodic acid	vol 30	E471

Iodic acid, thallium salt (aqueous)
+ iodic acid, lithium salt	vol 30	E271, 304

Iodic acid, titanium salt (aqueous)
+ iodic acid, lithium salt	vol 30	E271, E272 305, 306

Iodic acid, zinc salt (aqueous)
+ iodic acid	vol 30	E471, 482
+ iodic acid, rubidium salt	vol 30	E430, 439
+ iodic acid, potassium salt	vol 30	E379, 402

Iodic acid, zirconium salt (aqueous)
+ iodic acid, lithium salt	vol 30	E271, 307

Iodide
        see under individual iodidese
Iodine pentoxide
+ sulfuric acid	vol 30	E470, 479, 480

Iodobenzene
        see benzene,iodo-
1-Iodooctane
        see octane, 1-iodo-
2-Iodooctane
        see octane, 2-iodo-
IPTD
        see benzenesulfonamide, 4-amino-N-[5-(1-methylethyl)-
        1,3,4-thiadiazol-2-yl]-
Irgafene
        see benzamide, N-[(4-aminophenyl)sulfonyl]-3,4-dimethyl-
Irgafen
        see benzamide, N-[(4-aminophenyl)sulfonyl]-3,4-dimethyl-
Irgamide
        see 2-butenamide, N-[(4-aminophenyl)sulfonyl]-3-methyl-
Irgamid
        see 2-butenamide, N-[(4-aminophenyl)sulfonyl]-3-methyl-
Iridium
+ mercury	vol 25	E316
	vol 29	E214, 217

Iron
+ mercury	vol 25	E301, E302 303 - 306
	vol 29	E207, E214, 217

Iron (II) chloride (in hydrochloric acid)
+ hydrogen sulfide	vol 32	E26, 34

Iron(III) selenite
        see selenious acid, iron(III) salt
Iron sulfate
        see sulfuric acid, iron(II) salt, sulfuric acid, iron(III) salt
Iron sulfide (molten)
+ mercury	vol 29	E207, E208

Iron(II) sulfite
        see sulfurous acid, iron(II) salt
Iron, tris(4-amino-N-2-pyrimidinylbenzenesulfonamidato-NN,O)-
+ water	vol 36	214

Isarol
        see benzenesulfonamide, 4-amino-N-(1-phenyl-1H-pyrazol-5-yl)-

L-Isoleucine (aqueous)
                + mercury   vol 29   E50,   E51,   79
Isoxamin
            see benzenesulfonamide, 4-amino-N-(3,4-dimethyl-5-isoxazolyl)-

```
J

Jet fuel
          see fuel ,jet
JP-1, JP-4
          see fuel
Justamil
          see benzenesulfonamide, 4-amino-N-(4,5-dimethyl-2-oxazolyl)-
```

K

Kanamycin (or Kanamycin A)
          + mercury   vol 29   E87,   88 - 94
Kaoxidin
          see butanoic acid, 4-oxo-4-[[4-[(2-thiaxolylamino)sulfonyl]phenyl]
          amino]-
Kaoxidine
          see butanoic acid, 4-oxo-4-[[4-[(2-thiaxolylamino)sulfonyl]phenyl]
          amino]-
Kelametazine
          see benzenesulfonamide, 4-amino-N-(4,6-dimethyl-2-pyrimidinyl)-
Kelfizina
          see benzenesulfonamide. 4-amino-N-(3-methoxypryazinyl)-
Kelfizin
          see benzenesulfonamide. 4-amino-N-(3-methoxypryazinyl)-
Kerosene
          + ammonia   vol 21   18
Kerosene A-1
          + hydrogen sulfide   vol 32   E170, E171
          + methane   vol 27/28   575
Kerosene A-3
          + butane   vol 24   E414,   417
          + propane   vol 24   E414,   416, E422
                                                                          425
Kidney
          see rabbit kidney
Kinex
          see benzenesulfonamide, 4-amino-N-(6-methoxy-3-pyridazinyl)-
Kirocid
          see benzenesulfonamide, 4-amino-N-(5-methoxy-2-pyrimidinyl)-
Kiron
          see benzenesulfonamide, 4-amino-N-(5-methoxy-2-pyrimidinyl)-
Klucel MF
          see cellulose, ethers, 2-hydroxypropyl ether
Koffein
          see 1H-purine-2,6-dione, 3,7-dihydro-1,3,7-trimethyl-
Krypton
          + mercury   vol 29   E190, E191, 197
Kynex
          see benzenesulfonamide, 4-amino-N-(6-methoxy-3-pyridazinyl)-

L

Lactose
     see 4-O- -D-glucopyranosyl-D-glucose
Lanthanum
     + mercury   vol 25   E207, E208,
                                                                   209 - 213
Lanthanum acetate
     see acetic acid, lanthanum salt
Lanthanum ammonium nitrate
     see nitric acid, lanthanum ammonium salt
Lanthanum bromide
     + 1,4-dioxane   vol 22   94
     + 1,4-dioxane (ternary)   vol 22   95
     + methanol (ternary)   vol 22   95
     + 1-propanol (ternary)   vol 22   95
Lanthanum chloride
     + benzene (ternary)   vol 22   75,   76
     + 1,3-butadiene, hexachloro-   vol 22   74
     + 1,4-dioxane   vol 22   89
     + 1,4-dioxane (ternary)   vol 22   90,   91
     + 1,3-dioxolane   vol 22   88 -   89
     + 1,2-ethanediol   vol 22   81
     + ethane, 1-ethoxy-2-methoxy-   vol 22   88 -   89
     + ethanol   vol 22   78,   79
     + ethanol (aqueous)   vol 22   80
     + ethanol (ternary)   vol 22   76,   91
     + ethanol, 2-methoxy-   vol 22   86
     + ethanol, 2-ethoxy-   vol 22   87
     + heptane, 1-methoxy-   vol 22   88
     + methanol   vol 22   77
     + methanol (ternary)   vol 22   75,   76
                                                                   90,   91
     + octane, 1-methoxy-   vol 22   88
     + phosphoric triamide, hexamethyl-
                                            vol 22   92
     + phosphoryl chloride   vol 22   93
     + phosphoryl chloride (ternary)   vol 22   93
     + 1,2,3-propanetriol   vol 22   85
     + 1-propanol   vol 22   82
     + 1-propanol (ternary)   vol 22   76,   91
     + 2-propanol   vol 22   83 -   84
     + stannane, tetrachloro- (ternary)
                                            vol 22   93
     + water (ternary)   vol 22   80
Lanthanum chloride (aqueous)
     + butane   vol 24   E58 - E72,   95
Lanthanum cobalt nitrate
     see Nitric acid, cobalt lanthanum salt
Lanthanum fluoride
     + acetic acid, ethyl ester   vol 22   62
     + acidic nitrosyl fluoride (nitrosyl fluoride compound
       with hydrofluoric acid)
                                            vol 22   72
     + benzenamine (aniline)   vol 22   52
     + benzene (ternary)   vol 22   51
     + benzene, 1-chloro-2,4-dinitro-   vol 22   51
     + benzene, 1,3-dinitro-   vol 22   51
     + benzene, 1-fluoro-2,4-dinitro-   vol 22   51
     + butane, 1-(bromomethoxy)-   vol 22   50
     + butane, 1-(chloromethoxy)-   vol 22   50
     + butane, 1-methoxy-   vol 22   61
     + 1-butanesulfonyl fluoride, 1,1,2,2,3,3,4,4,4-nonafluoro-
                                            vol 22   50
     + 2-butanol   vol 22   59
     + 2-butanol, 2-methyl-   vol 22   60
     + decane, 2-methoxy-   vol 22   61
     + ethanamine, N,N-diethyl-   vol 22   66
     + 1,2-ethanediol (ethylene glycol)
                                            vol 22   57
     + ethane, 1,2-dimethoxy-   vol 22   61
     + ethane, 1-ethoxy-2-methoxy-   vol 22   61
     + ethanol   vol 22   55
     + ethanol (ternary)   vol 22   56

Lanthanum fluoride
        + ethanol, 2-amino-   vol 22   65
        + heptane, 2-methoxy-   vol 22   61
        + octane, 2-methoxy-   vol 22   61
        + 1-octanesulfonyl fluoride,
          1,1,2,2,3,3,4,4,5,5,6,6,7,7,8,8,8-heptadecafluoro-
                                        vol 22   50
        + pentane, 1-(bromomethoxy)-   vol 22   50
        + pentane, 1-(chloromethoxy)-   vol 22   50
        + pentane, 2-methoxy-   vol 22   61
        + petroleum ether (ternary)   vol 22   49
        + phosphoric acid, bis-(2-ethylhexyl) ester (ternary)
                                        vol 22   49
        + phosphoric acid, tributyl ester
                                        vol 22   63
        + 2-propanamine   vol 22   67
        + 1-propanamine, 2-methyl-N-(2-methylpropyl)-
                                        vol 22   68
        + propane, 1-(chloromethoxy)-   vol 22   50
        + propane, 1-methoxy-2-methyl- vol 22   61
        + 1-propanol, 2-methyl-   vol 22   58
        + methane, 1,1´-sulfinylbis-   vol 22   64
        + methanol   vol 22   53
        + methanol (ternary)   vol 22   54
        + pyridine   vol 22   69, 70
        + pyridine (ternary)   vol 22   71
        + urea (ternary)   vol 22   54, 56, 71
        + urea (aqueous)   vol 22   73
        + water (ternary)   vol 22   73
Lanthanum iodide
        + formamide, N,N-dimethyl-   vol 22   96
Lanthanum manganese nitrate
        see nitric acid, lanthanum manganese salt
Lanthanum nickel nitrate
        see nitric acid, lanthanum nickel salt
Lanthanum nitrate
        see nitric acid, lanthanum salt
Lauric acid
        see dodecanoic acid
Lead
        + mercury   vol 25   E157 -E159,
                                                          160 - 171
        + water   vol 29   E228, 234
Lead nitrate
        see nitric acid, lead salt
Lead oxide (molten)
        + mercury   vol 29   E207
Lead selenite
        see selenious acid, lead salt
Lead(II) sulfite
        see sulfurous acid, lead salt
Lead tellurite
        see tellurous acid, lead salt
Lederkin
        see benzenesulfonamide, 4-amino-N-(6-methoxy-3-pyridazinyl)-
Lederkyn
        see benzenesulfonamide, 4-amino-N-(6-methoxy-3-pyridazinyl)-
L-Leucine (aqueous)
        + mercury   vol 29   E50, 78
Leuconeoprontosil
        see 2,7-naphthalenedisulfonic acid, 6-acetylamino-
            3-[[4-(Aminosulfonyl)phenyl]azo]-4-hydroxy-, disodium salt
Levulose
        see D-fructose
Lignosulfonic acid (aqueous and multicomponent)
        + glucose   vol 26   E191, E192, 235
        + sulfurous acid, calcium salt vol 26   E191, E192, 235
        + xylose   vol 26   E191, E192, 235
Limonene
        see cyclohexane, 1-methyl-4-(1-methylethenyl)-(R)-
Linoleic acid
        see (zz)-9-12-Octadecadienoic acid
Lipids
        see under phospholipids

Lipo-Diazine
        see benzenesulfonamide, 4-amino-N-2-pyrimidinyl-
Lipo-Gantrisin
        see acetamide, N-[(4-aminophenyl)sulfonyl]-N-(3,4-dimethyl-
            5-isoxazolyl)-
Lipo-Gantrisin acetyl
        see acetamide, N-[(4-aminophenyl)sulfonyl]-N-(3,4-dimethyl-
            5-isoxazolyl)-
Lipo-Levazine
        see benzenesulfonamide, 4-amino-N-2-pyrimidinyl-
Liposorb SQO
        see sorbitan (z)-9-octadecenoate (2:3)
Liquadiazine
        see benzenesulfonamide, 4-amino-N-2-pyrimidinyl-
Liquid paraffin
                + arsine   vol 24   E302, 304, 305,
                                                                           308
                + germane   vol 24   E315, 321, 322
                + phosphine   vol 24   289, 298, 299
                + silane   vol 24   E315, 316, 317
                + stannane   vol 24   E315, 325, 326
                + stibine   vol 24   E310, 313, 314
Lisulfen
        see benzenesulfonamide, 4-amino-N-(6-methoxy-3-pyridazinyl)-
Lithium
                + mercury   vol 25   E1-E3,   4 -   12
Lithium carbonate
        see Carbonic acid, lithium salt
Lithium chloride
                + mercury (II) oxide   vol 23   335
Lithium chloride (aqueous)
                + butane   vol 24   E58 - E72,   98
                + mercury   vol 29   E207, 210
                + mercury (II) oxide   vol 23   335
                + methane   vol 27/28   E68, E69,
                                                                    114 - 116
                + propane   vol 24   E58 - E72,   97
Lithium chloride (multicomponent)
                + copper (II) hydroxide   vol 23   E10 - E26,   79
Lithium hydroxide (aqueous)
                + phosphoric acid, lithium salt vol 31   E4,   9,   10
                + mercury (II) oxide   vol 23   322
Lithium hydroxide (multicomponent)
                + cadmium hydroxide   vol 23   300
                + copper (II) hydroxide   vol 23   E10 - E26,   79
Lithium iodate
        see iodic acid, lithium salt
Lithium nitrate
        see nitric acid, lithium salt
Lithium oxide (aqueous)
                + phosphorus pentoxide   vol 31   E3-E5,   9,   10
Lithium perchlorate
        see perchloric acid, lithium salt
Lithium phosphate
         see phosphoric acid, trilithium salt
Lithium selenite
        see selenious acid, dilithium salt
Lithium sulfate
        see sulfuric acid, lithium salt
Lithium tellurite
        see tellurous acid, lithium salt
Lithium thiocyanate
        see thiocyanic acid, lithium salt
Longasulf
        see benzenesulfonamide, 4-amino-N-(5-methoxy-2-pyrimidinyl)-
Longin
        see benzenesulfonamide, 4-amino-N-(6-methoxy-3-pyridazinyl)-
Lucosil
        see benzenesulfonamide, 4-amino-N-(5-methyl-
            1,3,4-thiadiazol-2-yl)-
Lusil
        see benzenesulfonamide, 4-amino-
Lysococcine
        see benzenesulfonamide, 4-amino-

```
Lutetium
            + mercury                        vol 25                    E256,  257
Lutetium bromide
            + furan, tetrahydro-             vol 22                           381
Lutetium chloride
            + ethanol (aqueous)              vol 22                           378
            + phosphoric triamide, hexamethyl-
                                             vol 22                           379
          + phosphoryl chloride (ternary)    vol 22                           380
            + stannane tetrachloro- (ternary)
                                             vol 22                           380
            + water (ternary)                vol 22                           378
Lutetium nitrate
        see nitric acid, lutetium salt
Lymph nodes
        see guinea pig lymph nodes
L-Lysine, monohydrochloride (aqueous)
            + mercury                        vol 29                    E50,    85
L-Lysine, monohydrochloride (in Hank's balanced salt solution)
            + mercury                        vol 29            E50,   E87,    93
```

M

Mafenide
         see benzenesulfonamide, 4-(aminomethyl)-
Magnesium
              + mercury   vol 25   E57,   E58,
                                                                        59 -   65
Magnesium, bis-[4-amino-N-(2,6-dimethoxy-4-pyrimidinyl)-
benzenesulfonamidato]-, hydrate, (T-4)-
              + hydrochloric acid   vol 36   431
              + water   vol 36   431
Magnesium, bis-[4-amino-N-(6-methoxy-3-pyridazinyl)benzenesulfonamidato]-,
hydrate, (T-4)-
              + hydrochloric acid   vol 36   117
              + water   vol 36   117
Magnesium, bis-4-amino-N-2-thiazolylbenzenesulfonamidato-NN,O)-,
hydrate, (T-4)-
              + hydrochloric acid   vol 35   210
              + water   vol 35   210
Magnesium bisulfite
         see sulfurous acid, magnesium salt
Magnesium bromate
         see bromic acid, magnesium salt
Magnesium chlorate
         see chloric acid, magnesium salt
Magnesium chloride (aqueous)
              + methane   vol 27/28   E64,   E65,   104
              + sulfurous acid, magnesium salt
                                                vol 26   E157, E158,
                                                                    177,   186
              + zinc hydroxide   vol 23   E180,   256
Magnesium chloride (ternary)
              + methane   vol 27/28   E66,   E68,   E74,
                                                             E80,   E81,   106
                                                                    113,   146
Magnesium hydrogen sulfite
         see sulfurous acid, magnesium salt (2:1)
Magnesium hydrogen phosphate
         see phosphoric acid, magnesium salt (1:1)
Magnesium iodate
         see iodic acid, magnesium salt
Magnesium lanthanum nitrate
         see nitric acid, magnesium lanthanum salt
Magnesium nitrate
         see nitric acid, magnesium salt
Magnesium selenite
         see selenious acid, magnesium salt
Magnesium sulfate
         see sulfuric acid, magnesium salt
Magnesium sulfite
         see sulfurous acid, magnesium salt
Malfamin
         see benzenesulfonamide, 4-(aminomethyl)-
Maltose
         see 4-O- -D-glucopyranoxyl-D-glucose
Manganese
              + mercury   vol 25   E285, E286,
                                                                    287 -   299
                                                vol 29   E213,   220
Manganese acetate
         see acetic acid, manganese salt
Manganese, bis[4-amino-N-(2,6-dimethoxy-4-pyrimidinyl)benzenesulfonamidato]-
hydrate, (T-4)-
              + hydrochloric acid   vol 36   432
              + water   vol 36   432
Manganese, bis[4-amino-N-(4,6-dimethyl-2-pyrimidinyl)benzenesulfonamidato-
NN,01]-, hydrate
              + hydrochloric acid   vol 36   350
              + water   vol 36   350
Manganese, bis-[4-amino-N-(6-methoxy-3-pyridazinyl)benzenesulfon-amidato]-,
hydrate
              + hydrochloric acid   vol 36   118
              + water   vol 36   118

Manganese, bis-4-amino-N-2-thiazolylbensenesulfonamidato-NN,O)-,
hydrate
                + hydrochloric acid   vol 35   211
                + water   vol 35   211
Manganese nitrate
        see nitric acid, manganese salt
Manganese (ll) selenite
        see selenious acid, manganese (l) salt
Manganese sulfate
        see sulfate acid, manganese salt
Manganese sulfite
        see sulfurous acid, manganese salt
Manganous sulfate
        see sulfuric acid, manganese (2+) salt (1:1)
D-Mannitol (aqueous)
                + bromic acid, potassium salt   vol 30   245
Maphenid
        see benzenesulfonamide, 4-(aminomethyl)-
Maphenide
        see benzenesulfonamide, 4-(aminomethyl)-
Marprontil
        see benzenesulfonamide, 4-(aminomethyl)-
Mateina
        see 1H-purine-2,6-dione, 3,7-dihydro-1,3,7-trimethyl-
Mebacid
        see benzenesulfonamide, 4-amino-N-(4-methyl-2-pyrimidinyl)-
Medicel
        see benzenesulfonamide, 4-amino-N-(6-methoxy-3-pyridazinyl)-
Mefenal
        see benzenesulfonamide, 4-amino-N-(2,6-dimethyl-4-pyrimidinyl)-
Mercury
                + actinium   vol 25   E421
                + l-alanine   vol 29   E49,   72
                + aluminium   vol 25   E84,   E85
                                                                    86 - 92
                                                 vol 29   E214, 217
                + aluminium chloride   vol 29   E207, E208, 209
                + ammonia   vol 29   E191
                + antimony   vol 25   E173, E174
                                                                  175 - 181
                + L-arginine, monochloride (aqueous)
                                                 vol 29   E50, E51,   86
                + L-arginine, monohydrochloride
                   (in Hank´s balanced salt solution)
                                                 vol 29   E50, E51, E87
                                                                   88 - 94
                + argon   vol 29   E190 -E192
                                                                  195, 196
                + arsenic   vol 25   E172
                + L-aspartic acid (aqueous)   vol 29   E49,   73
                + barium   vol 25   E76, E77
                                                                   78 - 82
                + barium chloride   vol 29   E47,   57
                + beryllium   vol 25   E55,   56
                + benzene   vol 29   E149, 154 - 156
                + benzene, bromo-   vol 29   E178, 186
                + benzene, chloro-   vol 29   E179, 185
                + benzene, 1,2-dimethyl-   vol 29   E151, 160
                + benzene, nitro-   vol 29   E187, 188 - 189
                + benzene, methyl-   vol 29   E150, 157 - 159
                + benzene, (1-methylethyl)-   vol 29   E151, E152, 161
                + benzene, (2-methylpropyl)-   vol 29   E151, E153, 162
                + bismuth   vol 25   E182, E183
                                                                  198 - 193
                + blood   vol 29   E51,   97
                + boron   vol 25   E83
                + butane   vol 29   E191, E192
                                                                  201, 202
                + butane, 2,2-dimethyl-   vol 29   E107 -E109
                                                                  120, 121
                + butane, 2,3-dimethyl-   vol 29   E107 -E109, 122
                + butane, methyl-   vol 29   E107 -E109, 112
                + butane, 1,1-oxybis-   .   vol 29   E172, E173, 175

Mercury	+ cadmium	vol 25	E402 - E404
			405 - 420
	+ calcium	vol 25	E66, E67
			68 - 70
	+ calcium chloride	vol 29	E87, 88 - 94
	+ carbon	vol 25	E134
	+ carbonic acid, monosodium salt	vol 29	E87, 88 - 94
	+ cerium	vol 25	E214, E215
			216 - 220
	+ cesium	vol 25	E50, E51
			52 - 54
	+ chromium	vol 25	E277, 278 - 281
		vol 29	E214, 217
	+ cinnabar	vol 29	E47, 56
	+ cobalt	vol 25	E310, E311
			312, 313
		vol 29	E214, 217
	+ copper	vol 25	E335 - E337
			338 - 356
	+ cyclobutane, hexafluorobis(trifluoromethyl)-		
		vol 29	E49, 71, E177
			E178, 180, 181
	+ cyclohexane	vol 29	E133, E134
			138 - 141
	+ cyclohexane, 1,2-dimethyl-, cis-		
		vol 29	E134 -E137, 143
	+ cyclohexane, 1,2-dimethyl-, trans-		
		vol 29	E134 -E137, 144
	+ cyclohexane, 1,4-dimethyl-, cis-		
		vol 29	E134 -E137, 145
	+ cyclohexane, 1,4-dimethyl-, trans-		
		vol 29	E134 -E137, 146
	+ cyclohexane, methyl-	vol 29	E134 -E137, 142
	+ cyclohexene	vol 29	E147, 148
	+ L-cysteine monohydrochloride (aqueous)		
		vol 29	E50, 82
	+ L-cysteine monohydrochloride		
	(in Hank´s balanced salt solution)		
		vol 29	E50, E87, 91
	+ decane	vol 29	E102 -E104, E106
			131, 132
	+ dysprosium	vol 25	E245, 246
	+ erbium	vol 25	E249, 250
	+ ethanamine, N,N,N-triethyl-, bromide		
		vol 29	E47, 55
	+ ethanol	vol 29	E165
	+ europium	vol 25	E234, 235 - 238
	+ francium	vol 25	E421
	+ gadolinium	vol 25	E239, 240, 241
	+ gallium	vol 25	E93 - E95
			96 - 102
		vol 29	E213, E214, 216
	+ germanium	vol 25	E135, 136 - 138
	+ D-glucose	vol 29	E87, 88 - 94
	+ L-glutamine (aqueous)	vol 29	E50, 76
	+ gold	vol 25	E139 - E141, 142
		vol 29	E213, E214
			226, 227
	+ hafnium	vol 25	E267
	+ Hank´s balanced salt solution	vol 29	E87, 88
	+ helium	vol 29	E190, E191, 193
	+ hemoglobin (aqueous)	vol 29	E51, 99
	+ heptane	vol 29	E102 -E106
			123, 124
	+ hexane	vol 29	E102 -E106
			113 - 118, E165
	+ holmium	vol 25	E247, 248
	+ humus (aqueous)	vol 29	E51, 95, 96
	+ hydrochloric acid	vol 29	E47, 52
	+ hydrogen	vol 29	E190, E191, 198
	+ L-hystidine, monohydrochloride (aqueous)		
		vol 29	E50, E51, 83

Mercury	+ L-hystidine, monohydrochloride			
	(in Hank´s balanced salt solution)			
		vol 29	E50, E87,	90
	+ indium	vol 25	E103	-E105
			106 -	118
	+ iridium	vol 25		E316
		vol 29	E214,	217
	+ iron	vol 25	E301,	E302
			303 -	306
		vol 29	E207, E214,	217
	+ iron sulfide	vol 29	E207,	E208
	+ L-isoleucine	vol 29	E50, E51,	79
	+ Kanamycin (or Kanamycin A)	vol 29	E87, 88	- 94
	+ krypton	vol 29	E190, E191,	197
	+ lanthanum	vol 25	E207,	E208
			209 -	213
	+ lead	vol 25	E157	-E159
			160 -	171
	+ lead oxide (molten)	vol 29		E207
	+ L-leucine (aqueous)	vol 29	E50,	78
	+ lithium	vol 25	E1 - E3,	12
	+ lithium chloride (molten)	vol 29	E207,	210
	+ lutetium	vol 25	E256,	257
	+ L-lysine, monohydrochloride (aqueous)			
		vol 29	E50,	85
	+ L-lysine, monohydrochloride			
	(in Hank´s balanced salt solution)			
		vol 29	E50, E87,	93
	+ magnesium	vol 25	E57,	E58
			59 -	65
	+ manganese	vol 25	E285,	E286
			287 -	299
		vol 29	E213,	220
	+ mercury (II) chloride (molten)			
		vol 29	E207,	211
	+ mercury (II) oxide (molten)	vol 29	E207,	E208
	+ mercury (II) sulfide (aqueous)	vol 29		E47
	+ mercury telluride	vol 29	E207, E208,	212
	+ methanamine, N,N,N-trimethyl-, bromide (aqueous)			
		vol 29	E47,	54
	+ methane, tetrachloro-	vol 29	E178, 182	- 184
	+ methanol	vol 29	E163, E164,	E165
			166 - 168,	E192
			203,	204
	+ L-methionine (aqueous)	vol 29	E50,	77
	+ L-methionine (in Hank´s balanced salt solution)			
		vol 29	E50, E87,	89
	+ molybdenum	vol 25	E282,	283
		vol 29	E214,	217
	+ neodymium	vol 25	E225, 226	- 229
	+ neon	vol 29	E190, E191,	194
	+ nickel	vol 25	E317,	E318
			319 -	325
		vol 29	E214,	217
	+ niobium	vol 25	E271, 272,	273
		vol 29	E214,	217
	+ nitric acid	vol 29		29
	+ nitric acid, ammonium salt (aqueous)			
		vol 29	E47,	53
	+ nitric acid, sodium salt (aqueous)			
		vol 29	E49,	69
	+ nitrogen	vol 29	E190, E191,	199
	+ octane	vol 29	E102	-E106
			125 -	127
	+ 1-octanol	vol 29	E165, 170,	171
	+ L-ornithine, monohydrochloride (aqueous)			
		vol 29	E50,	84
	+ L-ornithine, monohydrochloride			
	(in Hank´s balanced salt solution)			
		vol 29	E50, E87,	92
	+ osmium	vol 25		E309
		vol 29	E214,	217
	+ palladium	vol 25	E326, 327	- 329

Mercury	+ pentane	vol 29	E102 -E106, 110, 111
	+ pentane, 3-methyl-	vol 29	E107 -E109, 119, E165
	+ pentane, 2,2,4-trimethyl-	vol 29	E107 -E109, 128 - 130
	+ perchloric acid, sodium salt	vol 29	E49, 67
	+ L-phenylalanine (aqueous)	vol 29	E50, 80
	+ phosphinic acid	vol 29	29
	+ phosphoric acid, disodium salt, dihydrate		
		vol 29	E87, 88 - 94
	+ phosphoric acid, monosodium salt		
		vol 29	E87, 88 - 94
	+ phosphorus	vol 29	E213, E214, 219
	+ plasma	vol 29	E51, 98
	+ platinum	vol 25	E330, 331 - 334
		vol 29	E213, E214, 225
	+ plutonium	vol 25	E434. 435
	+ polonium	vol 25	E421
	+ potassium	vol 25	E33, E34, 35 - 43
		vol 29	E213, E214, 218
	+ potassium chloride (aqueous)	vol 29	E49, 71
			E87, 88 - 94
	+ potassium chloride (molten)	vol 29	E207, 210
	+ praseodymium	vol 25	E221, 222 - 224
	+ L-proline	vol 29	E50, 75
	+ promethium	vol 25	E421
	+ propane	vol 29	E192, 200
	+ propane, 2,2-oxybis-	vol 29	E172, 174
	+ 2-propanol	vol 29	E164, E165, 169
	+ 2-propanone	vol 29	E192, 205, 206
	+ protactinium	vol 25	E421
	+ radium	vol 25	E421
	+ rhenium	vol 25	E300
		vol 29	E214, 217
	+ rhodium	vol 25	E314, 315
		vol 29	E214, 217
	+ rubidium	vol 25	E44, E45, 46 - 49
	+ ruthenium	vol 25	E307, 308
		vol 29	E214, 217
	+ samarium	vol 25	E230, E231, 232, 233
	+ scandium	vol 25	E206
	+ seawater	vol 29	45
	+ silicon	vol 25	E134
		vol 29	E213, E214, 219
	+ silver	vol 25	E357- E359, 360 - 368
		vol 29	E213, E214, 222
	+ sodium	vol 25	E13 - E18, 19 - 32
		vol 29	E213, E214, 218
	+ sodium bromide (aqueous)	vol 29	E49, 66
	+ sodium chloride (aqueous)	vol 29	E48, 59 - 65
			E87, 88 - 94
	+ sodium chloride (molten)	vol 29	E207, 209
	+ sodium fluoride (aqueous)	vol 29	E47, E48, 58
	+ sodium iodide (aqueous)	vol 29	E49
	+ strontium	vol 25	E71, E72, 73 - 75
	+ sulfuric acid, magnesium salt (aqueous)		
		vol 29	E87, 88 - 94
	+ sulfuric acid, sodium salt (aqueous)		
		vol 29	E49, 68
	+ sunflower seed oil	vol 29	176
	+ tantalum	vol 25	E274, 275, 276
		vol 29	E214, 217
	+ technetium	vol 25	E421
	+ tellurium	vol 25	E194, E195, 196 - 205
		vol 29	E213, 224
	+ terbium	vol 25	E242, 243 - 244

Mercury	+ thallium	vol 25	E119 -E121, 122 - 133
	+ thiocyanic acid, sodium salt (aqueous) vol 29		E49, 70
	+ thorium	vol 25	E422, 423, 424
		vol 29	E214, 217
	+ L-threonine (aqueous)	vol 29	E49, 74
	+ thulium	vol 25	E251, 252
	+ tin	vol 25	E139, E141, 142 - 156
		vol 29	E213, 223
	+ tissue culture medium	vol 29	E51, 100, 101
	+ titanium	vol 25	E258, E259, 260 - 262
		vol 29	E214, 217
	+ tungsten	vol 25	E284
		vol 29	E214, 217
	+ L-tyrosine	vol 29	E50, 81
	+ uranium	vol 25	E425- E427, 428 - 433
		vol 29	E214, 217
	+ vanadium	vol 25	E268, 269, 270
		vol 29	E214, 217
	+ water	vol 29	E1 - E24, 25-44, E46-E51, 52-55, 57 - 96, 99, 170, E228, 229 - 234
	+ water vapor	vol 29	E190
	+ xenon	vol 29	E191
	+ ytterbium	vol 25	E253, 254, 255
	+ yttrium	vol 25	E206
	+ zinc	vol 25	E385- E387, 388 - 401
		vol 29	E213, 221
	+ zirconium	vol 25	E263, E264, 265, 267
		vol 29	E214, 217
Mercury acetate see acetic acid, mercury salt			
Mercury (II) chloride (molten)	+ mercury	vol 29	E207, 211
Mercury (II) oxide	+ seawater	vol 23	E315, 318
	+ water	vol 23	E307 -E317, 318
Mercury (II) oxide (aqueous)	+ barium hydroxide	vol 23	320
	+ hydrochloric acid	vol 23	328, 329
	+ hydrofluoric acid	vol 23	E313, 319, 333
	+ lithium chloride	vol 23	335
	+ nitric acid	vol 23	326, 327
	+ potassium chloride	vol 23	335
	+ potassium hydroxide	vol 23	E307 -E313, 323, 334
	+ lithium hydroxide	vol 23	322
	+ perchloric acid, sodium salt	vol 23	331
	+ sodium hydroxide	vol 23	E307 -E313, 321, 324
	+ sodium sulfide	vol 23	E313, 332
Mercury (II) oxide (multicomponent)	+ acetic acid	vol 23	339
	+ acetic acid, sodium salt	vol 23	339
	+ carbon dioxide	vol 23	E316, 340, 341
	+ carbonic acid, sodium salt	vol 23	342
	+ nitric acid, potassium salt	vol 23	338
	+ nitric acid, sodium salt	vol 23	337
	+ perchloric acid, sodium salt	vol 23	342
	+ sodium chloride	vol 23	E316, 340, 341, 335
	+ sodium hydroxide	vol 23	336 - 338
	+ sulfuric acid, sodium salt	vol 23	336
Mercury (II) oxide (molten)	+ mercury	vol 29	E207, E208
Mercury (I) selenite			

```
Methane, tetrachloro-
          + methanamine                        vol 21                    E81,   E90,
                                                                          128 -  130
          + methane                             vol 27/28   E685,   686,   687
          + N-methylmethanamine                 vol 21                    E83,   E91,
                                                                          171 -  173
          + 2-methylpropane                     vol 24      E257,  E258,   278
          + phosphine                           vol 21      E281,  E282,   294
          + propane                             vol 24      E257,  E258,   260,
                                                                                261
Methane, tribromo-
          + hydrogen sulfide                    vol 32                   E178,  E179,
                                                                          279,   283
Methane, trichloro-
          + ammonia                             vol 21                     E9,    16,
                                                                           67,    68
          + arsine                              vol 21                   E302,   309
          + butane                              vol 24        33,  E257,  E258,
                                                                          269,   353,   398
          + N,N-dimethylmethanamine             vol 21       E87,   223,   224
          + hydrogen sulfide                    vol 32                   E178,  E179,
                                                                          279,   282
          + methanamine                         vol 21                    E81,   128
          + N-methylmethanamine                 vol 21                    E83,     3
          + methane                             vol 27/28                        580
          + 2-methylpropane                     vol 24        38,  E257,  E258
                                                                          277,   366,   406
          + propane                             vol 24        15,   228,   257,
                                                             E358,   259,   338,
                                                                          386,   415
Methanol
          + ammonia                             vol 21                     E4,    E5,
                                                               35 -   38,   E90
          + butane                              vol 24      E331-  E334,   350
          + N,N-dimethylmethanamine             vol 21                    E84,   186
          + iodic acid, potassium salt          vol 30                   417,   418
          + iodic acid, rubidium salt           vol 30                          447
          + iodic acid, sodium salt             vol 30                          373
          + hydrogen sulfide                    vol 32                   E171,  E172,
                                                                          238 -  241
          + mercury                             vol 29                   E163,  E164,  E165,
                                                                          166 -  168,  E192,
                                                                          203,   204
          + methane                             vol 27/28   580,  E582,  -E585,
                                                             596 -  600,  E629,
                                                                          639 -  634
          + N-methylmethanamine                 vol 21      E82,   142,   143
          + 2- methylpropane                    vol 24      E331  -E334,   363
          + 6,7,10,17,18,20,21-octahydrodibenzo[b,k]-
            [1,4,7,10,13,16]hexaoxacyclooctadecin
                                                vol 30      373,   418,   477
          + propane                             vol 24      E331-  E334,   335
Methanol (aqueous)
          + bromic acid, potassium salt         vol 30                          243

          + phosphoric acid, disodium salt
                                                vol 31      E98,  E101,   126
Methanol (multicomponent)
          + copper (II) hydroxide               vol 23      E10 -  E26,    68
L-Methionine (aqueous)
          + mercury                             vol 29                   E50,    77
L-Methionine (in Hanks balanced salt solution)
          + mercury                             vol 29      E50,   E87,    89
Methocel 54 HG
          see cellulose, ethers, 2-hydroxypropyl, methyl ether
Methoxybenzene
          see benzene, methoxy-
2-Methoxyethanol
          see ethanol, 2-methoxy-
3-Methoxypyrazine sulfanilamide
          see benzenesulfonamide, 4-amino-N-(3-methoxypryrazinyl)-
N1-(3-Methoxy-2-pyrazinyl)sulfanilamide
          see benzenesulfonamide, 4-amino-N-(3-methoxypryrazinyl)-
```

```
Mercury (I) selenite
        see selenious acid, mercury (I) salt
Mercury (II) selenite
        see selenious acid, mercury (II) salt
Mercury (II) sulfide (aqueous)
        + mercury                        vol 29                        E47
Mercury (I) sulfite
        see sulfurous acid, mercury salt
Mercury telluride
        + mercury                        vol 29        E207, E208,   212
Merian
        see benzenesulfonamide, 4-amino-N-(1-phenyl-1H-pyrazol-5-yl)-
Mermeth
        see benzenesulfonamide, 4-amino-N-(4,6-dimethyl-2-pyrimidinyl)-
Mesitylene
        see benzene, 1,3,5-trimethyl-
Mesudin
        see benzenesulfonamide, 4-(aminomethyl)-
Mesudrin
        see benzenesulfonamide, 4-(aminomethyl)-
Mesulfa
        see benzenesulfonamide, 4-amino-N-(4-methyl-2-pyrimidinyl)-
Metanilamide
        see benzenesulfonamide, 3-amino-
Metaphosphoric acid, sodium salt
        + water                          vol 31                         46
Methanamine (aqueous)
        + silver (I) oxide               vol 23               E83,  E84,
                                                              E93,   110
Methanamine, N,N-dimethyl-
        + methanamine                    vol 21          81,  115,  116
        + methanamine-d2                 vol 21        E234,  250,  251
Methanaminium, N,N,N-trimethyl-, bromide (aqueous)
        + butane                         vol 24        E58 - E72,   85
        + mercury                        vol 29              E47,   54
        + methane                        vol 27/28     E59, E60,    91
        + propane                        vol 24        E58 - E72,   84
Methane (multicomponent)
        + hydrogen sulfide               vol 32               75 -  77,
                                                              213,  214
Methane, chlorodifluoro-
        + methane                        vol 27/28           689,  690
Methane, dichloro-
        + arsine                         vol 21             E302,  309
Methane, dichlorodifluoro-
        + methane                        vol 27/28                 688
Methane, nitro-
        + N,N-diethylethanamine          vol 21             E252,  279
        + N,N-dimethylmethanamine        vol 21              E86,  208
Methane, 1,1-sulfinylbis-
        + iodic acid, potassium salt     vol 30                    419
        + hydrogen sulfide               vol 32             E185,  319
        + methane                        vol 27/28     E702,  705,  706
        + propane                        vol 24        E375- E378,  380
Methane, 1,1-sulfinylbis- (aqueous)
        + iodic acid, cesium salt        vol 30                    460
        + iodic acid, rubidium salt      vol 30                    446
        + iodic acid, potassium salt     vol 30                    416
        + methane                        vol 27/28                 182
Methanesulfonic acid, [[4-(aminosulfonyl)phenyl]amino]-, monosodium salt
        + 2-propanone                    vol 34                    195
Methanesulfonic acid, (p-sulfamylanilino)-, sodium salt
        see methanesulfonic acid, [[4-(aminosulfonyl)phenyl]amino]-,
            monosodium salt
Methane, tetrachloro-
        + ammonia                        vol 21               E9,   67
        + arsine                         vol 21             E302,  309
        + butane                         vol 24        E257, E258,  270
                                                                    271
        + N,N-dimethylmethanamine        vol 21              E87,  E91,
                                                              225,  226
        + hydrogen sulfide               vol 32        E178, E179,  279
        + mercury                        vol 29        E178,  182 - 184
```

Nl-(3-Methoxypyrazinyl)sulfanilamide
        see benzenesulfonamide, 4-amino-N-(3-methoxypryrazinyl)-
Nl-(6-Methoxy-3-pyridazinyl)sulfanilamide, monosodium salt
        see benzenesulfonamide, 4-amino-N-(6-methoxy-3-pyridoxinyl-,
            monosodium salt
[Nl-(6-Methoxy-3-pyridazinyl)sulfanilamido]sodium
        see benzenesulfonamide, 4-amino-N-(6-methoxy-3-pyridoxinyl-,
            monosodium salt
N-[4-[[(6-Methoxy-2-pyridazinyl)amino]sulfonyl]phenyl]-acetamide
        see acetamide, N-[4-[[(6-methoxy-2-pyridazinyl)amino]sulfonyl]-
            phenyl]-
4´-[(6-Methoxy-3-pyridazinyl)sulfamoyl]acetanilide
        see acetamide, N-[4-[[(6-methoxy-2-pyridazinyl)amino]sulfonyl]-
            phenyl]-
Nl-(6-Methoxy-3-pyridazinyl)sulfanilamide
        see benzenesulfonamide, 4-amino-N-(6-methoxy-3-pyridazinyl)-
6-Methoxy-3-pyridazinylsulfanilamide
        see benzenesulfonamide, 4-amino-N-(6-methoxy-3-pyridazinyl)-
Methoxypyrimal
        see benzenesulfonamide, 4-amino-N-(5-methoxy-2-pyrimidinyl)-
Nl-(4-Methoxy-2-pyrimidinyl)sulfanilamide
        see benzenesulfonamide, 4-amino-N-(4-methoxy-2-pyrimidinyl)-
Nl-(5-Methoxy-2-pyrimidinyl)sulfanilamide
        see benzenesulfonamide, 4-amino-N-(5-methoxy-2-pyrimidinyl)-
Nl-(6-Methoxy-4-pyrimidinyl)sulfanilamide
        see benzenesulfonamide, 4-amino-N-(6-methoxy-4-pyrimidinyl)-
3-Methoxy-2-sulanilamidopyrazine
        see benzenesulfonamide, 4-amino-N-(3-methoxypryrazinyl)-
3-Methoxy-6-sulfanilamidopyridazine
        see benzenesulfonamide, 4-amino-N-(6-methoxy-3-pyridazinyl)-
6-Methoxy-3-sulfanilamidopyridazine
        see benzenesulfonamide, 4-amino-N-(6-methoxy-3-pyridazinyl)-
4-Methoxy-6-sulfanilamidopyrimidine
        see benzenesulfonamide, 4-amino-N-(6-methoxy-4-pyrimidinyl)-
5-Methoxy-2-sulfanilamidopyrimidine
        see benzenesulfonamide, 4-amino-N-(5-methoxy-2-pyrimidinyl)-
6-Methoxy-4-sulfanilamidopyrimidine
        see benzenesulfonamide, 4-amino-N-(6-methoxy-4-pyrimidinyl)-
3-Methoxy-2-sulfapyrazine
        see benzenesulfonamide, 4-amino-N-(3-methoxypryrazinyl)-
3-Methyoxysulfolane
        see thiophene, tetrahydro-3-methoxy-, 1,1-dioxide
N-[4-[[(4-Methoxy-1,2,5-thiadiazol-3-yl)amino]sulfonyl]-phenyl]-
acetamide
        see acetamide, N-[4-[[(4-methoxy-1,2,5-thiadiazol-3-yl)amino]-
            sulfonyl]phenyl]-
Nl-(4-Methoxy-1,2,5-thiadiazol-3-yl)sulfanilamide
        see benzenesulfonamide, 4-amino-N-(4-methoxy-
            1,2,5-thiadiazol-3-yl)-
Nl-(4-Methoxy-1,3,4-thiadiazol-2-yl)sulfanilamide,
monosilver salt (1+) salt
        see benzenesulfonamide, 4-amino-N-(5-methyl-
            1,3,4-thiadiazol-2-yl)-, monosilver salt
Methoxy triethylene glycol acetate
        see ethanol, 2-(2-(methoxyethoxy)-ethoxy)-acetate
Methylacetate
        see acetic acid, methyl ester
N-Methylacetamide
        see acetamide, N-methyl-
Methylal
        see methane, dimethoxy-
2-Methyl aminobenzene
        see benzenamine, 2-methyl-
N-Methyl-p-aminobenzenesulfonamide
        see benzenesulfonamide, 4-amino-N-methyl-
3-Methyl-N-[(4-aminophenyl)sulfonyl]crotonamide
        see 2-butenamide, N-[(4-aminophenyl)sulfonyl]-3-methyl-
N-Methylaniline
        see benzenamine, N-methyl-
2-Methylaniline
        see benzenamine, 2-methyl-
3-Methylaniline
        see benzenamine, 3-methyl-

```
2-Methylanthracene
        see anthracene, 2-methyl-

9-Methylanthracene
        see anthracene, 9-methyl-
1-Methylbenz [a] anthracene
        see benz [a] anthracene, 1-methyl-
9-Methylbenz [a] anthracene
        see benz [a] anthracene, 9-methyl-
10-Methylbenz [a] anthracene
        see benz [a] anthracene, 10-methyl-
Methylbenzene
        see benzene, methyl-
N-Methylbenzenamine
        see benzenamine, N-methyl-
5-Methylbenzo [a] pyrene
        see benzo [a] pyrene, 5-methyl-
Methyl benzoate
        see benzoic acid, methyl ester
2-Methyl-1,3-butadiene
        see 1,3-butadiene, 2-methyl-
2-Methylbutane
        see butane, 2-methyl-
2-Methyl-1-butanol
        see 1-butanol, 2-methyl-
3-Methyl-1-butanol
        see 1-butanol, 3-methyl-
2-Methyl-2-butanol
        see 2-butanol, 2-methyl-
3-Methyl-2-butanol
        see 2-butanol, 3-methyl-
2-Methyl-2-butene
        see 2-butene, 2-methyl-
3-Methyl-1-butene
        see 1-butene, 3-methyl-
Methyl-butyl ethanoate
        see 1-butanol, 3-methyl-, acetate
Methyl butyl ketone
        see 2-hexanone
Methyl isobutyl ketone
        see 2-butanone, 3,3-dimethyl-
N-Methyl- -caprolactam
        see 2-H-azepin-2-one, hexahydro-1-methyl-
Methyl cellosolve
        see ethanol, 1-methoxy-
3-Methylcholanthrene
        see benz [j] aceanthrylene, 1,2-dihydro-3-methyl-
5-Methylchrysene
        see chrysene, 5-methyl-
N1-(3-Methylcrotonoyl)sulfanilamide
        see 2-butenamide, N-[(4-aminophenyl)sulfonyl]-3-methyl-
Methylcyclohexane
        see cyclohexane, methyl-
Methylcyclohexane, tetrafluorodeca-
        see cyclohexane, undecafluoro (trifluoromethyl)-
1-Methylcyclohexene
        see cyclohexene, 1-methyl-
Methylcyclopentane
        see cyclopentane, methyl-
2-Methyldecalin
        see naphthalene, decahydro-3-methyl-
Methyldiethanolamine
        see ethanol, 2,2´-(methylimino)di-
4-Methyl-1,3-dioxolan-2-one
        see 1,3-dioxolane-2-one, 4-methyl-
1,1´-Methylenebisbenzene
        see benzene, 1,1´-methylenebis-
Methylene dichloride
        see methane, dichloro-
1-Methylethylcyclohexane
        see cyclohexane, 1-methylethyl-
Methylethylketone
        see 2-butanone
```

```
N-Methylformamide
          see formamide, N-methyl-
Methyl formate
          see formic acid, methyl ester
2-Methylheptane
          see heptane, 2-methyl-
3-Methylheptane
          see heptane, 3-methyl-
2-Methyl-2-heptanol
          see 2-heptanol, 2-methyl-
3-Methyl-3-heptanol
          see 3-heptanol, 3-methyl-
2-Methylhexane
          see hexane, 2-methyl-
3-Methylhexane
          see hexane, 3-methyl-
2-Methyl-2-hexanol
          see 2-hexanol, 2-methyl-
5-Methyl-2-hexanol
          see 2-hexanol, 5-methyl-
2-Methyl-3-hexanol
          see 3-hexanol, 2-methyl-
3-Methyl-3-hexanol
          see 3-hexanol, 3-methyl-
Methylhexylketone,
          see 2-octanone
Methylhydrazine
          see hydrazine, methyl-
Methyl hydroxybenzoate
          see benzoic acid, 2-hydroxy-, methyl ester
N-[4-[[(5-Methyl-3-isoxaxolyl)amino]sulfonyl]phenyl]acetamide
          see acetamide, N-[4-[[(5-methyl-3-isoxaxolyl)amino]sulfonyl]-
               phenyl]-
4´-[(5-Methyl-3-isoxazolyl)sulfamoyl]acetanilide
          see acetamide, N-[4-[[(5-methyl-3-isoxaxolyl)amino]sulfonyl]-
               phenyl]-
N1-(5-Methyl-3-isoxazolyl)sulfanilamide
          see benzenesulfonamide, 4-amino-N-(5-methyl-3-isoxazolyl)-
1-Methyl-4-(1-methylethyl)benzene
          see benzene, 1-methyl-4-(1-methylethyl)-
5-Methyl-N-methylpyrrolidinone
          see 2-pyrrolidinone, 1,5-dimethyl-
1-Methylnaphthalene
          see naphthalene, 1-methyl-
1-Methyl-2-nitrobenzene
          see benzene,1-methyl-2-nitro-
N-Methyl-N-nitrosomethanamine
          see methanamine, N-methyl-N-nitroso-
4-Methyloctane
          see octane, 4-methyl-
7-Methyl-1-octanol
          see 1-octanol, 7-methyl-
p-(3-Methyl-5-oxo-2-pyrazolin-1-yl)benzenesulfonamide
          see benzenesulfonamide, 4-(4,5-dihydro-3-methyl-5-oxo-
               1H-pyrazol-1-yl)-
1-Methylphenanthrene
          see  phenanthrene, 1-methyl-
N1-(3-Methyl-1-phenylpyrazol-5-yl)sulfanilamide
          see benzenesulfonamide, 4-amino-N-(3-methyl-1-phenyl-
               1H-pyrazol-5-yl)-
1-Methyl-4-isopropylbenzene
          see benzene, 1-methyl-4-(1-methylethyl)-
2-Methylpentane
          see pentane, 2-methyl-
3-Methylpentane
          see pentane, 3-methyl-
4-Methyl-1-pentanol
          see 1-pentanol, 4-methyl-
2-Methyl-1-pentanol
          see 1-pentanol, 2-methyl-
2-Methyl-2-pentanol
          see 2-pentanol, 2-methyl-
3-Methyl-2-pentanol
          see 2-pentanol, 3-methyl-
4-Methyl-2-pentanol
          see 2-pentanol, 4-methyl-
```

```
2-Methyl-3-pentanol
        see 3-pentanol, 2-methyl-
3-Methyl-3-pentanol
        see 3-pentanol, 3-methyl-
4-Methyl-2-pentanone
        see 2-pentanone, 4-methyl-
2-Methyl-1-pentene
        see 1-pentene, 2-methyl-
4-Methyl-1-pentene
        see 1-pentene, 4-methyl-
4-Methyl-1-penten-3-ol
        see 1-penten-3-ol, 4-methyl-
3-Methylphenol
        see phenol, 3-methyl-
Methyl phenyl ketone
        see ethanone, 1-phenyl-
1-Methylpiperidine
        see piperidine, 1-methyl-
2-Methylpropane
        see propane, 2-methyl-
2-Methylpropanoic acid
        see propanoic acid, 2-methyl-
2-Methyl-1-propanol
        see 1-propanol, 2-methyl-
2-Methyl-2-propanol
        see 2-propanol, 2-methyl-
2-Methyl-1-propene
        see 1-propene, 2-methyl-
(1-Methylpropyl) benzene
        see benzene, (1-methylpropyl)-
1-Methyl-4-propylbenzene
        see benzene, 1-methyl-4-propyl-
Methyl-N-propyl carbinol
        see 2-pentanol
Methylisopropyl carbinol
        see 2-butanol, 3-methyl-
Methyl-propyl ethanoate
        see acetic acid, 2-methylpropyl ester
Methylpropylketone
        see 2-pentanone
3-Methyl-1H-pyrazole
        see 1H-pyrazole, 3-methyl-
N1-(6-Methyl-3-pyridazinyl)sulfanilamide
        see benzenesulfonamide, 4-amino-N-(6-chloro-3-pyridazinyl)-
Methyl pyridine
        see pyridine, methyl-
N-4-[[(4-Methyl-2-pyrimidinyl)amino]sulfonyl]phenyl]acetamide
        see acetamide, N-4-[[(4-methyl-2-pyrimidinyl)amino]sulfonyl]-
            phenyl]-
4´-[(4-Methyl-2-pyrimidinyl)sulfamoyl]acetanilide
        see acetamide, N-4-[[(4-methyl-2-pyrimidinyl)amino]sulfonyl]-
            phenyl]-
N1-(4-Methyl-2-pyrimidinyl)sulfanilamide
        see benzenesulfonamide, 4-amino-N-(4-methyl-2-pyrimidinyl)-
N-(4-Methyl-2-pyrimidyl)sulfanilamide
        see benzenesulfonamide, 4-amino-N-(4-methyl-2-pyrimidinyl)-
N1-(2-Methyl-4-pyrimidinyl)sulfanilamide
        see benzenesulfonamide, 4-amino-N-(2-methyl-4-pyrimidinyl)-
N1-Methyl-N1-2-pyridylsulfanilamide
        see benzenesulfonamide, 4-amino-N-methyl-N-2-pyridinyl-
1-Methylpyrrolidine
        see pyrrolidine, 1-methyl-
1-Methyl-2-pyrrolidone
        see 2-pyrrolidone, 1-methyl-
1-Methyl-2-pyrrolidinone
        see 2-pyrrolidinone, 1-methyl-
N-Methyl-2-pyrrolidinone
        see 2-pyrrolidinone, 1-methyl-
Methyl salicylate
        see benzoic acid, 2-hydroxy-, methyl ester
Methylstyrene
        see benzene, (1-methylethenyl)-
```

3-Methyl-1-(4-sulfamoylphenyl)-5-pyrazolone
        see benzenesulfonamide, 4-(4,5-dihydro-3-methyl-5-oxo-1H-
            pyrazol-1-yl)-
4´-(Methylsulfamoyl)sulfanilanilide
        see benzenesulfonamide, 4-amino-N-[4-[(methylamino)sulfonyl]-
            phenyl]-
N1-Methylsulfanilamide
        see benzenesulfonamide, 4-amino-N-methyl-
5-Methyl-3-sulfanilamidoisoxazole
        see benzenesulfonamide, 4-amino-N-(5-methyl-3-isoxazolyl)-

2-Methyl-5-sulfanilamido-1,3,4-thiadiazole
        see benzenesulfonamide, 4-amino-N-(5-methyl-
            1,3,4-thiadiazol-2-yl)-
4-Methyl-2-sulfanilamidothiazole
        see benzenesulfonamide, 4-amino-N-(4-methyl-2-thiazolyl)-

3-Methyl-N-sulfanilylcrotonamide
        see 2-butenamide, N-[(4-aminophenyl)sulfonyl]-3-methyl-
N1-Methylsulfapyridine
        see benzenesulfonamide, 4-amino-N-methyl-N-2-pyridinyl-
Methylsulfazine
        see benzenesulfonamid, $-amino-N-(4-methyl-2-pyrimidinyl)-
3-Methylsulfolane
        see thiophene, tetrahydro-3-methyl-, 1,1-dioxide
Methyltheobromine
        see 1H-purine-2,6-dione, 3,7-dihydro-1,3,7-trimethyl-
N-[4-[[(5-Methyl-1,3,4-thiadiazol-2-yl)amino]sulfonyl]-phenyl]acetamide
        see acetamide, N-[4-[[(5-methyl-1,3,4-thiadiazol-2-yl)amino]-
            sulfonyl]phenyl]-
4´-[(5-Methyl-1,3,4-thiadiazol-2-yl)sulfamoyl]acetanilide
        see acetamide, N-[4-[[(5-methyl-1,3,4-thiadiazol-2-yl)-amino]
            sulfonyl]phenyl]-
N1-(5-Methyl-1,3,4-thiadiazol-2-yl)sulfanilamide
        see benzenesulfonamide, 4-amino-N-(5-methyl-
            1,3,4-thiadiazol-2-yl)-
N-[4-[[(4-Methyl)-2-thiazolylamino]sulfonyl]phenyl]acetamide
        see acetamide, N-[4-[[(4-methyl)-2-thiazolylamino]sulfonyl]-
            phenyl]-
4´-[4-[(4-Methyl-2-thiazolyl)sulfamoyl]acetanilide
        see acetamide, N-[4-[[(4-methyl)-2-thiazolylamino]sulfonyl]-
            phenyl]-
N1-(4-Methyl-2-thiazolyl)sulfanilamide
        see benzenesulfonamide, 4-amino-N-(4-methyl-2-thiazolyl)-
N1-[6-(Methylthio)-3-pyridazinylbenzenesulfonamide
        see benzenesulfonamide, 4-amino-N-[6-(methylthio)-3-pyridazinyl)-
3-Methyltricyclo [4.4.0] decane
        see naphthalene, decahydro-2-methyl-
Metilsulfadiazin
        see benzenesulfonamide, 4-amino-N-(4-methyl-2-pyrimidinyl)-
Metilsulfazin
        see benzenesulfonamide, 4-amino-N-(4-methyl-2-pyrimidinyl)-
Microsul
        see benzenesulfonamide, 4-amino-N-(5-methyl-
            1,3,4-thiadiazol-2-yl)-
Microsulfon
        see benzenesulfonamide, 4-amino-N-2-pyrimidinyl-
Microtan pirazolo
        see benzenesulfonamide, 4-amino-N-(1-phenyl-1H-pyrazol-5-yl)-
Midicel
        see benzenesulfonamide, 4-amino-N-(6-methoxy-3-pyridazinyl)-
Midikel
        see benzenesulfonamide, 4-amino-N-(6-methoxy-3-pyridazinyl)-
Mineral oil
        see oil, mineral
Molybdenum
            + mercury   vol 25   E282, 283
                                               vol 29   E214, 217
N1-Monoacetylsulfisoxazole
        see acetamide, N-[(4-aminophenyl)sulfonyl]-N-
            (3,4-dimethyl-5-isoxazolyl)-
Monoethanolamine
        see ethanol, 2-amino-

```
Monosodium glutamate
         see DL-glutamic acid, monosodium salt
Monosodium glycinate
         see glycine, monosodium salt
Myasul
         see benzenesulfonamide, 4-amino-N-(6-methoxy-3-pyridazinyl)-
```

N

Nadisan
          see benzenesulfonamide, 4-amino-N-[(butylamino)carbonyl]-
Na disulfan
          see benzenesulfonamide, 4-amino-N-[4-(aminosulfonyl)phenyl]-,
          monosodium salt
Na disulfan monohydrate
          see benzenesulfonamide, 4-amino-N-4-[(aminosulfonyl)phenyl]-,
          monosodium salt monohydrate
Nadizan
          see benzenesulfonamide, 4-amino-N-[(butylamino)carbonyl]-
Nafcillin sodium
          see 4-thia-1-azabicyclo[3,2,0]heptane-2-carboxylic acid,
          6-[[(2-ethoxy-1-naphthalenyl)carbonyl]amino]-
          3,3-dimethyl-7-oxo-, monosodium salt
Naphthacene
          + water   vol 38   E493, 494 - 496
Naphthalene
          + water   vol 38   E236 -E238,
                                                                  239 - 249,
                                                                  251,   252,
                                                                  256,   257
Naphthalene (aqueous)
          + sodium chloride   vol 38   251,   252,
                                                                  256,   257
          + seawater   vol 38   E250, 253 - 255
Naphthalene (multicomponent)
          + methane   vol 27/28   579
Naphthalene, 1-bromo-
          + N,N-dimethylmethanamine   vol 21   E87,   231, 232
          + methanamine   vol 21   E81,   133
          + N-methylmethanamine   vol 21   177
Naphthalene, 1-chloro-
          + N,N-dimethylmethanamine   vol 21   E87,   233
          + methane   vol 27/28   701
          + 2-methylpropane   vol 24   E257, E258, 298
          + propane   vol 24   E257, E258, 289
Naphthalene, decahydro-
          + butane   vol 24   E218- E221, 244
          + hydrogen sulfide   vol 32   E166- E169, 222
          + methane   vol 27/28   E497, 510
          + phosphine   vol 21   E281, E282, 285
          + propane   vol 24   E218- E221, 236
          + water   vol 38   E287, 288 - 290
Naphthalene, decahydro-,cis-
          + ammonia   vol 21   E3,   26
Naphthalene, decahydro-,trans-
          + ammonia   vol 21   E3,   26
Naphthalene, decahydro-2-methyl-
          + water   vol 38   324
Naphthalene, 1,3-dimethyl-
          + water   vol 38   355
Naphthalene, 1,4-dimethyl-
          + water   vol 38   356
Naphthalene, 1,5-dimethyl-
          + water   vol 38   E357, 358, 359
Naphthalene, 2,3-dimethyl-
          + water   vol 38   E360, 361, 362
Naphthalene, 2,6-dimethyl-
          + water   vol 38   E363, 364, 365
2,7-Naphthalenedisulfonic acid, 6-acetylamino-3-[[4-(aminosulfonyl)-
phenyl]azo]-4-hydroxy-, disodium salt
          + 2-propanone   vol 34   199
Naphthalene, 1-ethyl-
          + water   vol 38   E366, E367,
                                                                  368 - 370
Naphthalene, 1-ethyl- (aqueous)
          + sodium chloride   vol 38   371
Naphthalene, 2-ethyl-
          + water   vol 38   372

```
Naphthalene, 1-methyl-
                + ammonia                     vol 21                    34,   E90
                + hydrogen sulfide            vol 32         E170, E171,  236
                + methane                     vol 27/28            E496,  508,
                                                                   545 - 547
                + 2-methylpropane             vol 24         E218- E221,  248
                + propane                     vol 24         E218- E221,  237
                + water                       vol 38              E311, E312,
                                                                   313 - 317
Naphthalene, 1-methyl- (aqueous)
                + sodium chloride             vol 38                          318
Naphthalene, 2-methyl-
                + water                       vol 38         E319,  320,  321
Naphthalene, 1-methyl- (multicomponent)
                + methane                     vol 27/28       488,  553 - 555,
                                                                           579
Naphthalene, 1-phenyl-(multicomponent)
                + methane                     vol 27/28                       579
Naphthalene, 1,2,3,4-tetrahydro-
                + arsine                      vol 21              E302,  308
                + methane                     vol 27/28           543,  544
                + phosphine                   vol 21         E281, E282,  285
                + propane                     vol 24              E251,  255
Naphthalene, 1,2,3,4-tetrahydro- (ternary)
                + propane                     vol 24              E251,  256
Naphthalene, 1,4,5-trimethyl-
                + water                       vol 38                          392
2-Naphthalenol, sodium salt (aqueous)
                + sulfurous acid, disodium salt
                                              vol 26              E5,   58
2-Naphthol (aqueous and multicomponent)
                + sodium hydroxide            vol 26              E5,   71
                + sulfurous acid, disodium salt
                                              vol 26              E5,   71
Neasina
        see benzenesulfonamide, 4-amino-N-(4,6-dimethyl-2-pyrimidinyl)-
Neazina
        see benzenesulfonamide, 4-amino-N-(4,6-dimethyl-2-pyrimidinyl)-
Neazine
        see benzenesulfonamide, 4-amino-N-2-pyrimidinyl-
Neodymium
                + mercury                     vol 25         E225,  226 - 229
Neazolin
        see benzenesulfonamide, 4-amino-N-(3,4-dimethyl-5-isoxazolyl)-
Neococcyl
        see benzenesulfonamide, 4-amino-
Neodisept
        see benzenesulfonamide, 4-amino-N-(4-methyl-2-thiazolyl)-
Neodymium bromide
                + 1-butanamine                vol 22                          193
                + 2-butanamine                vol 22                          193
                + 2-butanamine, N-(1-methylpropyl)-
                                              vol 22                          193
                + butane, 1-methoxy-          vol 22                          189
                + decane, 1-methoxy-          vol 22                          189
                + 1,4-dioxane                 vol 22                          191
                + 1,2-ethanediamine           vol 22                          192
                + ethane, 1,2-diethoxy-       vol 22                          188
                + ethanol, 2-amino-           vol 22                          194
                + furan, tetrahydro-          vol 22                          190
                + heptane, 1-methoxy-         vol 22                          189
                + morpholine                  vol 22                          195
                + nonane, 1-methoxy-          vol 22                          189
                + octane, 1-methoxy-          vol 22                          189
                + pentane, 1-methoxy-         vol 22                          189
                + 2-propanamine               vol 22                          193
Neodymium chloride
                + 1,3-butadiene, hexachloro-  vol 22                          151
                + 1-butanol                   vol 22              169 - 170
                + 1,4-dioxane                 vol 22                          176
                + 1,3-dioxolane               vol 22                          176
                + ethane, 1-ethoxy-2-methoxy- vol 22                          176
                + ethane, 1,1´-oxybis-        vol 22                          175
                + 1,2-ethanediol              vol 22              161,  165
```

Neodymium chloride
        + ethanol   vol 22   E156, 157 - 139
        + ethanol (aqueous)   vol 22   159
        + ethanol, 2-ethoxy-   vol 22   165, 174
        + ethanol, 2-methoxy-   vol 22   165, 173
        + furan, tetrahydro-   vol 22   178
        + gallium chloride (ternary)   vol 22   182, 183
        + heptane, 1-methoxy-   vol 22   177
        + methanol   vol 22   E152, 153 - 155
        + nonane, 1-methoxy-   vol 22   177
        + pentane, 1-methoxy-   vol 22   176
        + 1-pentanol   vol 22   171, 172
        + phosphoric acid, tributyl ester
                                            vol 22   179
        + phosphoric triamide, hexamethyl-
                                            vol 22   181
        + phosphoryl chloride (ternary)
                                            vol 22   182 - 187
        + 1-propanamine   vol 22   180
        + 2-propanamine   vol 22   180
        + 1,2,3-propanetriol   vol 22   168
        + 1-propanol   vol 22   E162, 163, 165
        + 2-propanol   vol 22   166 - 167
        + 2-propen-1-amine   vol 22   180
        + stannane, tetrachloro- (ternary)
                                            vol 22   184 - 186
        + water (ternary)   vol 22   160
        + zinc chloride (ternary)   vol 22   187
Neodymium cobalt nitrate
        see nitric acid, cobalt neodymium salt
Neodymium fluoride
        + butane, 1-(chloromethoxy)-   vol 22   147
        + decane, 1-methoxy-   vol 22   147
        + ethanol   vol 22   146
        + methane, 1,1´-sulfinylbis-   vol 22   149
        + methanol   vol 22   146
        + phosphoric acid, tributyl ester
                                            vol 22   148
        + pyridine   vol 22   150
Neodymium iodide
        + 1-butanamine   vol 22   199
        + 1-butanol (aqueous)   vol 22   196
        + 1,4-dioxane   vol 22   197
        + 1,2-ethanediamine   vol 22   198
        + ethanol, 2-amino-   vol 22   200
        + formamide, N,N-dimethyl-   vol 22   202
        + 1-propanamine   vol 22   199
        + 2-propanamine   vol 22   199
        + morpholine   vol 22   201
        + water (ternary)   vol 22   196
Neodymium manganese nitrate
        see nitric acid, manganese neodymium salt
Neodymium nickel nitrate
        see nitric acid, neodymium nickel salt
Neodymium nitrate
        see nitric acid, neodymium salt
Neofamid
        see benzenesulfonamide, 4-(aminomethyl)-
Neon
        + mercury   vol 29   E190, E191   194
Neoprontosil disodium salt
        see 2,7-naphthalenedisulfonic acid, 6-acetylamino-
            3-[[4-(aminosulfonyl)phenyl]azo]-4-hydroxy-, disodium salt
Neoprontosil sodium
        see 2,7-naphthalenedisulfonic acid, 6-acetylamino-
            3-[[4-(aminosulfonyl)phenyl]azo]-4-hydroxy-, disodium salt
Neostrpsan
        see benzenesulfonamide, 4-amino-N-2-thiazolyl-
Neotherapol
        see acetamide, N-[4-(aminosulfonyl)phenyl]-
Neo-Uliran
        see benzenesulfonamide, 4-amino-N-[4-[(methylamino)sulfonyl]
            phenyl]-

```
Neouliron
        see benzenesulfonamide, 4-amino-N-[4-[(methylamino)sulfonyl]-
        phenyl]-
Neoxazol
        see benzenesulfonamide, 4-amino-N-(3,4-dimethyl-5-isoxazolyl)-
Niamine
        see 3-pyridinecarboxamide, N,N-diethyl-
Nicamide
        see 3-pyridinecarboxamide, N,N-diethyl-
Nicetamide
        see 3-pyridinecarboxamide, N,N-diethyl-
Nicethamide
        see 3-pyridinecarboxamide, N,N-diethyl-
Nickel
        + mercury                              vol 25              E317, E318,
                                                                   319 - 325
        + mercury                              vol 29              E214, 217
Nickel acetate
        see acetic acid, nickel salt
Nickel, bis[4-amino-N-(2,6-dimethoxy-4-pyrimidinyl)-
benzenesulfonamidato]-, hydrate
        + hydrochloric acid                    vol 36                  433
        + water                                vol 36                  433
Nickel, bis[4-amino-N-(4,6-dimethyl-2-pyrimidinyl)-
benzenesulfonamidato-NN,01] diaqua-
        + hydrochloric acid                    vol 36                  351
        + water                                vol 36                  351
Nickel, bis-[4-amino-N-(6-methoxy-3-pyridazinyl)benzenesulfona-
midato]-, hydrate
        + hydrochloric acid                    vol 36                  119
        + water                                vol 36                  119
Nickel, bis(4-amino-N-2-thiazolyl-benzenesulfonamidato-NN,O)-, hydrate
        + hydrochloric acid                    vol 35                  212
        + water                                vol 35                  212
Nickel nitrate
        see nitric acid, nickel salt
Nickel (ll) selenite
        see selenious acid, nickel (ll) salt
Nickel sulfate
        see sulfuric acid, nickel salt
Nickel (II) sulfite
        see sulfurous acid, nickel (ll) salt
Nickel tellurite
        see tellurous acid, nickel salt
Nickelous sulfate
        see sulfuric acid, nickel salt
Ni-Cor
        see 3-pyridinecarboxamide, N,N-diethyl-
Nicorine
        see 3-pyridinecarboxamide, N,N-diethyl-
Nicotinic acid diethylamide
        see 3-pyridinecarboxamide, N,N-diethyl-
Nikardin
        see 3-pyridinecarboxamide, N,N-diethyl-
Niketamide
        see 3-pyridinecarboxamide, N,N-diethyl-
Nikethamide
        see 3-pyridinecarboxamide, N,N-diethyl-
Niketharol
        see 3-pyridinecarboxamide, N,N-diethyl-
Nikkol SO 15
        see sorbitan (z)-9-octadecenoate (2:3)
Nikorin
        see 3-pyridinecarboxamide, N,N-diethyl-
Niobium
        + mercury                              vol 25         E271, 272, 273
        + mercury                              vol 29              E214, 217
Nissan Nonion OP 83RAT
        see sorbitan (z)-9-octadecenoate (2:3)
Nitrate barium
        see nitric acid, barium salt
Nitrate calcium
        see nitric acid, calcium salt
```

```
Nitrate cesium
        see nitric acid, cesium salt
Nitrate lithium
        see nitric acid, lithium salt
Nitrate magnesium
        see nitric acid, magnesium salt
Nitrate sodium
        see nitric acid, sodium salt
Nitrate, lead
        see nitric acid, lead salt
Nitrate, silver
        see nitric acid, silver salt
Nitrate, sodium
        see nitric acid, sodium salt
Nitric acid (aqueous)
        + gold (III) hydroxide          vol 23              E148,  155
        + iodic acid                    vol 30      E470,  475 - 477
        + iodic acid, rubidium salt     vol 30              E428,  435
        + mercury                       vol 29                      29
        + mercury (II) oxide            vol 23               326,  327
        + selenious acid, cadmium salt  vol 26                     392
        + selenious acid, calcium salt    vol 26                   357
        + selenious acid, cobalt (II) salt
                                        vol 26                     371
        + selenious acid, copper (II) salt
                                        vol 26                     378
        + selenious acid, iron (III) salt
                                        vol 26                     368
        + selenious acid, magnesium salt
                                        vol 26      E348, E349,  355
        + selenious acid, mercury (I) salt
                                        vol 26               394,  395
        + selenious acid, nickel salt
                                        vol 26                     375
        + selenious acid, silver salt
                                        vol 26                     380
        + selenious acid, strontium salt
                                        vol 26      E348, E349,  359
        + selenious acid, zinc salt     vol 26                     389
        + tellurous acid, barium salt
                                        vol 26               E404,  420
Nitric acid (aqueous and multicomponent)
        + selenious acid, mercury (II) salt
                                        vol 26                     399
        + selenious acid, lead (II) salt
                                        vol 26               401,  427
Nitric acid, ammonium salt (aqueous)
        + hydrogen sulfide              vol 32                      39
        + mercury                       vol 29                E47,   53
        + sulfurous acid, calcium salt  vol 26      E191, E192,  215
Nitric acid, ammonium salt (multicomponent)
        + zinc hydroxide                vol 23                     264
Nitric acid, barium salt (aqueous)
        + silver (I) oxide              vol 23                     122
Nitric acid, calcium salt (aqueous)
        + chloric acid, sodium salt     vol 30                      78
Nitric acid, cesium salt
        + acetic acid, cesium salt      vol 33      E134,  134,  135
Nitric acid, cesium salt (aqueous)
        + iodic acid, cesium salt       vol 30              E448, 454
Nitric acid, lithium salt
        + acetic acid, lithium salt     vol 33      E210,  211 - 213
        + butanoic acid, lithium salt   vol 33                     217
        + formic acid, lithium salt     vol 33                     208
        + iodic acid, lithium salt      vol 30      E272,  294,  295
        + propanoic acid, lithium salt
                                        vol 33                     215
Nitric acid, potassium salt
        + acetic acid, potassium salt   vol 33               170,  171
        + bromic acid, potassium salt   vol 30                     230
        + butanoic acid, potassium salt vol 33              E187,  188
        + butanoic acid, 2-methyl-, potassium salt
                                        vol 33      E199,  200,  201
        + chloric acid, potassium salt  vol 30                     122
```

```
Nitric acid, potassium salt
            + formic acid, potassium salt      vol 33      E159,  151,   152
            + iodic acid, rubidium salt        vol 30             E428,   436
            + iodic acid, potassium salt       vol 30      E380,  388,   389
            + pentanoic acid, potassium salt
                                               vol 33             E195,   196
            + propanoic acid, 2-methyl-, potassium salt
                                               vol 33             E187,   188
Nitric acid, potassium salt (aqueous)
            + hydrogen sulfide                 vol 32                      39
            + nitric acid, sodium salt         vol 31              92  -   95
            + phosphoric acid, monopotassium salt
                                               vol 31        92 -  95, E208,
                                                           E210,  224,   232,
                                                            233,  253,   254,
                                                                  260,   261
            + phosphoric acid, monosodium salt
                                               vol 31              92  -   95
            + silver (I) oxide                 vol 23       E83,  E84,   E93,
                                                            112,  113,   120
            + sulfurous acid, dipotassium salt
                                               vol 26       E92,   96,    97
            + zinc hydroxide                   vol 23                     247
            + zinc oxide                       vol 23                     245
Nitric acid, potassium salt (aqueous and multicomponent)
            + selenious acid, ammonium salt
                                               vol 26                     373
            + selenious acid, cobalt (ll) salt
                                               vol 26                     373
Nitric acid, potassium salt (multicomponent)
            + mercury (II) oxide               vol 23                     338
            + silver (I) oxide                 vol 23             142 -  144
Nitric acid, rubidium salt
            + acetic acid, rubidium salt       vol 33      E332,  333 - 335
Nitric acid, rubidium salt (aqueous)
            + chloric acid, rubidium salt      vol 30      E164, 169, 170
Nitric acid, sodium salt
            + acetic acid, sodium salt         vol 33      E267,  268 - 272
            + butanoic acid, sodium salt       vol 33      E289,  290,   291
            + butandioic acid, 2-methyl-, sodium salt
                                               vol 33      E321,  322,   323
            + formic acid, sodium salt         vol 33             E326, E237,
                                                                  238 - 241
            + hexanoic acid, sodium salt       vol 33                     329
                                                                  346 - 353
            + pentanoic acid, sodium salt      vol 33      E308,  309,   310
            + propanoic acid, 2-methyl-, sodium salt
                                               vol 33             301,   302
            + propanoic acid, sodium salt      vol 33                     276
            + water                            vol 31        59 -  64,    79
Nitric acid, sodium salt (aqueous)
            + bromic acid, potassium salt      vol 30                     228
            + chloric acid, sodium salt        vol 30              E30,    41
            + hydrogen sulfide                 vol 32                      39
            + iodic acid, sodium salt          vol 30             E336, E337
            + mercury                          vol 29              E49,    69
            + nitric acid, potassium salt
                                               vol 31              92  -   95
            + phosphoric acid, disodium salt
                                               vol 31              E98,   110,
                                                                  121 -  123
            + phosphoric acid, monopotassium salt
                                               vol 31              92  -   95
            + phosphoric acid, monosodium salt
                                               vol 31             E47,   E48,
                                                                   59  -  64,
                                                             79,   92  -  95
            + tellurous acid, barium salt      vol 26            E404,   420
Nitric acid, sodium salt (aqueous and multicomponent)
            + acetic acid                      vol 26             301,   302
            + acetic acid, sodium salt         vol 26             301,   302
            + selenious acid, mercury (ll) salt
                                               vol 26             399,   400
            + selenious acid, sodium salt      vol 26                     400
```

Nitric acid, sodium salt (aqueous and multicomponent)
          + sulfurous acid, lead salt   vol 26   301,   302
          + tellurous acid, lead salt   vol 26   E404,   427
Nitric acid, sodium salt (multicomponent)
          + mercury (II) oxide   vol 23   337
          + silver (I) oxide   vol 23   138
Nitrobenzene
        see benzene, nitro-
Nitrogen
          + mercury   vol 29   E190, E191,   199
Nitrogen (ternary and multicomponent)
          + hydrogen sulfide   vol 32   75 -   77
          + methane   vol 27/28   313,   334,   335,
                                                         347,   348,   420-
                                                         435,   663 - 665,
                                                                667 - 671
Nitromethane
        see methane, nitro-
Nitropropane
        see propane, nitro-
N1-(5-Nitro-2-pyridyl)sulfanilamide
        see benzenesulfonamide, 4-amino-N-(5-nitro-2-pyridinyl)-
N-Nitroso-dimethylamine
        see methanamine,N-methyl-N-nitroso-
Nitrotoluene
        see benzene, 1-methyl-2-nitro-
2-Nitrotoluene
        see Benzene, 1-methyl-2-nitro-
Nitrous acid, cesium salt
          + acetic acid, cesium salt   vol 33   133
Nitrous acid, potassium salt
          + acetic acid, potassium salt   vol 33   E172,   173
          + butanoic acid, potassium salt   vol 33
E185,   186
          + butanoic acid, 2-methyl-, potassium salt
                                           vol 33   E197,   198
          + formic acid, potassium salt   vol 33   149
          + heptanoic acid, potassium salt
                                           vol 33   203
          + hexanoic acid, potassium salt
                                           vol 33   202
          + nonanoic acid, potassium salt
                                           vol 33   205
          + octanoic acid, potassium salt
                                           vol 33   204
          + pentanoic acid, potassium salt
                                           vol 33   E193,   194
          + propanoic acid, potassium salt
                                           vol 33   181,   182
          + propanoic acid, 2-methyl-, potassium salt
                                           vol 33   E189,   190
Nitrous acid, potassium salt (aqueous)
          + phosphoric acid, tripotassium salt
                                           vol 31   E297,   303
          + sulfurous acid, dipotassium salt
                                           vol 26   E92,   98,   99
Nitrous acid, sodium salt
          + acetic acid, sodium salt   vol 33   264 - 266
          + formic acid, sodium sal   vol 33   235
          + butandioic acid, 2-methyl-, sodium salt
                                           vol 33   E319,   320
          + butanoic acid, sodium salt   vol 33   E287,   288
          + pentanoic acid, sodium salt   vol 33   E306,   307
          + propanoic acid, sodium salt   vol 33   275
          + propanoic acid, 2-methyl-, sodium salt
                                           vol 33   300
          + water   vol 31   150
          + phosphoric acid, trisodium salt
                                           vol 31   E127,   150
Nitrous oxide
        see nitrogen oxide (N2O)
Nitrous oxide
          + methane   vol 27/28   761,   762

```
No-Doz
          see 1H-purine-2,6-dione, 3,7-dihydro-1,3,7-trimethyl-
1,1,2,2,3,3,4,4,4-Nonafluoro-N,N-bis (nonafluorobutyl)-1-butanamine
          see 1-butanamine, 1,1,2,2,3,3,4,4,4-nona-fluoro-N,N-bis-
              (nonafluorobutyl)-
1-Nonanamine, N-nonyl-
              + phosphine                     vol 21         E281, E282,   293

Nonadecane, 18-methyl-2,5,8,11,14,17-hexaoxa-
              + hydrogen sulfide              vol 32         E173, E174,   261
1,8-Nonadiyne
              + water                         vol 38                       211
Nonane
              + butane                        vol 24        E130 -E134,   174
              + hydrogen sulfide              vol 32               E166 -E169,
                                                                     188,   203
              + methanamine                   vol 21               E80,  E90,
                                                                      96,   97
              + methane                       vol 27/28     E202 -E204, E207
                                                                     230, E349,
                                                                     355 - 359
              + 2-methylpropane               vol 24        E130- E134,   194
              + propane                        vol 24        E130- E134,   149
              + seawater                      vol 38                       235
              + water                         vol 38               E225- E227,
                                                                     228 - 234
Nonane (ternary)
              + methane                       vol 27/28                    552
Nonane, 2,2,4,4,6,8,8-heptamethyl-
              + butane                        vol 24        E130 -E134,   181
              + methane                       vol 27/28           E213,   242
              + propane                        vol 24        E130 -E134,   160
1-Nonanol
              + methane                       vol 27/28     E582 -E584, E593,
                                                                           626
Nonanoic acid, potassium salt
              + nitrous acid, potassium salt  vol 33                       205
1-Nonyne
              + water                         vol 38                       212
Norboral
          see benzenesulfonamide, 4-amino-N-[(butylamino)carbonyl]-
2,5-Norbornadiene
              + water                         vol 37                       432
Norilgan-S
          see benzenesulfonamide, 4-amino-N-(3,4-dimethyl-5-isoxazolyl)-
Norsulfasol
          see benzenesulfonamide, 4-amino-N-2-thiazolyl-
Norsulfazole sodium
          see benzenesulfonamide, 4-amino-N-2-thiazolyl-, monosodium salt
Norsulfazole soluble
          see benzenesulfonamide, 4-amino-N-2-thiazolyl-, monosodium salt
Nosulfazol
          see benzenesulfonamide, 4-amino-N-2-thiazolyl-
Novoseptale
          see benzenesulfonamide, 4-amino-N-(4-methyl-2-thiazolyl)-
Nuprin
          see benzenesulfonamide, 4-amino-N-(4,5-dimethyl-2-oxazolyl)-
```

```
O
Octacosane
        + butane                        vol 24          E130 -E134,  185
        + methane                       vol 27/28       E202 -E204, E211,
                                                         E212,   241
        + propane                       vol 24          E130 -E134,  166
Octadecafluorodecahydronaphthalene
        see naphthalene, octadecafluorodecahydro-
Octadecafluorooctane
        see octane,octadecafluoro-
Octadecane
        + butane                        vol 24          E130 -E134,  182
        + methane                       vol 27/28       E202 -E204, E210,
                                                         240,   243
        + propane                       vol 24          E130 -E134,
                                                         159,   161
        + seawater                      vol 38                       507
        + water                         vol 38          E501, E502,
                                                         503 - 506
Octadecanoic acid, sodium salt
        + benzoic acid, sodium salt     vol 33          E330,   331
        + butanoic acid, sodium salt    vol 33          E283,   284
        + butanoic acid, 2-methyl-, sodium salt
                                        vol 33          E315,   316
        + hexanoic acid, sodium salt    vol 33          E326,   327
        + propanoic acid, 2-methyl-, sodium salt
                                        vol 33          E297,   298
Octadecanoic acid (in N,N-dimethylformamide)
        + propane                       vol 24          E125,   129
Octadecanoic acid (in 2-propanone)
        + propane                       vol 24          E125,   128
(Z)-9-Octadecenoic acid (aqueous)
        + butane                        vol 24          E58 - E72,   113
Octamethylcyclotetrasiloxane
        see cyclotetrasiloxane, octamethyl-
1-Octanamine
        + ammonia                       vol 21                 16,  E90
        + butane                        vol 24          E299 -E302,  327
        + N,N-dimethylmethanamine       vol 21                E86,  220
        + methanamine                   vol 21                E90,  125
        + N-methylmethanamine           vol 21                      169
        + propane                       vol 24          E299 -E302,  318
Octane
        + ammonia                       vol 21          E2,    13,   16
        + butane                        vol 24          E130 -E134,  172,
                                                         173, E196, E197,
                                                         213 - 215
        + hydrogen sulfide              vol 32          E166 -E169,  188,
                                                         190,   202
        + mercury                       vol 29          E102 -E106,
                                                         125 - 127
        + methane                       vol 27/28       E202 -E204, E206,
                                                         225,   227,   228,
                                                         E349 , 350 - 352
        + 2-methylpropane               vol 24          E130 -E134,  192
        + propane                       vol 24          E130 -E134,  147,
                                                         148, E196, E197,
                                                         206
        + seawater                      vol 38                158 - 160
        + water                         vol 38          E135 -E141,
                                                         142 - 157
Octane (ternary)
        + hydrogen sulfide              vol 32                213,   214
        + methane                       vol 27/28             397 - 399,
                                                         407 - 409,
                                                         412 - 417,
                                                         444 - 446
Octane (multicomponent)
        + methane                       vol 27/28                    579
iso-Octane
        see pentane, 2,2,4-trimethyl-
```

```
Octane, 1-bromo-
          + hydrogen sulfide              vol 32        E178, E179,   284
          + methane                       vol 27/28                   697
          + 2-methylpropane               vol 24           E257, E258,
                                                              282,   294
          + propane                       vol 24        E257, E258,   290
Octane, 1-chloro-
          + ammonia                       vol 21             E9,    69
          + methane                       vol 27/28                   698
          + 2-methylpropane               vol 24        E257, E258,   293
          + propane                       vol 24        E257, E258,   285
Octane, 2-cyclopentyl-
          + water                         vol 38                      394
Octane, 2,7-dimethyl-
          + water                         vol 38                      293
Octane, 1-iodo-
          + butane                        vol 24           E257, E258,
                                                              276,   295
          + methane                       vol 27/28                   699
          + propane                       vol 24           E257, E258,
                                                              268,   286
Octane, 2-iodo-
          + 2-methylpropane               vol 24        E257, E258,   296
          + methane                       vol 27/28                   700
          + propane                       vol 24        E257, E258,   287
Octane, 2-methyl-
          + water                         vol 38                      221
Octane, 3-methyl-
          + water                         vol 38                      222
Octane, 4-methyl-
          + water                         vol 38             223,   224
Octane, 1,1´-oxybis-
          + ammonia                       vol 21         E8,   16,    59
          + butane                        vol 24        E375 -E378,   404
          + N,N-dimethylmethanamine       vol 21        E85,  E91,   206
          + hydrogen sulfide              vol 32        E173, E174,   255
          + methanamine                   vol 21        E81,  E90,   112
          + N-methylmethanamine           vol 21                     158
          + 2-methylpropane               vol 24        E375 -E378,   409
          + propane                       vol 24        E375 -E378,   393
1-Octanol
          + ammonia                       vol 21         E6,   16,    48
          + butane                        vol 24        E331 -E334,   362
          + N,N-dimethylmethanamine       vol 21        E85,  E91,   191
          + hydrogen sulfide              vol 32        E171, E172,   244
          + mercury                       vol 29        E165,  170,   171
          + methanamine                   vol 21              E90,   109
          + methane                       vol 27/28     E582 -E584, E592,
                                                              623 -  625
          + N-methylmethanamine           vol 21        E82,  E91,   148
          + 2-methylpropane               vol 24        E331 -E334,   374
          + 1-propanamine                 vol 21             E252,   267
          + propane                       vol 24        E331 -E334,   348,
                                                                     349
Octanoic acid, potassium salt
          + nitrous acid, potassium salt  vol 33                     204
1-Octene
          + water                         vol 38             111,   112
2-Octene
          + water                         vol 38                     113
Octylamine
        see 1-octanamine
Octyl bromide
        see octane, 1-bromo-
Octyl chloride
        see octane, 1-chloro-
Octyl formate
        see formic acid, octyl ester
1-Octyne
          + water                         vol 38                      94
Oil, apiezon GW
        see Apiezon oil GW
Oil, aromatic, simulated light
          + methane                       vol 27/28          578,   579
```

```
Oil, gas (multicomponent)
          + methane                      vol 27/28              489,   560
Oil, hydrocarbon
          + methane                      vol 27/28              557,   558
Oil, mineral
          + methane                      vol 27/28                     569
Oil, olive
          + methane                      vol 27/28                     734
Oil, spray
          + propane                      vol 24               E422,   424
Oil
          see also colza oil, creozote oil, harp seal oil, hydrocarbon oil,
               jojoba oil, paraffin oil, silicone oil, solvent refined coal
               oil, sperm whale oil and vaseline oil
Oleic acid
          see (z)-9-octadecenoic acid
Oleum
          see disulfuric acid
Oranil
          see benzenesulfonamide, 4-amino-N-[(butylamino)carbonyl]-
Oranyl
          see benzenesulfonamide, 4-amino-N-[(butylamino)carbonyl]-
Orasulin
          see benzenesulfonamide, 4-amino-N-[(butylamino)carbonyl]-
Orgaguanidon
          see benzenesulfonamide, 4-amino-N-[(aminoiminomethyl)-
Orgaseptine
          see benzenesulfonamide, 4-amino-
Orisul
          see benzenesulfonamide, 4-amino-N-(1-phenyl-1H-pyrazol-5-yl)-
L-Ornithine, monohydrochloride (aqueous)
          + mercury                      vol 29               E50,    84
L-Ornithine, monohydrochloride (in Hank´s balanced salt solution)
          + mercury                      vol 29        E50,  E87,    92
Osmium
          + mercury                      vol 25                      E309
                                         vol 29              E214,   217
Orthanilamide
          see benzenesulfonamide, 2-amino-
Oxalic acid
          see 1,2-ethanedioic acid
Oxasulfa
          see benzenesulfonamide, 4-amino-N-(4,5-dimethyl-2-oxazolyl)-
Nl-2-Oxazolylsulfanilamide
          see benzenesulfonamide, 4-amino-N-2-oxazolyl-
Oxide
          see under individual oxides
3-Oxo-butanoic acid ethyl ester
          see butanoic acid, 3-oxo-ethyl ester
2-Oxopropanoic acid, sodium salt
          see propanoic acid, 2-oxo, sodium salt
4-Oxo-4-[[[4-(2-pyrimidinylamino)sulfonyl]phenyl]amino]-butanoicacid,
disilver(1+) salt
          see butanoic acid, 4-oxo-4-[[[4-(2-pyrimidinylamino)sulfonyl]-
               phenyl]-amino]-, disilver(1+) salt
4-Oxo-4-[[4-[(2-thiaxolylamino)sulfonyl]phenyl]amino]-butanoic acid
          see butanoic acid, 4-oxo-4-[[4-[(2-thiaxolylamino)sulfonyl]-
               phenyl]amino]-
1,1´-Oxybisbutane
          see butane, 1,1´-oxybis-
1,1´-Oxybis-2-chloroethane
          see ethane, 1,1´-oxybis(2-chloro-
1,1´-Oxybisethane
          see ethane, 1,1´-oxybis-
2,2´-Oxybisethanol
          see ethanol, 2,2´-oxybis-
1,1´-Oxybisoctane
          see octane, 1,1´-oxybis-
2,2´-Oxybispropane
          see propane, 2,2´-oxybis-
Oxybispropanol
          see propanol, oxybis-
Ox blood
          see bovine blood
```

```
P

Pabiamid
          see benzenesulfonamide, 4-amino-, monohydrochloride
PABS
          see benzenesulfonamide, 4-amino-
Paidazolo
          see benzenesulfonamide, 4-amino-N-(1-phenyl-1H-pyrazol-5-yl)-
Pancid
          see benzenesulfonamide, 4-amino-N-(3,4-dimethyl-5-isoxazolyl)
Palladium
          + mercury                          vol 25        E326,  327 - 329
Paraffin
          see kerosene in addition to entries given below
Paraffin (liquid)
          + hydrogen sulfide                 vol 32               E169, E170,
                                                                   277,  278
          + hydrogen selenide                vol 32               E330,  332
Paraffin oil
          + methane                          vol 27/28                   580
Paraffin wax
          + methane                          vol 27/28                   570
Paramenyl
          see benzenesulfonamide, 4-(aminomethyl)-
Paramid
          see benzenesulfonamide, 4-amino-N-(6-methoxy-3-pyridazinyl)-
Paramid Supra
          see benzenesulfonamide, 4-amino-N-(6-methoxy-3-pyridazinyl)-
Parazol
          see benzenesulfonamide, 4-[(2,4-diaminophenyl)azo]-
PASIT
          see benzenesulfonamide, 4-amino-N-[5-(1-methylethyl)-
              1,3,4-thiadiazol-2-yl]-
PEG-400-lauryl ether
          see poly(oxy-1,2-ethanediyl), a-(1-dodecyl)-w-hydroxy-
PEG-400-laurate
          see poly(oxy-1,2-ethanediyl), a-(1-oxododecyl)-w-hydroxy-
PEG-900-stearate
          see poly(oxy-1,2-ethanediyl), a-9-octadecenyl-w-hydroxy-,(Z)-
Pentachlorobenzene
          see benzene, pentachloro-
Pentachlorophenol
          see phenol, pentachloro-
Pentadecane
          + hydrogen sulfide                 vol 32        E166 -E169,  189
          + methane                          vol 27/28     E202 -E204, E208,
                                                                        235
1,4-Pentadiene
          + water                            vol 37                      11
Pentaethylene glycol methyl isopropyl ether
          see 2,5,8,11,14,17-hexaoxanonadecane, 18-methyl-
Pentane
          + butane                           vol 24        E130 -E134,  167
          + hydrogen sulfide                 vol 32        E166 -E169,  188,
                                                                   196,  197
          + methane                          vol 27/28     E202- E204,  216,
                                                           E303,  304 - 313,
                                                                   410 - 413
          + 2-methylpropane                  vol 24        E130 -E134,  189
          + mercury                          vol 29        E102 -E106,  110,
                                                                        111
          + phosphine                        vol 21        E281, E282,  283
          + propane                          vol 24        E130 -E134,  135,
                                                           E196, E197,
                                                                   202 - 205
          + water                            vol 37        E42 - E46,
                                                            47 -  62
Pentane (aqueous)
          + sodium chloride                  vol 37                      62
Pentane (ternary)
          + methane                          vol 27/28            383,  384,
                                                                  390 - 393,
                                                                  423 - 426
```

```
Pentane, 2,2-dimethyl-
             + water                      vol 37                      465,   466
Pentane, 2,3-dimethyl-
             + water                      vol 37                      467,   468
Pentane, 2,4-dimethyl-
             + water                      vol 37              E469, E470,
                                                                471 - 475
Pentane, 3,3-dimethyl-
             + water                      vol 37                      476,   477
Pentane, 2-methyl-
             + propane                    vol 24          E130 -E134,  141
             + water                      vol 37              E296 -E299,
                                                                300 - 306
Pentane, 3-methyl-
             + methane                    vol 27/28                  336,   337
             + mercury                    vol 29          E107 -E109,  119,
                                                                     E165
             + propane                    vol 24          E130 -E134,  142
             + water                      vol 37              E307, E308,
                                                                309 - 313
Pentane, 1,1´-oxybis-
             + butane                     vol 24          E374 -E377,  403
             + methanamine                vol 21               E81,   111
             + N-methylmethanamine        vol 21                      157
             + propane                    vol 24          E374 -E377,  393
Pentane, 2,2,4-trimethyl-
             + ammonia                    vol 21                E2,    15
             + methane                    vol 27/28           E213,   229,
                                                                353,   354
             + mercury                    vol 29          E107 -E109,
                                                                128 - 130
             + phosphine                  vol 21         E281, E282,   283
             + water                      vol 38         E114 -E117,
                                                                118 - 125
Pentane, 2,3,4-trimethyl-
             + water                      vol 38         E126, E127,
                                                                128 - 130
1-Pentanol
             + methane                    vol 27/28      E582- E584, E590,
                                                         E591,  615 - 618
iso-Pentanol
        see 1-butanol, 3-methyl-
tert-Pentanol
        see 2-butanol, 2-methyl-
3-Pentanone
             + N,N-dimethylmethanamine    vol 21                      181,   182
2,5,8,11,14-Pentaoxapentadecane
             + hydrogen sulfide           vol 32              E173, E175,
                                                                257 - 260
             + methane                    vol 27/28                  681
1-Pentene
             + water                      vol 37                       28
2-Pentene
             + water                      vol 37                       29
1-Pentene, 2-methyl-
             + water                      vol 37                      273
1-Pentene, 4-methyl-
             + water                      vol 37                      274
Pentyl acetate
        see acetic acid, pentyl ester
iso-Pentyl alcohol
        see 1-butanol, 3-methyl-
tert-Pentylbenzene
        see benzene (1,1-dimethylpropyl)-
tert-Pentyl carbinol
        see 1-butanol, 2,2-dimethyl-
Pentylmethyl carbinol
        see 2-heptanol
1-Pentyne
             + water                      vol 37                       12
Perchloric acid (aqueous)
             + hydrogen sulfide           vol 32               E26,    38
             + tellurous acid, sodium salt vol 26        E404,  415,   416
```

Perchloric acid (aqueous and multicomponent)
        + selenious acid, silver salt   vol 26   382 - 387
        + selenious acid   vol 26   382 - 387
Perchloric acid (multicomponent)
        + cadmium hydroxide   vol 23   301
Perchloric acid, copper (II) salt (aqueous)
        + copper (II) oxide   vol 23   E10 - E26,   81
Perchloric acid, lithium salt (aqueous)
        + hydrogen sulfide   vol 32   298
Perchloric acid, potassium salt (aqueous)
        + chloric acid, potassium salt   vol 30   E109, E110
                                                                    137, 138
        + potassium chloride   vol 30   E110, 137, 138
Perchloric acid, sodium salt (aqueous)
        + cadmium hydroxide   vol 23   290
        + hydrogen sulfide   vol 32   E26,   38,   40
        + mercury   vol 29   E49,   67
        + mercury (II) oxide   vol 23   331
        + phosphoric acid, disodium salt
                                           vol 31   125
        + phosphoric acid, monosodium salt
                                           vol 31   E51,   69,   97
        + phosphoric acid, trisodium salt
                                           vol 31   E127, 146
        + zinc hydroxide   vol 23   250, 251
        + zinc oxide   vol 23   248
Perchloric acid, sodium salt (aqueous and multicomponent)
        + sulfurous acid, calcium salt   vol 26   E189, E190
                                                                    207 - 210
                                                                    222, 223
        + tellurous acid, silver salt   vol 26   E404, 425
        + tellurous acid, sodium salt   vol 26   E404, 425
Perchloric acid, sodium salt (multicomponent)
        + cadmium hydroxide   vol 23   303
        + mercury (II) oxide   vol 23   342
        + silver (I) oxide   vol 23   139, 141
Perchloryl fluoride
        see chlorine, hexafluoride
Percoccide
        see benzenesulfonamide, 4-amino-N-(4-methyl-2-pyrimidinyl)-
Percoral
        see 3-pyridinecarboxamide, N,N-diethyl-
Perfluorobenzene
        see benzene, hexafluoro-
Perfluorobutyltetrahydrofuran
        see furan, heptafluorotetrahydro(nonafluorobutyl)-
Perfluorodecalin
        see naphthalene, octadecafluorodecahydro-
Perfluorodimethylcyclohexane
        see cyclohexane, decafluorobis(trifluoromethyl)-
Perfluoro-1,4-diisopropoxybutane
        see butane, 1,1,2,2,3,3,4,4-octafluoro-
            1,4-bis(1,2,2-tetrafluoro-1-(trifluoromethyl)-ethoxy)-
Perfluorodimethylcyclobutane
        see cyclobutane, hexafluorobis(trifluoromethyl)-
Perfluoroheptane
        see heptane, hexadecafluoro-
Perfluorohexane
        see hexane, tetradecafluoro-
Perfluoromethylcyclohexane
        see cyclohexane, undecafluoro(trifluoromethyl)-
Perfluoromethyldecalin
        see naphthalene, heptadecafluorodecahydro(trifluoromethyl)-
Perfluorononane
        see nonane, eicosafluoro-
Perfluorooctane
        see octane, octadecafluoro-
Perfluoro-1-isopropoxyhexane
        see hexane, 1,1,1,2,2,3,3,4,4,5,5,6,6-tridecafluoro-6-(1,2,2,2-
            tetrafluoro-1-(trifluoromethylll)ethoxy)-
Perfluorotributylamine
        see 1-butanamine,
            1,1,2,2,3,3,4,4,4-nonafluoro-N,N-bis(nonafluorobutyl)-

```
Permanganate, potassium
        see permanganic acid, potassium salt
Perylene
        + water                         vol 38                          520

Petrcleum
        + methane                       vol 27/28       568,  572 - 574
                                                        580,   581
Petroleum, crude oils
        + butane                        vol 24                  428,  429
        + propane                       vol 24                  E422,  423
Petroleum ether
        see ligroin
Petroleum ether
        + phosphine                     vol 21                          289
Phenanthrene (ternary)
        + methane                       vcl 27/28               541,  542

Phenanthrene
        + water                         vol 38                  E409, E410
                                                                411 - 418
Phenanthrene (aqueous)
        + sodium chloride               vol 38                  421 - 423
        + seawater                      vol 38                  E419,  420
Phenanthrene, 9,10-dihydro-
        + hydrogen                      vol 5/6                 432,  433
        + methane                       vol 27/28                       550
Phenanthrene, 1-methyl-
        + water                         vol 38                          438
Phenanthrene, 1-methyl (aqueous)
        + sodium chloride               vol 38                          439
Phenetole
        see benzene, ethoxy-
Phenol
        + butane                        vol 24          E331 -E334,  360
        + hydrogen sulfide              vol 32          E176, E178,  275
        + methane                       vol 27/28                       678
        + 2-methylpropane               vol 24          E331 -E334,  371
        + propane                       vol 24          E331 -E334,  346
Phenol (aqueous)
        + bromic acid, potassium salt   vol 30                          249
        + hydrogen sulfide              vol 32                  E20,   78
Phenol, 4-bromo-
        + water                         vol 20          E150,  151,  152
Phenol, 2-chloro-
        + water                         vol 20                  E183, E184
                                                                185 - 189
Phenol, 3-chloro-
        + water                         vol 20          E190,  191 - 194
Phenol, 4-chloro-
        + water                         vol 20                  E195, E196
                                                                197 - 201
Phenol, 4-chloro-2,5-dimethyl-
        + water                         vol 20                  E242,  243
Phenol, 4-chloro-2,6-dimethyl-
        + water                         vol 20                  E244,  245
Phenol, 4-chloro-3,5-dimethyl-
        + water                         vol 20                  E246,  247
Phenol, 4-chloro-3,5-dimethyl-2-(phenylmethyl)-
        + water                         vol 20                  E256,  257
Phenol, 2-chloro-6-methyl-
        + water                         vol 20                  E235,  236
Phenol, 4-chloro-2-methyl-
        + water                         vol 20                  E231,  232
Phenol, 4-chloro-3-methyl-
        + water                         vol 20                  E233,  234
Phenol, 4-chloro-3-methyl-5-ethyl-
        + water                         vol 20                  E250,  251
Phenol, 2,4-dibromo-
        + water                         vol 20                  E65,   66
Phenol, 2,4-dichloro-
        + water                         vol 20                  E116,  117
Phenol, 2,4-dichloro-6-butyl-
        + water                         vol 20                  E252,  253
```

```
Phenol, 2,4-dichloro-6-ethyl-
            + water                         vol 20              E240,   241
Phenol, 2,4-dichloro-6-methyl-
            + water                         vol 20              E228,   229

Phenol, 2,6-dichloro-4-methyl-
            + water                         vol 20              E226,   227
Phenol, 2,4-dichloro-6-(phenylmethyl)-
            + water                         vol 20              E254,   255
Phenol, 2,4-dichloro-6-propyl-
            + water                         vol 20              E248,   249
Phenol, methyl-
            + methane                       vol 27/28           580,    680
Phenol, pentachloro-
            + water                         vol 20              E6,     E7
                                                                  8 -   10
Phenol, pentachloro-(ternary)
            + water                         vol 20              E6,     E7
                                                                 11 -   15
Phenol, 2,4,5,6-tetrachloro-3-methyl-
            + water                         vol 20              E221,   222
Phenol, 2,4,6-tribromo-
            + water                         vol 20      E30,    31,     32
Phenol, 2,4,5-trichloro-
            + water                         vol 20              E43,    44
Phenol, 2,4,6-trichloro-
            + water                         vol 20      E45,    46 -    48
Phenol, 2,4,6-trichloro-3,5-dimethyl-
            + water                         vol 20              E238,   239
7-Phenylacetamidodeacetoxycephalosporanic acid
        see 5-thia-1-azabicyclo[4,2,0]oct-2-ene-2-carboxylic acid,
            3-methyl-8-oxo-7-(phenylacetamido)-
L-Phenylalanine(aqueous)
            + mercury                       vol 29              E50,    80
1-Phenylethanone
        see ethanone, 1-phenyl-
N,N-(p-Phenylenesulfonyl)bisacetamide
        see acetamide, N-[[(4-acetylamino)phenyl]sulfonyl]-
1-Phenylhexane
        see benzene hexyl-
Phenylmethane
        see benzene, methyl-
N-Phenyl-N-methyl amine
        see methanamine, N-phenyl-
Phenyl cellosolve
        see ethanol,2-phenoxy-
1-Phenyl benzoate
        see benzoic acid, 1-phenyl ester
Phenyl ethyl ether
        see benzene, ethoxy-
N1-(1-Phenylpyrazol-5-yl)sulfanilamide
        see benzenesulfonamide, 4-amino-N-(1-phenyl-1H-pyrazol-5-yl)-
2-Phenyl-3-sulfanilamidopyrazole
        see benzenesulfonamide, 4-amino-N-(1-phenyl-1H-pyrazol-5-yl)-
Phosphate, tricresyl-
        see phosphoric acid, tris(methylphenyl) ester
Phosphinic acid
            + mercury                       vol 29                      29
Phospholipids
            + butane                        vol 24              E414,   419
            + methane                       vol 27/28                   730
            + 2-methylpropane               vol 24              E414,   420
            + propane                       vol 24              E414,   418
Phosphoric acid
            + ammonia                       vol 31              269,    270
            + diphosphoric acid             vol 31              269,    270
            + iodic acid, lithium salt      vol 30                      320
            + phosphoric acid, disodium salt
                                            vol 31      23,     37,     38
            + phosphoric acid, lithium salt
                                            vol 31      E4,     9,      10
            + phosphoric acid, monocesium salt
                                            vol 31              E320,   E322
                                                                329 -   331
```

Phosphoric acid
          + phosphoric acid, monolithium salt
                                        vol 30   320
          + phosphoric acid, monopotassium salt
                                        vol 31   174,   178 - 181
          + phosphoric acid, monorubidium salt
                                        vol 31   E308, E310
                                                     313,   316,   317
          + phosphoric acid, monosodium salt
                                        vol 31   16,   20,   21
                                                                   24,   25
          + phosphoric acid, trisodium salt
                                        vol 31   E13,   E14
                                                                   26 -   36
          + phosphoric acid, tripotassium salt
                                        vol 31   E168 -E171
                                                                  175 - 177
                                                                  184 - 202
          + potassium hydroxide   vol 31   E168 -E171
                                                                  184 - 202
          + potassium oxide   vol 31   269,   270
          + zinc oxide   vol 23   E177,   258
Phosphoric acid, calcium salt (aqueous)
          + calcium chloride   vol 31   271
          + phosphoric acid   vol 31   264 - 266
          + phosphoric acid, monopotassium salt
                                        vol 31   264 - 266,   271
          + potassium chloride   vol 31   271
          + sulfurous acid, calcium salt   vol 26   E191, E192,   231
Phosphoric acid, copper(2+) salt
          + water   vol 31   148
Phosphoric acid, copper(2+) salt (aqueous)
          + phosphoric acid, trisodium salt
                                        vol 31   E127,   148
Phosphoric acid, diammonium salt
          + water   vol 31   313
Phosphoric acid, diammonium salt (aqueous)
          + ammonium chloride   vol 31   119,   120
          + phosphoric acid, disodium salt
                                        vol 31   117,   119,   120
          + phosphoric acid, monoammonium salt
                                        vol 31   86,   87
          + phosphoric acid, monorubidium salt
                                        vol 31   E308, E309,   313
          + phosphoric acid, monosodium salt
                                        vol 31   E51,   86,   87
          + sodium chloride   vol 31   119,   120
Phosphoric acid, dipotassium salt
          + water   vol 31   172 - 174
Phosphoric acid, dipotassium salt (aqueous)
          + phosphoric acid   vol 31   174
          + phosphoric acid, disodium salt
                                        vol 31   108,   109
Phosphoric acid, disodium salt
          + water   vol 31   23,   E98 -E102
                                                                  103 - 126
Phosphoric acid, disodium salt, dihydrate
          + mercury   vol 29   E87,   88 -   94
Phosphoric acid, disodium salt (aqueous)
          + ammonium chloride   vol 31   119,   120
          + boric acid   vol 31   E98,   118
          + boric acid, sodium salt   vol 31   E98
          + N,N´-1,2-ethanediylbis(N-carboxymethyl)glycine,
            disodium salt   vol 31   E98,   116
          + hydrogen peroxide   vol 31   E98, 107,   113
          + methanol   vol 31   E98, E101,   126
          + nitric acid, sodium salt   vol 31   E98, E101,   110
                                                                  121 - 123
          + perchloric acid, diammonium salt
                                        vol 31   117,   119,   120
          + phosphoric acid   vol 31   23,   37,   38
          + phosphoric acid, dipotassium salt
                                        vol 31   E98, E101
                                                                  108,   109

```
Phosphoric acid, disodium salt (aqueous)
        + phosphoric acid, magnesium salt
                                          vol 31              E101,   115
        + 2-propanone                     vol 31               E98,  E101
                                                        124,   125,  E129
        + silicic acid, sodium salt       vol 31               E98,   114
        + sodium chloride                 vol 31       E98,  E101,   111
                                                        119,   120,   121
        + sodium hydroxide                vol 31       E13,   E14,    18
                                                                19,    22
                                                                37,    38
        + sulfuric acid, sodium salt      vol 31               E98,   112
                                                                122,   123
Phosphoric acid, lithium salt
        + water                           vol 31       E1  -  E6,     7
Phosphoric acid, lithium salt (aqueous)
        + ammonia                         vol 31                E3,     8
        + lithium hydroxide               vol 31       E4,     9,    10
        + lithium oxide                   vol 31               9,    10
        + phosphoric acid                 vol 31       E4,     9,    10
Phosphoric acid, magnesium salt (1:1) (aqueous)
        + phosphoric acid, disodium salt
                                          vol 31              E101,   115
Phosphoric acid, monoammonium salt
        + water                           vol 31               55  -   58
Phosphoric acid, monoammonium salt (aqueous)
        + ammonium chloride               vol 31                       90
        + phosphoric acid, diammonium salt
                                          vol 31                86,    87
        + phosphoric acid, monocesium salt
                                          vol 31              E318,  E319
                                                             E322,   326
        + phosphoric acid, monopotassium salt
                                          vol 31      E208,  E209,   227
                                                      230,   231,   236
                                                      244  - 247,   256
                                                             262,   263
                                                             274,   275
        + phosphoric acid, monosodium salt
                                          vol 31               E51,    54
                                                                55  -   58
                                                       86,    87,    90
        + sodium chloride                 vol 31                       90
        + urea                            vol 31              274,   275
Phosphoric acid, monocesium salt (aqueous)
        + cesium chloride                 vol 31                     E318
        + cesium oxide                    vol 31      E320,  E322
                                                             329  - 331
Phosphoric acid, monocesium salt (aqueous)
        + phosphoric acid                 vol 31      E320,  E322
                                                             329  - 331
        + phosphoric acid, monoammonium salt
                                          vol 31              E318,  E319
                                                             E322,   326
        + phosphoric acid, monopotassium salt
                                          vol 31      E318  -E321,   328
Phosphoric acid, mcnolithium salt (aqueous)
        + iodic acid, lithium salt        vol 30                      320
        + phosphoric acid                 vol 30                      320
Phosphoric acid, monopotassium salt
        + water
                                          vol 31       70  -  72,    75
                                                             E206  -E212
                                                      213  - 296,   328
Phosphoric acid, monopotassium salt (aqueous)
        + acetic acid, potassium salt     vol 31                      221
        + ammonium chloride               vol 31              262,   263
                                                             267,   268
        + arsenic acid, monopotassium salt
                                          vol 31              E208,   220
        + boric acid, potassium salt      vol 31              E208,  E210
                                                             224,   232
                                                      233,   253,   254
                                                             260,   261
```

Phosphoric acid, monopotassium salt (aqueous)
          + calcium chloride   vol 31   271
          + carbonic acid, dipotassium salt
                                          vol 31   E212,   222
          + ethanamine, N,N-diethyl-, phosphate
                                          vol 31   273
          + formamide   vol 31   E212,   272,
                                                                     276,   277
          + hydrogen peroxide   vol 31   E212,   225
          + nitric acid, potassium salt   vol 31   92 -   95
          + phosphoric acid   vol 31   174,   178 -   181,
                                                                     203 -   205,
                                                                     264 -   266
          + phosphcric acid, calcium salt
                                          vol 31   264 -   266,   271
          + phosphoric acid, monoammonium salt
                                          vol 31   E208, E209,   227,
                                                       230,   231,   236,
                                                       244 -   247,   256,
                                                       262,   263,   267,
                                                       268,   274,   275
          + phosphoric acid, monocesium salt
                                          vol 31   E318 -E321,   328
          + phosphoric acid, monosodium salt
                                          vol 31   E47,   E50 - E52,
                                                               67,   68,
                                                       70 -   72,   75,
                                                       80 -   85,   88,
                                                       89,   91 -   95
          + potassium chloride   vol 31   80 -   85,   88,
                                                       89,   91,   E210,
                                                  E212,   226,   228,
                                                   229,   240 -   243,
                                                   251,   252,   255,
                                                   257,   262,   263,
                                                   267,   268,   271
          + potassium hydroxide   vol 31   182,   183,
                                                               203 -   205
          + selenic acid, potassium salt vol 31   E208,   248
          + sodium chloride   vol 31   80 -   85,
                                                       88,   89,   91
          + sulfuric acid, potassium salt
                                          vol 31   E210, E212,   223,
                                                               237 -   239,
                                                               249,   250
          + urea   vol 31   E208,   234,   235,
                                                               274,   275
Phosphoric acid, monorubidium salt
          + water   vol 31   E307 -E311,
                                                               312 -   317
Phosphoric acid, monorubidium salt (aqueous)
          + phosphcric acid   vol 31   E308, E310,
                                                               316,   317
          + phosphoric acid, ammonium salt
                                          vol 31   E308, E309,   313
          + rubidium chloride   vol 31   E307,   314
          + rubidium oxide   vol 31   E308, E310,   315,
                                                               316,   317
Phosphoric acid, monosodium salt
          + mercury   vol 29   E87,   88   -   94
          + water   vol 31   20,   24,   25,
                                                               E39 -   E42,
                                                                43 -   45,
                                                               E47 -   E52,
                                                                53 -   97,
                                                              E318 -E320,
                                                               321 -   331
          + water-d2   vol 31   332,   333
Phosphoric acid, monosodium salt (aqueous)
          + ammonium chloride   vcl 31   90
          + boric acid sodium salt   vol 31   E47,   73,   74
          + nitric acid, potassium salt   vol 31   92   -   95

```
Phosphoric acid, monosodium salt (aqueous)
          + nitric acid, sodium salt       vol 31              E47,   E48
                                                       59 -   64,    79
                                                              92 -   95
          + perchloric acid, sodium salt   vol 31      E51,   69,    97
          + phosphoric acid                vol 31      16,    20,    21
                                                              24,    25
          + phosphoric acid, diammonium salt
                                           vol 31      E51,   86,    87
          + phosphoric acid, monoammonium salt
                                           vol 31              E51,   54
                                                              55 -   58
                                                       86,    87,    90
          + phosphoric acid, monopotassium salt
                                           vol 31      E47,   E50 - E52
                                                              67,    68
                                                       70 -   72,    75
                                                       80 -   85,    88
                                                       89,    91 -   95
          + potassium chloride             vol 31              E51,   E52
                                                       80 -   85,    88
                                                              89,    91
          + 2-propanone                    vol 31      E51,   96,    97
                                           E129
          + sodium chloride                vol 31      E47,   E49,   E51
                                                       E52,   65,    66
                                                              76 -   78
                                                              80 -   85
                                                              88,    89
                                                              90,    91
          + sulfuric acid, disodium salt   vol 31              E51,   53
Phosphoric acid, sodium salt (aqueous and multicomponent)
          + acetic acid, sodium salt       vol 26      E191, E192,   234
          + formic acid, sodium salt       vol 26      E191, E192,   234
          + sulfurous acid, calcium salt   vol 26      E191, E192,   234
Phosphoric acid, tributyl ester
          + ammonia                        vol 21               E8,   64
          + butane                         vol 24      E375 -E378,   402
          + hydrogen sulfide               vol 32      E183 -E185,   260
                                                              323 - 325
          + methane                        vol 27/28            727,  728
          + propane                        vol 24      E375 -E378,   392
Phosphoric acid, triethyl ester
          + butane                         vol 24      E375 -E378,   402
          + hydrogen sulfide               vol 32              E183 -E185
          + methane                        vol 27/28                 727
          + propane                        vol 24      E375 -E378,   392
Phosphoric acid, trilithium salt (aqueous)
          + iodic acid, lithium salt       vol 30              E272,  296
Phosphoric acid, trimethyl ester
          + butane                         vol 24      E375 -E378,   402
          + hydrogen sulfide               vol 32              E183 -E185
          + propane                        vol 24      E375 -E378,   392
Phosphoric acid, tripotassium salt
          + water                          vol 31      E168 -E171, E297
                                                              298 - 306
Phosphoric acid, tripotassium salt (aqueous)
          + ammonia                        vol 31      E297,  300,   301
          + boric acid, potassium salt     vol 31      E297,         304
          + nitrous acid, potassium salt   vol 31      E297,         303
          + phosphoric acid                vol 31      E168 -E171
                                                              175 - 177
                                                              184 - 202
          + phosphoric acid, trisodium salt
                                           vol 31      E129,  149,   165
          + potassium hydroxide            vol 31      E168 -E171
                                                              184 - 202
Phosphoric acid, tripotassium salt (aqueous)
          + sulfuric acid, dipotassium salt
                                           vol 31              E297,  302
                                                              305,   306
          + vanadic acid, potassium salt   vol 31      E297,  305,   306
Phosphoric acid, tripropyl ester
          + butane                         vol 24      E375 -E378,   402
```

```
Phosphoric acid, tripropyl ester
          + hydrogen sulfide           vol 32            E183 -E185
          + methane                    vol 27/28                727
          + propane                    vol 24       E375 -E378,  392
Phosphoric acid, tris(2-methylpropyl) ester
          + butane                     vol 24       E375 -E378,  402
          + hydrogen sulfide           vol 32            E183 -E185
          + methane                    vol 27/28                727
          + propane                    vol 24       E375 -E378,  392
Phosphoric acid, trisodium salt
          + water                      vol 31        E11 - E15,   17,
                                                          26 -   36,
                                                      E127 -E130,
                                                         131 - 165
Phosphoric acid, trisodium salt (aqueous)
          + boric acid, sodium salt    vol 31            E127,   147
          + carbonic acid, disodium salt  vol 31    E127, E129,  139,
                                                     152,  157,  158
          + hydrogen peroxide          vol 31       E127,  141,  142
          + nitrous acid, sodium salt  vol 31            E127,  150
          + nitrous oxide              vol 9              E35,   92
          + perchloric acid, sodium salt  vol 31         E127,  146
          + phosphoric acid            vol 31            E13,  E14,
                                                          26 -   36
          + phosphoric acid, copper(2+) salt
                                       vol 31            E127,  148
          + phosphoric acid, tripotassium salt
                                       vol 31       E129,  149,  165
          + 2-propanone                vol 31       E129,  166,  167
          + silicic acid, dipotassium salt
                                       vol 31                    165
          + silicic acid, sodium salt  vol 31       E129,  143,
                                                     153,  165
          + sodium aluminium oxide     vol 31       E129,  159 - 162
          + sodium chloride            vol 31       E127,  134,  135
          + sodium fluoride            vol 31            E127,  154
          + sodium hydroxide           vol 31        E11,  E12,  E14,
                                                      17,   26 -   36
          + sodium hydroxide           vol 31       E129,  155 - 162
          + sulfuric acid, disodium salt
                                       vol 31            E128, E129,
                                                     136 - 138,  140,
                                                     145,  155,  156,
                                                     161 - 164
          + tungstic acid, disodium salt  vol 31         E127,  151
          + vanadic acid, sodium salt  vol 31       E129,  144,  159,
                                                     160,  163,  164
Phosphoric acid-d3 (aqueous)
          + phosphoric acid-d3, monocesium salt
                                       vol 31             332,  333
Phosphoric acid-d3, monocesium salt (aqueous)
          + cesium oxide               vol 31             332,  333
          + phosphoric acid-d3         vol 31             332,  333
Phosphoric triamide, hexamethyl-
          + butane                     vol 24       E299 -E302,  321
          + hydrogen sulfide           vol 32       E183 -E185,  326
          + methane                    vol 27/28                729
          + propane                    vol 24       E299 -E302,  314
Phosphorous acid, dibutylester
          + ammonia                    vol 21                    62
Phosphorous acid, triethyl ester
          + ammonia                    vol 21                    61
Phosphorous acid, triphenyl ester
          + ammonia                    vol 21                    63
Phosphorus
          + mercury                    vol 29       E213, E214,  299
Phosphorus (ternary)
          + phosphine                  vol 21                   300
Phosphorus (V) oxide (aqueous)
          + zinc oxide                 vol 23       E177,  195,  196
Phosphorus pentoxide (aqueous)
          + lithium oxide              vol 31              E3 -   E5,
                                                            9,   10
          + rubidium oxide             vol 31                   315
```

```
Phosphorus pentoxide (aqueous)
            + sodium cxide                    vol 31                      Ell
Phosphorus pentoxide (aqueous-d2)
            + rubidium oxide                  vol 31                      315
Phosphorus pentoxide (aquecus and multicomponent)
            + sulfurous acid, calcium salt    vol 26            E191, E192
                                                                  229,  230
Picene
            + water                           vol 38                      540
Picoline
        see pyridine,methyl-
Pinene
        see hept-2-ene, 2,6,6-trimethyl-, bicyclo(3,1,1)
Piperidine
            + methane                         vcl 27/28                   707
Piperidine (aqueous)
            + bromic acid, potassium salt     vol 30                      251
Piridazol
        see benzenesulfonamide, 4-amino-N-2-pyridinyl-
Piridisir
        see benzenesulfonamide, 4-amino-N-2-pyrimidinyl-
Piridolo
        see benzenesulfonamide, 4-amino-N-(6-methoxy-3-pyridazinyl)-
Pirimal-M
        see benzenesulfonamide, 4-amino-N-(4-methyl-2-pyrimidinyl)-
Pirimal
        see benzenesulfonamide, 4-amino-N-2-pyrimidinyl-
Pirmazin
        see benzenesulfonamide, 4-amino-N-(4,6-dimethyl-2-pyrimidinyl)-
Placenta
        see human placenta
Planomide
        see benzenesulfonamide, 4-amino-N-2-thiazolyl-
Plasma
            + mercury                         vol 29            E51,    98
Plasma
        see also human plasma, human hyperlipidemic plasma
Platinum
            + mercury                         vol 25            E330,  331 - 334
            + mercury                         vol 29            E213, E214,  225
Plutonium
            + mercury                         vol 25            E434,  435
Plisulfan
        see benzenesulfonamide, 4-amino-N-(1-phenyl-1H-pyrazol-5-yl)-
Poliseptil
        see benzenesulfonamide, 4-amino-N-2-thiazolyl-
Polonium
            + mercury                         vol 25                     E421
Polycidal
        see benzenesulfonamide. 4-amino-N-(3-methoxypryrazinyl)-
Polyethylene glycol
        see hydro- -hydroxy-poly(oxy-1,2-ethanediyl)-
Potassium
            + mercury                         vol 25            E33,  E34
                                                                 35 -  43
            + mercury                         vol 29            E213, E214,  225
Potassium acetate
        see acetic acid, potassium salt
Potassium bicarbonate
        see carbonic acid, monopotassium salt
Potassium borate
        see boric acid, potassium salt
Potassium bromate
        see bromic acid, potassium salt
Potassium bromide (aqueous)
            + bromic acid, potassium salt     vol 30                     233 - 235
            + chloric acid, potassium salt    vol 30            E108, E109,  141
            + iodic acid, potassium salt      vol 30            E379,  392,  393
Potassium bromide (ternary)
            + butane                          vol 24            E58 - E72,  117
            + propane                         vol 24            E58 - E72,  116
Potassium bromide (in liquid ammonia)
            + iodic acid, potassium salt      vol 30                     425
```

```
Potassium carbonate
        see carbonic acid, dipotassium salt
Potassium chlorate
        see chloric acid, potassium salt
Potassium chloride (aqueous)
        + ammonium chloride          vol 31               267,    268
        + bromic acid, potassium salt vol 30              232,    243
        + butane                     vol 24       E58 - E72,    115
        + calcium chloride           vol 31                       271
        + chloric acid, potassium salt vol 30            E108 -E110,
                                                         125 - 140,
                                                         153, E164
        + chloric acid, rubidium salt vol 30             E110,  139,
                                                         140, E164
        + chloric acid, sodium salt  vol 30              E28,     29
        + hydrogen sulfide           vol 32       E25,    39,     41
        + iodic acid, potassium salt vol 30              E379,   391
        + mercury                    vol 29       E49,    71,   E87,
                                                          88  - 94
        + methane                    vol 27/28            E79  -E81,
                                                         162  -165
        + perchloric acid, potassium salt
                                     vol 30       E110,  137,    138
        + phosphoric acid, calcium salt (aqueous)
                                     vol 31                       271
        + phosphoric acid, monoammonium salt
                                     vol 31               267,    268
        + phosphoric acid, monopotassium salt
                                     vol 31        80 -   85,     88,
                                                   89,   91, E210,
                                                E212,  226,    228,
                                                 229,  240 -  243,
                                                 251,  252,    255,
                                                 257,  262,    263,
                                                 267,  268,    271
        + phosphoric acid, monosodium salt
                                     vol 31              E51,    E52,
                                                          80 -   85,
                                                   88,   89,     91
        + propane                    vol 24       E58 - E72,    114
        + rubidium chloride          vol 30       E110,  139,    140,
                                                               E164
        + sodium chloride            vol 31        80 -   85,     88,
                                                          89,     91
Potassium chloride (molten)
        + mercury                    vol 29              E207,   210
Potassium chloride (ternary)
        + methane                    vol 27/28    E80,   E83,   E84,
                                                         113,    166
        + water                      vol 20               E6,     E7,
                                                          11,     12
Potassium chloride (multicomponent)
        + copper (I) oxide           vol 23       E1  -  E3,      8
        + copper (II) oxide          vol 23       E10 - E26,     80
        + mercury (II) oxide         vol 23                      335
Potassium chloride (in liquid ammonia)
        + iodic acid, potassium salt vol 30              421,    423
Potassium chromate
        see chromic acid, potassium salt
Potassium dicyanoargentate
        see argentate(1)-, bis(cyano-C)-potassium
Potassium dihydrogen arsenate
        see arsenic acid, monopotassium salt
Potassium dihydrogen phosphate
        see phosphoric acid, monopotassium salt
Potassium ferrocyanide
        see ferrate (4-), hexakis(cyano-C-), tetrapotassium
Potassium fluoride (multicomponent)
        + copper (II) oxide          vol 23       E10 - E26,     63
Potassium hydrogen selenate
        see selenic acid, monopotassium salt
Potassium hydrogen sulfite
        see sulfurous acid, monopotassium salt
```

```
Potassium hydroxide (aqueous)
          + chloric acid, potassium salt    vol 30              E108,   147
          + iodic acid, lithium salt        vol 30              E272,  E380
          + iodic acid, potassium salt      vol 30       E378,  E380,   405
                                                         406,   409,   410
          + lithium hydroxide               vol 30              E273,  E380
          + methane                         vol 27/28    E79,   160,   161
          + mercury (II) oxide              vol 23             E307  -E313
                                                                323,   334
          + phosphoric acid                 vol 31             E168  -E171
          + phosphoric acid, monopotassium salt
                                            vol 31              182,   183
          + potassium iodide                vol 30       E380,  409,   410
          + silver (I) oxide                vol 23       E83,   E84,   E86
                                                         E87,   118,   129
          + silver (II) oxide               vol 23             E144,  E145
                                                                146,   147
          + sulfurous acid, dipotassium salt
                                            vol 26              E92,   102
          + zinc hydroxide                  vol 23       E162,  222 -  224
          + zinc oxide                      vol 23       E162,  205,   206
                                                         228,   242,   243
                                                         246,   249,   253
Potassium hydroxide (multicomponent)
          + cadmium hydroxide               vol 23                     300
          + silver (I) oxide                vol 23                     140
          + zinc hydroxide                  vol 23                     265
Potassium iodate
      see iodic acid, potassium salt
Potassium iodide (aqueous)
          + bromic acid, potassium salt     vol 30                     236
          + butane                          vol 24       E58 - E72,    119
          + chloric acid, potassium salt    vol 30       E379,  E380,   394
                                                                409,   410
          + methane                         vol 27/28           E81,   167
          + potassium hydroxide             vol 30       E380,  409,   410
          + propane                         vol 24       E58 - E72,    118
Potassium iodide (in liquid ammonia)
          + iodic acid, potassium salt      vol 30                     426
Potassium oxide (aqueous)
          + ammonia                         vol 31              269,   270
          + diphosphoric acid               vol 31              269,   270
          + phosphoric acid                 vol 31              269,   270
Potassium nitrate
      see nitric acid potassium salt
Potassium nitrite
      see nitrous acid, potassium salt
Potassium perchlorate
      see perchloric acid, potassium salt
Potassium periodate
      see periodic acid, potassium salt
Potassium permanganate
      see permanganic acid, potassium salt
Potassium phosphate
      see phosphoric acid, potasium salt
Potassium pyroselenite
      see diselenious acid, dipotassium salt
Potassium pyrosulfite
      see disulfurous acid, dipotassium salt
Potassium selenite
      see selenious acid, dipotassium salt
Potassium sulfate
      see sulfuric acid, potassium salt
Potassium sulfide
      see sulfurous acid, dipotassium salt
Potassium thiocyanate
      see thiocyanic acid, potassium salt
Potassium trihydrogen diselenite
          + water                           vol 26       E331,  340,   341
Potassium vanadate
      see vanadic acid, dipotassium salt
Praseodymium
          + mercury                         vol 25       E221,  222 - 224
```

```
Praseodymium bromide
            + 1-butanamine                      vol 22                    142
            + 2-butanamine                      vol 22                    142
            + 2-butanamine, N-(1-methylpropyl)-
                                                vol 22                    142
            + 1,4-dioxane                       vol 22                    141
            + ethane, 1,2-diethoxy-             vol 22                    140
            + 1-propanamine                     vol 22                    142
            + 2-propanamine                     vol 22                    142
Praseodymium chloride
            + 1,3-butadiene, hexachloro-        vol 22                    123
            + 1-butanol                         vol 22                    130
            + 1,4-dioxane                       vol 22                    132
            + 1,3-dioxolane                     vol 22                    132
            + 1,2-ethanediol                    vol 22                    128
            + ethane, 1-ethoxy-2-methoxy-       vol 22                    132
            + ethanol                           vol 22                    125
            + ethanol (aqueous)                 vol 22                    126
            + ethanol, 2-methoxy-               vol 22                    133
            + ethanol, 2-ethoxy-                vol 22                    133
            + furan, tetrahydro-                vol 22                    135
            + heptane, 1-methoxy-               vol 22                    134
            + methanol                          vol 22                    124
            + nonane, 1-methoxy-                vol 22                    134
            + octane, 1-methoxy-                vol 22                    134
            + 1-pentanol                        vol 22                    131
            + phosphoric acid, tributyl ester
                                                vol 22                    136
            + phosphoric triamide, hexamethyl-
                                                vol 22                    138
            + phosphoryl chloride               vol 22                    139
            + 1-propanamine                     vol 22                    137
            + 2-propanamine                     vol 22                    137
            + 1-propanol                        vol 22           127 - 129
            + 2-propen-1-amine                  vol 22                    137
            + 2-propen-1-ol                     vol 22                    128
            + stannane, tetrachloro-            vol 22                    139
            + water (ternary)                   vol 22                    126
Praseodymium cobalt nitrate
         see nitric acid, cobalt praseodymium nitrate
Praseodymium copper nitrate
         see nitric acid copper praseodymium nitrate
Praseodymium fluoride
            + acidic nitrosyl fluoride (nitrosyl fluoride compound
              with hydrofluoric acid)          vol 22                    122
            + butane, 1-(chloromethoxy)-        vol 22                    116
            + decane, 1-methoxy-                vol 22                    117
            + furan, tetrahydro-                vol 22                    118
            + methane, 1,1´sulfinylbis-         vol 22                    120
            + methanol                          vol 22                    115
            + phosphoric acid, tributyl ester
                                                vol 22                    119
            + pyridine                          vol 22                    121
Praseodymium iodide
            + 1-butanamine                      vol 22                    144
            + 2-butanamine                      vol 22                    144
            + 1,4-dioxane                       vol 22                    143
            + formamide, N,N-dimethyl-          vol 22                    145
            + 1-propanamine                     vol 22                    144
            + 2-propanamine                     vol 22                    144
Praseodymium manganese nitrate
         see nitric acid, praseodymium manganese salt
Praseodymium nickel nitrate
         see ntric acid, nickel praseodymium nitrate
Praseodymium nitrate
         see nitric acid, praseodymium salt
Praseodymium zinc nitrate
         see nitric acid praseodymium zinc salt
Polyvinylpyrrolidone
         see 2-Pyrrolidinone, 1-ethenyl-, polymers
Pratonal
         see benzamide, N-[(4-aminophenyl)sulfonyl]-3,4-dimethyl-
L-Proline(aqueous)
            + mercury                           vol 29                E50,  75
```

```
Promethium
            + mercury                          vol 25                      E421
Pronotosil soluble
        see 2,7-naphthalenedisulfonic acid, 6-acetylamino-
            3-[[4-(Aminosulfonyl)phenyl]azo]-4-hydroxy-, disodium salt
Pronotosil S
        see 2,7-naphthalenedisulfonic acid, 6-acetylamino-
            3-[[4-(aminosulfonyl)phenyl]azo]-4-hydroxy-, disodium salt
Prontalbin
        see benzenesulfonamide, 4-amino-
Prontosil Album
        see benzenesulfonamide, 4-amino-
Prontosil flavum
        see benzenesulfonamide, 4-[(2,4-diaminophenyl)azo]-
Prontosil I
        see benzenesulfonamide, 4-amino-
Prontosil rubrum
        see benzenesulfonamide, 4-[(2,4-diaminophenyl)azo]-
Prontosil white
        see benzenesulfonamide, 4-amino-
Prontylin
        see benzenesulfonamide, 4-amino-
Pronzin Album
        see benzenesulfonamide, 4-amino-
Pronzin rubrum
        see benzenesulfonamide, 4-[(2,4-diaminophenyl)azo]-
Propanaminium bromide, N,N,N-tripropyl- (aqueous)
            + butane                           vol 24       E58 - E72,    91
            + propane                          vol 24             E58 - E72
                                                                  89,    90
Propanaminium, N,N,N-tripropyl-, bromide
            + methane                          vol 27/28          E61,    95
Propane
            + hydrogen sulfide                 vol 32            E166,    191
            + mercury                          vol 29            E192,    200
            + methane                          vol 27/28   E266,  267 - 280
Propane (multi component)
            + methane                          vol 27/28         194,    195
                                                                 378 - 382
                                                                 387 - 406
                                                                 553 - 556
Propane, 2,2-dimethyl-
            + methane                          vol 27/28   E314,  315 - 316
            + water                            vol 37                     30
Propane, 2,2-dimethyl- (ternary)
            + methane                          vol 27/28                 410
Propane, 2-methyl-
            + hydrogen sulfide                 vol 32            E166, E169
                                                                 194,    195
            + methane                          vol 27/28   E297,  298 - 302
Propane, 2-methyl-(multicomponent)
            + methane                          vol 27/28         485 - 490
Propane, nitro-
            + phosphine                        vol 21      E281, E282,  290
Propane, 1,1'-oxybis-
            + mercury                          vol 29            E172,   174
            + methane                          vol 27/28         E641,   647
Propanenitrile
            + arsine                           vol 21            E302,   303
1,2,3-Propanetriol
            + ammonia                          vol 21              E7,    53
            + butane                           vol 24      E331 -E334,   354
            + N,N-dimethylmethanamine          vol 21            E85,   E86
                                                                 E91,   197
            + methanamine                      vol 21      E80,  102,   103
            + N-methylmethanamine              vol 21      E82,  150,   151
            + 1-propanamine                    vol 21            E252,   267
1,2,3-Propanetriol (aqueous)
            + hydrogen sulfide                 vol 32            E20,     24
1,2,3-Propanetriol, triacetate
            + ammonia                          vol 21              E9,    65
            + bromic acid, potassium salt      vol 30                    243
            + chloric acid, potassium salt     vol 30            E108,   153
            + methane                          vol 27/28                 679
```

```
1-Propanol
          + ammonia                          vol 21              E5,     38,
                                                                 43,     E90
          + bromic acid, potassium salt      vol 30                      243
          + N,N-dimethylmethanamine          vol 21                      188
          + N-methylmethanamine              vol 21      E82,   146,     147
1-Propanol (aqueous)
          + methane                          vol 27/28                   175
1-Propanol (multicomponent)
          + copper (II) hydroxide            vol 23      E10 - E26,      70
2-Propanol
          + ammonia                          vol 21       E5,    38,     44
          + butane                           vol 24      E331 -E334,    356
          + mercury                          vol 29      E164, E165,    169
          + methane                          vol 27/28         E588,  E589,
                                                                610,    636
          + 2-methylpropane                  vol 24      E331 -E334,    368
          + propane                          vol 24      E331 -E334,    340
iso-Propanol
       see 2-propanol
iso-Propanolamine
       see 2-propanol, 1-amino-
2-Propanol, 1,1´-iminobis- (aqueous)
          + hydrogen sulfide                 vol 32              E80,    116
2-Propanol, 1,1´-iminobis- (multicomponent)
          + hydrogen sulfide                 vol 32              E82,    117,
                                                                160 -   165
1-Propanol, 2-methyl-
          + ammonia                          vol 21       E6,    45,     46
          + methane                          vol 27/28         E590,    614
2-Propanol, 2-methyl-
          + ammonia                          vol 21                      46
2-Propanol, 1,1´-oxybis-
          + methane                          vol 27/28   E641, E642,    676
1-Propanol, oxybis-
          + butane                           vol 24      E375 -E378,    395
          + hydrogen sulfide                 vol 32      E173, E174,    250
          + 2-methylpropane                  vol 24      E331 -E334,    372
          + propane                          vol 24      E375 -E378,    388
2-Propanone
          + ammonia                          vol 21               E7,    E8,
                                                                 53,     55
          + arsine                           vol 21             E302,    308
          + chloric acid, lithium salt       vol 30                      23
          + chloric acid, sodium salt        vol 30                      100
          + N,N-dimethylmethanamine          vol 21              E85,    196
          + hydrogen sulfide                 vol 32      E176, E178,    274
          + iodic acid, lithium salt         vol 30                      329
          + mercury                          vol 29      E192,  205,    206
          + methane                          vol 27/28         580,  E655,
                                                                656 -   659
          + phosphine                        vol 21      E281, E282,    290
          + propane                          vol 24      E125,  126,    128,
                                                                E375 -E378,
                                                                381,    382
2-Propanone (aqueous)
          + bromic acid, potassium salt      vol 30                      247
          + chloric acid, potassium salt     vol 30      E108,  154 -   156
          + perchloric acid, sodium salt     vol 31              97,    125
          + phosphoric acid, disodium salt
                                             vol 31             124,    125
          + phosphoric acid, monosodium salt
                                             vol 31      E51,    96,     97
          + phosphoric acid, trisodium salt
                                             vol 31      E129,  166,    167
2-Propanone (multicomponent)
          + copper (II) hydroxide            vol 23      E10 - E26,      67
Propene, 1,2-epoxy-
          + phosphine                        vol 21      E281, E282,    290
1-Propene (ternary)
          + propane                          vol 24                      256
1-Propenylcyclopentane see cyclopentane, 1-propenyl-
Propionic acid
       see propanoic acid
```

```
Propionitrile
        see propanenitrile
iso-Propanol
        see 2-propanol
Propyl acetate
        see acetic acid, propyl ester
iso-Propyl alcohol
        see 2-propanol
N-Propylamine
        see 1-propanamine
Propylbenzene
        see benzene, propyl-
iso-Propylbenzene
        see benzene, (1-methylethyl)-
Propylene
        see 1-propene
Propylene carbonate
        see 1,3-dioxolan-2-one, 4-methyl-
Isopropylcyclohexane
        see cyclohexane, (1-methylethyl)-
Propylcyclohexane
        see cyclohexane, propyl-
Propylcyclopentane
        see cyclopentane,propyl-
Isopropylcyclopentane
        see cyclopentane, methylethyl-
Propylene glycol
        see 1,2-Propanediol
iso-Propyl ether
        see propane, 2,2´-oxybis-
Propylethyl carbinol
        see 3-hexanol
iso-Propylethyl carbinol
        see 3-pentanol, 2-methyl-
Propyl nitrate
        see nitric acid, propyl ester
N1-(5-Propyl-1,3,4-thiadiazol-2-yl)sulfanilamide
        see benzenesulfonamide, 4-amino-N-(5-propyl-
              1,3,4-thiadiazol-2-yl)-
N1-(5-Isopropyl-1,3,4-thiadiazol-2-yl)sulfanilamide
        see benzenesulfonamide, 4-amino-N-[5-(1-methylethyl)-
              1,3,4-thiadiazol-2-yl]-
2-Propyl-1,3,5-trimethylbenzene
        see benzene, 1,3,5-trimethyl-2-propyl-
Proseptal
        see benzenesulfonamide, 4-amino-
Proseptine
        see benzenesulfonamide, 4-amino-
Proseptol
        see benzenesulfonamide, 4-amino-
Protonsil red
        see benzenesulfonamide, 4-[(2,4-diaminophenyl)azo]-
Protonsil
        see benzenesulfonamide, 4-[(2,4-diaminophenyl)azo]-
Protactinium
            + mercury                         vol 25                    E421
Pyralcid
        see benzenesulfonamide, 4-amino-N-(4-methyl-2-pyrimidinyl)-
Pyran, 2,3-dihydro-
            + methane                         vol 27/28         E643, 653
2H-Pyran, tetrahydro-
            + methane                         vol 27/28         E643, 653
N1-(Pyrazinyl)sulfanilamide
        see benzenesulfonamide, 4-amino-N-pyrazinyl-
N1-(Pyrazinyl)sulfanilamide, monosodium salt
        see benzenesulfonamide, 4-amino-N-pyrazinyl-, monosodium salt
N1-2-Pyrazinylsulfanilamide
        see benzenesulfonamide, 4-amino-N-pyrazinyl-
(N1-Pyrazinylsulfanilamido)sodium
        see benzenesulfonamide, 4-amino-N-pyrazinyl-, monosodium salt
1-H-Pyrazole, 1,3-dimethyl-
            + hydrogen sulfide                vol 32         E180, E182, 301
1-H-Pyrazole, 3-methyl-
            + hydrogen sulfide                vol 32         E180 -E182, 301
```

```
Pyrene
            + water                              vol 38              E446 -E448,
                                                                     449 - 455
Pyrene (aqueous)
            + sodium chloride                    vol 38                  458,   459
            + salt water                         vol 38                         457
            + seawater                           vol 38              E456,   460
Pyriamid
        see benzenesulfonamide, 4-amino-N-2-pyridinyl-
Pyricardyl
        see 3-pyridinecarboxamide, N,N-diethyl-
N1-3-Pyridazinylsulfanilamide
        see benzenesulfonamide, 4-amino-N-3-pyridazinyl-
Pyridazol
        see benzenesulfonamide, 4-amino-N-2-pyridinyl-
Pyridine
            + N,N-dimethylmethanamine            vol 21                         210
            + hydrogen sulfide                   vol 32              E180 -E182, 301
            + methane                            vol 27/28                      707
            + methanamine                        vol 21              E81,   117
            + N-methylmethanamine                vol 21              E83,   160
Pyridine (aqueous)
            + bromic acid, potassium salt        vol 30                         251
Pyridine (multicomponent)
            + copper (II) hydroxide              vol 23              E10 - E26,   77
Pyridine, 2,5-bis[[(4-aminophenyl)sulfonyl]amino]-
        see benzenesulfonamide, N,N´-(2,5-pyridinediyl)bis[4-amino-
1(2H)-Pyridineacetic acid, 2-[[(4-aminophenyl)sulfonyl]amino]-
            + water                              vol ??                          82
Pyridine-3-carboxdiethylamide
        see 3-pyridinecarboxamide, N,N-diethyl-
Pyridine-3-carboxylic acid diethylamide
        see 3-pyridinecarboxamide, N,N-diethyl-
N-[4-[(2-Pyridinylamino)sulfonyl]phenyl]acetamide
        see acetamide, N-[4-[(2-pyridinylamino)sulfonyl]phenyl]-
4-[(2-Pyridylamino)sulfonyl]aniline
        see benzenesulfonamide, 4-amino-N-2-pyridinyl-
N1-2(1H)-Pyridylidenesulanilamide
        see benzenesulfonamide, 4-amino-N-2-pyridinyl-
4´-(2-Pyridylsulfamoyl)acetanilide
        see acetamide, N-[4-[(2-pyridinylamino)sulfonyl]phenyl]-
N1-3-Pyridylsulfanilamide
        see benzenesulfonamide, 4-amino-N-3-pyridinyl-
N1-2-Pyridylsulfanilamide
        see benzenesulfonamide, 4-amino-N-2-pyridinyl-
(N1-2-Pyridylsulfanilamido)sodium
        see benzenesulfonamide, 4-amino-N-2-pyridinyl-, monosodium salt
N1-2-Pyridylsulfapyridine, monosodium salt
        see benzenesulfonamide, 4-amino-N-2-pyridinyl-, monosodium salt
Pyrimal
        see benzenesulfonamide, 4-amino-N-2-pyrimidinyl-
Pyrimal M
        see benzenesulfonamide, 4-amino-N-(4-methyl-2-pyrimidinyl)-
N-[4-[(2-Pyrimidinylamino)sulfonyl]phenyl]acetamide
        see acetamide, N-[4-[(2-pyrimidinylamino)sulfonyl]phenyl]-
N1-2(1H)-Pyrimidinylidenesulfanilamide
        see benzenesulfonamide, 4-amino-N-2-pyrimidinyl-
4´-(2-Pyrimidinylsulfamoyl)acetanilide
        see acetamide, N-[4-[(2-pyrimidinylamino)sulfonyl]phenyl]-
N1-2-Pyrimidinylsulfanilamide, monosilver(1+) salt
        see silver, (4-amino-N-2-pyrimidinylbenzenesulfonamidato-NN-01)-
N1-2-Pyrimidinylsulfanilamide
        see benzenesulfonamide, 4-amino-N-2-pyrimidinyl-
N1-4-Pyrimidinylsulfanilamide
        see benzenesulfonamide, 4-amino-N-4-pyrimidinyl-
N1-5-Pyrimidinylsulfanilamide
        see benzenesulfonamide, 4-amino-N-5-pyrimidinyl-
N1-2-Pyrimidinylsulfanilamide
        see benzenesulfonamide, 4-amino-N-2-pyrimidinyl-
(N1-2-Pyrimidinylsulfanilamido)sodium
        see benzenesulfonamide, 4-amino-N-2-pyrimidinyl-, monosodium salt
N1-2-Pyrimidylsulfanilamide
        see benzenesulfonamide, 4-amino-N-2-pyrimidinyl-
```

```
Pyrone, dimethyl-
          + bromic acid, potassium salt     vol 30                            252
Pyrrolidine
          + methane                         vol 27/28                         707
Pyrrolidine, 1-methyl-
          + methane                         vol 27/28      E712,   713 - 718
Pyrrolidone, 1-methyl-(ternary)
          + methane                         vol 27/28        97,   716 - 718
2-Pyrrolidinone, 1-methyl-
          + ammonia                         vol 21          E8,    56 -  58
          + butane                          vol 24        E299 -E302,   320
          + hydrogen sulfide                vol 32        E180, E181,   259
                                                                      305 - 313
          + 2-methylpropane                 vol 24        E299 -E302,   328
          + propane                         vol 24        E299 -E302,   306
Pysococcine
        see benzenesulfonamide, 4-amino-
```

Q

Quinoline

+ ammonia	vol 21	E11,	75
+ butane	vol 24	E299 -E302,	322
+ N,N-dimethylmethanamine	vol 21		210
+ hydrogen sulfide	vol 32	E180 -E182,	299
+ methanamine	vol 21	E81,	117
+ methane	vol 27/28	710,	710
+ N-methylmethanamine	vol 21	E83,	160
+ propane	vol 24	E299 -E302,	313

Quinoseptyl

      see benzenesulfonamide, 4-amino-N-(6-methoxy-3-pyridazinyl)-

R

Ro 4-3506
        see benzenesulfonamide, 4-amino-N-(6-methoxy-4-pyrimidinyl)-
Ro 4-4426
        see benzenesulfonamide, 4-amino-N-(5,6-dimethoxy-4-pyrimidinyl)-
RP 146
        see benzenesulfonamide, 4-amino-N-(4-methyl-2-thiazolyl)-
RP 2145
        see benzenesulfonamide, 4-amino-N-(5-methyl-
            1,3,4-thiadiazol-2-yl)-
RP 2254
        see benzenesulfonamide, 4-amino-N-[5-(1-methylethyl)-
            1,3,4-thiadiazol-2-yl]-
RP 2255
        see benzenesulfonamide, 4-amino-N-[(aminothioxomethyl)-
RP 2275
        see benzenesulfonamide, 4-amino-N-[(aminoiminomethyl)-
RP 2652
        see benzenesulfonamide, 4-amino-N-(4-methyl-2-pyrimidinyl)-
Rabbit brain and blood and saline solution
            + methane   vol 27/28   731
Rabbit blood
        see also blood,rabbit
Rabbit brain
        see also brain,rabbit
Radium
            + mercury   vol 25   E421
Radonil
        see benzenesulfonamide, 4-amino-N-(5-methyl-3-isoxazolyl)-
Rat abdominal muscle
            + methane   vol 27/28   735
Rat blood (venous)
        see Blood, rat (venous)
Raziosulfa
        see benzenesulfonamide, 4-amino-N-(1-phenyl-
            1H-pyrazol-5-yl)-
Red blood cells
        see Human red blood cells
Red cell membrane
        see Human red cell membrane
Red streptocide
        see benzenesulfonamide, 4-[(2,4-diaminophenyl)azo]-
Reformin
        see 3-pyridinecarboxamide, N,N-diethyl-
Refresh´n
        see 1H-purine-2,6-dione, 3,7-dihydro-1,3,7-trimethyl-
Region
        see acetamide, N-[(4-aminophenyl)sulfonyl]-
Rehormin
        see 3-pyridinecarboxamide, N,N-diethyl-
Relbapiridine
        see benzenesulfonamide, 4-amino-N-2-pyridinyl-
Relbazon
        see benzenesulfonamide, 4-[(2,4-diaminophenyl)azo]-
Resorcinol dimethyl ester
        see Benzene, 1,3-dimethoxy-
Resulfon
        see benzenesulfonamide, 4-amino-N-[(aminoiminomethyl)-
Retamid
        see benzenesulfonamide, 4-amino-N-(6-methoxy-3-pyridazinyl)-
Retasulfin
        see benzenesulfonamide, 4-amino-N-(6-methoxy-3-pyridazinyl)-
Retasulfine
        see benzenesulfonamide, 4-amino-N-(6-methoxy-3-pyridazinyl)-
Retina
        see Rabbit retina
Rhenium
            + mercury   vol 25   E300
                                                vol 29   E214, 217
Rhodium
            + mercury   vol 25   E314, 315
                                                vol 29   E214, 217

Rolsul
          see butanoic acid, 4-oxo-4-[[4-[(2-thiaxolylamino)-
          sulfonyl]phenyl]amino]-
Romezin
          see benzenesulfonamide, 4-amino-N-(4-methyl-2-pyrimidinyl)-
Ronin
          see benzenesulfonamide, 4-amino-N-2-pyridinyl-
Roxosul
          see benzenesulfonamide, 4-amino-N-(3,4-dimethyl-5-isoxazolyl)-
Rubiazol A
          see benzenesulfonamide, 4-amino-
Rubiazol I
          see benzenesulfonamide, 4-[(2,4-diaminophenyl)azo]-
Rubidium
               + mercury   vol 25   E44,   E45,
                                                                   46 -   49
Rubidium bromate
          see bromic acid, rubidium salt
Rubidium chloride (aqueous)
               + cesium chloride   vol 30   E164,   174,
                                                                   175, E178
               + chloric acid, cesium salt   vol 30   E164,   174,
                                                                   175, E178
               + chloric acid, rubidium salt vol 30   E110, 139,   140,
                                                         E164, 171,   172,
                                                          174, 175, E178
               + chloric acid, potassium salt vol 30   E164
               + chloric acid, sodium salt   vol 30   E164
               + phosphoric acid, monorubidium salt
                                               vol 31   E307,   314
               + potassium chloride   vol 30   E110,   139
                                                                   140, E164
               + sodium chloride   vol 30   E164
Rubidium chlorate
          see chloric acid, rubidium salt
Rubidium dihydrogen phosphate
          see phosphoric acid, monorubidium salt
Rubidium hydroxide (aqueous)
               + iodic acid, rubidium salt   vol 30   E428, E430,   437
Rubidium iodate
          see iodic acid, rubidium salt
Rubidium nitrate
          see nitric acid, rubidium salt
Rubidium oxide (aqueous)
               + phosphoric acid, monorubidium salt
                                               vol 31   E308, E310,
                                                          315,   316,   317
               + phosphorus pentoxide   vol 31   315
Rubidium oxide (aqueous-d2)
               + phosphorus pentoxide   vol 31   315
Rubidium perchlorate
          see perchloric acid, rubidium salt
Rubidium sulfate
          see Sulfuric acid, rubidium salt
Rufol
          see benzenesulfonamide, 4-amino-N-(5-methyl-
          1,3,4-thiadiazol-2-yl)-
Ruocid
          see benzenesulfonamide, 4-amino-N-[(aminoiminomethyl)-
Ruthenium
               + mercury   vol 25   E307,   308
                                               vol 29   E214,   217

```
S
Salazodimethoxine
        see benzoic acid, 5-[[4-[[2,6-dimethoxy-4-pyrimidinyl)amino]-
            sulfonyl]-phenyl]azo]-2-hydroxy-
Saline solution
        see also sodium chloride (aqueous)
Salvacard
        see 3-pyridinecarboxamide, N,N-diethyl-
Salvoseptyl
        see benzenesulfonamide, 4-amino-N-[(aminothioxomethyl)-
Samarium
        + mercury                       vol 25              E230, E231,
                                                              232, 233
Samarium bromate
        see bromic acid, samarium salt
Samarium bromide
        + 1-butanamine                  vol 22                   228
        + 2-butanamine                  vol 22                   228
        + 2-butanamine, N-(1-methylpropyl)-
                                        vol 22                   228
        + butane, 1-methoxy-            vol 22                   225
        + decane, 1-methoxy-            vol 22                   225
        + 1,4-dioxane                   vol 22                   227
        + ethane, 1,2-diethoxy-         vol 22                   224
        + furan, tetrahydro-            vol 22                   226
        + heptane, 1-methoxy-           vol 22                   225
        + nonane, 1-methoxy-            vol 22                   225
        + octane, 1-methoxy-            vol 22                   225
        + pentane, 1-methoxy-           vol 22                   225
        + 1-propanamine                 vol 22                   228
        + 2-propanamine                 vol 22                   228
Samarium chloride
        + 1,3-butadiene, hexachloro-    vol 22                   208
        + methanol                      vol 22                   209
        + 1,4-dioxane                   vol 22                   218
        + 1,3-dioxolane                 vol 22                   218
        + ethane, 1-ethoxy-2-methoxy-   vol 22                   218
        + ethane, 1,1´-oxybis-          vol 22                   217
        + 1,2-ethanediol                vol 22                   213
        + ethanol                       vol 22                   210
        + ethanol (aqueous)             vol 22                   211
        + ethanol, 2-ethoxy-            vol 22              215, 216
        + ethanol, 2-methoxy-           vol 22              214, 216
        + furan, tetrahydro-            vol 22                   219
        + phosphoric acid, tributyl ester
                                        vol 22                   220
        + phosphoric triamide, hexamethyl-
                                        vol 22                   222
        + phosphoryl chloride           vol 22                   223
        + 2-propanamine                 vol 22                   221
        + 1-propanol                    vol 22                   213
        + 2-propanol                    vol 22                   212
        + 2-propen-1-amine              vol 22                   221
        + 2-propen-1-ol                 vol 22                   213
        + stannane, tetrachloro-        vol 22                   223
        + water (ternary)               vol 22                   211
Samarium cobalt nitrate
        see nitric acid, cobalt samarium salt
Samarium fluoride
        + acidic nitrosyl fluoride (nitrosyl fluoride compound
          with hydrofluoric acid )      vol 22                   207
        + butane, 1-(chloromethoxy)-    vol 22                   204
        + decane, 1-methoxy-            vol 22                   204
        + ethanol                       vol 22                   203
        + methanol                      vol 22                   203
        + methane, 1,1-sulfinylbis-     vol 22                   205
        + pyridine                      vol 22                   206
Samarium iodide
        + formamide, N,N-dimethyl-      vol 22                   229
Samarium manganese nitrate
        see nitric acid, manganese samarium salt
Samarium nickel nitrate
        see nitric acid, nickel samarium salt
```

```
Samarium nitrate trihydrate
         see nitric acid, samarium salt (trihydrate)
Samarium nitrate
         see nitric acid, samarium salt
Samarium zinc nitrate
         see nitric acid, samarium zinc salt
Sanamid
         see benzenesulfonamide, 4-amino-
Sanasil
         see benzenesulfonamide, 4-amino-N-(5,6-dimethoxy-4-pyrimidinyl)-
Sancora
         see 3-pyridinecarboxamide, N,N-diethyl-
Sanodiazine
         see benzenesulfonamide, 4-amino-N-2-pyrimidinyl-
Sanotiazol
         see benzenesulfonamide, 4-amino-N-2-thiazolyl-
Santowax R
                 + methane                    vol 27/28              567
Scandium
                 + mercury                    vol 25               E206
Scandium bromide
                 + 1,4-dioxane                vol 22                19
Scandium chloride
                 + acetic acid, ethyl ester   vol 22                14
                 + acetonitrile               vol 22                18
                 + 2-butanamine               vol 22                16
                 + butane, 1-methoxy-         vol 22                12
                 + 1-butanol                  vol 22                 4
                 + ethanamine, N,N-diethyl-   vol 22                15
                 + ethanamine, N-ethyl-       vol 22                15
                 + ethane, 1,2-diethoxy-      vol 22                11
                 + ethanol                    vol 22                 2
                 + ethanol, 2-ethoxy-         vol 22                10
                 + formamide, N,N-dimethyl-   vol 22                17
                 + furan, tetrahydro-         vol 22                13
                 + 1-heptanol                 vol 22                 7
                 + 1-hexanol                  vol 22                 6
                 + methanol                   vol 22                 1
                 + 1-nonanol                  vol 22                 9
                 + 1-octanol                  vol 22                 8
                 + octanamine, N,N-dioctyl-   vol 22                15
                 + pentanamine, N-pentyl-     vol 22                15
                 + 1-pentanol                 vol 22                 5
                 + 1-propanamine, 2-methyl-N-(2-methylpropyl)-
                                              vol 22                16
                 + 1-propanol                 vol 22                 3
Scandium nitrate
         see nitric acid, scandium salt
SDA
         see benzenesulfonamide, 4-amino-N-2-pyrimidinyl-
SDS
         see sulfuric acid, monododecyl ester sodium salt
Sea water
                 + butane                     vol 24      E70 - E72,  120
                 + hydrogen sulfide           vol 32      E51,  52,   53
                 + mercury                    vol 29                  45
                 + methane                    vol 27/28        E50,  E51,
                                                              52 - 55
                 + sulfurous acid, calcium salt  vol 26    E191, E192,
                                                           226 - 228
Selenic acid, disodium salt (aqueous)
                 + selenious acid, disodium salt vol 26     311 - 313
Selenic acid, potassium salt (aqueous)
                 + phosphoric acid, monopotassium salt
                                              vol 31      E208,  248

Selenious acid (aqueous)
                 + selenious acid, dipotassium salt
                                              vol 26       329,  330
                 + selenious acid, disodium salt vol 26    318,  319
Selenious acid (aqueous and multicomponent)
                 + perchloric acid, sodium salt  vol 26    382 - 387
                 + selenious acid, silver salt   vol 26    382 - 387
```

Selenious acid, ammonium salt (2:1)
            + water   vol 26   E331,   344,
                                                                     346,   347
Selenious acid, barium salt
            + water   vol 26   E348, E349,
                                                                     362 - 366
Selenious acid, barium salt (aqueous)
            + selenium dioxide   vol 26   365,   366
Selenious acid, beryllium salt
            + water   vol 26   E348, E349,   353
Selenious acid, cadmium salt
            + water   vol 26   E348, E350,
                                                                     391 - 393
Selenious acid, cadmium salt (aqueous)
            + nitric acid   vol 26   392
            + sulfuric acid   vol 26   392
Selenious acid, calcium salt
            + water   vol 26   E348, E349,
                                                                     356,   357
Selenious acid, calcium salt (aqueous)
            + hydrochloric acid   vol 26   357
            + nitric acid   vol 26   357
Selenious acid, cobalt (II) salt
            + water   vol 26   E348, E350,
                                                                     370 - 373
Selenious acid, cobalt (II) salt (aqueous and multicomponent)
            + nitric acid   vol 26   371
            + nitric acid, potassium salt   vol 26   373
            + selenious acid, diammonium salt
                                            vol 26   373
            + sulfuric acid   vol 26   371
Selenious acid, copper (II) salt
            + water   vol 26   E348, E350,
                                                                     377 - 379
Selenious acid, copper (II) salt (aqueous)
            + acetic acid   vol 26   379
            + hydrochloric acid   vol 26   379
            + nitric acid   vol 26   378
            + sulfuric acid   vol 26   378,   379
Selenious acid, copper (II) salt (aqueous and multicomponent)
            + ammonia   vol 26   379
Selenious acid, diammonium salt
            + water   vol 26   E331,   342,   343

Selenious acid, diammonium salt (aqueous and multicomponent)
            + nitric acid, potassium salt   vol 26   373
            + selenious acid, cobalt (II) salt
                                            vol 26   373
Selenious acid, dipotassium salt
            + water   vol 26   E322 -E324,
                                                                     325 - 330
Selenious acid, dipotassium salt (aqueous)
            + ethanol   vol 26   328
            + selenious acid   vol 26   329,   330
            + sulfurous acid, dipotassium salt
                                            vol 26   327
Selenious acid, disodium salt
            + water   vol 26   E303 -E305,
                                                                     306 - 319
Selenious acid, disodium salt (aqueous)
            + carbonic acid, disodium salt vol 26   314,   315
            + ethanol   vol 26   310
            + selenic acid, disodium salt   vol 26   311 - 313
            + selenious acid   vol 26   318,   319
            + sulfurous acid, disodium salt vol 26   309
            + tellurous acid, sodium salt   vol 26   316,   317
Selenious acid, disodium salt (aqueous and multicomponent)
            + nitric acid, sodium salt   vol 26   400
            + selenious acid, mercury (II) salt
                                            vol 26   398,   400
Selenious acid, iron (III) salt
            + water   vol 26   E348, E350,   369

Selenious acid, iron (III) salt (aqueous)
            + nitric acid   vol 26   368
            + sulfuric acid   vol 26   368
Selenious acid, lead (II) salt
            + water   vol 26   E348, E351,
                                                                     401,   402
Selenious acid, lead (II) salt (aqueous)
            + hydrochloric acid   vol 26   401
            + nitric acid   vol 26   401
Selenious acid, lithium salt
        + water   vol 26   E331,   332
Selenious acid, magnesium salt
            + water   vol 26   E348, E349,
                                                                     354,   355
Selenious acid, magnesium salt (aqueous)
            + hydrochloric acid   vol 26   E348, E349,   355
            + nitric acid   vol 26   E348, E349,   355
Selenious acid, manganese (II) salt (aqueous)
            + hydrochloric acid   vol 26   367
            + sulfuric acid   vol 26   367
Selenious acid, mercury (I) salt
            + water   vol 26   E348, E350,
                                                                     394 - 396
Selenious acid, mercury (I) salt (aqueous)
            + nitric acid   vol 26   394,   395
            + sulfuric acid   vol 26   394,   395
Selenious acid, mercury (II) salt
            + water   vol 26   E348, E351,
                                                                     397 - 400
Selenious acid, mercury (II) salt (aqueous and multicomponent)
            + nitric acid   vol 26   399
            + nitric acid, sodium salt   vol 26   399
            + selenious acid, disodium salt vol 26   398,   400
Selenious acid, nickel salt
            + water   vol 26   E348, E350,
                                                                     374 - 376
Selenious acid, nickel salt (aqueous)
            + nitric acid   vol 26   375
            + selenium dioxide   vol 26   376
            + sulfuric acid   vol 26   375
Selenious acid, silver salt (aqueous)
            + nitric acid   vol 26   380
            + sulfuric acid   vol 26   380
Selenious acid, silver salt (aqueous and multicomponent)
            + perchloric acid, sodium salt vol 26   382 - 387
            + selenious acid   vol 26   382 - 387
Selenious acid, silver salt
            + water   vol 26   E348, E350,
                                                                     380 - 388
Selenious acid, strontium salt
            + water   vol 26   E348, E349,
                                                                     358 - 361
Selenious acid, strontium salt (aqueous)
            + hydrochloric acid   vol 26   E348, E349,   359
            + nitric acid   vol 26   E348, E349,   359
Selenious acid, zinc salt
            + water   vol 26   E348, E350,
                                                                     389,   390
Selenious acid, zinc salt (aqueous)
            + nitric acid   vol 26   389
            + sulfuric acid   vol 26   389
Selenium dioxide
            + water   vol 26   320,   321
Selenium dioxide (aqueous)
            + selenious acid, barium salt   vol 26   365,   366
            + selenious acid, nickel salt   vol 26   376
            + sodium hydroxide   vol 26   320,   321
Selenium (IV) oxide (aqueous)
            + silver (I) oxide   vol 23   E83, E84, E92,
                                                                            134
Selenium oxide (aqueous)
            + copper (II) hydroxide   vol 23   E10 - E26,   62
N1-Senecioylsulfanilamide
        see 2-butenamide, N-[(4-aminophenyl)sulfonyl]-3-methyl-

```
Sepasolv MPE
            + hydrogen sulfide              vol 32      E173 -E175,  262
Septacil
        see benzenesulfonamide, 4-amino-N-(4-methyl-2-pyrimidinyl)-
Septamide album
        see benzenesulfonamide, 4-amino-
Septanilam
        see benzenesulfonamide, 4-amino-
Septicid-6
        see benzenesulfonamide, 4-(aminomethyl)-
Septinal
        see benzenesulfonamide, 4-amino-
Septipulmon
        see benzenesulfonamide, 4-amino-N-2-pyridinyl-
Septolix
        see benzenesulfonamide, 4-amino-
Septoplex
        see benzenesulfonamide, 4-amino-
Septoplix
        see benzenesulfonamide, 4-amino-
Serum albumen
        see bovine serum albumen
SETD
        see benzenesulfonamide, 4-amino-N-(5-methyl-
          1,3,4-thiadiazol-2-yl)-

Sethadil
        see benzenesulfonamide, 4-amino-N-(5-methyl-
          1,3,4-thiadiazol-2-yl)-
SH 613
        see benzenesulfonamide, 4-amino-N-(5-methoxy-2-pyrimidinyl)-
Silane, triethoxy-
            + hydrogen sulfide              vol 32      E185,   277,  278
            + hydrogen selenide             vol 32      E330,   334,  335
Silane, tetramethyl-
            + ammonia                       vol 21              E11,   79
Silane, triethoxy-
            + ammonia                       vol 21               E9,   66
            + arsine                        vol 21      E302,   306,  307
            + phosphine                     vol 21      E281, E282,  296,
                                                                     297
            + stannane                      vol 21      E315,   327,  328
            + stibine                       vol 21      E310,   311,  312
Silicic acid, disodium salt
            + water                         vol 31                    143
Silicic acid, disodium salt (aqueous)
            + phosphoric acid, disodium salt
                                            vol 31              E98,  114
            + phosphoric acid, trisodium salt
                                            vol 31             E129,  143,
                                                                     153,  165
Silicic acid, tetraethyl ester
            + arsine                        vol 21      E302,   306,  307
            + germane                       vol 21      E315,   323,  324
            + hydrogen sulfide              vol 32      E185,   277,  278
            + hydrogen selenide             vol 32      E330,   334,  335
            + phosphine                     vol 21      E281, E282,  296,
                                                                     297
            + silane                        vol 21      E315,   318,  319
            + stannane                      vol 21      E315,   327,  328
            + stibine                       vol 21      E310,   311,  312
Silicon
            + mercury                       vol 25       +             E134
                                            vol 29      E213, E214,  219
Silicone oils
            + hydrogen sulfide              vol 32      E185,   277,  278
Silicone PFMS-4F
            + arsine                        vol 21      E302,   304,  305
            + germane                       vol 21      E315,   321,  322
            + phosphine                     vol 21              298,  299
            + silane                        vol 21      E315,   316,  317
            + stannane                      vol 21      E315,   325,  326
```

```
Silicone VKZH-94B
        + arsine                    vol 21        E302,  306,   307
        + germane                   vol 21        E315,  323,   324
        + phosphine                 vol 21               296,   297
        + silane                    vol 21        E315,  318,   319
        + stannane                  vol 21        E315,  327,   328
        + stibine                   vol 21        E310,  311,   312
Silicone 702-DF
        + arsine                    vol 21        E302,  306,   307
        + germane                   vol 21        E315,  323,   324
        + phosphine                 vol 21               296,   297
        + silane                    vol 21        E315,  318,   319
        + stannane                  vol 21        E315,  327,   329
Silicon tetrachloride
        see silane, tetrachloro-
Silvadene
        see silver, (4-amino-N-2-pyrimidinylbenzenesulfonamidato-NN-01)-

Silver
        + mercury                   vol 25              E357 -E359,
                                                         360 - 368
                                    vol 29        E228, E214,  222
        + water                     vol 29        E228,  231 - 233
Silver, (4-amino-N-2-pyrimidinylbenzene-sulfonamidato-NN,O)-
bis(1-imidazole-N3)-,(T-4)-
        + water                     vol 36                     223
Silver (I) hydroxide
        + water                     vol 23                     106
Silver nitrate
        see nitric acid, silver salt
Silver (I) oxide
        + water                     vol 23              E83 - E97,
                                                  98 - 105,  107
Silver (I) oxide (aqueous)
        + ammonium hydroxide        vol 23        E83,  E84,  E93,
                                                         109,  116
        + barium hydroxide          vol 23        E83,  E84,  E88,
                                                               119
        + boron (III)               vol 23        E83,  E84,  E92,
                                                         130 - 133
        + chromium (VI)             vol 23        E83,  E84,  E92,
                                                               123
        + ethanol                   vol 23        E83,  E84,  E94,
                                                               111
        + ethanol, 2-amino-         vol 23        E83,  E84,  E93,
                                                               124
        + ethanol, 2,2´-imino-      vol 23        E83,  E84,  E93,
                                                               125
        + ethanol, 2,2´,2"-nitrilotri-  vol 23   E83,  E84,  E93,
                                                               126
        + methanamine               vol 23        E83,  E84,  E93,
                                                               110
        + nitric acid, barium salt     vol 23                  122
        + nitric acid, potassium salt  vol 23    E83,  E84,  E93,
                                                  112,  113,  120
        + potassium hydroxide       vol 23        E83,  E84,  E86,
                                                         118,  129
        + selenium (IV) oxide       vol 23        E83,  E84,  E92,
                                                               134
        + sodium hydroxide          vol 23        E83,  E84,  E86,
                                                  108,  114,  115,
                                                  117,  127,  128,
                                                         135 - 137
        + sulfuric acid, potassium salt vol 23                 121
Silver (I) oxide (multicomponent)
        + ammonium hydroxide        vol 23              142 - 144
        + nitric acid, potassium salt  vol 23           142 - 144
        + nitric acid, sodium salt  vol 23                     138
        + perchloric acid, sodium salt  vol 23          139,  141
        + potassium hydroxide       vol 23                     140
        + sodium hydroxide          vol 23        138,  139,  141
        + zinc oxide                vol 23                     140
Silver (II) oxide
        + water                     vol 23              E144, E145
```

```
Silver (II) oxide (aqueous)
        + potassium hydroxide          vol 23              E144, E145,
                                                            146,  147
Silver selenite
        see selenious acid, silver salt
Silver sulfadiazine
        see silver, (4-amino-N-2-pyrimidinylbenzenesulfonamidato-NN-01)-

Silver sulfamethazine
        see benzenesulfonamide, 4-amino-N-(4,6-dimethyl-2-pyrimidinyl)-,
            monosilver(1+) salt
Silver sulfamethiazole
        see benzenesulfonamide, 4-amino-N-(5-methyl-
            1,3,4-thiadiazol-2-yl)-, monosilver salt
Silver sulfamethizole
        see benzenesulfonamide, 4-amino-N-(5-methyl-
            1,3,4-thiadiazol-2-yl)-, monosilver salt
Silver sulfite
        see sulfurous acid, silver salt
Silver tellurite
        see tellurous acid, silver salt
Sinomin
        see benzenesulfonamide, 4-amino-N-(5-methyl-3-isoxazolyl)-
Slosul
        see benzenesulfonamide, 4-amino-N-(6-methoxy-3-pyridazinyl)-
SMOP
        see benzenesulfonamide, 4-amino-N-(6-methoxy-3-pyridazinyl)-
SMP2
        see benzenesulfonamide. 4-amino-N-(3-methoxypryrazinyl)-
Sodium
        + mercury                       vol 25              E13 - E18,
                                                             19 -  32
                                        vol 29        E213, E214,  218
Sodium acetate
        see acetic acid, sodium salt
Sodium aluminate
        see sodium aluminium oxide
Sodium aluminium oxide
        + water                         vol 31              159,  160
Sodium aluminium oxide (aqueous)
        + phosphoric acid, trisodium salt
                                        vol 31        E129, 159 - 162
        + sodium hydroxide              vol 31              159 - 162
        + sulfuric acid, disodium salt  vol 31              161,  162
        + vanadic acid, sodium salt     vol 31              159,  160
Sodium aminoacetate
        see glycine, monosodium salt
Sodium aminopropionate
        see DL-alanine, monosodium salt
Sodium bicarbonate
        see carbonic acid, monosodium salt
Sodium bisulfate
        see sulfuric acid, monosodium salt
Sodium bisulfite
        see sulfurous acid, monosodium salt
Sodium borate
        see boric acid, sodium salt
Sodium bromate
        see bromic acid, sodium salt
Sodium bromide
        + acetic acid, sodium salt      vol 33              219,  220
        + formic acid, sodium salt      vol 33                    218
Sodium bromide (aqueous)
        + bromic acid, sodium salt      vol 30        E196, 210 - 212,
                                                            216,  217
        + carbonic acid, disodium salt  vol 30                    216
        + carbonic acid, monosodium salt
                                        vol 30                    217
        + chloric acid, sodium salt     vol 30        E28,  E31,  E32,
                                                                   61
        + iodic acid, sodium salt       vol 30              E335, E336,
                                                            359 - 363
```

```
Sodium bromide (aqueous)
          + mercury                        vol 29                  E49,    66
          + methane                        vol 27/28       E75,   148,   149
          + propane                        vol 24          E58 - E72,   112
Sodium bromoacetate
       see acetic acid, bromo-, sodium salt
Sodium butanoate
       see butanoic acid, sodium salt
Sodium butyrate
       see butanoic acid, sodium salt
Sodium carbonate
       see carbonic acid, disodium salt
Sodium chlorate
       see chloric acid, sodium salt
Sodium chloride
          + acetic acid, sodium salt       vol 33                  257,   258
          + formic acid, sodium salt       vol 33                         228
          + water                          vol 31                   76 -   78,
                                                                   134,   135
Sodium chloride (aqueous)
          + ammonium chloride              vol 31          90,   119,   120
          + anthracene                     vol 38                407,   407
          + anthracene, 2-methyl-          vol 38                        436
          + benz [a] anthracene            vol 38                484,   485
          + benzene                        vol 37                190 - 192
          + benzo [e] pyrene               vol 38                        519
          + 1,1´-biphenyl                  vol 38                        353
          + bromic acid, potassium salt    vol 30                        229
          + bromic acid, sodium salt       vol 30                208,   209
          + butane                         vol 24          E58 - E72,
                                                                   106 - 110
          + carbonic acid, sodium salt     vol 30     E31,   E32,    82
          + cesium chloride                vol 30                E31, E178
          + chloric acid, cesium salt      vol 30                E31, E178
          + chloric acid, potassium salt   vol 30                E31, E108,
                                                                   E109,   121
          + chloric acid, rubidium salt    vol 30                E31, E164
          + chloric acid, sodium salt      vol 30     E28, E31, 51-58,
                                                            82 - 84, 86-89,
                                                                   E164, E178
          + chlorous acid, sodium salt     vol 30     E31,   E32,    83
                                                            84,    86 -   89
          + chrysene                       vol 38                        492
          + disulfurous acid, disodium salt
                                           vol 26                E74,    86
          + fluoranthene                   vol 38                        445
          + fluorene                       vol 38                        390
          + hexane                         vol 37                        368
          + hydrogen sulfide               vol 32     E25,   E26,    36,
                                                                    37,    39,
                                                                    41,    42
          + iodic acid, sodium salt        vol 30     E335, E336,   357,
                                                                          358
          + mercury                        vol 29     E48,    59 -   65
                                                      E87,    88 -   94
          + mercury (II) oxide             vol 23                        335
          + methane                        vol 27/28             E69 - E74,
                                                                   117 - 145
          + methylbenzene                  vol 37          427,   428,   430
          + methylcyclopentane             vol 37                        220
          + naphthalene                    vol 38                251 - 254,
                                                                   256,   257
          + naphthalene, 1-ethyl-          vol 38                        371
          + naphthalene, 1-methyl-         vol 38                        318
          + nitric acid, sodium salt       vol 31                        121
          + pentane                        vol 37                         62
          + phenanthrene                   vol 38                421 - 423
          + phenanthrene, 1-methyl-        vol 38                        439
          + phosphoric acid, disodium salt vol 31          E98, E101,   111,
                                                                   119 - 121
          + phosphoric acid, monoammonium salt
                                           vol 31                         90
```

```
Sodium chloride (aqueous)
        + phosphoric acid, monopotassium salt
                                        vol 31          80 -  85,   88,
                                                             89,   91
        + phosphoric acid, monosodium salt
                                        vol 31          E47,  E49,  E51,
                                                        E52,   65,   66,
                                                               76 -  78,
                                                               80 -  85,
                                                               88 -  91
        + phosphoric acid, trisodium salt
                                        vol 31          E127,  134,  135
        + potassium chloride            vol 30                      E31
                                        vol 31          80 -  85,   88,
                                                             89,   91
        + propane               vol 24                  E58 - E72
                                                        99 - 105
        + pyrene                vol 38                  459,  458
        + rubidium chloride     vol 30                  E31, E164
        + sulfurous acid, barium salt   vol 26  E240, E241,  246
        + sulfurous acid, strontium salt
                                vol 26                  E236,  239
        + tellurous acid, barium salt   vol 26          E404,  420
Sodium chloride (aqueous and multicomponent)
        + ammonium chloride     vol 26          E5,   66 -  70
                                vol 26          E74,   87,   88
        + disulfurous acid, diammonium salt
                                vol 26          E74,   87,   88
        + disulfurous acid, disodium salt
                                vol 26          E74,   87 -  89
        + sulfuric acid, disodium salt
                                vol 26          E5,   59 -  62
        + sulfurous acid, calcium salt
                                vol 26                  E191, E192,
                                                        224 - 228
        + sulfurous acid, diammonium salt
                                vol 26          E5,   66 -  70
        + sulfurous acid, disodium salt
                                vol 26                  E4,   E5,
                                                        34 -  38,
                                                        59 -  62,
                                                        66 -  70
Sodium chloride (in liquid ammonia)
        + iodic acid, potassium salt    vol 30                  423
Sodium chloride (in hydrochloric acid)
        + hydrogen sulfide      vol 32          E26,   36,   37
Sodium chloride (ternary and multicomponent)
        + cadmium oxide         vol 23          E280,  305,  306
        + copper (I) oxide      vol 23          E1 -  E3,    8
        + copper (II) oxide     vol 23          E10 - E26,   80
        + mercury (II) oxide    vol 23          E316,  340,  341
        + methane               vol 27/28       E74,  E75,  E77,
                                                E81,  113,  146,
                                                147,  156,  166
        + zinc oxide            vol 23                  E181,  268
                                                        231 - 234
Sodium chloride (molten)
        + mercury               vol 29                  E207,  209
Sodium cholate
        see cholan-24-oic acid, 3,7,12-trihydroxy-, monosodium salt,
        (3a,5B,7a,12a)
Sodium cyano-acetate
        see acetic acid, cyano-, sodium salt
Sodium dihydrogen phosphate
        see phosphoric acid, monosodium salt
Sodium dithionite (aqueous)
        + sulfurous acid, monosodium salt
                                vol 26                          85
Sodium dodecyl sulfate
        see sulfuric acid, monododecyl ester sodium salt
Sodium fluoride
        + water                 vol 31                          154
```

Sodium fluoride (aqueous)
          + chloric acid, sodium salt   vol 30   E28,   E31,
                                                                   E32,   50
          + mercury   vol 29   E47,   E48,   58
          + phosphoric acid, trisodium salt
                                               vol 31   E127, 154
Sodium gluconate
       see D-gluconic acid, monosodium salt
Sodium glutamate
       see DL-glutamic acid, disodium salt
Sodium glycolate
       see acetic acid, hydroxy-, monosodium salt
Sodium glycyl glycinate
       see glycine, N-glycyl-, sodium salt
Sodium hexanoate
       see hexanoic acid, sodium salt
Sodium hexylsulfate
       see sulfuric acid, monohexyl ester, sodium salt
Sodium hydrogen carbonate
       see carbonic acid, monosodium salt
Sodium hydrogen sulfite
       see sulfurous acid, monosodium salt
Sodium hydroxide (aqueous)
          + cadmium hydroxide   vol 23   E270 -E277,   286,
                                                                   287,   288
          + cadmium oxide   vol 23   E270 -E277,   286,
                                                            287,   296,   297
          + carbonic acid, disodium salt
                                               vol 31   157,   158
          + chloric acid, sodium salt   vol 30   E30,   42 -   49
          + gold (III) hydroxide   vol 23   E148 -E150,   153,
                                                                          154
          + mercury (II) oxide   vol 23   E307 -E313,   321,
                                                            324,   325,   330
          + phosphoric acid, disodium salt
                                               vol 31   E13,   E14,   18,
                                                                    19,   22,
                                                                    37,   38
          + phosphoric acid, monosodium salt
                                               vol 31   24,   25
          + phosphoric acid, trisodium salt
                                               vol 31   E11,   E12,   E14,
                                                             17,   26 -   36,
                                                           E129,   155 - 162
          + selenium dioxide   vol 26   320,   321
Sodium hydroxide (aqueous)
          + silver (I) oxide   vol 23   E83,   E84,   E86,
                                                            E87,   108,   114,
                                                            115,   117,   127,
                                                            128,   135 - 137
          + sodium aluminium oxide   vol 31   159,   160,
                                                                   161,   162
          + sulfuric acid, disodium salt   vol 31   155,   156,
                                                                   161,   162
          + sulfurous acid, disodium salt   vol 26   E5,   56,   57
          + tellurous acid, sodium salt   vol 26   E403,   407 - 410
          + vanadic acid, sodium salt   vol 31   159,   160
          + zinc hydroxide   vol 23   E159 -E162,   192,
                                                            193,   198,   199,
                                                            202,   203,   208,
                                                            209,   213,   229,
                                                                   260,   261
          + zinc oxide   vol 23   E159 -E162,   197,
                                                                   201,   207,
                                                            215 - 219,   227,
                                                                          224
Sodium hydroxide (multicomponent)
          + cadmium hydroxide   vol 23   300,   301
          + copper (II) hydroxide   vol 23   E10 - E26,   65,
                                                                           66
          + copper (II) oxide   vol 23   E10 - E26,   81
          + mercury (II) oxide   vol 23   336 - 338
          + 2-naphthol   vol 26   E5,   71
          + silver (I) oxide   vol 23   138,   139,   141

Sodium hydroxide (multicomponent)
          + sulfurous acid, calcium salt   vol 26   E191, E192,   232
          + sulfurous acid, disodium salt   vol 26   E5,   71
          + zinc oxide   vol 23   231 - 234,   265,
                                                                           266
Sodium hydroxyacetate
        see acetic acid, hydroxy-, monosodium salt
Sodium B-hydroxybutyrate
        see butanoic acid, 3-hydroxy-, monosodium salt
Sodium iodate
        see iodic acid, sodium salt
Sodium iodide
          + acetic acid, sodium salt   vol 33   202,   263
Sodium iodide (aqueous)
          + bromic acid, sodium salt   vol 30   213
          + chloric acid, sodium salt   vol 30   E28,   E31, E32,
                                                                            61
          + iodic acid, sodium salt   vol 30   E335, E336,   364,
                                                                           365
          + mercury   vol 29   E49
          + methane   vol 27/28   E75,   E76, 150,
                                                                           151
Sodium iodide (in N,N-dimethylformamide)
          + propane   vol 24   E125,   127
Sodium iodide (in 2-propanone)
          + propane   vol 24   E125,   126
Sodium lactate
        see propanoic acid, 2-hydroxy-, monosodium salt
Sodium malate
        see butanedioic acid, hydroxy-, monosodium salt
Sodium mandelate
        see mandelic acid, monosodium salt
Sodium metaphosphate
        see metaphosphoric acid, sodium salt
Sodium methoxyacetate
        see acetic acid, methoxy-, sodium salt
Sodium molybdate
        see disodium (I-4) tetroxomolydate (2-)
Sodium 2-naphtholate
        see 2-naphthalenol, sodium salt
Sodium nitrate
        see nitric acid, sodium salt
Sodium nitrite
        see nitrous acid, sodium salt
Sodium norsulfazole
        see benzenesulfonamide, 4-amino-N-2-thiazolyl-, monosodium
            salt
Sodium octanoate
        see octanoic acid, sodium salt
Sodium oxide (aqueous)
          + phosphorus pentoxide   vol 31   E11
Sodium pentanoate
        see pentanoic acid, sodium salt
Sodium perchlorate
        see perchloric acid, sodium salt
Sodium phosphate
        see phosphoric acid, trisodium salt
Sodium propanoate
        see propanoic acid, sodium salt
Sodium propionate
        see propanoic acid, sodium salt
Sodium pyroselenite
        see diselenious acid, disodium salt
Sodium pyrosulfite
        see disulfurous acid, disodium salt
Sodium pyruvate
        see propanoic acid, 2-oxo-, sodium salt
Sodium salycilate
        see benzoic acid, 2-hydroxy-, monosodium salt
Sodium selenite
        see selenious acid, sodium salt
Sodium streptocide
        see methanesulfonic acid, [[4-(aminosulfonyl)phenyl]amino]-,
            monosodium salt

```
Sodium sulfadiazine
        see benzenesulfonamide, 4-amino-N-2-pyrimidinyl-, monosodium salt
Sodium sulfadicramide
        see 2-butenamide, N-[(4-aminophenyl)sulfonyl]-3-methyl-,
            monosodium salt
Sodium sulfamethoxypyridazine
        see benzenesulfonamide, 4-amino-N-(6-methoxy-3-pyridoxinyl-
            monosodium salt
Sodium p-sulfamoylanilinoacetate
        see glycine,  N-[(4-aminosulfonyl)phenyl]-, monosodium salt
Sodium sulfamyd
        see benzenesulfonamide, 4-(aminomethyl)-, monosodium salt
Sodium sulfanilamide
        see benzenesulfonamide, 4-amino-, monosodium salt
Sodium, sulfanilamido-
        see benzenesulfonamide, 4-amino-, monosodium salt
Sodium 2-sulfanilamidothiazole
        see benzenesulfonamide, 4-amino-N-2-thiazolyl-, monosodium
            salt
Sodium sulfapyridazine
        see benzenesulfonamide, 4-amino-N-(6-methoxy-3-pyridoxinyl-
            monosodium salt
Sodium sulfapyridine
        see benzenesulfonamide, 4-amino-N-2-pyridinyl-, monosodium salt
Sodium sulfapyrimidine
        see benzenesulfonamide, 4-amino-N-2-pyrimidinyl-, monosodium salt
Sodium sulfate
        see sulfuric acid, sodium salt
Sodium sulfathiazole
        see benzenesulfonamide, 4-amino-N-2-thiazolyl-, monosodium
            salt
Sodium sulfide (aqueous)
        + cadmium hydroxide                vol 23                       293
        + hydrogen sulfide                 vol 32                        48
        + mercury (II) oxide               vol 23            E313,      332
        + zinc oxide                       vol 23            E181,      252
Sodium sulfite
        see sulfurous acid, sodium salt
Sodium p-sulfonamidophenylaminomethanesulfonate
        see methanesulfonic acid, [[4-(aminosulfonyl)phenyl]amino]-,
            monosodium salt
Sodium tartarate
        see butanedioic acid, 2,3-dihydroxy-, disodium salt
Sodium taurocholate
        see ethanesulfonic acid,2-[[(3a,5B,7a,12a)-3,7,12-trihydroxy-,
            24-oxocholan-24-yl]amino]-, monosodium salt
Sodium tauroglycocholate
        see tauroglycocholic acid, sodium salt
Sodium tellurite
        see tellurous acid, sodium salt
Sodium thiocyanate
        see thiocyanic acid, sodium salt
Sodium thiosulfate
        see thiosulfuric acid, disodium salt
Sodium trihydrogen diselenite
        + water                            vol 26            E331,  336,  337
Sodium vanadate
        see vanadic acid, sodium salt
Soluble sulfadiazine
        see benzenesulfonamide, 4-amino-N-2-pyrimidinyl-, monosodium salt
Soluble sulfathiazole
        see benzenesulfonamide, 4-amino-N-2-thiazolyl-, monosodium salt
Solu Cortef
        see pregn-4-ene-3,20-dione,21(3-carboxy-1-oxopropoxy)-11,
            17-dihydroxy-monosodium salt (11-B)
Soludagenan
        see benzenesulfonamide, 4-amino-N-2-pyridinyl-, monosodium salt
Soludiazine
        see benzenesulfonamide, 4-amino-N-2-pyrimidinyl-, monosodium salt
Solufontamide
        see benzenesulfonamide, 4-amino-N-[(aminothioxomethyl)-
Soluthiazomide
        see benzenesulfonamide, 4-amino-N-2-thiazolyl-, monosodium salt
```

```
Solyacord
        see 3-pyridinecarboxamide, N,N-diethyl-
Sorbitol
        see D-glucitol
Sonilyn
        see benzenesulfonamide, 4-amino-N-(6-chloro-3-pyridazinyl)-
Soxisol
        see benzenesulfonamide, 4-amino-N-(3,4-dimethyl-5-isoxazolyl)-
Span 20
        see sorbitan monolaurate
Spanbolet
        see benzenesulfonamide, 4-amino-N-(4,6-dimethyl-2-pyrimidinyl)-
Spleen
        see guinea pig spleen and human fetal spleen
Spofadazine
        see benzenesulfonamide, 4-amino-N-(6-methoxy-3-pyridazinyl)-
Sporfadrizine
        see benzenesulfonamide, 4-amino-N-2-pyrimidinyl-
Spray oil
        see oil, spray
Squalane
        see tetracosane, 2,6,10,15,19,23-hexamethyl-
SRC recyle solvent
            + methane                    vol 27/28              566
Stansin
        see benzenesulfonamide, 4-amino-N-(3,4-dimethyl-5-isoxazolyl)-
Staphylamid
        see benzenesulfonamide, 4-amino-N-(4-methyl-2-thiazolyl)-
Stearic acid
        see octadecanoic acid
Steramide
        see acetamide, N-[(4-aminophenyl)sulfonyl]-
Sterazine
        see benzenesulfonamide, 4-amino-N-2-pyrimidinyl-
(E)-Stilbene
        see benzene, 1,1´-(1,2-ethenediyl)bis-
Stim
        see 1H-purine-2,6-dione, 3,7-dihydro-1,3,7-trimethyl-
Stimulin
        see 3-pyridinecarboxamide, N,N-diethyl-
Stomach
        see guinea pig stomach
Stopton album
        see benzenesulfonamide, 4-amino-
Stramid
        see benzenesulfonamide, 4-amino-
Strepamide
        see benzenesulfonamide, 4-amino-
Streptocide
        see benzenesulfonamide, 4-[(2,4-diaminophenyl)azo]-
Streptocid Rubrum
        see 2,7-naphthalenedisulfonic acid, 6-acetylamino-
            3-[[4-(Aminosulfonyl)phenyl]azo]-4-hydroxy-, disodium salt
Streptocid soluble
        see methanesulfonic acid, [[4-(aminosulfonyl)phenyl]amino]-,
            monosodium salt
Streptocide album soluble
        see methanesulfonic acid, [[4-(aminosulfonyl)phenyl]amino]-,
            monosodium salt
Streptocide white soluble
        see methanesulfonic acid, [[4-(aminosulfonyl)phenyl]amino]-,
            monosodium salt
Streptosilpyridine
        see benzenesulfonamide, 4-amino-N-2-pyridinyl-
Streptosilthiazole
        see benzenesulfonamide, 4-amino-N-2-thiazolyl-
Streptozon
        see benzenesulfonamide, 4-[(2,4-diaminophenyl)azo]-
Streptozon II
        see 2,7-naphthalenedisulfonic acid, 6-acetylamino-
            3-[[4-(aminosulfonyl)phenyl]azo]-4-hydroxy-, disodium salt
Streptozon S
        see 2,7-naphthalenedisulfonic acid, 6-acetylamino-
            3-[[4-(Aminosulfonyl)phenyl]azo]-4-hydroxy-, disodium salt
```

```
Strontium
            + mercury                         vol 25                E71,  E72,
                                                                    73 -   75
Strontium bromate
        see bromic acid, strontium salt
Strontium chlorate
        see chloric acid, strontium salt
Strontium iodate
        see iodic acid, strontium salt
Strontium nitrate
        see nitric acid, strontium salt
Strontium selenite
        see selenious acid, strontium salt
Strontium sulfite
        see sulfurous acid, strontium salt
Styrene
        see benzene, ethenyl-
2-(N4-Succinylsulfanilamido)thiazole
        see butanoic acid, 4-oxo-4-[[4-[(2-thiaxolylamino)-
            sulfonyl]phenyl]amino]-
Succinylsulfathiazole
        see butanoic acid, 4-oxo-4-[[4-[(2-thiaxolylamino)-
            sulfonyl]phenyl]amino]-
Sucrose
        see also a-D-glucopyranoside,b-D-fructofuranosyl
Sucrose (aqueous)
            + sulfurous acid, barium salt
                                            vol 26        E240, E241,   244
Sucrose (aqueous and multicomponent)
            + copper (II) hydroxide         vol 23        E10 - E26,    72
            + sulfurous acid, calcium salt
                                            vol 26               E191, E192,
                                                                 216 - 221,   231
Suganyl
        see benzenesulfonamide, 4-amino-N-[(aminoiminomethyl)-
Sulamyd
        see acetamide, N-[(4-aminophenyl)sulfonyl]-
Sulanilamidothiazole
        see benzenesulfonamide, 4-amino-N-2-thiazolyl-
Sulanilsulfanilmethylamide
        see benzenesulfonamide, 4-amino-N-[4-[(methylamino)sulfonyl]-
            phenyl]-
Sulbio
        see benzenesulfonamide, 4-amino-N-(3,4-dimethyl-5-isoxazolyl)-
Sulfabenzamide
        see benzamide, N-[(4-aminophenyl)sulfonyl]-
Sulfabenzide
        see benzamide, N-[(4-aminophenyl)sulfonyl]-
Sulfabenzothiazole
        see benzenesulfonamide, 4-amino-N-2-benzothiazolyl-
Sulfabenzoylamide
        see benzamide, N-[(4-aminophenyl)sulfonyl]-
Sulfabid
        see benzenesulfonamide, 4-amino-N-(1-phenyl-1H-pyrazol-5-yl)-
Sulfabutin
        see benzenesulfonamide, 4-amino-N-(4,5-dimethyl-2-oxazolyl)-
Sulfacarbamide
        see benzenesulfonamide, 4-amino-N-(aminocarbonyl)-
Sulfacet
        see acetamide, N-[(4-aminophenyl)sulfonyl]-
Sulfacetamide
        see acetamide, N-[(4-aminophenyl)sulfonyl]-
Sulfacetimide
        see acetamide, N-[(4-aminophenyl)sulfonyl]-
Sulfachloropyridazine
        see benzenesulfonamide, 4-amino-N-(6-chloro-3-pyridazinyl)-
Sulfachlorpyridazine
        see benzenesulfonamide, 4-amino-N-(6-chloro-3-pyridazinyl)-
Sulfachlorazine
        see benzenesulfonamide, 4-amino-N-(6-chloro-3-pyridazinyl)-
Sulfacyl
        see acetamide, N-[(4-aminophenyl)sulfonyl]-
Sulfadiazin
        see benzenesulfonamide, 4-amino-N-2-pyrimidinyl-
```

```
Sulfadiazine
        see benzenesulfonamide, 4-amino-N-2-pyrimidinyl-
Sulfadiazine silver
        see silver, (4-amino-N-2-pyrimidinylbenzenesulfonamidato-NN-O1)-
Sulfadiazine sodium
        see benzenesulfonamide, 4-amino-N-2-pyrimidinyl-, monosodium salt
Sulfadicramide
        see 2-butenamide, N-[(4-aminophenyl)sulfonyl]-3-methyl-
Sulfadigesin
        see butanoic acid, 4-oxo-4-[[4-[(2-thiaxolylamino)sulfonyl]-
            phenyl]amino]-
Sulfadimerazine
        see benzenesulfonamide, 4-amino-N-(4,6-dimethyl-2-pyrimidinyl)-
Sulfadimesin
        see benzenesulfonamide, 4-amino-N-(4,6-dimethyl-2-pyrimidinyl)-
Sulfadimesine
        see benzenesulfonamide, 4-amino-N-(4,6-dimethyl-2-pyrimidinyl)-
Sulfadimethoxypyrimidine
        see benzenesulfonamide, 4-amino-N-(4,6-dimethoxy-2-pyrimidinyl)-
Sulfadimethyldiazine
        see benzenesulfonamide, 4-amino-N-(4,6-dimethyl-2-pyrimidinyl)-
Sulfadimethylisoxazole
        see benzenesulfonamide, 4-amino-N-(3,4-dimethyl-5-isoxazolyl)-
Sulfadimethyloxazole
        see benzenesulfonamide, 4-amino-N-(4,5-dimethyl-2-oxazolyl)-
Sulfadimethylpyrimidine
        see benzenesulfonamide, 4-amino-N-(4,6-dimethyl-2-pyrimidinyl)-
Sulfadimetine
        see benzenesulfonamide, 4-amino-N-(2,6-dimethyl-4-pyrimidinyl)-
Sulfadimezin
        see benzenesulfonamide, 4-amino-N-(4,6-dimethyl-2-pyrimidinyl)-
Sulfadimezine
        see benzenesulfonamide, 4-amino-N-(4,6-dimethyl-2-pyrimidinyl)-
Sulfadimidin
        see benzenesulfonamide, 4-amino-N-(4,6-dimethyl-2-pyrimidinyl)-
Sulfadimidine
        see benzenesulfonamide, 4-amino-N-(4,6-dimethyl-2-pyrimidinyl)-
Sulfadine
        see benzenesulfonamide, 4-amino-N-(4,6-dimethyl-2-pyrimidinyl)-
Sulfadoxin
        see benzenesulfonamide, 4-amino-N-(5,6-dimethoxy-4-pyrimidinyl)-
Sulfadoxine
        see benzenesulfonamide, 4-amino-N-(5,6-dimethoxy-4-pyrimidinyl)-
Sulfaethidol
        see benzenesulfonamide, 4-amino-N-(5-methyl-
            1,3,4-thiadiazol-2-yl)-
Sulfaethylthiadiazole
        see benzenesulfonamide, 4-amino-N-(5-methyl-
            1,3,4-thiadiazol-2-yl)-
Sulfafurazole
        see benzenesulfonamide, 4-amino-N-(3,4-dimethyl-5-isoxazolyl)-
Sulfagan
        see benzenesulfonamide, 4-amino-N-(3,4-dimethyl-5-isoxazolyl)-
Sulfaguanidin
        see benzenesulfonamide, 4-amino-N-[(aminoiminomethyl)-
Sulfaguanidine
        see benzenesulfonamide, 4-amino-N-[(aminoiminomethyl)-
Sulfaguanidine monohydrate
        see benzenesulfonamide, 4-amino-N-[(aminoiminomethyl)-
            monohydrate
Sulfaguanil
        see benzenesulfonamide, 4-amino-N-[(aminoiminomethyl)-
Sulfaguine
        see benzenesulfonamide, 4-amino-N-[(aminoiminomethyl)-
Sulfaisodimerazine
        see benzenesulfonamide, 4-amino-N-(2,6-dimethyl-4-pyrimidinyl)-
Sulfaisodimidine
        see benzenesulfonamide, 4-amino-N-(2,6-dimethyl-4-pyrimidinyl)-
Sulfaisomidine
        see benzenesulfonamide, 4-amino-N-(2,6-dimethyl-4-pyrimidinyl)-
Sulfaisopropylthiadiazole
        see benzenesulfonamide, 4-amino-N-[5-(1-methylethyl)-
            1,3,4-thiadiazol-2-yl]-
```

```
Sulfaisoxazole
        see benzenesulfonamide, 4-amino-N-(3,4-dimethyl-5-isoxazolyl)-
Sulfalene
        see benzenesulfonamide. 4-amino-N-(3-methoxypryazinyl)-
Sulfalex
        see benzenesulfonamide, 4-amino-N-(6-methoxy-3-pyridazinyl)-
Sulfamelazine
        see acetamide, N-4-[[(4-methyl-2-pyrimidinyl)amino]sulfonyl]-
            phenyl]-
Sulfameradine
        see benzenesulfonamide, 4-amino-N-(4-methyl-2-pyrimidinyl)-
Sulfamerazin
        see benzenesulfonamide, 4-amino-N-(4-methyl-2-pyrimidinyl)-
Sulfamerazine
        see benzenesulfonamide, 4-amino-N-(4-methyl-2-pyrimidinyl)-
Sulfameter
        see benzenesulfonamide, 4-amino-N-(5-methoxy-2-pyrimidinyl)-
Sulfamethalazole
        see benzenesulfonamide, 4-amino-N-(5-methyl-3-isoxazolyl)-
Sulfamethazine
        see benzenesulfonamide, 4-amino-N-(4,6-dimethyl-2-pyrimidinyl)-
Sulfamethazine hemihydrate
        see benzenesulfonamide, 4-amino-N-(4,6-dimethyl-2-pyrimidinyl)-,
            hemihydrate
Sulfamethiazine
        see benzenesulfonamide, 4-amino-N-(4,6-dimethyl-2-pyrimidinyl)-
Sulfamethiazole
        see benzenesulfonamide, 4-amino-N-(5-methyl-
            1,3,4-thiadiazol-2-yl)-
Sulfamethine
        see benzenesulfonamide, 4-amnio-N-(4,6-dimethyl-2-pyrimidinyl)-
Sulfamethizole
        see benzenesulfonamide, 4-amino-N-(5-methyl-
            1,3,4-thiadiazol-2-yl)-
Sulfamethizole N4-acetate
        see acetamide, N-[4-[[(5-methyl-1,3,4-thiadiazol-2-yl)amino]-
            sulfonyl]-phenyl]-
Sulfamethoxazole
        see benzenesulfonamide, 4-amino-N-(5-methyl-3-isoxazolyl)-
Sulfamethopyrazine
        see benzenesulfonamide, 4-amino-N-(3-methoxypryazinyl)-
Sulfamethoxine
        see benzenesulfonamide, 4-amino-N-(5-methoxy-2-pyrimidinyl)-
Sulfamethoxydiazine
        see benzenesulfonamide, 4-amino-N-(5-methoxy-2-pyrimidinyl)-
Sulfamethoxydin
        see benzenesulfonamide, 4-amino-N-(5-methoxy-2-pyrimidinyl)-
Sulfamethoxypyrazine
        see benzenesulfonamide, 4-amino-N-(3-methoxypryazinyl)-
Sulfamethoxypyridazine
        see benzenesulfonamide, 4-amino-N-(6-methoxy-3-pyridazinyl)-
Sulfamethoxypyrimidine
        see benzenesulfonamide, 4-amino-N-(5-methoxy-2-pyrimidinyl)-
2-Sulfa-5-methoxypyrimidine
        see benzenesulfonamide, 4-amino-N-(5-methoxy-2-pyrimidinyl)-
Sulfamethyldiazine
        see benzenesulfonamide, 4-amino-N-(4-methyl-2-pyrimidinyl)-
Sulfamethylisothiourea
        see benzenesulfonamide, 4-amino-N-[imino(methylthio)methyl]-
Sulfamethylmercaptopyridazine
        see benzenesulfonamide, 4-amino-N-[6-(methylthio)-3-pyridazinyl]-
Sulfamethylphenazole
        see benzenesulfonamide, 4-amino-N-(3-methyl-1-phenyl-
            1H-pyrazol-5-yl)-
Sulfamethylpyridazine
        see benzenesulfonamide, 4-amino-N-(6-chloro-3-pyridazinyl)-
2-Sulfa-4-methylpyrimidine
        see benzenesulfonamide, 4-amino-N-(4-methyl-2-pyrimidinyl)-
Sulfamethylthiadiazole
        see benzenesulfonamide, 4-amino-N-(5-methyl-
            1,3,4-thiadiazol-2-yl)-
Sulfamethylthiazole
        see benzenesulfonamide, 4-amino-N-(4-methyl-2-thiazolyl)-
```

```
Sulfametin
        see benzenesulfonamide, 4-amino-N-(5-methoxy-2-pyrimidinyl)-
Sulfametopyrazine
        see benzenesulfonamide, 4-amino-N-(3-methoxypryrazinyl)-
Sulfametoyl
        see benzamide, N-[(4-aminophenyl)sulfonyl]-3,4-dimethyl-
Sulfametrole
        see benzenesulfonamide, 4-amino-N-
            (4-methoxy-1,2,5-thiadiazol-3-yl)-
Sulfamezathine
        see benzenesulfonamide, 4-amino-N-(4,6-dimethyl-2-pyrimidinyl)-
Sulfamidochrysoidin
        see benzenesulfonamide, 4-[(2,4-diaminophenyl)azo]-
Sulfamidochrysoidine
        see benzenesulfonamide, 4-[(2,4-diaminophenyl)azo]-
Sulfamonomethoxine
        see benzenesulfonamide, 4-amino-N-(5-methoxy-2-pyrimidinyl)-
Sulfamonomethoxine complex with 18-crown-6 (1:1)
        see benzenesulfonamide, 4-amino-N-(6-methoxy-4-pyrimidinyl)-,
            comp. with 1,4,7,10,13,16-hexaoxacyclooctadecane (1:1)
Sulfamonomethoxine-18-crown-6 complex
        see benzenesulfonamide, 4-amino-N-(6-methoxy-4-pyrimidinyl)-,
            comp. with 1,4,7,10,13,16-hexaoxacyclooctadecane (1:1)
Sulfamoprine
        see benzenesulfonamide, 4-amino-N-(4,6-dimethoxy-2-pyrimidinyl)-
Sulfamoxolum
        see benzenesulfonamide, 4-amino-N-(4,5-dimethyl-2-oxazolyl)-
4´-Sulfamoylacetanilide
        see acetamide, N-[4-(aminosulfonyl)phenyl]-
p-Sulfamoylacetanilide
        see acetamide, N-[4-(aminosulfonyl)phenyl]-
4-Sulfamoylbenzylamine
        see benzenesulfonamide, 4-(aminomethyl)-
p-Sulfamoylbenzylamine
        see benzenesulfonamide, 4-(aminomethyl)-
N-(4-Sulfamoylphenyl)acetamide
        see acetamide, N-[4-(aminosulfonyl)phenyl]-
N-(p-Sulfamoylphenyl)glycine
        see glycine, N-[(4-Aminosulfonyl)phenyl]-
4´-Sulfamoylsulfanilanilide
        see benzenesulfonamide, 4-amino-N-[4-(aminosulfonyl)phenyl]-
Sulfamyd
        see benzenesulfonamide, 4-(aminomethyl)-
p-Sulfamylacetanilide
        see acetamide, N-[4-(aminosulfonyl)phenyl]-
(p-Sulfamylanilino)methanesulfonic acid sodium salt
        see methanesulfonic acid, [[4-(aminosulfonyl)phenyl]amino]-,
            monosodium salt
N-Sulfamylbenzamide
        see benzamide, N-[(4-aminophenyl)sulfonyl]-
Sulfamylon
        see benzenesulfonamide, 4-(aminomethyl)-
4´-Sulfamylsulfanilanilide
        see benzenesulfonamide, 4-amino-N-[4-(aminosulfonyl)phenyl]-
Sulfanilamide
        see benzenesulfonamide, 4-amino-
o-Sulfanilamide
        see benzenesulfonamide, 2-amino-
Sulfanilamide-N4-acetate
        see acetamide, N-[4-(aminosulfonyl)phenyl]-
Sulfanilamide hydrochloride
        see benzenesulfonamide, 4-amino-, monohydrochloride
Sulfanilamide monohydrate
        see benzenesulfonamide, 4-amino-, monohydrate
Sulfanilamide, monosodium salt
        see benzenesulfonamide, 4-amino-, monosodium salt
Sulfanilamide sodium
        see benzenesulfonamide, 4-amino-, monosodium salt
Sulfanilamide soluble
        see benzenesulfonamide, 4-amino-, monosodium salt
5-Sulfanilamido-2-aminopyridine
        see benzenesulfonamide, 4-amino-N-(6-amino-3-pyridinyl)-
2-Sulfanilamido-5-aminopyridine
        see benzenesulfonamide, 4-amino-N-(5-amino-2-pyridinyl)-
```

5-Sulfanilamido-5-bromopyridine
        see benzenesulfonamide, 4-amino-N-(2-bromo-5-pyridinyl)-
2-Sulfanilamido-5-bromopyridine
        see benzenesulfonamide, 4-amino-N-(5-bromo-2-pyridinyl)-
6-Sulfanilamido-3-chloropyridazine
        see benzenesulfonamide, 4-amino-N-(6-chloro-3-pyridazinyl)-
5-Sulfanilamido-2-chloropyridine
        see benzenesulfonamide, 4-amino-N-(2-chloro-5-pyridinyl)-
4-Sulfanilamido-5,6-dimethoxypyrimidine
        see benzenesulfonamide, 4-amino-N-(5,6-dimethoxy-4-pyrimidinyl)-
6-(Sulfanilamido)-2,4-dimethyl-1,3-diazine
        see benzenesulfonamide, 4-amino-N-(2,6-dimethyl-4-pyrimidinyl)-
5-Sulfanilamido-3,4-dimethylisoxazole
        see benzenesulfonamide, 4-amino-N-(3,4-dimethyl-5-isoxazolyl)-
2-Sulfanilamido-4,6-dimethylpyrimidine
        see benzenesulfonamide, 4-amino-N-(4,6-dimethyl-2-pyrimidinyl)-
4-Sulfanilamido-2,6-dimethylpyrimidine
        see benzenesulfonamide, 4-amino-N-(2,6-dimethyl-4-pyrimidinyl)-
6-Sulfanilamido-2,4-dimethylpyrimidine
        see benzenesulfonamide, 4-amino-N-(2,6-dimethyl-4-pyrimidinyl)-
2-Sulfanilamido-5-ethylthiadiazole
        see benzenesulfonamide, 4-amino-N-(5-methyl-
            1,3,4-thiadiazol-2-yl)-
2-Sulfanilamidoimidazole
        see benzenesulfonamide, 4-amino-N-1H-imidazol-2-yl-
2-Sulfanilamido-5-isopropylthiadiazole
        see benzenesulfonamide, 4-amino-N-[5-(1-methylethyl)-
            1,3,4-thiadiazol-2-yl]-
2-Sulfanilamido-3-methoxypyrazine
        see benzenesulfonamide, 4-amino-N-(3-methoxypryrazinyl)-
6-Sulfanilamido-3-methoxypyridazine
        see benzenesulfonamide, 4-amino-N-(6-methoxy-3-pyridazinyl)-
2-Sulfanilamido-5-methoxypyrimidine
        see benzenesulfonamide, 4-amino-N-(5-methoxy-2-pyrimidinyl)-
3-Sulfanilamido-6-methoxypyridazine
        see benzenesulfonamide, 4-amino-N-(6-methoxy-3-pyridazinyl)-
6-Sulfanilamido-4-methoxypyrimidine
        see benzenesulfonamide, 4-amino-N-(6-methoxy-4-pyrimidinyl)-
3-Sulfanilamido-5-methylisoxazole
        see benzenesulfonamide, 4-amino-N-(5-methyl-3-isoxazolyl)-
2-Sulfanilamido-4-methylpyrimidine
        see benzenesulfonamide, 4-amino-N-(4-methyl-2-pyrimidinyl)-
2-Sulfanilamido-4-methylthiazole
        see benzenesulfonamide, 4-amino-N-(4-methyl-2-thiazolyl)-
5-Sulfanilamido-1-phenylpyrazole
        see benzenesulfonamide, 4-amino-N-(1-phenyl-1H-pyrazol-5-yl)-
Sulfanilamidopyrazine
        see benzenesulfonamide, 4-amino-N-pyrazinyl-
3-Sulfanilamidopyridazine
        see benzenesulfonamide, 4-amino-N-3-pyridazinyl-
2-Sulfanilamidopyridine
        see benzenesulfonamide, 4-amino-N-2-pyridinyl-
3-Sulfanilamidopyridine
        see benzenesulfonamide, 4-amino-N-3-pyridinyl-
2-Sulfanilamidopyridine sodium salt
        see benzenesulfonamide, 4-amino-N-2-pyridinyl-, monosodium salt
2-Sulfanilamidopyrimidine
        see benzenesulfonamide, 4-amino-N-2-pyrimidinyl-
4-Sulfanilamidopyrimidine
        see benzenesulfonamide, 4-amino-N-4-pyrimidinyl-
5-Sulfanilamidopyrimidine
        see benzenesulfonamide, 4-amino-N-5-pyrimidinyl-
2-Sulfanilamidopyrimidine sodium salt
        see benzenesulfonamide, 4-amino-N-2-pyrimidinyl-, monosodium salt
2-Sulfanilamidothiazole
        see benzenesulfonamide, 4-amino-N-2-thiazolyl-
2-Sulfanilamidothiazole sodium salt
        see benzenesulfonamide, 4-amino-N-2-thiazolyl-, monosodium salt
3-Sulfanilamido-1,2,4-triazole
        see benzenesulfonamide, 4-amino-N-1H-1,2,4-triazol-3-yl-
4-Sulfanilamido-1,2,4-triazole
        see benzenesulfonamide, 4-amino-N-1H-1,2,4-triazol-4-yl-
2-Sulfanilamidoxazole
        see benzenesulfonamide, 4-amino-N-2-oxazolyl-

```
Sulfanilcarbamid
        see benzenesulfonamide, 4-amino-N-(aminocarbonyl)-
Sulfanildimethylacroylamide
        see 2-butenamide, N-[(4-aminophenyl)sulfonyl]-3-methyl-
Sulfanilguanidine
        see benzenesulfonamide, 4-amino-N-[(aminoiminomethyl)-
Sulfanilthiocarbamide
        see benzenesulfonamide, 4-amino-N-[(aminothioxomethyl)-
Sulfanilthiourea
        see benzenesulfonamide, 4-amino-N-[(aminothioxomethyl)-
N-Sulfanilylacetamide
        see acetamide, N-[(4-aminophenyl)sulfonyl]-
2-Sulfanilylamidopyrimidine
        see benzenesulfonamide, 4-amino-N-2-pyrimidinyl-
2-(Sulfanilylamino)thiazole
        see benzenesulfonamide, 4-amino-N-2-thiazolyl-
N-Sulfanilylbenzamide
        see benzamide, N-[(4-aminophenyl)sulfonyl]-
N-Sulfanilyl-N´-butylurea
        see benzenesulfonamide, 4-amino-N-[(butylamino)carbonyl]-
N-Sulfanilyl-N,N-dimethylacrylamide
        see 2-butenamide, N-[(4-aminophenyl)sulfonyl]-3-methyl-
Sulfanilylguanidine
        see benzenesulfonamide, 4-amino-N-[(aminoiminomethyl)-
N-Sulfanilylseneciamide
        see 2-butenamide, N-[(4-aminophenyl)sulfonyl]-3-methyl-
N4-Sulfanilylsulfanilamide
        see benzenesulfonamide, 4-amino-N-[4-(aminosulfonyl)phenyl]-
1-Sulfanilyl-2-thiourea
        see benzenesulfonamide, 4-amino-N-[(aminothioxomethyl)-
Sulfanilylurea
        see benzenesulfonamide, 4-amino-N-(aminocarbonyl)-
Sulfanilyl-3,4-xylamide
        see benzamide, N-[(4-aminophenyl)sulfonyl]-3,4-dimethyl-
N-Sulfanilyl-3,4-xylamide
        see benzamide, N-[(4-aminophenyl)sulfonyl]-3,4-dimethyl-
Sulfano
        see benzenesulfonamide, 4-amino-N-(4,5-dimethyl-2-oxazolyl)-
Sulfa-Perlongit
        see benzenesulfonamide, 4-amino-N-(5-methyl-
             1,3,4-thiadiazol-2-yl)-
Sulfaphenazol
        see benzenesulfonamide, 4-amino-N-(1-phenyl-1H-pyrazol-5-yl)-
Sulfaphenazole
        see benzenesulfonamide, 4-amino-N-(1-phenyl-1H-pyrazol-5-yl)-
Sulfaphenazon
        see benzenesulfonamide, 4-amino-N-(1-phenyl-1H-pyrazol-5-yl)-
Sulfaphenylpipazol
        see benzenesulfonamide, 4-amino-N-(1-phenyl-1H-pyrazol-5-yl)-
Sulfaphenylpyrazole
        see benzenesulfonamide, 4-amino-N-(1-phenyl-1H-pyrazol-5-yl)-
Sulfapiridazin
        see benzenesulfonamide, 4-amino-N-(6-methoxy-3-pyridazinyl)-
Sulfapyrazine
        see benzenesulfonamide, 4-amino-N-pyrazinyl-
Sulfapyrazinemethoxine
        see benzenesulfonamide, 4-amino-N-(3-methoxypyrazinyl)-
Sulfapyrazine sodium
        see benzenesulfonamide, 4-amino-N-pyrazinyl-, monosodium salt
Sulfapyrazole
        see benzenesulfonamide, 4-amino-N-(3-methyl-1-phenyl-
             1H-pyrazol-5-yl)-
Sulfapyridazine sodium
        see benzenesulfonamide, 4-amino-N-(6-methoxy-3-pyridoxinyl-
             monosodium salt
Sulfasol
        see benzenesulfonamide, 4-amino-N-(3,4-dimethyl-5-isoxazolyl)-
Sulfasuccidin
        see butanoic acid, 4-oxo-4-[[4-[(2-thiaxolylamino)sulfonyl]-
             phenyl]amino]-
Sulfasuccidine
        see butanoic acid, 4-oxo-4-[[4-[(2-thiaxolylamino)sulfonyl]-
             phenyl]amino]-
```

```
Sulfasuccinil
         see butanoic acid, 4-oxo-4-[[4-[(2-thiaxolylamino)sulfonyl]-
              phenyl]amino]-
Sulfasuccithiazole
         see butanoic acid, 4-oxo-4-[[4-[(2-thiaxolylamino)sulfonyl]-
              phenyl]amino]-
Sulfasuxidine
         see butanoic acid, 4-oxo-4-[[4-[(2-thiaxolylamino)-
              sulfonyl]phenyl]amino]-
Sulfates
         see under sulfuric acid
Sulfathiazole
         see benzenesulfonamide, 4-amino-N-2-thiazolyl-
2-Sulfathiazole
         see benzenesulfonamide, 4-amino-N-2-thiazolyl-
Sulfathiazole sodium
         see benzenesulfonamide, 4-amino-N-2-thiazolyl-, monosodium salt
Sulfathiocarbamid
         see benzenesulfonamide, 4-amino-N-[(aminothioxomethyl)-
Sulfathiocarbamide
         see benzenesulfonamide, 4-amino-N-[(aminothioxomethyl)-
Sulfathiourea
         see benzenesulfonamide, 4-amino-N-[(aminothioxomethyl)-
Sulfaurea
         see benzenesulfonamide, 4-amino-N-(aminocarbonyl)-
Sulfaxylamide
         see benzamide, N-[(4-aminophenyl)sulfonyl]-3-methyl-
Sulfavigor
         see benzenesulfonamide, 4-amino-N-(4,5-dimethyl-2-oxazolyl)-
Sulfazamet
         see benzenesulfonamide, 4-amino-N-(3-methyl-1-phenyl-
              1H-pyrazol-5-yl)-
Sulfazin
         see acetamide, N-[(4-aminophenyl)sulfonyl]-N-
              (3,4-dimethyl-5-isoxazolyl)-
Sulfazol
         see benzenesulfonamide, 4-amino-N-(4-methyl-2-thiazolyl)-
Sulfazole
         see benzenesulfonamide, 4-amino-N-(4-methyl-2-thiazolyl)-
Sulfenterone
         see butanoic acid, 4-oxo-4-[[4-[(2-thiaxolylamino)-
              sulfonyl]phenyl]amino]-
1-Sulfentidine
         see benzenesulfonamide, 4-amino-N-[(aminoiminomethyl)-
Sulfethidiole
         see benzenesulfonamide, 4-amino-N-(5-methyl-1,3,4-thiadiazol-2-yl)-
Sulfinylbismethane
         see methane, sulfinylbis-
Sulfirgamid
         see 2-butenamide, N-[(4-aminophenyl)sulfonyl]-3-methyl-
Sulfirgamide
         see 2-butenamide, N-[(4-aminophenyl)sulfonyl]-3-methyl-
Sulfismezole-N4-acetate
         see acetamide, N-[4-[[(5-methyl-3-isoxaxolyl)amino]sulfonyl]-
              phenyl]-
Sulfisomezole
         see benzenesulfonamide, 4-amino-N-(5-methyl-3-isoxazolyl)-
Sulfisoxazole acetyl
         see acetamide, N-[(4-aminophenyl)sulfonyl]-N-
              (3,4-dimethyl-5-isoxazolyl)-
Sulfisoxazole dialamine
         see benzenesulfonamide, 4-amino-N-(3,4-dimethyl-5-isoxazolyl)-
Sulfmidil
         see benzenesulfonamide, 4-amino-N-(4,5-dimethyl-2-oxazolyl)-
Sulfocerol
         see benzenesulfonamide, 4-amino-N-2-thiazolyl-
2-Sulfocerol
         see benzenesulfonamide, 4-amino-N-2-thiazolyl-
Sulfoguanidine
         see benzenesulfonamide, 4-amino-N-[(aminoiminomethyl)-
Sulfoguanil
         see benzenesulfonamide, 4-amino-N-[(aminoiminomethyl)-
Sulfoguanyl
         see benzenesulfonamide, 4-amino-N-[(aminoiminomethyl)-
```

```
Sulfoguenil
        see benzenesulfonamide, 4-amino-N-[(aminoiminomethyl)-
Sulfolane
        see thiophene, tetrahydro-1,1-dioxide
Sulfono
        see benzenesulfonamide, 4-amino-N-(4,5-dimethyl-2-oxazolyl)-
Sulfstat
        see benzenesulfonamide, 4-amino-N-(5-methyl-
        1,3,4-thiadiazol-2-yl)-
Sulfune
        see benzenesulfonamide, 4-amino-N-(4,5-dimethyl-2-oxazolyl)-
Sulfuno
        see benzenesulfonamide, 4-amino-N-(4,5-dimethyl-2-oxazolyl)-
Sulfur
        + hydrogen sulfide                  vol 32                E185,   317
Sulfur dioxide
        + methane                           vol 27/28                     763
Sulfur dioxide (aqueous)
        + ammonia                           vol 26                   124, 125
        + sulfurous acid, barium salt   vol 26            E240, E241, 245
        + sulfurous acid, copper (I, II) salt
                                            vol 26            E262,  263, 264
Sulfur dioxide (aqueous)
        + sulfurous acid, diammonium salt
                                            vol 26            E115, 124,  125,
                                                                  129 - 131
        + sulfurous acid, iron (II) salt
                                            vol 26                   254, 255
        + sulfurous acid, lead salt     vol 26                   297, 298
        + zinc oxide                        vol 23            E177,   236
Sulfur dioxide (aqueous and multicomponent)
        + sulfuric acid, magnesium salt
                                            vol 26            E157, E158,  180,
                                                                          181
        + sulfurous acid, calcium salt
                                            vol 26            E189, E190,  196-
                                                              211,  229,  230
        + sulfurous acid, magnesium salt
                                            vol 26            E155, E156, E157,
                                                              E158,  165 - 167,
                                                                     180,  181
        + sulfurous acid, manganese salt
                                            vol 26                  E248, 251
        + sulfurous acid, zinc salt     vol 26                  E271 -E274,
                                                                  276 - 282
Sulfur trioxide (aqueous)
        + zinc oxide                        vol 23                  E177,  210
Sulfuric acid
        + hydrogen sulfide                  vol 32            E26,    32,  33
        + methane                           vol 27/28         E58,  E59,  87,
                                                                           88
Sulfuric acid (aqueous)
        + butane                            vol 24            E58 -  E72,  76
        + gold (III) hydroxide              vol 23            E148, E150, 151,
                                                                          152
        + 2-methylpropane                   vol 24            E58 - E72,  77
        + propane                           vol 24            E58 - E72,  75
        + selenious acid, cadmium salt  vol 26                         392
        + selenious acid, cobalt (II) salt
                                            vol 26                         371
        + selenious acid, copper (II) salt
                                            vol 26                   378,  379
        + selenious acid, iron (III) salt
                                            vol 26                         368
        + selenious acid, manganese (II) salt
                                            vol 26                         367
        + selenious acid, mercury (I) salt
                                            vol 26                   394, 395
        + selenious acid, nickel salt   vol 26                         375
        + selenious acid, silver salt   vol 26                         380
        + selenious acid, zinc salt     vol 26                         389
        + tellurous acid, cobalt salt   vol 26                  E404, 422
        + tellurous acid, copper salt   vol 26                  E404, 423
        + tellurous acid, nickel salt   vol 26                  E404, 422
```

```
Sulfuric acid (aqueous)
        + tellurous acid, sodium salt    vol 26              E403, 414
Sulfuric acid (ternary)
        + methane                        vol 27/28           E72, 154
Sulfuric acid (multicomponent)
        + hydrogen sulfide               vol 32          52,   53,   67
Sulfuric acid, ammonium salt (aqueous)
        + cadmium oxide                  vol 23                      295
        + hydrogen sulfide               vol 32                 39,  67
Sulfuric acid, ammonium salt (multicomponent)
        + copper (II) hydroxide          vol 23          E10 - E26,  64,
                                                                     75
        + zinc hydroxide                 vol 23                      264
        + zinc oxide                     vol 23              E178, 269
Sulfuric acid, cadmium salt (aqueous)
        + cadmium hydroxide              vol 23                      289
        + sulfurous acid, cadmium salt
                                         vol 26                      288
Sulfuric acid, calcium salt (aqueous and multicomponent)
        + sulfurous acid, calcium salt   vol 26          E191, E192, 212-
                                                         214,  220 -233
Sulfuric acid, copper (II) salt (aqueous)
        + sulfurous acid, copper (I, II) salt
                                         vol 26                      265
Sulfuric acid, diammonium salt (aqueous)
        + disulfurous acid, diammonium salt
                                         vol 26              E144, E145,
                                                             147 - 151
        + sulfuric acid, diammonium salt
                                         vol 26          E115, 141 - 143
        + sulfurous acid, diammonium salt
                                         vol 26          E115, 132 - 137,
                                                             141 - 143
Sulfuric acid, dipotassium salt
        + water                          vol 31              305, 306
Sulfuric acid, dipotassium salt (aqueous and multicomponent)
        + phosphoric acid, monopotassium salt
                                         vol 31          E210, E212, 223,
                                                         237 - 239, 249,
                                                                     250
        + phosphoric acid, tripotassium salt
                                         vol 31          E297,  302, 305,
                                                                     306
        + sulfurous acid, dipotassium salt
                                         vol 26          E92,  104, 105,
                                                                     107
Sulfuric acid, disodium salt
        + water                          vol 31          145,  161 -164
Sulfuric acid, disodium salt (aqueous)
        + disulfurous acid, disodium salt
                                         vol 26          E74,   78 - 83
        + phosphoric acid, disodium salt
                                         vol 31          E98,  112,  122,
                                                                     123
        + phosphoric acid, monosodium salt
                                         vol 31              E51,  53
        + phosphoric acid, trisodium salt
                                         vol 31              E128, E129,
                                                         136 - 138,  140,
                                                         145,  155,  156,
                                                             161 - 164
        + sodium aluminium oxide         vol 31              161,  162
        + sodium hydroxide               vol 31          155,  156,  161,
                                                                     162
        + sulfurous acid, disodium salt
                                         vol 26          E3, E4, 16 -  33
        + vanadic acid, sodium  salt     vol 31              163,  164
Sulfuric acid, disodium salt (aqueous and multicomponent)
        + disulfurous acid, disodium salt
                                         vol 26          E5,   63 -  65
        + sulfuric acid, zinc salt       vol 26              284,  285
        + sulfurous acid, cadmium salt   vol 26              289,  290
```

```
Sulfuric acid, disodium salt (aqueous and multicomponent)
          + sulfurous acid, disodium salt
                                          vol 26           E5,     59 - 65,
                                                               284,   285,
                                                               289,   290
          + sulfurous acid, zinc salt    vol 26            284,   285
Sulfuric acid, lithium salt (aqueous)
          + bromic acid, lithium salt    vol 30                      193
          + iodic acid, lithium salt     vol 30           E272,   281
Sulfuric acid, magnesium salt (aqueous)
          + mercury                      vol 29           E87,    88 - 94
          + sulfurous acid, magnesium salt
                                          vol 26                E157, E158,
                                                               173 - 176
Sulfuric acid, magnesium salt (aqueous and multicomponent)
          + magnesium chloride           vol 26           E157, E158,  186
          + sulfur dioxide               vol 26           E157, E158,  180,
                                                                       181
          + sulfurous acid, magnesium salt
                                          vol 26           E157, E158,  173-
                                                               176,  180 - 185
          + magnesium hydrogen sulfite   vol 26           E157, E158,
                                                               182 - 185
Sulfuric acid, manganese salt (aqueous)
          + methane                      vol 27/28        E65,  E66,  105

Sulfuric acid, manganese salt (ternary)
          + methane                      vol 27/28        E66,  E77, E82,
                                                               106,  155
Sulfuric acid, monododecyl ester, sodium salt (aqueous and multicomponent)
          + methane                      vol 27/28              189 - 191
Sulfuric acid, monohexyl ester, sodium salt (aqueous)
          + methane                      vol 27/28                    188
Sulfuric acid, potassium salt (aqueous)
          + bromic acid, potassium salt  vol 30                      231
          + chloric acid                 vol 30                 122,   123
          + hydrogen sulfide             vol 32                        39
          + iodic acid, potassium salt   vol 30           E380,   390
          + methane                      vol 27/28        E81,  E82,  168
          + silver (I) oxide             vol 23                      121
Sulfuric acid, potassium salt (ternary)
          + methane                      vol 27/28        E82,  155,  159
Sulfuric acid, sodium salt
          + water                        vol 30                  42 - 49
Sulfuric acid, sodium salt (aqueous)
          + bromic acid, sodium salt     vol 30                 204 - 207
          + chloric acid, sodium salt    vol 30           E30,   42 -  49
          + copper (II) hydroxide        vol 23           E10 - E26,   64
          + hydrogen sulfide             vol 32           E25,  E26,   39,
                                                                40,   43
          + iodic acid, sodium salt      vol 30           E337,  354 - 356
          + mercury                      vol 29           E49,   68
          + mercury (II) oxide           vol 23                      336
          + methane                      vol 27/28        E76,  E77, E79,
                                                               152,  153
Sulfuric acid, sodium salt (ternary)
          + methane                      vol 27/28        E77,  E82, 154,
                                                               155,  156, 159
Sulfuric acid, zinc salt (aqueous)
          + zinc oxide                   vol 23                      212
Sulfuric acid, zinc salt (aqueous and multicomponent)
          + sulfur dioxide               vol 26                 278 - 282
          + sulfuric acid, disodium salt vol 26            284,   285
          + sulfurous acid, zinc salt    vol 26            278 - 282
Sulfurine
          see benzenesulfonamide, 4-amino-N-(5-methyl-
          1,3,4-thiadiazol-2-yl)-
Sulfurous acid (aqueous)
          + sulfurous acid, dipotassium salt
                                          vol 26           E92,   103
Sulfurous acid, barium salt
          + water                        vol 26           E240, E241,
                                                               242 - 247
```

```
Sulfurous acid, barium salt (aqueous)
          + sodium chloride              vol 26       E240, E241,  246
          + sucrose                      vol 26       E240, E241,  244
          + sulfur dioxide               vol 26       E240, E241,  245
Sulfurous acid, barium salt (aqueous and multicomponent)
          + carbonic acid, disodium salt vol 26       E240, E241,  244
          + sulfurous acid, disodium salt vol 26      E240, E241,  244
Sulfurous acid, beryllium salt
          + water                        vol 26                    E152
Sulfurous acid, cadmium salt
          + water                        vol 26       E286,  287 - 290
Sulfurous acid, cadmium salt (aqueous)
          + sulfuric acid, cadmium salt   vol 26                    288
Sulfurous acid, cadmium salt (aqueous and multicomponent)
          + sulfuric acid, disodium salt  vol 26                289, 290
          + sulfurous acid, disodium salt
                                         vol 26                289, 290
Sulfurous acid, calcium salt
          + water                        vol 26       E190-E193,
                                                      194 - 235
Sulfurous acid, calcium salt (aqueous)
          + nitric acid, ammonium salt   vol 26       E191, E192, 215
Sulfurous acid, calcium salt (aqueous and multicomponent)
          + acetic acid                  vol 26       E191, E192,
                                                      232, 233
          + acetic acid, sodium salt     vol 26       E191, E192, 234
          + citric acid                  vol 26       E191, E192, 233
Sulfurous acid, calcium salt (aqueous and multicomponent)
          + formic acid, sodium salt     vol 26       E191, E192, 234
          + glucose                      vol 26       E191, E192, 220,
                                                      221, 235
          + hydrochloric acid            vol 26       E191, E192, 224,
                                                      225
          + lignosulfonic acid           vol 26       E191, E192, 235
          + perchloric acid, sodium salt vol 26       E189, E190, 207-
                                                      210,  229, 230
          + phosphoric acid, sodium salt vol 26       E191, E192, 234
          + seawater                     vol 26       E191, E192, 226-
                                                      228
          + sodium chloride              vol 26        E191, E192,
                                                      224 - 228
          + sodium hydroxide             vol 26       E191, E192, 232
          + sucrose                      vol 26        E191, E192,
                                                      216 - 221,  231
          + sulfur dioxide               vol 26       E189, E190, E191,
                                                      196 - 211,  229,
                                                      230
          + sulfuric acid, calcium salt  vol 26       E191, E192, 212-
                                                      214,  220 -223
          + xylose                       vol 26       E191, E192, 235
Sulfurous acid, cobalt (II) salt
          + water                        vol 26       E256, 257
Sulfurous acid, copper (I) salt
          + water                        vol 26       E260, 261
Sulfurous acid, copper (I, II) salt
          + water                        vol 26       E262, 263-266
Sulfurous acid, copper (I, II) salt (aqueous)
          + sulfur dioxide               vol 26       E262, 263, 264
          + sulfuric acid, copper (II) salt vol 26                 265
Sulfurous acid, diammonium salt
          + water                        vol 26       E113-E116,
                                                      117 - 123
Sulfurous acid, diammonium salt (aqueous)
          + ammonia                      vol 26       E115, 121-123
                                                      126
          + ammonium chloride            vol 26       E115, 132-135
          + disulfurous acid, diammonium salt
                                         vol 26       E115, 127, 128
          + sulfur dioxide               vol 26       E115, 124, 125
                                                      129-131
          + sulfuric acid, diammonium salt vol 26     E115,  132-135
          + sulfurous acid, disodium salt  vol 26     E5,   46 - 49
          + sulfurous acid, monoammonium salt
                                         vol 26       E115, 141-143
```

Sulfurous acid, diammonium salt (aqueous)
        + thiosulfuric acid, ammonium salt
                                        vol 26   E115, 140
Sulfurous acid, diammonium salt (aqueous and multicomponent)
        + ammonium chloride   vol 26   E5, 66-70
        + sodium chloride   vol 26   E5, 66-70
        + sulfurous acid, disodium salt vol 26   E5, 66-70, E115
                                                                  138, 139
Sulfurous acid, dipotassium salt
        + water   vol 26   E90-E92, 93-107
Sulfurous acid, dipotassium salt (aqueous)
        + ethanol   vol 26   E92, 95
        + nitric acid, potassium salt   vol 26   E92, 96, 97
        + nitrous acid, potassium salt vol 26   E92, 98, 99
        + potassium hydroxide   vol 26   E92, 102
        + selenious acid, dipotassium salt
                                        vol 26   327
        + sulfurous acid   vol 26   E92, 103
        + urea   vol 26   E92, 100, 101
Sulfurous acid, dipotassium salt (aqueous and multicomponent)
        + carbonic acid, dipotassium salt
                                        vol 26   E92, 106, 107
        + sulfuric acid, dipotassium salt
                                        vol 26   E92, 104, 105
                                                                      107
Sulfurous acid, disodium salt
        + water   vol 26   E1-E7, 8-71

Sulfurous acid, disodium salt (aqueous)
        + ethanol   vol 26   E5, 54, 55
        + 2-naphthalenol, sodium salt   vol 26   E5, 58
        + selenious acid, sodium salt   vol 26   309
        + sodium chloride   vol 26   E4, E5, 34-38
        + sodium hydroxide   vol 26   E5, 56, 57
        + sulfuric acid, disodium salt vol 26   E3, E4, 16-33
        + sulfurous acid, diammonium salt
                                        vol 26   E5, 46-49
        + sulfurous acid, magnesium salt vol 26   E157, E158
                                                               178, 179
        + thiosulfurous acid, disodium salt
                                        vol 26   E5, 50-53
Sulfurous acid, disodium salt (aqueous and multicomponent)
        + 2-naphthol   vol 26   E5, 71
        + ammonium chloride   vol 26   E5, 66-70
        + disulfurous acid, disodium salt vol 26   E5, 39-45, 63-65
        + sodium chloride   vol 26   E5, 59-62, 66-70
        + sodium hydroxide   vol 26   E5, 71
        + sulfuric acid, disodium salt
                                        vol 26   E5, 59-70, 284
                                                              285, 289, 290
        + sulfuric acid, zinc salt   vol 26   284, 285
        + sulfurous acid, barium salt   vol 26   E240, E241, 247
        + sulfurous acid, cadmium salt vol 26   289, 290
        + sulfurous acid, zinc salt   vol 26   283 -285
Sulfurous acid, iron (II) salt
        + water   vol 26   E252, 253-255
Sulfurous acid, iron (II) salt (aqueous)
        + sulfur dioxide   vol 26   254, 255
Sulfurous acid, lead salt
        + water   vol 26   E293, E294, 295-
                                                                      302
Sulfurous acid, lead salt (aqueous)
        + acetic acid, ammonium salt   vol 26   299, 300
        + sulfur dioxide   vol 26   297, 298
Sulfurous acid, lead salt (aqueous and multicomponent)
        + acetic acid   vol 26   301, 302
        + acetic acid, sodium salt   vol 26   301, 302
        + nitric acid, sodium salt   vol 26   301, 302
Sulfurous acid, magnesium salt
        + water   vol 26   E153-E159, 160
                                                                      186
Sulfurous acid, magnesium salt (2:1) (aqueous)
        + magnesium chloride   vol 26   E157, E158, 186
        + sulfuric acid, magnesium salt vol 26   182- 186

```
Sulfurous acid, magnesium salt (2:1) (aqueous)
          + sulfurous acid, magnesium salt    vol 26      E155, E156, 168-
                                                          172, 182-186
Sulfurous acid, magnesium salt (aqueous)
          + magnesium chloride                vol 26      E157, E158, 177
          + magnesium hydrogen sulfite        vol 26      E155, E156, 168-
                                                          172, 182 -186
          + sulfur dioxide                    vol 26        E155, E156,
                                                          165 - 167
          + sulfuric acid, magnesium salt     vol 26         E157, E158
                                                          173 - 176
                                                          180 - 185
          + sulfurous acid, disodium salt     vol 26         E157, E158,
                                                          178, 179
Sulfurous acid, magnesium salt (aqueous and multicomponent)
          + sulfur dioxide                    vol 26         E157, E158,
                                                          180,  181
          + sulfuric acid, magnesium salt     vol 26         E157, E158,
                                                          180,  181
Sulfurous acid, manganese salt (aqueous and multicomponent)
          + carbon dioxide                    vol 26         E248, 251
          + sulfur dioxide                    vol 26         E248, 251
Sulfurous acid, mercury salt
          + water                             vol 26         E291, 292
Sulfurous acid, monoammonium salt (aqueous)
          + sulfuric acid, diammonium salt    vol 26      E115, 141-143
          + sulfurous acid, diammonium salt   vol 26      E115, 141-143
Sulfurous acid, monopotassium salt
          + water                             vol 26                112
Sulfurous acid, monosodium salt (aqueous)
          + dithionous acid, sodium salt      vol 26                 85
Sulfurous acid, nickel (II) salt
          + water                             vol 26         E258, 259
Sulfurous acid, silver salt
          + water                             vol 26      E267, 268-270
Sulfurous acid, strontium salt
          + water                             vol 26      E236, 237-239
Sulfurous acid, strontium salt (aqueous)
          + sodium chloride                   vol 26         E236, 239
Sulfurous acid, zinc salt
          + water                             vol 26         E271-E274,
                                                          275 -285
Sulfurous acid, zinc salt (aqueous and multicomponent)
          + sulfur dioxide                    vol 26         E271-E274,
                                                          276 -282
          + sulfuric acid, disodium salt      vol 26         284, 285
          + sulfurous acid, disodium salt     vol 26         283- 285
Sulgin
          see benzenesulfonamide, 4-amino-N-[(aminoiminomethyl)-
Sulgin ASG
          see acetamide, N-[4-[[(aminoiminomethyl)amino]sulfonyl]phenyl]-
Sulisoxazole
          see benzenesulfonamide, 4-amino-N-(3,4-dimethyl-5-isoxazolyl)-
Sulphacetamide
          see acetamide, N-[(4-aminophenyl)sulfonyl]-
Sulphafuazole
          see benzenesulfonamide, 4-amino-N-(3,4-dimethyl-5-isoxazolyl)-
Sulphaphenazole
          see benzenesulfonamide, 4-amino-N-(1-phenyl-1H-pyrazol-5-yl)-
Sulphmethoxazole
          see benzenesulfonamide, 4-amino-N-(5-methyl-3-isoxazolyl)-
Sulphocidine album
          see methanesulfonic acid, [[4-(aminosulfonyl)phenyl]amino]-
              monosodium salt
Sul-Spansion
          see benzenesulfonamide, 4-amino-N-(5-methyl-
              1,3,4-thiadiazol-2-yl)-
Sul-Spantab
          see benzenesulfonamide, 4-amino-N-(5-methyl-
              1,3,4-thiadiazol-2-yl)-
Sulzol
          see benzenesulfonamide, 4-amino-N-2-thiazolyl-
Sunflower seed oil
          + mercury                            vol 29                 176
```

T

Tantalum
                + mercury   vol 25   E274,   275, 276
                                                     vol 29   E214,   217
Tardamide
        see benzenesulfonamide, 4-amino-N-(4,5-dimethyl-2-oxazolyl)-
Tardamid
        see benzenesulfonamide, 4-amino-N-(4,5-dimethyl-2-oxazolyl)-
Tartaric acid
        see butanedioic acid, 2,3-dihydrooxy-
Technetium
                + mercury   vol 25   E421
Telluric (IV) acid
        see tellurous acid
Tellurium
                + mercury   vol 25   E194, E195,
                                                                   196 - 205
                                                     vol 29   E213,   224
Tellurous acid, barium salt (aqueous)
                + hydrochloric acid   vol 26   E404,   420
                + nitric acid, sodium salt   vol 26   E404,   420
                + nitric acid   vol 26   E404,   420
                + sodium chloride   vol 26   E404,   420
Tellurous acid, cesium salt
                + water   vol 26   E404,   419
Tellurous acid, cobalt salt (aqueous)
                + sulfuric acid   vol 26   E404,   421
Tellurous acid, copper salt (aqueous)
                + hydrochloric acid   vol 26   E404,   423
                + sulfuric acid   vol 26   E404,   423
Tellurous acid, lead salt (aqueous and multicomponent)
                + nitric acid, sodium salt   vol 26   E404,   427
                + nitric acid   vol 26   E404,   427
Tellurous acid, lithium salt
                + water   vol 26   E403,   406
Tellurous acid, nickel salt (aqueous)
                + hydrochloric acid   vol 26   E404,   422
                + sulfuric acid   vol 26   E404,   422
Tellurous acid, silver salt
                + water   vol 26   E404,   424 - 426
Tellurous acid, silver salt (aqueous and multicomponent)
                + perchloric acid, sodium salt vol 26   E404,   425
                + tellurous acid, sodium salt   vol 26   E404,   425
Tellurous acid, sodium salt
                + water   vol 26   E403,   407 - 418
Tellurous acid, sodium salt (aqueous)
                + carbonic acid, disodium salt vol 26   E403,   411 - 413
                + ethanol   vol 26   E404,   417, 418
                + perchloric acid   vol 26   E404,   415, 416
                + selenious acid, disodium salt vol 26   316,   317
                + sodium hydroxide   vol 26   E403,   407 - 410
                + sulfuric acid   vol 26   E403,   414
Tellurous acid, sodium salt (aqueous and multicomponent)
                + perchloric acid, sodium salt vol 26   E404,   425
                + tellurous acid, silver salt   vol 26   E404,   425
Terbium
                + mercury   vol 25   E242,   243 - 244
Terbium bromide
                + 1-butanamine   vol 22   283
                + 2-butanamine   vol 22   283
                + 1,4-dioxane   vol 22   282
                + furan, tetrahydro-   vol 22   281
                + 1-propanamine   vol 22   283
Terbium chloride
                + 2-butanamine   vol 22   278
                + 2-butanamine, N-(1-methylpropyl)-
                                                     vol 22   278
                + butane, 1-methoxy-   vol 22   276
                + decane, 1-methoxy-   vol 22   276
                + 1,4-dioxane   vol 22   275
                + ethane, 1,2-diethoxy-   vol 22   274
                + ethane, 1-ethoxy-2-methoxy-   vol 22   275
                + ethanol (aqueous)   vol 22   272

```
Terbium chloride
            + ethanol, 2-ethoxy-               vol 22                        273
            + ethanol, 2-methoxy-              vol 22                        273
            + heptane, 1-methoxy-             vol 22                        276
            + nonane, 1-methoxy-              vol 22                        276
            + octane, 1-methoxy-              vol 22                        276
            + pentane, 1-methoxy-            vol 22                        276
            + phosphoric acid, tributyl ester
                                              vol 22                        277
            + phosphoric triamide, hexamethyl-
                                              vol 22                        279
            + phosphoryl chloride            vol 22                        280
            + stannane, tetrachloro-         vol 22                        280
            + water (ternary)                vol 22                        272
Terbium fluoride
            + butane, 1-(chlorometyhoxy)-    vol 22                        270
            + decane, 1-methoxy-             vol 22                        270
            + methane, 1,1´-sulfinylbis-     vol 22                        271
Terbium iodide
            + furan, tetrahydro-             vol 22                        284
Terbium nitrate
         see nitric acid, terbium salt
Ternary systems involving salt plus water are listed under the salt
1,2,4,5-Tetrabromobenzene
         see benzene, 1,2,4,5-tetrabromo-
1,1,2,2-Tetrabromoethane
         see ethane, 1,1,2,2-tetrabromo-
Tetrabutylammonium bromide
         see 1-butanaminium, N,N,N-tributyl-, bromide
1,2,3,4-Tetrachlorobenzene
         see benzene 1,2,3,4-tetrachloro-
1,2,3,5-Tetrachlorobenzene
         see benzene 1,2,3,5-tetrachloro-
1,2,4,5-Tetrachlorobenzene
         see benzene 1,2,4,5-tetrachloro-
1,1,2,2-Tetrachloroethane
         see ethane, 1,1,2,2-tetrachloro-
1,1,2,2-Tetrachloroethene
         see ethene, 1,1,2,2-tetrachloro-
Tetrachloroethylene
         see ethene, 1,1,2,2-tetrachloro-
Tetrachloroethene
         see ethene, 1,1,2,2-tetrachloro-
Tetrachloromethane
         see methane, tetrachloro-
2,4,5,6-Tetrachloro-3-methylphenol
         see phenol, 2,4,5,6-tetrachloro-3-methyl-
2,3,4,6-Tetrachloro-3-methylphenol
         see phenol, 2,3,4,6-tetrachloro-3-methyl-
Tetrachlorosilane
         see silane, tetrachloro-
Tetracid
         see benzenesulfonamide, 4-amino-N-(5-methyl-1,3,4-thiadiazol-2-yl)-
Tetracosane
            + butane                         vol 24          E130 -E134,   185
Tetracosane,2,6,10,15,19,23-hexamethyl-
            + butane                         vol 24          E130- E134,   186
            + methane                        vol 27/28       E213, E214,   247
            + propane                        vol 24          E130 -E134,   165
Tetradecafluoromethylcyclohexane
         see cyclohexane, undecafluoro (trifluoromethyl)-
Tetradecane
            + hydrogen sulfide               vol 32                  E166 -E169,
                                                                     189,   209
            + methane                        vol 27/28              E202 -E204,
                                                                     E208,   234
            + propane                        vol 24          E130 -E134,   153
            + seawater                       vol 38                        432
            + water                          vol 38                 E426, E427,
                                                                     428 - 431
Tetradecane, 7,8-dimethyl-
            + water                          vol 38                        463
```

```
Tetraethanolammonium bromide
          see ethanaminium, 2-hydroxy-, N,N,N-tris (2-hydroxyethyl)-,
               bromide
Tetraethylammonium bromide
          see 1-ethanaminium, N,N,N-triethyl-, bromide
Tetraethylene glycol dimethyl ether
          see 2,5,8,11,14-pentaoxapentadecane
Tetraethoxysilane
          see silicic acid, tetraethyl ester
Tetrafluorohydrazine
          see Hydrazine, tetrafluoro-
Tetraglyme
          see pentadecane, 2,5,11,14-pentaoxa-
Tetrahydrofuran
          see Furan, tetrahydro-
Tetrahydrofurfuryl alcohol
          see 2-Furanmethanol, tetrahydro-
1,2,3,4-Tetrahydronaphthalene
          see Naphthalene, 1,2,3,4-tetrahydro-
Tetrahydro-2H-pyran
          see 2H-pyran, tetrahydro-
Tetrahydrothiophene, 1,1-dioxide
          see thiophene, tetrahydro-, 1,1-dioxide
Tetrahydroxyethylammonium bromide
          see ethanaminium, 2-hydroxy-N,N,N-tris (2-hydroxyethyl)-,
               bromide
Tetralin
          see Naphthalene, 1,2,3,4-tetrahydro-
Tetramethylammonium bromide
          see methanamine, N,N,N-trimethyl-, bromide
Tetramethylammonium hydroxide
          see methanaminium, N,N,N-trimethyl-, hydroxide
Tetramethylammonium iodide,
          see methanaminium, N,N,N-trimethyl-, iodide
Tetramethylene glycol dimethyl ester
          see 2,5,8,11,15-pentaoxapentadecane
Tetramethylene sulfone
          see thiophene,tetrahydro-, 1,1-dioxide
2,5,8,11-Tetraoxadodecane
          + hydrogen sulfide                   vol 32        E173 -E175,  257
3,6,9,12-Tetraoxahexadecan-1-ol
          + hydrogen sulfide                   vol 32        E173, E174,  257
Tetrapropylammonium bromide
          see propanaminium, N,N,N-tripropyl-, bromide
1,2,4,5-Tetramethylbenzene
          see benzene, 1,2,4,5-tetramethyl-
2,2,4,4-Tetramethyl-3-pentanone
          see 3-pentanone,2,2,4,4-tetramethyl-
Tetramethylsilane
          see silane, tetramethyl-
Tetramethylsulfone
          see thiophene,tetrahydro-,1,1-dioxide
Tetraphenylarsonium nitrate
          see nitric acid, tetraphenylarsonium salt
Tetratriacontane
          + butane                             vol 24        E130 -E134,  188
Thallium
          + mercury                            vol 25        E119 -E121,
                                                             122 - 133
Thallium silver cyanide
          see thallium dicyanoargentate
Theradiazine
          see benzenesulfonamide, 4-amino-N-2-pyrimidinyl-
Thiacoccine
          see benzenesulfonamide, 4-amino-N-2-thiazolyl-
Thiacyl
          see butanoic acid, 4-oxo-4-[[4-[(2-thiaxolylamino)sulfonyl]-
               phenyl]amino]-
1,2,5-Thiadiazole, acetamide derivative
          see acetamide, N-[(4-aminophenyl)sulfonyl]-N-(4-methoxy-
               1,2,5-thiadiazol-3-yl)-
1,2,5-Thiadiazole, benzenesulfonamide derivative
          see benzenesulfonamide, 4-amino-N-(4-methoxy-
               1,2,5-thiadiazol-3-yl)-
```

Thiaminokal
         see benzenesulfonamide, 4-amino-, monosodium salt
Thianaphthene
         see benzo(b)thiophene
Thiasin
         see benzenesulfonamide, 4-amino-N-(3,4-dimethyl-5-isoxazolyl)-
Thiasulfol
         see benzenesulfonamide, 4-amino-N-2-thiazolyl-
Thiazamide
         see benzenesulfonamide, 4-amino-N-2-thiazolyl-
N1-4-Thiazolin-2-ylidenesulfanilamide
         see benzenesulfonamide, 4-amino-N-2-thiazolyl-
N-[4-[(2-Thiazolylamino)sulfonyl]phenyl]acetamide
         see acetamide, N-[4-[(2-thiazolylamino)sulfonyl]phenyl]-
4´-(2-Thiazolylsulfamoyl)acetanilide
         see acetamide, N-[4-[(2-thiazolylamino)sulfonyl]phenyl]-
4´-(2-Thiazolylsulfamoyl)succinanilic acid
         see butanoic acid, 4-oxo-4-[[4-[(2-thiaxolylamino)sulfonyl]-
                 phenyl]amino]-
N1-(2-Thiazolyl)sulfanilamide
         see benzenesulfonamide, 4-amino-N-2-thiazolyl-
N1-2-Thiazolylsulfanilamide
         see benzenesulfonamide, 4-amino-N-2-thiazolyl-
N1-2-Thiazolylsulfanilamide, monosodium salt
         see benzenesulfonamide, 4-amino-N-2-thiazolyl-, monosodium salt
(N1-2-Thiazolylsulfanilamido)sodium
         see benzenesulfonamide, 4-amino-N-2-thiazolyl-, monosodium salt
p-2-Thiazoylsulfamylsuccinanilic acid
         see butanoic acid, 4-oxo-4-[[4-[(2-thiaxolylamino)sulfonyl]-
                 phenyl]amino]-
Thidicur
         see benzenesulfonamide, 4-amino-N-(5-methyl-
                 1,3,4-thiadiazol-2-yl)-
Thioacetamide
         see ethanethioamide
Thioacetimidic acid
         see ethanethioamide
N1-(Thiocarbamoyl)sulfanilamide
         see benzenesulfonamide, 4-amino-N-[(aminothioxomethyl)-
Thiocyanate, ammonium
         see thiocyanic acid, ammonium salt
Thiocyanate, potassium
         see thiocyanic acid, potassium salt
Thiocyanate, sodium
         see thiocyanic acid, sodium salt
Thioocyanic acid, ammonium salt (multicomponent)
                 + copper (II) hydroxide   vol 23   E10 - E26,   76
Thiocyanic acid, lithium salt
                 + acetic acid, lithium salt   vol 33   209
                 + butanoic acid, lithium salt vol 33   216
                 + formic acid, lithium salt   vol 33   207
Thiocyanic acid, potassium salt
                 + acetic acid, potassium salt vol 33   E167, 168, 169
                 + butanoic acid, potassium salt   vol 33
E183,   184
                 + formic acid, potassium salt   vol 33   E145, 146, 147
                 + propanoic acid, potassium salt
                                                 vol 33   180
Thiocyanic acid, sodium salt
                 + acetic acid, sodium salt   vol 33   259 - 261
                 + butanoic acid, 2-methyl-, sodium salt
                                                 vol 33   E317, 318
                 + butanoic acid, sodium salt   vol 33   285 - 286
                 + formic acid, sodium salt   vol 33   E229, 230 - 233
                 + hexanoic acid, sodium salt   vol 33   328
                 + mercury   vol 29   E49,   70
                 + pentanoic acid, sodium salt   vol 33   E303, 304, 305
                 + propanoic acid, 2-methyl-, sodium salt
                                                 vol 33   299
                 + propanoic acid, sodium salt   vol 33   273 - 274
Thiophene, tetrahydro-,1,1-dioxide
                 + hydrogen sulfide   vol 32   E185, 320, 321
Thiophene, tetrahydro-, 1,1-dioxide (multicomponent)
                 + hydrogen sulfide   vol 32   117

Thioseptal
        see benzenesulfonamide, 4-amino-N-2-pyridinyl-
Thiosulfil
        see benzenesulfonamide, 4-amino-N-(5-methyl-
            1,3,4-thiadiazol-2-yl)-
Thiosulfuric acid, diammonium salt (aqueous)
            + sulfurous acid, diammonium salt
                                        vol 26   E115,   140
Thiosulfuric acid, disodium salt (aqueous)
            + sulfurous acid, disodium salt vol 26   E5,   50 - 53
Toriseptin M
        see benzenesulfonamide, 4-amino-N-(4-methyl-2-thiazolyl)-
Thorium
            + mercury   vol 25   E422, 423, 424
                                            vol 29   E214, 217
L-Threonine (aqueous)
            + mercury   vol 29   E49, 74
Thulium
            + mercury   vol 25   E251, 252
Thulium bromide
            + furan, tetrahydro-   vol 22   356
Thulium chloride
            + ethane, 1,2-diethoxy-   vol 22   351
            + ethane, 1,1´-oxybis-   vol 22   350
            + ethanol (aqueous)   vol 22   348
            + ethanol, 2-methoxy-   vol 22   349
            + phosphoric acid, tributyl ester
                                            vol 22   352
            + phosphoric triamide, hexamethyl-
                                            vol 22   354
            + phosphoryl chloride   vol 22   355
            + 2-propanamine   vol 22   353
            + 2-propen-1-amine   vol 22   353
            + stannane, tetrachloro-   vol 22   355
            + water (ternary)   vol 22   348
Thulium nitrate
        see nitric acid, thulium salt
Tin
            + mercury   vol 25   E139, E141,
                                                            142 - 156
                                            vol 29   E213, 223
Tissue culture medium
            + mercury   vol 29   E51, 100, 101
Titanium
            + mercury   vol 25   E258, E259,
                                                            260 - 262
                                            vol 29   E214, 217
Toluene
        see benzene, methyl-
Toluene, m-bromo-
        see benzene,1-bromo-3-methyl-
Toluidine
        see Benzenamine, ar-methyl-
Triacetin
        see 1,2,3-propanetriol, triacetate
Triacontane
            + butane   vol 24   E130 -E134, 187
Trianon
        see benzenesulfonamide, 4-amino-N-2-pyridinyl-
N1-4H-1,2,4-Triazol-4-ylsulfanilamide
        see benzenesulfonamide, 4-amino-N-1H-1,2,4-triazol-4-yl-
1,2,4-Tribromobenzene
        see benzene, 1,2,4-tribromo-
1,3,5-Tribromobenzene
        see benzene, 1,3,5-tribromo-
Tribromomethane
        see methane, tribromo-
2,4,6-Tribromophenol
        see phenol, 2,4,6-tribromo-
Tributylamine
        see 1-Butanamine,N,N-dibutyl-
Tributylamine, perfluoro-
        see 1-Butanamine,1,1,2,2,3,3,4,4,4-nonafluoro-N,N-bis-
            (nonafluorobutyl)-

N,N,N-Tributylbutanaminium bromide
        see Butanaminium bromide, N,N,N-tributyl-
Tributyl phosphate
        see Phosphoric acid, tributyl ester
Triisobutyl phosphate
        see Phosphoric acid, tris(2-methylpropyl) ester
Tributyrin
        see Butanoic acid, 1,2,3-propanetriyl ester
1,2,3-Trichlorobenzene
        see benzene, 1,2,3-trichloro-
1,2,4-Trichlorobenzene
        see benzene, 1,2,4-trichloro-
1,3,5-Trichlorobenzene
        see benzene, 1,3,5-trichloro-
2,4,6-Trichloro-3,5-dimethylphenol
        see phenol, 2,4,6-trichloro-3,5-dimethyl-
1,1,1-Trichloroethane
        see ethane, 1,1,1-trichloro-
1,1,2-Trichloroethane
        see ethane, 1,1,2-trichloro-
2,2,2-Trichloroethanol
        see ethanol, 2,2,2-trichloro-
1,1,1-Trichloroethanol
        see ethanol, 2,2,2-trichloro-
Trichloroethene
        see ethene, 1,1,2-trichloro-
Trichloroethylene
        see ethene, 1,1,2-trichloro-
1,1,2-Trichloroethene
        see ethene, 1,1,2-trichloro-
1,1,2-Trichloroethylene
        see ethene, 1,1,2-trichloro-
Trichlorofluoromethane
        see methane, trichlorofluoro-
1,1,1-Trichloro-2-hydroxyethane
        see ethanol, 2,2,2-trichloro-
Trichloromethane
        see methane, trichloro-
(Trichloromethyl)benzene
        see benzene, (trichloromethyl)-
2-(Trichloromethyl)pyridine
        see pyridine, 2-(trichloromethyl)-
2,4,6-Trichlorophenol
        see phenol, 2,4,6-trichloro-
3,4,5-Trichlorophenol
        see phenol, 3,4,5-trichloro-
1,2,3-Trichloropropane
        see propane, 1,2,3-trichloro-
1,1,1-Trichloro-2-hydroxyethane
        see ethanol, 2,2,2-trichloro
1,1,2-Trichloro-1,2,2-trifluoroethane
        see ethane, 1,1,2-trichloro-1,2,2-trifluoro-
3,4,5-Trichloro-2-(dichloromethyl)pyridine
        see pyridine, 3,4,5-trichloro-2-(dichloromethyl)-
3,4,5-Trichloro-2-(trichloromethyl)pyridine
        see pyridine, 3,4,5-trichloro-2-(trichloromethyl)-
            1,1,1,2,4,4,5,7,7,8,10,10,11,13,13,14,16,16,17,17,18,18,18-
Tricosafluoro-5,8,11,14-tetrakis(trifluoromethyl)-3,6,9,12,15-
pentaoxaoctadecane
        see 3,6,9,12,15-pentaoxaoctadecane, 1,1,1,2,4,4,5,7,7,8,10,10,
            11,13,13,14,16,16,17,17,18,18,18-tricosafluoro-5,8,11,14-
            tetrakis(trifluoromethyl)-
Tricresyl phosphate
        see phosphoric acid, tris(methyl phenyl) ester
Tricyclodecanedimethane
        + hydrogen sulfide   vol 32   258
Tridecane
        + hydrogen sulfide   vol 32   E166 -E169, 189
        + methane   vol 27/28   E202 -E204,
                                        E208, 234
        + water   vol 38   395
Triethanolamine
        see ethanol, 2,2´,2´´-nitrilotris-

Triethoxysilane
          see silane, triethoxy-
Triethylamine
          see ethanamine, N,N-diethyl-
Triethylamine nitrate
          see ethanamine, N,N-diethyl-, nitrate
Triethylene glycol
          see ethanol, 2,2´-(1,2-ethanediylbis(oxy))bis-
Triethylene glycol dimethyl ether
          see 2,5,8,11-tetraoxadodecane
Triethylene phosphate
          see ethanamine, N,N-diethyl-, phosphate
Triethylenetetramine
          see 1,2-ethanediamine, N,N-bis(2-aminoethyl)-
N,N,N-Triethylethanaminium bromide
          see ethanaminium bromide, N,N,N-triethyl-
Triethyl phosphate
          see phosphoric acid, triethyl ester
(Trifluoromethyl)benzene
          see benzene, (trifluoromethyl)-
Triglyme
          see 2,5,8,11-tetraoxadodecane
Trihexanoin
          see hexanoic acid, 1,2,3-propanetriyl ester
2,2,3-Trimethylbutane
          see butane, 2,2,3-trimethyl-
1,3,5-Trimethylbenzene
          see benzene, 1,3,5-trimethyl-
2,2,3-Trimethylbutane
          see butane, 2,2,3-trimethyl-
2,3,3-Trimethyl-2-butanol
          see 2-butanol, 2,3,3-trimethyl-
1,1,3-Trimethylcyclopentane
          see cyclopentane, 1,1,3-trimethyl-
1,3,7-Trimethyl-2,6-dioxopurine
          see 1H-purine-2,6-dione, 3,7-dihydro-1,3,7-trimethyl-
1,4,5-Trimethylnaphthalene
          see naphthalene, 1,4,5-trimethyl-
2,2,4-Trimethylpentane
          see pentane, 2,2,4-trimethyl-
2,3,4-Trimethylpentane
          see pentane, 2,3,4-trimethyl-
2,3,3-Trimethyl-3-pentanol
          see 3-pentanol, 2,3,3-trimethyl-
2,4,6-Trimethyl-2-phenylheptene
          see heptene, 2,4,6-trimethyl-2-phenyl-
N,N,N-Trimetylmethanamine hydroxide
          see methanamine, N,N,N-trimethyl-, hydroxide
N,N,N-Trimethylmethanaminium bromide
          see methanaminium bromide, N,N,N-trimethyl-
Trioctanoin
          see octanoic acid, 1,2,3-propanetriyl ester
Triphenylene
                + water   vol 38   E497, 498 - 500
Triphenyl methane
          see benzene, 1,1´,1´´-methylidynetris-
Tripotassium phosphate
          see phosphoric acid, potassium salt
Tripropylphosphate
          see phosphoric acid, tripropyl ester
N,N,N-Tripropylpropanaminium bromide
          see propanaminium bromide, N,N,N-tripropyl-
Triisobutyl phosphate
          see phosphoric acid, tris(2-methyl propyl) ester
Tris(4-amino-N-2-pyrimidinylbenzenesulfonamidato-NN,O)aluminium
          see aluminium, tris(4-amino-N-2-pyrimidinylbenzene-
                sulfonamidato-NN,O)-
Tris(4-amino-N-2-pyrimidinylbenzenesulfonamidato-NN,O)cerium
          see cerium, tris(4-amino-N-2-pyrimidinyl-
                benzenesulfonamidato-NN,O)-
Tris(4-amino-N-2-pyrimidinylbenzenesulfonamidato-NN,O)iron
          see iron, tris(4-amino-N-2-pyrimidinylbenzenesulfonamidato-NN,O)-
Trisodium phosphate
          see phosphoric acid, sodium salt

```
          see phosphoric acid, sodium salt
Tris(methylphenyl)phosphate
          see phosphoric acid, tris(methylphenyl)ester
Tris (hydroxymethyl) aminomethane
          see 1,3-propanediol, 2-amino-2-(hydroxymethyl)-
Trisodium phosphate
          see phosphoric acid, trisodium salt
Tritolyl phosphate
          see phosphoric acid, tris(methyl phenyl) ester
Tungsten
          + mercury                        vol 25                              E284
                                           vol 29              E214,  217
Tungstic acid, disodium salt
          + water                          vol 31                         151
Tungstic acid, disodium salt (aqueous)
          + phosphoric acid, trisodium salt
                                           vol 31              E127,  151
Turpentine, oil
          + propane                        vol 24         15,  228,  259,
                                                        338, E414,  415
Tween 20
      see sorbitan, monododecanoate poly(oxy-1,2-ethanediyl)
          derivatives
Tween  40
      see sorbitan, monohexadecanoate poly(oxy-1,2-ethanediyl)
          derivatives
Tween 60
      see sorbitan, monooctadecanoate poly8oxy-1,2-ethanediyl)
          derivatives
L-Tyrosine (aqueous)
          + mercury                        vol 29                E50,   81
```

U

U 6987
         see benzenesulfonamide, 4-amino-N-[(butylamino)carbonyl]-
Uleron
         see benzenesulfonamide, 4-[[(4-aminophenyl)sulfonyl]amino]-
             N,N-dimethyl-
Uliran
         see benzenesulfonamide, 4-[[(4-aminophenyl)sulfonyl]amino]-
             N,N-dimethyl-
Uliron
         see benzenesulfonamide, 4-[[(4-aminophenyl)sulfonyl]amino]-
             N,N-dimethyl-
Uliron C
         see benzenesulfonamide, 4-amino-N-[4-(aminosulfonyl)phenyl]-
Uleron
         see benzenesulfonamide, 1-amino-N-[5-(1,1´dimethylethyl)-
             1,3,4-thiadiazol-2-yl)-
Uliran
         see benzenesulfonamide, 1-amino-N-[5-(1,1´dimethylethyl)-
             1,3,4-thiadiazol-2-yl)-
Uliron
         see benzenesulfonamide, 1-amino-N-[5-(1,1´dimethylethyl)-
             1,3,4-thiadiazol-2-yl)-
Ultrasul
         see benzenesulfonamide, 4-amino-N-(5-methyl-
             1,3,4-thiadiazol-2-yl)-
Ultraseptal
         see benzenesulfonamide, 4-amino-N-(4-methyl-2-thiazolyl)-

Ultraseptyl
         see benzenesulfonamide, 4-amino-N-(4-methyl-2-thiazolyl)-
Ultrax
         see benzenesulfonamide, 4-amino-N-(5-methoxy-2-pyrimidinyl)-
Umbilical cord
         see human umbilical cord
Undecafluoro (trifluoromethyl) cyclohexane
         see cyclohexane, undecafluoro (trifluoromethyl)-
Undecane
         + hydrogen sulfide   vol 32   E166 -E169,   189
         + methane   vol 27/28   232,   239
                                                       E202 -E204,
                                                       E207,   233
         + seawater   vol 38   331
         + water   vol 38   E326, E327,
                                                        328 - 330
1-Undecanol
         + methane   vol 27/28   E582 -E584,
                                                       E594,   628
Unisulf
         see benzenesulfonamide, 4-amino-N-(3,4-dimethyl-5-isoxazolyl)
Uractyl
         see benzenesulfonamide, 4-amino-N-(aminocarbonyl)-
Uramid
         see benzenesulfonamide, 4-amino-N-(aminocarbonyl)-
Uranium
         + mercury   vol 25   E425 -E427,
                                                        428 - 433
Urea (aqueous)
         + bromic acid, potassium salt   vol 30   253
         + butane   vol 24   E70 - E72,   123
         + chloric acid, sodium salt   vol 30   81
         + hydrogen sulfide   vol 32   E20,   22
         + methane   vol 27/28   177,   178
         + 2-methylpropane   vol 24   E70 - E72,   124
         + phosphoric acid, monoammonium salt
                                         vol 31   274,   275
         + phosphoric acid, monopotassium salt
                                         vol 31   E208, 234,   235,
                                                              274,   275
         + propane   vol 24   E70 - E72,
                                                              121,   122
         + sulfurous acid   vol 26   E92,   100,   101

```
Urea (multicomponent)
              + copper (II) hydroxide              vol 23         E10 - E26,    73
Urenil
          see benzenesulfonamide, 4-amino-N-(aminocarbonyl)-
Urethane
          see carbamic acid, ethyl ester
Uritrisin
          see benzenesulfonamide, 4-amino-N-(3,4-dimethyl-5-isoxazolyl)
Urocydal
          see benzenesulfonamide, 4-amino-N-(5-methyl-1,3,4-thiadiazol-2-yl)
Urodiaton
          see benzenesulfonamide, 4-amino-N-(5-methyl-1,3,4-thiadiazol-2-yl)
Urolucosil
          see benzenesulfonamide, 4-amino-N-(5-methyl-1,3,4-thiadiazol-2-yl)
Urosulfane
          see benzenesulfonamide, 4-amino-N-(aminocarbonyl)-
Urosulfan
          see benzenesulfonamide, 4-amino-N-(aminocarbonyl)-
Urosulfin
          see benzenesulfonamide, 4-amino-N-(5-methyl-1,3,4-thiadiazol-2-yl)
Urosulfon
          see acetamide, N-[(4-aminophenyl)sulfonyl]-
```

V

Valeric acid,
        see pentanoic acid
Vanadic acid
                + water   vol 31   305, 306
Vanadic acid, potassium salt
                + water   vol 31   305, 306
Vanadic acid, potassium salt (aqueous)
                + phosphoric acid, tripotassium salt
                                                vol 31   E297, 305, 306
Vanadic acid, sodium salt
                + water   vol 31   144, 159, 160
                                                                      163, 164
Vanadic acid, sodium salt (aqueous)
                + phosphoric acid, trisodium salt
                                                vol 31   E129, 144, 159
                                                                160, 163, 164
                + sodium aluminium oxide   vol 31   159, 160
                + sodium hydroxide   vol 31   159, 160
                + sulfuric acid, disodium salt vol 31   163, 164
Vanadium
                + mercury   vol 25   E268, 269, 270
Vegetable oil
        see Oil, vegetable
Ventramine
        see 3-pyridinecarboxamide, N,N-diethyl-
Vesulong
        see benzenesulfonamide, 4-amino-N-(3-methyl-1-phenyl-
            1H-pyrazol-5-yl)-
Veta-Merazine
        see benzenesulfonamide, 4-amino-N-(4-methyl-2-pyrimidinyl)-
Vitreous body
        see rabbit vitreous body
4-Vinyl-1-cyclohexene
        see 1-cyclohexene,4-vinyl-
Vk 53
        see benzenesulfonamide, 4-amino-N-(5-methyl-1,3,4-thiadiazol-2-yl)-
VK 55
        see benzenesulfonamide, 4-amino-N-(5-methyl-1,3,4-thiadiazol-2-yl)-
Vk 57
        see benzenesulfonamide, 4-amino-N-[5-(1-methylethyl)-
            1,3,4-thiadiazol-2-yl]-

```
W

Water : Aqueous solid-liquid and liquid-liquid systems are indexed under
        organic components not under water, only gas-liquid
        systems are indexed below.
Water
              + butane                        vol 24              E16,   E17,
                                                                   18  -  33,
                                                           269,   352,   397
              + hydrogen sulfide              vol 32      E1-E3,   4  -  19
              + methane                       vol 27/28           E1  -  E6,
                                                                    7  -  23
                                                                  E24 - E28,
                                                            29  -  44,   580
              + 2-methylpropane               vol 24      E34,    35  -  38,
                                                           E48,    49  -  53
                                                           277,   365,   405
              + propane                       vol 24              E1,    E2,
                                                                    3  -  15,
                                                           E39,    40  -  47,
                                                           228,   259,   338,
                                                                  385,   414
Water (multicomponent)
              + hydrogen sulfide              vol 32      E20,    21  -  24,
                                                                  E25 - E28,
                                                            29  -  50,   E51,
                                                            52,    53,   E54,
                                                           E55,    56  -  78,
                                                                  E79 - E83,
                                                            84 - 165,   298,
                                                                  E330, E331,
                                                                  336 - 339
Water-d2
              + butane                        vol 24      E54,    56,    57
              + chloric acid, potassium salt  vol 30             119,   120
              + deuterium sulfide             vol 32      E327,  328,   329
              + iodic acid, potassium salt    vol 30                    387
              + methane                       vol 27/28          E45 - E46,
                                                                   47  -  49
              + propane                       vol 24      E54,    55
Wintrazole
         see benzenesulfonamide, 4-amino-N-2-thiazolyl-
WR 4103
         see benzenesulfonamide, 4-amino-N-(5,6-dimethoxy-4-pyrimidinyl)-
```

X

Xylene
        see Benzene, dimethyl-
m-Xylene
        see benzene, 1,3-dimethyl-
o-Xylene
        see benzene, 1,2-dimethyl-
p-Xylene
        see benzene, 1,4-dimethyl-
Xylidine
        see benzenamine,ar,ar-dimethyl-
Xylose (aqueous and multicomponent)
                + glucose   vol 26   E191, E192,   235
                + lignosulfonic acid   vol 26   E191, E192,   235
                + sulfurous acid, calcium salt   vol 26   E191, E192,   235
Xyloylsulfamine
        see benzamide, N-[(4-aminophenyl)sulfonyl]-3,4-dimethyl-

Y

Ytterbium
                + mercury   vol 25   E253,   254,   255
Ytterbium bromide
                + 1,4-dioxane   vol 22   374
                + ethane, 1,2-diethoxy-   vol 22   372
                + furan, tetrahydro-   vol 22   373
Ytterbium chloride
                + 1-butanamine   vol 22   369
                + 2-butanamine   vol 22   369
                + 2-butanamine,N-(1-methylpropyl)-
                                                 vol 22   369
                + butane, 1-ethoxy-   vol 22   367
                + 1,4-dioxane   vol 22   367
                + ethanamine   vol 22   369
                + ethanamine, N-ethyl-   vol 22   369
                + ethane, 1,2-diethoxy-   vol 22   365
                + ethane, 1-ethoxy-2-methoxy-   vol 22   367
                + ethane, 1,1´-oxybis-   vol 22   364
                + ethanol   vol 22   360
                + ethanol (ternary)   vol 22   361
                + ethanol, 1-ethoxy-   vol 22   363
                + ethanol, 2-methoxy-   vol 22   363
                + furan, tetrahydro-   vol 22   368
                + hexane, 1-methoxy-   vol 22   366
                + methanol   vol 22   359
                + pentane, 1-methoxy-   vol 22   367
                + phosphoric triamide, hexamethyl-
                                                 vol 22   370
                + phosphoryl chloride   vol 22   371
                + 1-propanamine   vol 22   369
                + 2-propanamine   vol 22   369
                + propane, 1,1´-oxybis-   vol 22   367
                + 2-propanol   vol 22   362
                + 2-propen-1-amine   vol 22   369
                + stannane, tetrachloro-   vol 22   371
                + water (ternary)   vol 22   361
Ytterbium fluoride
                + methane, 1,1´-sulfinylbis-   vol 22   357
                + pyridine   vol 22   358
Ytterbium iodide
                + 1-butanamine   vol 22   377
                + 2-butanamine   vol 22   377
                + 1,4-dioxane   vol 22   376
                + 1-propanamine   vol 22   377
                + 2-propanamine   vol 22   377
Yttrium
                + mercury   vol 25   E206
Yttrium bromate
                see bromic acid, yttrium salt
Yttrium bromide
                + 1,4-dioxane   vol 22   47
                + ethane, 1,2-diethoxy-   vol 22   45
                + furan, tetrahydro-   vol 22   46
Yttrium chloride
                + benzene (ternary)   vol 22   20
                + 1-butanamine   vol 22   42
                + 2-butanamine   vol 22   43
                + butane, 1-ethoxy-   vol 22   31
                + butane, 1-methoxy-   vol 22   30
                + decane, 1-methoxy-   vol 22   34
                + 1,4-dioxane   vol 22   36
                + 1,4-dioxane (ternary)   vol 22   37
                + ethanamine, N-ethyl-   vol 22   38
                + ethane, 1-ethoxy-2-methoxy-   vol 22   29
                + ethanol   vol 22   21,   E22,   23
                                                                      24,   37
                + ethanol (ternary)   vol 22   20,   25
                + ethanol, 2-ethoxy-   vol 22   28
                + ethanol, 2-methoxy-   vol 22   27
                + furan, tetrahydro-   vol 22   35
                + heptane, 1-methoxy-   vol 22   34

Yttrium chloride
          + methanol   vol 22   21
          + methanol (ternary)   vol 22   20,   37
          + nonane, 1-methoxy-   vol 22   34
          + octane, 1-methoxy-   vol 22   34
          + pentane, 1-methoxy-   vol 22   32,   33,   34
          + 1-propanamine   vol 22   39
          + 2-propanamine   vol 22   40
          + 1-propanamine, 2-methyl-N-(2-methylpropyl)-
                                            vol 22   44
          + 1-propanol (ternary)   vol 22   20,   37
          + 2-propanol   vol 22   26
          + 2-propen-1-amine   vol 22   41
          + water   vol 22   25
Yttrium iodide
          + furan, tetrahydro-   vol 22   48
Yttrium nitrate
        see Nitric acid, yttrium salt

Z

Zinc
                + mercury   vol 25   E385 -E387
                                                                      388 - 401
                + mercury   vol 29   E213, 221
                + water   vol 29   E228, 230
Zinc acetate
          see acetic acid, zinc salt
Zinc bisulfite
          see sulfurous acid zinc salt
Zinc, bis(4-amino-N-2-pyrimidinylbenzenesulfonamidato-NN,O)-, (T-4)-
                + water   vol 36   58
Zinc, bis[4-amino-N-(2,6-dimethyl-4-pyrimidinyl)benzenesulfonamidato-,(T-4)-
                + water   vol 36   402
Zinc, bis[4-amino-N-(4,6-dimethyl-2-pyrimidinyl)-benzenesulfonamidato-
NN,O]-, (T-4)-
                + water   vol 36   357
Zinc, bis[4-amino-N-(4-methyl-2-pyrimidinylbenzenesulfonamidato-NN,O)-
                + water   vol 36   268
Zinc, bis(4-amino-N-2-thiazolylbenzenesulfamidato-NN,O)-, (T-4)-
                + water   vol 35   215
Zinc chloride (aqueous)
                + zinc oxide   vol 23   E179, 204
Zinc chloride (in hydrochloric acid)
                + hydrogen sulfide   vol 32   E26,   35
Zinc hydroxide
                + water   vol 23   E156 -E184, 187
                                                                      189, 191
Zinc hydroxide (aqueous)
                + ammonia   vol 23   E178, E179, 214
                + ammonium hydroxide   vol 23   E178, E179
                                                                      254, 255
                + magnesium chloride   vol 23   E180, 256
                + nitric acid, potassium salt   vol 23   247
                + perchloric acid, sodium salt   vol 23   250, 251
                + potassium hydroxide   vol 23   E162, 222 - 224
                + sodium hydroxide   vol 23   E159 -E162, 192
                                                                193, 198, 199
                                                                202, 203, 208
                                                                209, 213, 229
                                                                      260, 261
Zinc hydroxide (multi component)
                + ammonium chloride   vol 23   264
                + hydrochloric acid   vol 23   266, 267
                + nitric acid, ammonium salt   vol 23   264
                + potassium hydroxide   vol 23   265
                + sulfuric acid, ammonium salt   vol 23   264
Zinc nitrate
          see nitric acid, zinc salt
Zinc oxide
                + seawater   vol 23   E180, 188
                + water   vol 23   E156 -E184
                                                                185, 186, 190
Zinc oxide (aqueous)
                + ammonium hydroxide   vol 23   E178, E179, 200
                + arsenic (V) oxide   vol 23   E174, 262, 263
                + chromium (VI) oxide   vol 23   E174 -E176
                                                                      194, 230
                + ethanol 2,2´,2"-nitrilotri-   vol 23   E179, 259
                + glycine   vol 23   E179, 257
                + hydrogen peroxide   vol 23   E180, 237 - 241
                + nitric acid, potassium salt   vol 23   245
                + perchloric acid, sodium salt   vol 23   248
                + phosphoric acid   vol 23   E177, 258
                + phosphorus (V) oxide   vol 23   E177, 195, 196
                + potassium hydroxide   vol 23   E162, 205, 206
                                                                211, 220, 221
                                                                225, 226, 228
                                                                242, 243, 246
                                                                      249, 253

```
Zinc oxide (aqueous)
          + sodium hydroxide               vol 23        E159 -E162,  197
                                                          201,   207
                                                          215 -  219
                                                          227,   244
          + sodium sulfide                 vol 23             E181,  252
          + sulfur dioxide                 vol 23             E177,  236
          + sulfur trioxide                vol 23             E177,  210
          + sulfuric acid, zinc salt       vol 23                    212
          + zinc chloride                  vol 23             E179,  204
Zinc oxide (multi component)
          + ammonia                        vol 23        E178,  269,  303
          + carbon dioxide                 vol 23             E181,  268
          + silver (I) oxide               vol 23                    140
          + sodium chloride                vol 23        E181,  231 -  234
                                                                     268
          + sulfuric acid, ammonium salt   vol 23             E178,  269
          + sodium hydroxide               vol 23             231 -  234
                                                          265,   266
Zinc selenite
          see selenious acid, zinc salt
Zinc sulfadiazine
          see zinc, bis(4-amino-N-2-pyrimidinylbenzenesulfamido-NN,O)-,
              (T-4)-
Zinc sulfamerazine
          see zinc, bis[4-amino-N-(4-methyl-2-pyrimidinyl-
              benzenesulfonamidato-NN,O)-
Zinc sulfamethazine
          see zinc, bis[4-amino-N-(4,6-dimethyl-2-pyrimidinyl)-
              benzenesulfonamidato-NN,O]- (T-4)-
Zinc sulfate
          see sulfuric acid, zinc salt
Zinc sulfite
          see sulfurous acid, zinc salt
Zirconium
          + mercury                        vol 25        E263, E264,
                                                          265,   267
          + mercury                        vol 29        E214,  217
Zn(II) sulfamerazine
          see zinc, bis[4-amino-N-(4-methyl-2-pyrimidinyl-
              benzenesulfonamidato-NN,O)-
Zn(II) sulfamethazine
          see zinc, bis[4-amino-N-(4,6-dimethyl-2-pyrimidinyl)-
              benzenesulfonamidato-NN,O]- (T-4)-
Zn(II) sulfapyridine
          see zinc, bis-(4-amino-N-2-pyridinylbenzenesulfonamidato-NN,O)-,
              (T-4)-
Zn(II) sulfisomidine
          see zinc, bis[4-amino-N-(2,6-dimethyl-4-pyrimidinyl)-
              benzenesulfonamidato-,(T-4)-
Zn(II) sulfanilamide
          see benzenesulfonamide, 4-amino-, zinc salt (2:1)
Zn(II) sulfathiazole
          see zinc, bis(4-amino-N-2-thiazolylbenzenesulfamidato-NN,O)-,
              (T-4)-
```

REGISTRY NUMBER INDEX

Page numbers preceded by E refer to evaluation texts whereas those not preceded by E refer to compiled tables.

50-00-0	vol 27/28	172
	vol 30	246
50-01-1	vol 24	E61, 81-83
	vol 27/28	E64, 103
50-32-8	vol 38	E512, 513-515
50-70-4	vol 35	37
50-99-7	vol 23	71
	vol 27/28	183
	vol 29	E87, 88-94
	vol 30	245
	vol 34	149, 301, 314
	vol 35	39, 178
	vol 36	5
51-79-6	vol 30	253
	vol 34	126
	vol 35	174
52-89-1	vol 29	E50, 82, E87, 91
54-21-7	vol 34	E13-E15, 100-102
56-23-5	vol 21	E9, 67, E81, E83, E87-E91, 128-130, 171-173, 225, 226, E281, E282, 294, E302, 309
	vol 24	E257, E258, 260, 261, 270, 271, 278
	vol 27/28	E685, 686, 687
	vol 29	E178, 182-184
	vol 32	E178, E179, 279
56-40-6	vol 23	257
	vol 30	248, 413
	vol 34	91, 174, 175
	vol 35	31
54-41-7	vol 29	E49, 72
56-49-5	vol 38	E528, 529, 530
56-55-3	vol 38	E477, E478, 479-482, E483, 484, 485
56-56-4	vol 38	523
56-81-5	vol 21	E7, 53, E80, E82, E85, E88-E91, 102, 103, 150, 151, 197, E252, 267
	vol 22	85, 168
	vol 24	E331-E334, 354
	vol 30	153, 243
	vol 32	E20, 24
	vol 34	114, 127, 295, 296, 312, 313, 318, 319
	vol 35	171, 172
	vol 36	44, 45, 172, 173, 259, 260
56-84-8	vol 29	E49, 73
56-85-9	vol 29	E50, 76
57-09-0	vol 27/28	100
57-11-4	vol 24	E125, 128, 129
57-13-6	vol 22	54, 56, 71, 73
	vol 23	73

```
57-13-6      vol 24      E70, E71, 121-123
             vol 27/28   177, 178
             vol 30      81, 253
             vol 31      234, 235, 274, 275, 284, 285, 291, 292
             vol 32      E20, 22
             vol 34      116, 117, 127, 186, 238, 269, 296, 313, 318
             vol 35      76, 99, 172, 173, 273, 288, 300, 309
             vol 36      43, 45, 108, 121, 173, 179, 180, 260, 478, 479,
                         481, 482

57-14-7      vol 21      E10, 73, E88-E91

57-50-1      vol 23      72
             vol 27/28   180, 181
             vol 36      185

57-55-6      vol 34      113, 154

57-60-1      vol 35      41

57-67-0      vol 34      E302, 303-317

57-68-1      vol 36      E303, 304-345, 474, 476, 478, 479

57-97-6      vol 38      522

58-08-2      vol 35      175, 176
             vol 36      46, 184

58-70-3      vol 38      E534, 535-537

58-90-2      vol 20      E24, 25

59-01-8      vol 29      E50, 81

59-26-7      vol 34      119, 249

59-40-5      vol 36      444-446

59-50-7      vol 20      E233, 234

60-18-4      vol 29      E50, 81

60-29-7      vol 21      E85, 200
             vol 22      101, 175, 217, 254, 292, 350, 364
             vol 24      15, 33, 38, 228, 259, 269, 277, 338, 353, 366,
                         E375-E378, 386, 398, 406, 415
             vol 27/28   580, E640, E641, 645, 646
             vol 30      244
             vol 32      E173-E175, 254
             vol 35      48

60-34-4      vol 21      E10, 72, E88-E91

60-35-5      vol 30      253
             vol 34      241

61-73-4      vol 35      166

61-90-5      vol 29      E50, 78

62-53-3      vol 21      E10, E11, 16, 74, 75, E81, E83, E86, E88-E91, 119,
                         162, 213, E281, E282, 286
             vol 22      52
             vol 23      78
             vol 24      E299-E302, 309, 311, 313, 322, 323
             vol 27/28   580, 709
             vol 32      E180-E182, 290-293

62-55-5      vol 34      242

62-56-6      vol 34      125, 191
```

63-68-3	vol 29	E50, 77, E87, 89
63-74-1	vol 34	E13-E15, 16-166
	vol 36	450
63-91-2	vol 29	E50, 80
64-04-0	vol 21	16
64-17-5	vol 21	E5, 38-42, E82, E84, E88-E91, 144, 145, 187
	vol 22	2, 20, 37, 55, 56, 76, 78-80, 91, 99, 112, 125, 126, 146, E156, 157-160, 203, 210, 211, 234, E247, 248-250, 272, 285, 290, 304, 307, 311, 322, E328, 329-331, 348, 360, 361, 378
	vol 23	69, E83, E84, E94, 111
	vol 24	15, 33, 38, 228, 259, 269, 277, E331-E334, 336-338, 351-353, 355, 364-366, 386, 398, 406, 415
	vol 27/28	174, 580, E582-E585, 601-605, 635
	vol 29	E165
	vol 30	151, 152, 243, 254, 328, 411
	vol 32	E20, 23, E171, E172, 242
	vol 34	109-111, 114, 127, 176, 239, 270, 293-296, 311-313, 318, 319, 320
	vol 35	E14, E15, 33, 34, 82, 83, 90, 168-172, 186, 190, 275, 276
	vol 36	42, 44, 45, 110, 111, 172, 173, 188, 189, 260, 261, 263, 267, 338, 380, 381, 384-386, 401, 414, 415
64-18-6	vol 36	381
64-19-7	vol 21	E251, E282, 290
	vol 23	339
	vol 24	E375-E378, 379
	vol 30	248
	vol 32	E176, 270
	vol 34	288
	vol 36	163
64-20-0	vol 24	E61, E62, 84, 85
	vol 27/28	E59, E60, 91
	vol 29	E47, 54
65-85-0	vol 36	337, 341
67-56-1	vol 21	E4, E5, 35-38, E82, E88-E91, 142, 143, 186
	vol 22	1, 20, 21, E22, 23-25, 37, 53, 54, 75-77, 90, 91, 95, 115, 124, 146, E152, 153-155, 203, 209, 246, 307, 322, 327, 359
	vol 23	68
	vol 24	E331-E334, 335, 350, 363
	vol 27/28	173, 580, E582-E585, 596-600, E629, 630-634
	vol 29	E163, E164, E165, 166-168, E192, 203, 204
	vol 30	243, 373, 417, 418
	vol 31	126
	vol 32	E171, E172, 238-241
	vol 35	E14, E15, 43, 44, 47, 81
	vol 36	186, 187, 382, 383, 412, 413
67-63-0	vol 21	E5, 38, 44, E88-E91
	vol 22	26, 83, 84, 166, 167, 212, 251, 332, 362
	vol 24	E331-E334, 340, 356, 368
	vol 27/28	E582-E584, E588, E589, 610, 636
	vol 29	E164, E165, 169
	vol 34	150, 316, 322, 323
	vol 35	45, 187, 188, 230
	vol 36	6-8, 10, 14, 18, 49-51, 55, 56, 192, 225, 262, 439, 440
67-64-1	vol 21	E7, E8, 53, 55, E281, E282, 290, E302, 308
	vol 23	67
	vol 24	E125, 126, 128, E375-E378, 381, 382
	vol 27/28	580, E655, 656-659
	vol 29	E192, 205, 206
	vol 30	154-156, 194, 247, 329

67-64-1	vol 31	96, 97, 124, I25, 166, 167
	vol 32	E176, E178, 274
	vol 34	161, 177, 179, 192-199, 263, 272, 273, 281, 317
	vol 35	194, 238
	vol 36	343
67-66-3	vol 21	E9, 16, 67, 68, E81, E83, E89, 128, 173, 223, 224, E302
	vol 21	309
	vol 24	15, 33, 38, 228, E257, E258, 259, 269, 277, 338, 353, 366, 386, 398, 406, 415
	vol 27/28	580
	vol 32	E178, E179, 279, 282
	vol 34	159, 160, 201, 203, 262, 279, 283
	vol 35	7, 8, 10, 54-57, 60, 92, 93, 100, 103, 191-193, 218, 248, 301
	vol 36	47, 48, 91, 95, 99, 101, 113, 114, 123, 204, 205, 259, 265, 266, 278, 291, 344, 399, 400, 426, 427
67-68-5	vol 22	64, 120, 149, 205, 231, 245, 271, 288, 310, 325, 357
	vol 24	E375-E378, 380
	vol 27/28	182, E702, 705, 706
	vol 30	416, 419, 446
	vol 32	E185, 319
68-04-2	vol 34	97
68-12-2	vol 21	E16, E81, E82, E86, E88-E91, 113, 114, 159, 209
	vol 22	17, 96, 145, 202, 229, 269
	vol 24	E125, 127, 129, E299-E302, 303-305, 319
	vol 27/28	720
	vol 30	103, 160, 414, 415, 444, 445, 458, 459
	vol 32	E180, E183, 299, 302-304
	vol 34	245
	vol 36	176-178, 202
68-35-9	vol 36	E124, 125-205, 453-455, 459-461, 466-472, 474-476, 478, 479, 483-485
69-65-8	vol 30	245
69-79-4	vol 35	42
70-34-8	vol 22	51
71-23-8	vol 21	E5, 38, 43, E82, E84, E88-E91, 146, 147, 188
	vol 22	3, 20, 37, 76, 82, 91, 95, E127, 128, 129, E162, 163-165, 213, 233
	vol 23	70
	vol 27/28	175, E582-E564, E586, E587, 606-609
	vol 30	243
	vol 35	84
	vol 36	190, 191, 388, 389, 416, 417
71-36-3	vol 21	E5, 6, 45-47, E80, E82, E84, E88-E91, 105,
	vol 21	148, 189
	vol 22	4, 113, 130, 169, 170, 196, 305
	vol 24	E331-E334, 342, 357, 358, 369
	vol 27/28	E582-E584, E589, E590, 611-613
	vol 32	E171, E172, 243
	vol 34	151
	vol 35	85
	vol 36	193, 194, 390, 391, 418, 419
71-41-0	vol 22	5, 131, 171, 172
	vol 27/28	E582-E584, E590, E591, 615-618
	vol 34	152
	vol 35	86
	vol 36	195, 196, 392, 393, 420, 421, 485
71-43-2	vol 21	E3, E4, 13, 28-32, E80, E82, E84, E88-E91, 100, 140, 184, 185, E281, E282, 286

71-43-2	vol 22	20, 51, 75, 76
	vol 24	15, E218-E221, 225-229, 246, E251, 252-254, 259, 338, 386, 415
	vol 27/28	E493, E494, 498-501, E511, 512-521, 578-580
	vol 29	E149, 154-156
	vol 32	E170, E171, 224-226
	vol 34	158
	vol 35	50-53
	vol 36	290, 292, 293
	vol 37	E63-E85, 86-187, E188, 189-192
71-48-7	vol 36	116
71-91-0	vol 24	E62, 86-88
	vol 27/28	E60, 92-94
	vol 29	E47, 55
72-14-0	vol 35	E110-E112, 113-206
	vol 36	451, 453-458, 468-472, 483, 487
72-17-3	vol 34	93
72-19-5	vol 29	E49, 74
73-32-5	vol 29	E50, 79
74-82-8	vol 32	75, 76, 213, 214
74-84-0	vol 27/28	194, 195, E248, E249, 250-265, 378-386, 483, 484, 553-555
74-85-1	vol 27/28	E477, 478-490
74-89-5	vol 21	E80-E91, 92-134
	vol 23	E83, E84, E93, 110
74-90-8	vol 21	E302, E303
74-96-4	vol 32	E178, E179, 280
74-98-6	vol 24	E1, E2, 3-15, E39, 40-47, E54, 55, E58-E72, 73, 75, 78-81, 84, 86, 89, 90, 92, 94, 97, 99-105, 111, 112, 114, 116, 118, 121, 122, E125, 126-129, E130-E134, 135-166, E196, E197, 198-209, E218-E221, 222-229, E251, 252-256, E257, E258, 259-268, E299-E302, 303-318, E331-E334, 335-348, E375-E378, 379-394, E411, 412, E414, 415, 416, 418, 422-427
	vol 27/28	194, 195, E266, 267-280, 378-382, 387-406, 553-555, 556
	vol 29	E192, 200
	vol 32	E166, 191
75-04-7	vol 21	E252, 253-256
	vol 22	369
75-05-8	vol 21	E86, 207, E302, 303
	vol 22	18
	vol 23	E1-E3, 9, 82
	vol 32	E180, E182, 298, 299
75-09-2	vol 21	E302, 309
75-12-7	vol 30	253
	vol 31	272, 276, 277
	vol 34	240
75-15-0	vol 21	E281, E282, 295, 300
	vol 24	E218-E221, 222
	vol 27/28	580, E702, 703, 704
	vol 32	E185, 318
75-21-8	vol 24	E411, 412, 413
	vol 27/28	648, 649

75-25-2	vol 32	E178, E179, 279, 283
75-28-5	vol 24	E34, 35, E58-E72, 77, 83, 124, E130-E134, 189-195, E196, E197, 201, E218-E221, 246-250, E251, E257, E258, 277-282, E299-E302, 328-330, E331-E334, 363-374, E375-E378, 405-410, E414, 420
	vol 27/28	E297, 298-302, 485-490
	vol 32	E166, E167, 194, 195
75-31-0	vol 21	E253, 268-270, 274
	vol 22	40, 67, 137, 142, 144, 180, 193, 199, 221, 228, 239, 244, 259, 265, 268, 297, 303, 341, 347, 353, 369, 375, 377
75-39-8	vol 22	194, 200
75-45-6	vol 27/28	689, 690
75-50-3	vol 21	E80-E91, 115, 116, 178-233, E234, 250, 251
75-52-5	vol 21	E86, 208, E252, 279
75-56-9	vol 21	E281, E282, 290
75-65-0	vol 21	46
75-71-8	vol 27/28	688
75-76-3	vol 21	E11, 19
75-83-2	vol 24	E131, E133, E134, 143, 144, 170, 191
	vol 27/28	E212, E213, 223
	vol 29	E107-E109, 120, 121
	vol 37	E283, E284, 285-288
75-85-4	vol 22	60
	vol 30	243
76-01-7	vol 32	E178, E179, 280
76-05-1	vol 21	E281, E282, 290, 301
76-13-1	vol 27/28	684
76-64-1	vol 30	E4, 23, 100
77-92-9	vol 34	5, 11, 80-82, 96, 237, 291, 292
	vol 35	6, 77-79, 98, 161, 271, 272, 310-315
	vol 36	104, 166-168, 332, 336, 365, 378, 379, 441, 442, 476, 480
78-10-4	vol 21	E281, E282, 296, 297, E302, 306, 307, E310, 311, 312, E315, 318, 319, 323, 324, 327, 328
	vol 32	E185, 277, 278, E330, 334, 335
78-40-0	vol 24	E375-E378, 392, 402
	vol 27/28	727
	vol 32	E183, 322
78-78-4	vol 27/28	317, 318, 411
	vol 29	E107-E109, 112
	vol 37	E31-E34, 35-41
78-79-5	vol 37	E7, E8, 9, 10
78-81-9	vol 21	E252, 277
78-83-1	vol 21	E6, 45, 46
	vol 22	58
	vol 27/28	E590, 614
78-92-2	vol 21	E6, 46
	vol 22	59
	vol 35	190

78-93-3	vol 21	179, 180
79-01-6	vol 21	E302, 309
	vol 32	281
79-05-0	vol 34	243
79-16-3	vol 30	218, 256
	vol 34	246
79-20-9	vol 23	E1-E3, 9, 82
	vol 27/28	662
	vol 30	250
79-27-6	vol 32	E178, E179, 280
79-29-8	vol 24	E131, 145
	vol 29	E107-E109, 122
	vol 37	E289-E291, 292-295
79-31-2	vol 2	298
79-34-5	vol 21	E302, 309
	vol 32	E178, E179, 280
79-43-6	vol 24	E257, E258, 272
	vol 32	E178, E179, 282
80-32-0	vol 36	98, 99
80-34-2	vol 35	311
80-35-3	vol 36	102-114
80-56-8	vol 21	E281, E282, 285
	vol 27/28	E491, 492
80-74-0	vol 35	94, 95
83-32-9	vol 27/28	578, 579
	vol 38	E332-E334, 335-340
84-77-5	vol 21	E281, E282, 296, 297, E302, 306, 307, E315, 318, 319, 323, 324, 327, 328
	vol 32	E176, E178, 277, 278, E330, 332, 333
85-01-8	vol 27/28	541, 542, 578, 579
	vol 38	E408-E410, 411-423
86-73-7	vol 27/28	578, 579
	vol 38	E385, E386, 387-390
87-61-6	vol 20	E33, 34
87-64-9	vol 20	E235, 236
87-68-3	vol 22	74, 98, 123, 151, 208
87-78-5	vol 34	283, 284
	vol 35	38
	vol 36	258, 259, 277, 278
87-86-5	vol 20	E6, E7, 8-14
88-04-0	vol 20	E246, 247
88-06-2	vol 20	E45, 46-48
88-72-2	vol 21	E86, E88-E91, 121, 164, 216, 217
	vol 24	E299-E302, 324
90-11-9	vol 21	E81, E87, 133, 177, 231, 232

90-12-0	vol 27/28	545-547, 553-555, 578, 579
	vol 38	E311, E312, 313-318
90-13-1	vol 21	E87, 233
	vol 24	E257, E258, 289, 298
	vol 27/28	701
91-17-8	vol 21	E281, E282, 285
	vol 24	E218-E221, 236, 244
	vol 27/28	E497, 510
	vol 32	E166, E169, 222
	vol 38	E287, 288-290
91-20-3	vol 27/28	578, 579
	vol 38	E236-E238, 239-257
91-22-5	vol 21	E11, 75, E81, E83, E85, 117, 160, 210
	vol 24	E299-E302, 313, 322
	vol 27/28	710, 711
	vol 32	297
91-51-3	vol 27/28	E453, 464, 487
91-57-6	vol 38	E319, 320, 321
91-66-7	vol 21	16, E81, E83, E86, E88-E91, 127, 170, 222
	vol 32	E180-E182, 295
92-24-0	vol 38	E493, 494-496
92-51-3	vol 21	E2, E3, 24, 25, E88-E91
	vol 24	E218-E221, 238, 245, 249
	vol 32	E166, E169, 223
92-52-4	vol 27/28	578, 579
	vol 38	E341-E343, 344-354
93-89-0	vol 21	E82, E85, E88-E91, 110, 156, 204, 205
94-19-9	vol 35	E289, 290-301
	vol 36	452
95-46-5	vol 21	E81, E87, 132, 176, 230
95-47-6	vol 27/28	E495, E496, 504, 578, 579
	vol 29	E151, 160
	vol 32	E170, E171, 230
	vol 38	E9-E12, 13-22
95-50-1	vol 20	E73-E77, 78-88
95-57-8	vol 20	E183, E184, 185-189
95-63-6	vol 38	E169, E170, 171-176
95-75-0	vol 20	225
95-93-2	vol 38	258, 259
95-94-3	vol 20	E22, 23
95-95-4	vol 20	E43, 44
96-14-0	vol 24	E131, 142
	vol 27/28	336, 337
	vol 29	E107-E109, 119, E165
	vol 37	E307, E308, 309-313
96-22-0	vol 21	179, 180
96-31-1	vol 34	122, 188
96-37-7	vol 37	E212, E213, 214-220

97-00-7	vol 22	51
98-01-1	vol 21	E281, E282, 296, 297, E302, 306, 307, E310, E311, 312, E315, 318, 319, 323, 324, 327, 328
	vol 24	E375-E378, 387, 407
	vol 30	255
	vol 32	E176, E178, 277, 278, E330, 334, 335
98-06-6	vol 38	E265, E266, 267-270
98-18-0	vol 34	7-12
98-82-8	vol 21	E281, E282, 286
	vol 29	E151, E152, 161
	vol 32	E170, E171, 233
	vol 38	E187-E189, 190-200
98-86-2	vol 21	E82, E85, 107, 154, 201, 202
	vol 24	E299-E302, 312, 400
98-92-0	vol 34	118, 248
98-95-3	vol 21	E86, E88-E91, 118, 161, 211, 212, E281, E282, 292, 298, 299, E302, 304, E310, 313, 314, E315, 316, 317, 321, 322, 325, 326
	vol 24	E299-E302, 307, 308, 312, 324, 329, 330
	vol 27/28	580, 722
	vol 29	E187, 188, 189
	vol 32	E180, E182, 277, 278, 315, 316, E330, 332, 333
99-65-0	vol 22	51
99-87-5	vol 38	E260, 261-263
100-40-3	vol 38	92
100-41-4	vol 24	E218-E221, 232
	vol 32	E170, E171, 233
	vol 38	E59-E66, 67-91
100-42-5	vol 38	E1, E2, 3-8
100-44-7	vol 21	E10, 28
100-46-9	vol 21	E83, E86, E88-E91, 122, 166, 218
	vol 24	E299-E302, 317, 325
100-47-0	vol 21	E88-E91, 120, 163, 215
	vol 24	E299-E302, 315, 325
	vol 32	E180, E182, 300
100-51-6	vol 21	16, E80, E84, E88-E91, 106, 153, 192-194
	vol 24	E331, E334, 345, 347, 361, 373, 400
	vol 27/28	677
	vol 32	E171, E172, 245
100-61-8	vol 21	E81, E83, E86, E88-E91, 123, 165, 214
	vol 24	E299-E302, 316, 326
	vol 32	E180, E182, 293
100-66-3	vol 27/28	E643, E644, 654
100-97-0	vol 27/28	179
100-90-3	vol 36	358-362, 477, 480-482
101-02-0	vol 21	E8, 63
101-81-5	vol 24	E218-E221, 239
	vol 27/28	E496, 509, 548, 549
	vol 32	E170, E171, 235
	vol 38	391

102-71-6	vol 21	E302, 308
	vol 23	E179, 259
	vol 32	E80, E81, 112-115
	vol 36	175
102-76-1	vol 21	E9, 65
	vol 27/28	679
103-12-8	vol 34	198
103-30-3	vol 38	424
103-65-1	vol 32	E170, E171, 233
	vol 38	E201, E202, 203-210
103-69-5	vol 21	E81, E83, E86, E88-E91, 124, 167, 219
	vol 32	E180, E182, 295
103-73-1	vol 21	E81, E82, E85, 108, 155, 203
	vol 24	E374-E377, 390, 399
	vol 32	E173-E175, 256
103-84-4	vol 30	489
104-51-8	vol 27/28	539, 540, 542
	vol 38	E277, E278, 279-285
105-59-9	vol 32	E80, 109-111
106-37-6	vol 20	E62, 63, 64
106-38-7	vol 20	230
106-39-8	vol 20	E53, 54
106-41-2	vol 20	E150, 151, 152
106-42-3	vol 27/28	E495, E496, 507, 578, 579
	vol 38	E43-E46, 47-58
106-43-4	vol 24	E257, E258, 284, 292
	vol 27/28	696
106-46-7	vol 20	E99, E100, 101, 102, E103-E105, 106-115
106-48-9	vol 20	E195, E196, 197-201
106-93-4	vol 24	E218-E221, 272
	vol 27/28	E637, 638
	vol 32	E178, E179, 281, 283
106-97-8	vol 21	E80, E88-E91, 92, 93, E252, 253, 254
	vol 24	E16, E17, 18-33, E48, 49-53, E54, 56, 57, E58-E72, 74, 76, 82, 85, 87, 88, 91, 93, 95, 96, 98, 106-110, 113, 115, 117, 119, 120, 123, E130-E134, 167-188, E196, E197, 198-200, 210-217, E218-E221, 240-245, E251, E257, E258, 269-276, E299-E302, 319-327, E331-E334, 350-362, E375-E378, 395-404, E411, 413, E414, 417, 419, 421, 422, 428, 429
	vol 27/28	196-199, E281, 282-296, 387-389, 407-409, 420-422, 436-440, 447-449
	vol 29	E191, E192, 201, 202
	vol 32	E166, E168, 192, 193
107-06-2	vol 21	E9, 67, E281, 282, 294, 302, 309
	vol 32	281
107-10-8	vol 21	E252, 259, 263, 267
	vol 22	39, 137, 142, 144, 180, 199, 228, 244, 259, 268, 283, 318, 347, 369, 375, 377
107-11-9	vol 22	41, 137, 180, 221, 239, 259, 297, 341, 353, 369
107-12-0	vol 21	E302, 303

107-15-3	vol 22	192, 198
	vol 30	101, 159
107-18-6	vol 22	128, 213, 249
107-21-1	vol 21	E6, E7, 50, E80, E82, E85, E88-E91, 101, 102, 149, 195, E252, 267, 274, 278, E281, E282, 290
	vol 22	57, 81
	vol 24	E331-E334, 339, 354, 367
	vol 30	98, 157, 243
	vol 32	E171, E172, E175, 246-249
	vol 34	112, 153
107-83-5	vol 24	E131, 141
	vol 37	E296-E299, 300-306
107-88-0	vol 21	E252, 267
108-08-7	vol 37	E469, E470, 471-475
108-20-3	vol 29	E172, 174
108-24-7	vol 24	E375-E378, 383, 397
	vol 32	E176, 271
108-32-7	vol 24	E375-E378, 385, 405
	vol 27/28	E663, 664-675
	vol 32	E173-E176, E177, 260, 266-269
108-36-1	vol 20	E59, 60, 61
108-37-2	vol 20	E51, 52
108-38-3	vol 21	82, E88-E91, 100, 140, 185
	vol 24	E218-E221, 233, 234, 242
	vol 27/28	E495, E496, 505, 506, E533, 534-536
	vol 32	E170, E171, 231
	vol 38	E23-E28, 29-41
108-39-4	vol 27/28	680
108-43-0	vol 20	E190, 191-194
108-67-8	vol 21	16, E80, E82, 100, 141, 185
	vol 24	E218-E221, 247
	vol 27/28	537, 538, 578, 579
	vol 32	E170, E171, 232-234
	vol 38	E177, E178, 179-186
108-70-3	vol 20	E41, 42
108-86-1	vol 20	E132-E136, 137, 149
	vol 21	E10, 28, E81, E83, 117, 131, 175, 228
	vol 24	E257, E258, 281
	vol 29	E178, 186
	vol 32	E178, E179, 224, 286, 288
108-87-2	vol 21	E2, 23
	vol 29	E134-E137, 142
	vol 27/28	E450, E451, 457
	vol 32	E168, E169, 217
	vol 37	E445-E447, 448-453
108-88-3	vol 21	E4, 13, 28, 33, E80, E81, 100, 140, 185, E281, E282, 287
	vol 24	E218-E221, 230, 231, 240, 241
	vol 27/28	E494, 502, 503, E523, 524-532, 553-555, 556, 578, 579
	vol 29	E150, 157-159
	vol 32	E170, E171, 224, 227-229
	vol 34	284
	vol 36	258, 277
	vol 37	E369-E379, 380-424, E425, 426-431

108-90-7	vol 20	E153-E157, 158-182
	vol 21	E4, 10, 28, 70, E83, 175
	vol 24	E257, E258, 263, 272, 280, 283, 291
	vol 27/28	E692, 693-695
	vol 29	E179, 185
	vol 32	E178, E179, 199, 224, 285, 287
108-91-8	vol 24	E299-E302, 310
	vol 27/28	708
108-93-0	vol 21	E6, 49, E281, E282, 288, E315, 320
	vol 27/28	E637, 638, 639
108-94-1	vol 21	179, 180
	vol 27/28	661
108-95-2	vol 24	E331-E334, 346, 360, 371
	vol 27/28	678
	vol 30	249
	vol 32	E20, 78, E176, E178, 275
109-65-9	vol 32	E178, E179, 283
109-66-0	vol 21	E281, E282, 283
	vol 24	E130, E132-E134, 135, 167, 189, 202-205
	vol 27/28	E202-E204, 216, E303, 304-313, 383, 384, 390-393
		410-413, 423-426
	vol 29	E102-E106, 110, 111
	vol 32	E166-E168, 188, 196, 197
	vol 37	E42-E46, 47-62
109-67-1	vol 37	28
109-68-2	vol 37	29
109-73-9	vol 22	42, 142, 144, 193, 199, 228, 239, 259, 265, 268,
		283, 297, 341, 347, 369, 375, 377
109-74-0	vol 21	E302, 303
109-87-5	vol 22	27, 86, 100, 133, 165, 173, 214, 216, 233, 249,
		252, 273, 291, 312, 329, 349, 363
109-89-7	vol 21	E252, 278
	vol 22	15, 38, 341, 369
	vol 30	251
	vol 34	84
109-99-9	vol 21	E281, E282, 291
	vol 22	13, 35, 46, 48, 104, 109, 114, 118, 135, 178, 190,
		219, 226, 243, 257, 263, 266, 281, 284, 295, 301,
		306, 316, 321, 339, 345, 356, 368, 373, 381
	vol 27/28	E643, 653
	vol 32	E173-E175, 263
110-54-3	vol 21	E1, 13, 14, 15, E80, E82, E83, E84, E88-E91, 94,
		95, 135-138, 178, 181, 182, E234, 235-249, E252,
		255, 256, 259-266, 268-273, 280, E281, E282, 283
	vol 24	E130-E134, 135, 136-140, 167-169, 189, 190
	vol 27/28	E202-E205, 216-222, E320, 321-335, 414-417,
		427-435, 441-443
	vol 29	E102-E106, 113-118, E165
	vol 32	E166-E168, 188, 190, 198, 199
	vol 36	203, 398, 401
	vol 37	E314-E323, 324-365, E366, 367, 368
110-71-4	vol 22	61
110-80-5	vol 21	E281, E282, 298, 299, E302, 304, 305, E310, 313,
		314, E315, 316, 317, 321, 322, 325, 326
	vol 22	10, 28, 87, 133, 165, 174, 215, 216, 233, 249, 253,
		273, 291, 312, 329, 363
	vol 24	E331-E334, 341, 370
	vol 32	E171, E172, 277, 278, E330, 332, 333

110-80-5	vol 35	46, 89, 189
	vol 36	264, 282, 342, 387
110-82-7	vol 21	E2, 20-23, E88-E91, E281, E282, 285
	vol 24	E218-E221, 223, 224
	vol 27/28	E450, 455, 456, 463, E465, 466-474
	vol 29	E133, E134, 138-141, E147
	vol 32	E169, 190, 216
	vol 37	E221-E229, 230-271
	vol 38	544
110-83-8	vol 21	27, 31, 32, E281, E282, 285
	vol 27/28	463, E491
	vol 29	E147, 148
	vol 37	E197-E199, 200-208
110-85-0	vol 36	110-112
110-86-1	vol 21	E81, E83, E86, 117, 160, 210
	vol 22	69, 70, 71, 105, 110, 111, 121, 150, 206, 232, 289, 326, 358
	vol 23	77
	vol 27/28	707
	vol 30	251
110-89-1	vol 27/28	707
110-89-4	vol 30	251
110-91-8	vol 22	195, 201
110-96-3	vol 22	16, 44, 68
110-97-4	vol 32	E80, E82, 116, 117, 160-165, 308
110-98-5	vol 27/28	E641, E642
111-01-3	vol 24	E131-E134, 165, 186
	vol 27/28	E213, E214, 247
111-27-3	vol 22	6
	vol 27/28	E582-E584, E591, E592, 618, 620, 621
111-42-2	vol 23	E93, 125, E279, 304
	vol 26	E5, E74, 89
	vol 27/28	185, 717
	vol 32	E79, E80, E82, 95-107, 140-154
	vol 36	109, 174
111-43-3	vol 22	315, E334, 337, 338, 367
	vol 27/28	E641, 647
111-44-4	vol 21	E281, E282, 298, 299, E302, 304, 305, E310, 313, 314, E315, 316, 317, 321, 322, 325, 326
	vol 24	E257, E258, 262, 279
	vol 32	E178, E179, 277, 278, E330, 332, 333
111-46-6	vol 21	E7, 51, 52, E80, E88-E91, 102, E252, 267, 274, 275
	vol 34	115, 157
111-65-9	vol 21	E2, 13, 16
	vol 24	E130, E132-E134, 147, 148, 172, 173, 192, 206-207, 213-215
	vol 27/28	E202-E204, E206, 225, 227, 228, E349, 350-352, 397-399, 407-409, 412-417, 444-448, 578, 579
	vol 29	E102-E106, 125-127
	vol 32	E166-E169, 188, 190, 202, 213, 214
	vol 38	E135-E141, 142-157, E158, 159, 160
111-66-0	vol 38	111, 112
111-67-1	vol 38	113
111-69-3	vol 21	E11, 76-78

```
111-70-6      vol 22       7
              vol 24       E331-E334, 344
              vol 27/28    E582-584, E592, 622, 623

111-77-3      vol 32       E173-E175, 257

111-83-1      vol 24       E257, E258, 282, 290, 294
              vol 27/28    697
              vol 32       E178, E179, 284

111-84-2      vol 21       E80, E88-E91, 96, 97
              vol 24       E130, E132-E134, 149, 174, 194
              vol 27/28    E202-E204, E207, 230, E349, 355-359, 552
              vol 32       E166-E168, 188, 203
              vol 38       E225-E227, 228-235

111-85-3      vol 21       E9, 69
              vol 24       E257, E258, 285, 293
              vol 27/28     698

111-86-4      vol 21       16, E86, E88-E91, 125, 169, 220
              vol 24       E299-E302, 318, 327

111-87-5      vol 21       E6, 48, E81, E82, E84, E88-E91, 109, 148, 191,
                           E252, 267
              vol 22       8
              vol 24       E331-E334, 348, 349, 362, 374
              vol 27/28    E582-E584, E592, 623-625
              vol 29       E165, 170, 171
              vol 32       E171, E172, 244
              vol 35       87
              vol 36       197, 198, 394, 396, 422, 423

111-96-6      vol 32       E173-E175, 257

112-14-1      vol 21       E8, 60

112-24-3      vol 21       E253, 275

112-27-6      vol 21       E253, 267, 274, 275
              vol 27/28    669
              vol 32       E173-E175, 249, 251

112-30-1      vol 27/28    E582-E584, E593, E594, 626, 627
              vol 35       88
              vol 36       199, 200, 396, 397, 424, 425

112-40-3      vol 21       E2, 13, 16, E281, E282, 283
              vol 24       E130, E132, E134, 152, 177, 178
              vol 27/28    E202-E204, E208, 232, 233
              vol 32       E166-E168, 189, 190, 208
              vol 38       E376, E377, 378-384

112-42-5      vol 27/28    E582-E584, E594, 628

112-49-2      vol 32       E173-E175, 257

112-53-8      vol 27/28    E582-E584, E594, 628

112-60-7      vol 21       E253, 275

112-95-8      vol 24       E131-E134, 162, 163, 183, 184, 195
              vol 27/28    E202-204, E210, 244, 245, E370, 376
              vol 38       525, 526

114-63-6      vol 34       103

115-07-1      vol 24       E251, 256

115-20-8      vol 21       E7, 16, 54, E85, E88-E91, 134, 174, 227
              vol 24       E257, E258, 272
              vol 32       E178, E179, 282

115-68-4      vol 34       271, 272
```

116-43-8	vol 35	230
116-44-9	vol 36	439
118-74-1	vol 20	E1, 2
118-79-6	vol 20	E30, 31, 32
119-36-8	vol 24	E375-E398, 390, 401
	vol 32	E176, E178, 273
119-64-2	vol 21	E281, E282, 285, 286, E302, 308
	vol 24	E251, 255, 256
	vol 27/28	543, 544
120-12-7	vol 38	E396-E398, 399-404, E405, 406, 407
120-34-3	vol 34	277, 281
120-71-6	vol 23	E93, 126
120-82-1	vol 20	E35, 36-40
120-83-2	vol 20	E116, 117
120-92-3	vol 21	179, 180
	vol 27/28	660
121-44-8	vol 21	E11, 75, E252, 279, 280
	vol 22	15, 66
121-46-0	vol 37	432
121-61-9	vol 34	181-192
	vol 36	450
121-69-7	vol 21	E81, E83, E86, E88-E91, 126, 168, 221
	vol 24	E299-E302, 317, 326
	vol 32	E180-E182, 293, 294
122-11-2	vol 36	E404, 406-427
122-52-1	vol 21	61
122-62-3	vol 24	E375-E378, 394, 410
123-39-7	vol 27/28	719
123-51-3	vol 21	E6, 45, E84, 190
	vol 24	E331-E334, 343, 359
	vol 27/28	580, E582-E584, E591, 619
123-75-1	vol 27/28	707
123-91-1	vol 21	E2, 20, E81, E82, E85, 104, 152, 198
	vol 22	19, 36, 37, 47, 81, 90, 91, 94, 132, 141, 143, 176, 191, 197, 218, 227, 237, 256, 264, 267, 275, 282, 294, 302, 315, 338, 346, 367, 374, 376
	vol 24	E375-E378, 384, 396
	vol 27/28	176, E642, 650-653
	vol 30	412
	vol 32	E173, 264
124-09-4	vol 21	E252, 275
124-18-5	vol 21	16, E80, E82, E83, E84, E88-E91, 98, 99, 139, 183, E281, E282, 283
	vol 24	E130, E132-E134, 149-151, 174-176, 193, 194, 208, 209, 216, 217
	vol 27/28	E202-E204, 207, 230, E360, 361-369, 400-406, 418, 419, 578, 579
	vol 29	E102, E104, E106, 131, 132
	vol 32	E166-E168, 188, 204-207
	vol 38	E294-E296, 297-307, E308, 309, 310

124-38-9	vol 23	E1-E3, 8, 80, 268, 305, 306, 340, 341
	vol 27/28	200, 201, 436-446, E737, 738-748, 757-760
	vol 32	E20, 21, E27, 48, 50, 68-78, E79-E83, 84-165,
124-40-3	vol 21	E80-E91, 135-177
126-33-0	vol 30	102
	vol 32	E185, 117, 320, 321
126-71-6	vol 24	E375-E378, 392, 402
	vol 27/28	727
	vol 32	E183, 322
126-73-8	vol 21	E8, 64
	vol 22	63, 119, 136, 148, 179, 220, 238, 258, 277, 287, 296, 309, 317, 324, 340, 352
	vol 24	E375-E378, 392, 402
	vol 27/28	727, 728
	vol 32	E184, 259, 332-325
127-08-2	vol 31	221
	vol 33	E26, 27-30, E35, 36-39, E50, 51-55, 56, E57-E60, 61-71, 72, E73, 74-76, 137-139, 153, E154, 155, 156, E157, 158, E159, 160, E161, 162, E163, 164-166 E167, 168-171, E172, 173, 174, E175, 176-179
127-09-3	vol 23	339
	vol 26	E191, E192, 234, 301, 302
	vol 33	E31, 32, 33, E43, 44, 45, E57-E60, 61-71, E77, 78-80, E85, 86, 87, E88, 89, 90, E91, 92-94, 219-221, 242, E243, 244-247, E248, 249, E250, 251, 252, E253, 254-266, E267, 268-272
	vol 34	86, 288
	vol 36	163
127-18-4	vol 21	E302, 309
	vol 32	E178, E179, 281
127-19-5	vol 32	E180, E183, 299
	vol 34	247
	vol 36	201
127-57-1	vol 36	57
127-69-5	vol 35	E64, 65-93
127-71-9	vol 34	275, 276
	vol 36	466-468, 484
127-73-1	vol 36	E269, 270-278, 464, 465, 473, 477, 480-482, 485
127-74-2	vol 36	E228, 229-240, 463, 465, 473, 475, 477, 480-482,485
127-76-4	vol 35	219-229
	vol 36	451, 463, 464, 473, 485
127-79-7	vol 36	E243, 244-267, 456-461, 466-472, 474, 476, 478, 479 483-485
129-00-0	vol 38	E446-E448, 449-455, E456, 457-460
132-64-9	vol 32	E176, E178, 276
133-10-8	vol 34	104
133-60-8	vol 34	199
133-80-9	vol 34	216
135-19-3	vol 26	E5, 71
135-98-8	vol 38	E271, 272-276
137-40-6	vol 33	E99, 100-102, E103, 104, 222, 242, 273-276

137-40-6	vol 34	87
138-39-6	vol 34	322
138-43-2	vol 34	195
139-33-3	vol 31	116
141-43-5	vol 21	75, E302, 308
	vol 22	65
	vol 23	E93, 124, E279, 304
	vol 27/28	184, 716
	vol 30	99, 158
	vol 32	E79, E81, E82, 84-94, 118-139, 313
141-53-7	vol 26	E191, E192, 216-221, 231
	vol 33	4, 7, E15, 16, 17, 19-21, 23, 218, 221, 222, E223, 224, 225, E226, 227, 228, E229, 230-235, E236, E237, 238-241
	vol 34	85
	vol 36	381
141-78-6	vol 21	199
	vol 22	14, 62
	vol 32	199
	vol 35	91
141-97-9	vol 21	E281, E282, 290
142-29-0	vol 37	E2, E3, 4-6
142-62-1	vol 24	E375-E378, 389, 399, 408
	vol 32	E176, 272
142-68-7	vol 27/28	E643, 653
142-72-3	vol 33	56, E85, 86, 87
142-82-5	vol 21	E83, 179, 180, E281, E282, 283
	vol 24	E130, E132-E134, 146, 147, 171, 172, 192, 210-212
	vol 27/28	E202-E206, 219, 224-226, E338, 339-348, 385, 386, 394-396
	vol 29	E102-E106, 123, 124
	vol 32	E166-E168, 188, 200, 201
	vol 37	E485-E491, 492-515, E516, 517
142-96-1	vol 34	166
	vol 29	E172, E173, 175
143-08-8	vol 22	9
	vol 27/28	E582-E584, E593, 626
143-19-1	vol 24	E69, 113
143-24-8	vol 27/28	681
	vol 32	E173-E175, 257-260
143-74-8	vol 29	E87, 88-94
144-33-2	vol 35	107, 160
	vol 34	173
	vol 36	164, 165, 233, 234, 257, 273, 335, 362
144-55-8	vol 27/28	E77, E78, 157
	vol 29	E87, 88-94
	vol 30	202, 217
	vol 32	48
	vol 35	69, 71
	vol 34	56, 76
	vol 36	144, 151, 372, 374
144-74-1	vol 35	213
144-80-9	vol 34	E204, 205-215, 217-264

144-82-1	vol 35	E257, 258-276
144-83-2	vol 36	E19, E20, 21-53, 462
145-42-6	vol 35 vol 36	162 169
147-55-8	vol 29	E87, 88-94
151-21-3	vol 27/28 vol 35 vol 36	189-191 274 171
152-47-6	vol 36	441
155-91-9	vol 36	365
156-54-7	vol 33	E108, E109, 110-112, E113, 114, 115, E116, 117, 118, E223, 224, E243, 244-246, E277, E278, 279, E280, 281, 282, E283, 284, E285, 286, E287, 288, E289, 290, 291
191-07-1	vol 38	543
191-24-2	vol 38	531, 532
192-97-2	vol 38	516, E517, 518, 519
198-55-0	vol 38	520
206-44-0	vol 27/28 vol 38	578, 579 E440, 441-445
213-41-1	vol 38	392
213-46-7	vol 38	540
215-58-7	vol 38	533
217-59-4	vol 38	E497, 498-500
218-01-9	vol 38	E486, E487, 488-492
218-32-9	vol 27/28	578, 579
224-41-9	vol 38	538, 539
238-84-3	vol 38	475
243-17-4	vol 38	476
280-57-9	vol 27/28	186, 187
287-92-3	vol 37	E13-E15, 16-21
291-64-5	vol 37	454
292-64-8	vol 27/28 vol 38	E252, E253, 462 110
294-93-9	vol 32	E173-E176, 265
298-07-7	vol 22	49
298-14-6	vol 27/28	E83, 169
302-01-2	vol 21 vol 22 vol 30	E10, 71, E88-E91 107 104, 219, 427
307-35-7	vol 22	50
311-89-7	vol 27/28	721

327-62-8	vol 33	97, 98, E99, 100, 101, 140, 153, 180-182
333-20-0	vol 33	E145, 146, 147, E167, 168, 169, 180, E183, 184
	vol 34	65, 66, 225, 226
	vol 36	322, 323
335-57-9	vol 24	E256, E257, 267
	vol 27/28	682
339-43-5	vol 34	E285, 286-292
355-02-2	vol 21	E281, E282, 285
355-42-0	vol 24	E256, E257, 265
361-09-1	vol 35	165-167
367-11-3	vol 20	E118, 119
372-18-9	vol 20	E120, 121
375-72-4	vol 22	50
392-56-3	vol 21	E281, E282, 286
	vol 27/28	683
402-31-3	vol 20	237
420-05-3	vol 3	E38
462-06-6	vol 20	E202-E204, 205-209
463-58-1	vol 27/28	749
463-82-1	vol 27/28	E314, 315, 316, 410
	vol 37	30
464-06-2	vol 27/28	551, 552
	vol 37	464
471-34-1	vol 26	E191, E192, 216-221, 231
479-23-2	vol 38	521
493-01-6	vol 21	E3, 26
493-02-7	vol 21	E3, 26
	vol 27/28	541
496-11-7	vol 38	E161, 162, 163
497-19-8	vol 23	342
	vol 24	E69, 112
	vol 26	E240, 247, 314, 315, 411-413
	vol 27/28	E78, E79, 156, 158, 159
	vol 30	E32, 82, 85, 201, 216
	vol 31	139, 152, 157, 158
	vol 32	E26, E27, 45, 48
	vol 35	70, 71
	vol 34	57, 76
	vol 36	143, 151, 373, 374
504-63-2	vol 36	109, 112
506-87-6	vol 23	74
512-46-1	vol 24	E375-E378, 392
512-56-1	vol 24	E375-E378, 402
	vol 32	E183, 322
513-08-6	vol 24	E375-E378, 392, 402
	vol 27/28	727
	vol 32	E183, 322

513-35-9	vol 37	E22, E23, 24
515-49-1	vol 34	293-296
515-59-3	vol 35	E231, 232-238
515-62-8	vol 36	97
515-64-0	vol 36	E367, 368-401
515-67-3	vol 36	18
526-73-8	vol 38	E164, E165, 166-168
532-32-1	vol 33 vol 34	256, 282, 296, E313, 314, E324, 325, E330, 331 98, 99
535-65-9	vol 35	313
538-93-2	vol 29	E151, E153, 162
539-66-2	vol 33	E125, 126, 127, E226, 227, E250, 251, 252, E278, 279, E292, 293, E311, 312, E313, 314, E315, 316, E317, 318, E319, 320, E321, 322, 323
540-36-3	vol 20	E122, 123, 124
540-59-0	vol 21	E302, 309
540-72-7	vol 29 vol 33	E49, 70 E229, 230-233, 259-261, 273, 274, E285, 286, 299, 303, 304, 305, E317, 318, E328
540-84-1	vol 21 vol 27/28 vol 29 vol 38	E2, 15, E281, E282, 283 E213, 229 E134-E137, 145 E114-E117, 118-125
541-35-5	vol 34	244
541-43-5	vol 33	3-5
541-73-1	vol 20	E89-E91, 92-98
542-97-7	vol 37	1
543-90-8	vol 33	25, E26, 27-30, E31, 32-34
544-10-5	vol 27/28	691
544-17-2	vol 33	6-8
544-25-2	vol 37	E433, E434, 435-437
544-76-3	vol 21 vol 24 vol 27/28 vol 32 vol 38	E2, 13, 16, 17, E88-E91 E130-E134, 154-159, 166, 179, 180, 182 E202-E204, E208-E210, 235-241, E370, 371-375, 486 E166-E168, 189, 190, 210-212 E464, E465, 466-474
544-85-4	vol 24 vol 27/28	E131-E134, 187 E370, 377, 418, 419, 540
546-89-4	vol 33	E40, 41, 42, E50, 51-55, E77, 78-80, E81, 82-84, 206, 209, E210, 211-213
547-31-9	vol 36	440
547-32-0	vol 36	225, 226
547-44-4	vol 34	283, 284
547-52-4	vol 36	1-6

547-53-5	vol 36	11-14
551-76-8	vol 20	E223, 224
554-13-2	vol 30	293
556-37-6	vol 33	E123, 124
556-45-6	vol 33	E105, 106, 107, E116, 117, 118
556-63-8	vol 33	E9, 10, 11, 19, 20, 206-208
556-65-0	vol 33	207, 209, 214, 216
556-67-2	vol 24 vol 27/28	E218-E221, 235, 243, 250 E724, 725, 726
557-27-7	vol 33	98, E103, 104
557-34-6	vol 33	49, 76, 84, E91, 92-96
557-36-8	vol 24 vol 27/28	E257, E258, 287, 296 700
557-39-1	vol 33	E12, 13, 14, 21, 22
562-08-9	vol 35	2-8
562-49-2	vol 37	476, 477
563-45-1	vol 37	27
563-67-7	vol 33	34, E46, 47, 48, E73, 74, 75, E81, 82, 83, E88, 89, 90, 96, E332, 333-335
563-78-0	vol 37	272
564-02-3	vol 27/28	353, 354
565-59-3	vol 37	467, 468
565-75-3	vol 38	E126, E127, 128-130
571-58-4	vol 38	356
571-61-9	vol 38	E357, 358, 359
575-41-7	vol 38	355
581-40-8	vol 38	E360, 361, 362
581-42-0	vol 38	E363, 364, 365
583-53-9	vol 20	E57, 58
584-08-7	vol 26 vol 27/28 vol 31 vol 32	E92, 106, 107 E83, E84, 166, 170 222, 283, 291, 292 E26, E27, 46, 47, 49, 50
584-94-1	vol 38	131
589-34-4	vol 37	E481, 482-484
589-39-9	vol 33	E105, 106, 107, E108, E109, 110-112, E141, 141, E154, 155, 156, E183, 184, E185, 186, E187, 188
589-43-5	vol 38	132
589-46-8	vol 33	E125, 126, 127, E161, 162, 197, 198, E199, 200, 201
589-81-1	vol 38	133, 134
589-87-7	vol 20	E55, 56

589-90-2	vol 38	101, 103
590-29-4	vol 33	3, 6, E9, 10, 11, E12, 13, 14, E15, 16-18, 136, 139, 140, E141, E142, 143, 144, E145, 146, 147, 148, 149, E150, 151, 152
590-35-2	vol 37	465, 466
591-17-3	vol 24	E257, E258, 266, 275
591-49-1	vol 37	439
591-50-4	vol 20 vol 21 vol 24 vol 32	E210-E213, 214-220 E83, E87, 175, 229 E257, E258, 264, 274 E178, E179, 289
591-76-4	vol 37	478-480
591-93-5	vol 37	11
592-41-6	vol 37	E275, E276, 277-281
592-42-7	vol 37	209, 210
592-43-8	vol 37	282
592-76-7	vol 37	E455, E456, 457-459
592-77-8	vol 37	E460, 461-463
592-89-2	vol 33	18, 23, 24
593-45-3	vol 38	E501, E502, 503-507
593-45-3	vol 24 vol 27/28	E131-E134, 159, 161, 182 E202-E204, E210, 240, 243
598-38-0	vol 21	E7, 54
598-50-5	vol 34	120, 187
598-94-7	vol 34	123, 189
599-81-5	vol 36	54
599-82-6	vol 36	206
599-84-8	vol 36	366
605-02-7	vol 27/28	578, 579
608-93-5	vol 20	E3, 4, 5
613-12-7	vol 38	E433, 434-436
615-41-8	vol 20	E67, 68
615-42-9	vol 20	E125, 126
615-54-3	vol 20	E26, 27
615-58-7	vol 20	E65, 66
624-24-3	vol 27/28	E451, E452, 460
624-29-3	vol 29	E134-E137, 145
624-38-4	vol 20	E129, 130, 131
625-44-5	vol 22	61
625-52-5	vol 34	121

625-99-0	vol 20	E69, 70
626-00-6	vol 20	E127, 128
626-23-3	vol 22	142, 193, 228, 265, 278, 318, 341, 347, 369, 375
627-19-0	vol 37	12
628-28-4	vol 22	12, 30, 61, 189, 225, 242, 276
628-41-1	vol 37	E193, E194, 195, 196
628-71-7	vol 37	442
628-80-8	vol 22	E32, 33, 34, 61, 176, 189, 225, 236, 242, 256, 276, 294, 314, 315, E336, 337, 338, 367
628-81-9	vol 22	31, 315, E335, 337, 338, 367
628-92-2	vol 37	440
629-14-1	vol 22	11, 45, 140, 188, 224, 235, 255, 262, 274, 293, 300, 313, 333, 344, 351, 365, 372
629-27-6	vol 24	E257, E258, 268, 276, 286, 295
	vol 27/28	699
629-32-3	vol 22	34, 61, 88, 102, 134, 177, 189, 225, 236, 242, 276, 314
629-39-1	vol 20	E28, 29
629-50-5	vol 27/28	E202-E204, E208, 234
	vol 32	E166-E168, 189
	vol 38	395
629-59-4	vol 24	E130, E132, 153
	vol 27/28	E202-E204, E208, 234
	vol 32	E166-E168, 189, 209
	vol 38	E426, E427, 428-432
629-62-9	vol 27/28	E202-E204, E208, 235
	vol 32	E166-E168, 189
629-78-7	vol 24	E131-E134, 158
629-82-3	vol 21	E8, 16, 59, E81, E82, E85, E88-E91, 112, 158, 206
	vol 24	E375-E378, 393, 404, 409
	vol 32	E173-E175, 255
629-97-0	vol 24	E131-E134, 164, 184
	vol 27/28	E202-E204, E211, 246
630-01-3	vol 38	545
630-02-4	vol 24	E131-E134, 166, 185
630-06-8	vol 24	E131-E134, 166, 188
	vol 27/28	E202-E204, E212, 241
	vol 38	546, 547
630-08-0	vol 32	75, 76
631-61-8	vol 26	299, 300
632-22-4	vol 34	124, 190
634-66-2	vol 20	E17, 18
634-90-2	vol 20	E19, 20
636-28-2	vol 20	E15, 16
637-87-6	vol 20	E71, 72

638-04-0	vol 27/28	E451, E452, 461
638-68-6	vol 24	E131-E134, 187
643-90-8	vol 20	21
645-09-2	vol 30	252
645-35-2	vol 29	E50, 83, E87, 90
646-06-0	vol 22	89, 103, 132, 176, 218, 237
646-31-1	vol 24	E131-E134, 185
651-06-9	vol 36	280-283
655-78-7	vol 36	441
657-27-2	vol 29	E50, 85, E87, 93
657-84-1	vol 34	105
672-65-1	vol 24	E257, E258, 288, 297
680-31-9	vol 22	92, 106, 138, 181, 222, 240, 260, 279, 298, 319, 342, 354, 370, 379
	vol 24	E299-E302, 314, 321
	vol 27/28	729
	vol 32	E183, E185, 326
691-37-2	vol 37	274
693-02-7	vol 37	211
693-65-2	vol 21	E81, E82, 111, 157
	vol 24	E375-E378, 393, 403
694-48-4	vol 32	E180-E182, 301
694-80-4	vol 20	E49, 50
696-29-7	vol 32	E166, E169, 221
723-46-6	vol 35	E14, E15, 16-57
723-47-7	vol 35	108
729-99-7	vol 29	105-107
763-29-1	vol 37	273
764-71-6	vol 33	204
776-35-2	vol 27/28	550
779-02-2	vol 32	E170, E171, 237
	vol 38	437
781-43-1	vol 38	461
822-16-2	vol 33	E283, 284, E297, 298, E315, 316, E326, 327, E330, 331
832-69-6	vol 38	438, 439
852-19-7	vol 35	9, 10
868-18-8	vol 34	95
872-05-9	vol 38	292
872-50-4	vol 21	E8, 56-58
	vol 24	E257, E258, 306, 320, 328
	vol 27/28	E712, 713-717

872-50-4	vol 32	E180, 259, 305-313
875-83-2	vol 26	E5, 58
877-24-7	vol 34	6, 12, 83
	vol 35	95
917-72-6	vol 21	E234
929-06-6	vol 32	E80, E82, 108, 155-159
929-56-6	vol 22	34, 61, 88, 102, 134, 189, 225, 236, 242, 276, 314
929-73-7	vol 24	E61, 80
939-27-5	vol 38	372
992-98-3	vol 33	5, 8, 22, 24
996-30-5	vol 33	E119, 120, 225, 247, E277, E292, 293, E294, 295, 296, E297, 298-302
998-30-1	vol 21	E9, 66, E281, E282, 296, 297, E302, 306, 307, E310, 311, 312, E315, 318, 319, 323, 324, 327, 328
	vol 32	E185, 277, 278, E330, 334, 335
1004-36-0	vol 30	252
1037-51-0	vol 35	302-309
	vol 36	452
1072-05-5	vol 38	220
1072-16-8	vol 38	293
1077-16-3	vol 38	375
1116-76-3	vol 22	15
1119-34-2	vol 29	E50, E51, 86, E87, 94
1120-21-4	vol 27/28	E202-E204, E207, 233
	vol 32	E166-E168, 189
	vol 38	E326, E327, 328-331
1123-63-3	vol 20	E244, 245
1124-06-7	vol 20	E242, 243
1125-66-2	vol 20	E250, 251
1127-76-0	vol 38	E366, 367-371
1132-61-2	vol 35	277
	vol 36	60, 61, 356
1220-83-3	vol 36	285-293
1222-52-1	vol 21	E8
1301-96-8	vol 23	E144, E145, 146, 147
1302-42-7	vol 31	159, 160-162
1303-28-2	vol 23	E174, 262, 263, E279, 298, 299
1303-52-2	vol 23	E148-E150, 151-155
1303-86-2	vol 23	E92, 130-133
1306-19-0	vol 23	E270-E282, 283, 286, 287, 293-296, 298, 299, 304-306

```
1309-60-0    vol 29        E207

1310-51-0    vol 35        E14, E15, 24, 25, 94, 95, 142, 143, 222

1310-58-3    vol 23        E10, E11, E15-E18, 47, 48, E83-E88, 118, 129,
                           140, E144, E145, 146, 147, 205, 206, 211,
                           220-226, 228, 242, 243, 246, 249, 253, 265,
                           300, E314, 323, 334
             vol 26        E92, 102
             vol 27/28     E79, 160, 161
             vol 30        147, 325-327, 405, 406, 409, 410
             vol 31        E168-E171, 182-205
             vol 36        142

1310-65-2    vol 23        79, 300, 322
             vol 30        313, 325-327
             vol 31        E1-E6

1310-66-3    vol 31        E4

1310-73-2    vol 23        E10-E15, 34, 35-39, 46, 52, 58, 60, 61, 65,
                           66, 81, E83-E86, 108, 114, 115, 117, 127, 128,
                           135-139, 141, E148-E150, 153, 154, E163-E165,
                           192, 193, 197-199, 201-203, 207-209, 213,
                           215-219, 227, 229, 231-234, 244, 260, 261, 265,
                           266, 267, 268, 269, E270-E277, 286-288, 296,
                           297, 300-302, E307-E313, 321, 324, 325, 330,
                           336-338
             vol 26        E5, 56, 57, 71, E191, E192, 232, 320, 321, 407-410
             vol 30        E30, 67
             vol 31        E11-E15, 17-19, 22, 24-38, 155-162
             vol 34        4, 10, 79, 170, 175, 276, 290
             vol 36        32, 70, 139-141, 164, 165, 232-234, 250, 257, 273,
                           335, 347, 358, 362, 453, 454, 456, 457, 459, 460,
                           467, 469-471, 484

1310-82-3    vol 30        437

1313-82-2    vol 23        140, 252, 293, 332
             vol 32        48

1314-13-2    vol 23        E156-E184, 185, 186, 188, 190, 194-197, 201,
                           204-207, 210-212, 215, 216, 220, 221, 225, 227,
                           228, 230-236, 242-244, 246, 248, 249, 252, 253,
                           257-259, 261-263

1314-56-3    vol 23        E177, 195, 196
             vol 26        E191, E192, 229, 230
             vol 31        315

1317-37-9    vol 29        E207

1317-38-0    vol 23        E10-E26, 27-30, 32, 33, 35, 36, 40-42, 46-48,
                           53, 56, 58-60, 63, 80-82

1317-39-1    vol 23        E1-E3, 4-9

1319-77-3    vol 27/28     580

1321-94-4    vol 21        34, E88-E91
             vol 24        E218-E221, 237, 248
             vol 27/28     488, 508
             vol 32        E170, E171, 236

1330-20-7    vol 21        E281, E282, 286
             vol 24        231, 240
             vol 27/28     580

1330-43-4    vol 31        147
             vol 36        162

1331-22-2    vol 34        162, 163
             vol 35        195, 196
             vol 36        52, 53
```

```
1333-74-0     vol 27/28    663-665
              vol 29       E190, E191, 198
              vol 32       75, 76, 237, 240, 276

1333-82-0     vol 23       E92, 123, E174-E176, 194, 230, 235

1336-21-6     vol 23       E1-E3, 7, E10, E11, E18-E20, 49, 51, 59,
                           64-78, E83, E84, E93, 109, 116, 142, 143, E178,
                           E179, 200, 254, 255
              vol 26       379

1338-39-2     vol 34       164

1344-48-5     vol 29       E47

1344-81-0     vol 26       E258

1453-58-3     vol 32       301

1559-34-8     vol 32       E173-E175, 257

1570-64-5     vol 20       E231, 232

1570-65-6     vol 20       E228, 229

1640-89-7     vol 37       443, 444

1643-19-2     vol 24       E63, E64, 92, 93
              vol 27/28     E61, E62, 96-98

1678-91-7     vol 32       E168, E169, 218
              vol 38       E104-E106, 107-109

1678-92-8     vol 32       E166, E169, 219, 220

1694-09-2     vol 35       167

1704-78-3     vol 34       178

1709-39-7     vol 34       202, 203

1709-52-0     vol 34       200, 201

1762-95-4     vol 29       76

1867-85-2     vol 20       E226, 257

1941-30-6     vol 24       E62, E63, 89-91
              vol 27/28    E61, 95

2029-98-0     vol 30       476

2040-95-1     vol 38       213

2044-21-5     vol 21       E281, E282, 293

2049-95-8     vol 38       323

2050-92-2     vol 22       15

2082-84-0     vol 27/28    101

2099-72-1     vol 27/28    E63, E64

2207-01-4     vol 27/28    E451, E452, 458
              vol 29       E134-E137, 143

2207-03-6     vol 27/28    E451, E452, 461

2207-04-7     vol 27/28    E451, E452, 460
              vol 29       E134-E137, 146

2207-98-9     vol 27/28    188
```

2216-33-3	vol 38	222
2216-34-4	vol 38	223, 224
2319-96-2	vol 38	527
2351-69-1	vol 22	50, 116, 147, 204, 230, 270, 286, 308, 323
2381-15-9	vol 38	510
2381-16-0	vol 38	509
2396-63-6	vol 37	438
2396-65-8	vol 38	211
2406-52-2	vol 21	E309, 325-328
2432-12-4	vol 20	E226, 227
2447-57-6	vol 36	434
2466-09-3	vol 31	269, 270
2498-77-3	vol 38	508
2556-73-2	vol 32	E180, E184, 314
2577-32-4	vol 36	116
2614-35-9	vol 21	E234
2651-18-5	vol 34	297-301
2801-86-7	vol 38	463
2958-76-1	vol 38	324
2980-70-3	vol 38	425
3070-53-9	vol 37	441
3073-66-3	vol 38	214, 215
3163-31-3	vol 36	403
3184-13-2	vol 29	E50, 84, E87, 92
3213-22-7	vol 36	279
3221-61-2	vol 38	221
3306-62-5	vol 34	1-6
3396-11-0	vol 33	25, E35, 36-39, E40, 41, 42, E43, 44, 45, E46, 47-49, 133, E134, 134, 135
3452-09-3	vol 38	212
3522-94-9	vol 38	E216, E217, 218, 219
3587-57-3	vol 22	50
3697-24-3	vol 38	511
3741-00-2	vol 38	291
3767-37-1	vol 21	E234
3811-04-9	vol 30	E29, 68-70, 90-92, E105-E112, 113-161, 411
3875-51-2	vol 38	97
3982-67-0	vol 38	322

4206-74-0	vol 35	99, 100
4328-04-5	vol 24 vol 27/28	E64, 94 E62, E63, 99
4390-04-9	vol 24 vol 27/28	E131, E133, 160, 181 E213, 242
4432-31-9	vol 36	221, 222, 302
4457-00-5	vol 38	325
4485-46-6	vol 36	297
4516-69-2	vol 38	95, 96
4587-19-3	vol 3	E185, 186, E213, 216, 221, E223, 226-228
4747-07-3	vol 22	366
4810-01-9	vol 38	394
4810-04-2	vol 38	374
4810-05-3	vol 38	373
4810-06-4	vol 38	462
4810-07-5	vol 38	393
4917-19-8	vol 30	344, 345
5018-54-2	vol 36	435
5137-45-1	vol 22	29, 61, 89, 132, 176, 218, 237, 256, 275, 294, 315, 338, 367
5433-63-6	vol 36	100, 101
5581-55-5	vol 21	E234
5587-98-2	vol 21	E281, E282, 284
5623-78-9	vol 38	93
5626-90-4	vol 34	265-270
5852-45-9	vol 21	E234, E252, 257, 258
5989-27-5	vol 38	286
6101-31-1	vol 34	177
6101-41-8	vol 33	E121, 122, E123, 124, E248, 249, E303, 304, 305, E306, 307, E308, 309, 310
6106-41-8	vol 34	88, 92
6138-01-8	vol 35	250
6138-11-0	vol 34	193
6303-21-5	vol 29	29
6484-52-2	vol 23 vol 26 vol 29 vol 32	264 E191, E192, 215 E47, 53 39
6531-45-9	vol 33	97, 102, 214, 215
6834-92-0	vol 31	114, 143, 153, 165

6876-23-9	vol 27/28	E451, E452, 459
	vol 29	E134-E137, 144
6912-98-7	vol 36	438
6972-47-0	vol 20	E238, 239
7289-51-2	vol 22	34, 102, 134, 177, 189, 224, 236, 242, 276, 314
7289-52-3	vol 22	34, 61, 117, 147, 189, 204, 225, 230, 236, 242, 270, 276, 286, 308, 314, 323
7395-10-0	vol 21	E253, 271-273
7429-90-5	vol 25	E84, E85, 86-92
	vol 29	E214, 217
7429-91-6	vol 25	E245, 246
7439-88-5	vol 25	E316
	vol 29	E214, 217
7439-89-6	vol 25	E301, E302, E303-306
	vol 29	E207, E214, 217
7439-90-9	vol 29	E190-E192, 197
7439-91-0	vol 25	E207-E208, 209-213
7439-92-1	vol 25	E157-E159, 160-171
	vol 29	E228, 234
7439-93-2	vol 25	E1-E3, 4-12
7439-94-3	vol 25	E256, 257
7439-95-4	vol 25	E57, E58, 59-65
7439-96-5	vol 25	E285, E286, 287-299
	vol 29	E213, 220
7439-97-6	vol 25	Solute covered in this volume, see all pages
	vol 29	Solute covered in this volume, see all pages
7439-98-7	vol 25	E282, 283
	vol 29	E214, 217
7440-00-8	vol 25	E225, 226-229
	vol 29	E190, E191, 194
7440-02-0	vol 25	E317, E318, 319-325
	vol 29	E214, 217
7440-03-1	vol 25	E271, 272, 273
	vol 29	E214, 217
7440-04-2	vol 25	E309
	vol 29	E214, 217
7440-05-3	vol 25	E326, 327-329
7440-06-4	vol 25	E330, 331-334
	vol 29	E213, E214, 225
7440-07-5	vol 25	E434, 435
7440-08-6	vol 25	E421
7440-09-7	vol 25	E33-E34, 35-43,
	vol 29	E213, E214, 225
7440-10-0	vol 25	E221, 222-224
7440-12-2	vol 25	E421

7440-13-3	vol 25	E421
7440-14-4	vol 25	E421
7440-15-5	vol 25	E300
7440-16-6	vol 25	E314, 315
	vol 29	E214, 217
7440-17-7	vol 25	E44-E45, 46-49
7440-18-8	vol 25	E307, 308
	vol 29	E214, 217
7440-19-9	vol 25	E230, E231, 232, 233
7440-20-2	vol 25	E206
7440-21-3	vol 25	E134
	vol 29	E213, E214, 219
7440-22-4	vol 25	E357-E359, 360-368
	vol 29	E213, E214, 222, E228, 231-233
7440-23-5	vol 25	E13-E18, 19-32
	vol 29	E213, E214, 218
7440-24-6	vol 25	E71, E72, 73-75
7440-25-7	vol 25	E274, 275, 276
	vol 29	E214, 217
7440-26-8	vol 25	E421
7440-27-9	vol 25	E242, 243, 244
7440-28-0	vol 25	E119-E121, 122-133
7440-29-1	vol 25	E422, 423, 424
	vol 29	E214, 217
7440-30-4	vol 25	E251, 252
7440-31-5	vol 25	E139-E141, 142-156,
	vol 29	E213, 223
7440-32-6	vol 25	E258, E259, 260-262
	vol 29	E214, 217
7440-33-7	vol 25	E284
	vol 29	E214, 217
7440-34-8	vol 25	E421
7440-36-0	vol 25	E173, E174, 175-181
7440-37-1	vol 29	E190-E192, 195, 196
7440-38-2	vol 25	E172
7440-39-3	vol 25	E76, E77, 78-82
7440-41-7	vol 25	E55
7440-42-8	vol 25	E83
7440-43-9	vol 25	E402-E404, 405-420
7440-44-0	vol 25	E134
7440-45-1	vol 25	E214, E215, 216-220
7440-46-2	vol 25	E50, E51, 52-54

7440-47-3	vol 25	E277, 278-281
	vol 29	E214, 217
7440-48-4	vol 25	E310, E311, 312, 313
	vol 29	E214, 217
7440-50-8	vol 25	E335-E337, 338-356
	vol 29	E228, 229
7440-52-0	vol 25	E249, 250
7440-53-1	vol 25	E234, 235-238
7440-54-2	vol 25	E239, 240, 241
7440-55-3	vol 25	E93-E95, 96-103
	vol 29	E213, E214, 216
7440-56-4	vol 25	E135, 136-138
7440-57-5	vol 25	E369-E371, 372-384
	vol 29	E213, E214, 226, 227
7440-58-6	vol 25	E267
7440-59-7	vol 29	E190, E191, 193
7440-60-0	vol 25	E247, 248
7440-61-1	vol 25	E425-E427, 428-433
	vol 29	E214, 217
7440-62-2	vol 25	E268, 269, 270
	vol 29	E214, 217
7440-63-6	vol 29	E191
7440-64-4	vol 25	E253, 254, 255
7440-65-5	vol 25	E206
7440-66-6	vol 25	E385-E387, 388-401
	vol 29	E213, 221, E228, 230
7440-67-7	vol 25	E263, E264, 265, 266
	vol 29	E214, 217
7440-69-9	vol 25	E182, E183, 184-193
7440-70-2	vol 25	E66, E67, 68-70
7440-73-5	vol 25	E421
7440-74-6	vol 25	E103-E105, 106-118
7446-08-4	vol 23	E10, E11, E20, 62, E92, 134
	vol 26	320, 321, 365, 366, 376
7446-09-5	vol 23	E177, 236
	vol 26	E115, 124, 125, 129-131, E155, E156, 165-167, 180, 181, E188-E190, 196-211, 229, 230, E240, 245, 254, 255, 263, 264, 276, 278-282, 297, 298
	vol 27/28	763
7446-10-8	vol 26	E293, 294-302
7446-11-9	vol 23	E177, 210
7446-70-0	vol 29	E207, 209
7447-14-5	vol 34	238
	vol 38	251, 252, 256, 257, 318, 371, 390, 406, 407, 421-423, 436, 439, 445, 458, 459, 485, 492, 519

7447-40-7	vol 20	11, 12
	vol 23	E1-E3, 8, 80, 335
	vol 24	E69, 114, 115
	vol 27/28	E79, E80, E81, E83, E84, 113, 162-166
	vol 29	E49, 71, E87, 88-94, E207, 210
	vol 30	90-92, 125-140, 232, 391, 421, 422
	vol 32	39, 41
	vol 33	144, 165, 166
	vol 34	3, 4, 9, 10, 59, 60, 78, 79, 219, 220, 232, 269, 287-290
	vol 35	76, 99, 146, 149, 273, 288, 300, 309
	vol 36	41, 108, 121, 161, 315-317, 478, 479, 481, 482
7447-41-8	vol 23	79, 335
	vol 24	E65, 97, 98
	vol 27/28	E68, E69, 114-116
	vol 29	E207, E208, 210
	vol 30	20-22, 254, 282
	vol 36	313
7487-88-9	vol 26	E157, E158, 173-176, 180-185
	vol 29	E87, 88-94
7487-94-7	vol 29	E207, 211
7488-51-9	vol 26	E351, 401-402
7488-52-0	vol 26	E271
7550-35-8	vol 30	283
7557-82-6	vol 31	112
7558-79-4	vol 31	E11, 18, 19, 22, 23, 37, 38, E98-E102, 103-126
7558-80-7	vol 31	E11, 16, 20, 21, 24, 25, E39-E42, 43-45, E47-E52, 53-97
	vol 36	162, 163, 169, 170
7558-94-4	vol 35	4-6, 9, 27-30, 32, 35, 62, 63, 72-75, 78, 79, 96-98, 101, 106, 147, 150-159, 161-163, 223, 225-229, 235, 237, 243, 245, 252-255, 265, 267-269, 271, 272, 283, 285-287, 293, 295-299, 304, 306-308, 310-315
	vol 36	12, 13, 15, 17, E19, E20, 34, 36-40, 71, 73-76, 98, 100, 104, 105-107, 120, 122, 149, 152-160, 167-170, 236-242, 251-256, 258, 259, 288, 289, 294, 329-334, 336, 359-361, 363, 365, 375-378, 403, 410, 411, 434, 435, 441, 442, 455, 458, 461, 463-465, 472, 473, 474, 476, 477, 480, 485
7601-54-9	vol 26	E191, E192, 234
	vol 31	E11-E15, 17, 26-36, E127-E130, 131-167
	vol 36	150
7601-89-0	vol 23	56, 57, 139, 248, 250, 251, 290, 302, 303, 331, 342
	vol 26	E190, 207-210, 222, 382, 383-387, 425
	vol 29	E49, 67
	vol 31	69, 97, 125, 146
	vol 32	E26, 38, 40
7601-90-3	vol 23	141, 301
	vol 26	415, 416
	vol 32	E16, E26, 38
	vol 36	445
7631-14-5	vol 26	400
7631-90-5	vol 26	85
7631-95-0	vol 30	E30, 76, 77, 200, 343
7631-99-4	vol 23	337

7631-99-4	vol 26	301, 302, 399, 420, 427
	vol 29	E49, 69
	vol 30	E30, 41, 203, 228, 346-353
	vol 31	59-64, 79, 92-95, 110, 121, 122
	vol 32	39
	vol 33	E236, E237, 238-241, E267, 268-272, 276, E289, 290, 291, 301, 302, E308, 309, 310, E321, 322, 323, E329
7632-00-0	vol 31	150
	vol 33	235, 264-266, 275, E287, 288, 300, E306, 307, E319, 320
7646-69-7	vol 27/28	186, 187
7646-78-8	vol 22	93, 108, 139, 184, 186, 223, 241, 261, 280, 299, 320, 343, 355, 371, 380
7646-85-7	vol 23	E179, E180, 204
	vol 32	E26, 35
7647-01-0	vol 23	266, 267, 328, 329
	vol 24	E59, 73, 74
	vol 26	E191, E192, 224, 225, 355, 357, 359, 367, 379, 401, 420, 422, 423
	vol 27/28	E57, E58, E86
	vol 29	E47, 52
	vol 32	E25, E26, 29, 30, 34, 35-37
	vol 35	3, E14, E15, 22, 23, 31, 36, 58, 68, 105, 107, 160, 203-205, 208-212, 263, 264
	vol 36	103, 115, 117-119, 137, 138, 150, 283, 287, 295, 328, 348-351, 371, E404, 408, 409, 428-433, 444
7647-14-5	vol 23	E1-E3, 8, 80, 231-234, 268, 305, 306, 335, 340, 341
	vol 24	E65-E68, 99-110
	vol 26	E4, E5, 34-38, 59-62, 66-70, E74, 86-89, 224-228, E236, 239, E240, 246
	vol 27/28	E69-E75, E77, E80, E81, 113, 117-147, 156, 166
	vol 29	E48, 59-65, E87, 88-94, E207, 209
	vol 30	E31, E32, 82-84, 86-97, 121, 208, 209, 229, 357, 358, 423
	vol 31	65, 66, 76-78, 80-85, 88-91, 111, 119-121, 134, 135
	vol 32	E25, E26, 36, 37, 39, 41, 42
	vol 33	228, 257, 258
	vol 35	26, 31, 32, 35, 61, 76, 99, 105, 144, 145, 149, 203-205, 264, 270, 271, 273, 288, 292, 300, 309
	vol 36	33, 41, 108, 121, 138, 145-147, 161-163, 258, 259, 277, 278, 314, 328, 478, 479, 481, 482
	vol 37	62, 190-192, 220, 368, 427, 428, 430
	vol 38	251, 252, 256, 257, 318, 371, 390, 406, 407, 421-423, 436, 439, 445, 458, 459, 485, 492, 519
7647-15-6	vol 24	E68, 111
	vol 27/28	E75, 148, 149
	vol 29	E49, 66
	vol 30	E28, 61, 210-213, 216, 217, 359-363
	vol 33	218-220
7647-17-8	vol 27/28	E84, 171
	vol 30	95-97, 174, 175, 182, 183
	vol 31	327
7664-38-2	vol 23	E177, 258
	vol 30	320
	vol 31	E1-E6, 9, 10, E11-E15, 16, 20, 21, 23, 26-38, E168-E171, 174-181, 184-205, 264-266, 269, 270, 316, 317
	vol 35	270
	vol 36	163
7664-39-3	vol 23	E10, E11, 63, E313, 319, 333
	vol 30	478

7664-41-7	vol 21	E1-E12, 13-79
	vol 23	E178, E179, 214, 269, E277, E279, 294, 304
	vol 26	E115, 121-126
	vol 27/28	736
	vol 29	E191
	vol 30	161, 253, 257, 420-426
	vol 31	E1-E6, 8, 269, 270, 300, 301, 329-331
	vol 32	E54, E55, 56-78
7664-93-9	vol 23	40-44, E148, E150, 151, 152
	vol 24	E59, E60, 75-77
	vol 26	367, 368, 371, 375, 378, 380, 389, 392, 394, 395, 414, 421-423
	vol 27/28	E58, E59, E77, 87, 88, 154
	vol 30	479
	vol 32	E26, 32, 33, 52, 53
7681-11-0	vol 24	E70, 118
	vol 27/28	E81, 167
	vol 30	144, 236, 394, 409, 410, 426
	vol 33	148, 170, 171
	vol 36	320, 321
7681-49-4	vol 29	E47, E48, 58
	vol 30	E28, 50-58
	vol 31	154
7681-55-2	vol 30	65, 66, 285, E330-E339, 340-373
7681-57-4	vol 26	E5, 39-45, 63-65, E72-E75, 76-84, 86-89
7681-82-5	vol 24	E125, 126, 127
	vol 27/28	E75, E76, 150, 151
	vol 29	E49
	vol 30	E28, 64, 213, 364, 365
7681-82-5	vol 33	234, 262, 263
7697-37-2	vol 23	E10, E11, E24, 53, E148, E150, 155, E307-E313, 326, 327
	vol 26	355, 357, 359, 368, 371, 375, 378-380, 389, 392, 394, 395, 399, 401, 420, 427
	vol 29	29
	vol 30	435, 475-477
	vol 36	59, 62, 218, 220, 352-355, 446
7704-34-9	vol 32	E185, 317
7722-76-1	vol 31	54-58, 86, 87, 90, 227, 230, 231, 236, 244-247, 256, 262, 263, 267, 268, 274, 275, 293-296, 313, 326
	vol 35	76, 99, 273, 288, 300, 309
	vol 36	108, 121, 478, 479, 481, 482
7722-84-1	vol 31	107, 113, 141, 142, 225
7723-14-0	vol 21	300
	vol 29	E214, E214, 219
7727-37-9	vol 27/28	313, 334, 335, 347, 348, 420-435
	vol 29	E191, 199
	vol 32	75, 76
7729-20-0	vol 20	182
7732-18-5	vol 22	25, 73, 80, 99, 112, 113, 126, 160, 196, 211, 234, 250, 272, 290, 304, 305, 311, 331, 348, 361, 378
	vol 23	E1-E3, 6-8, E10-E26, 27, 29-81, E83-E97, 98-143, E144, E145, 146, 147, E148-E150, 151-155, E156-E184, 185-269, E270-E282, 283-306, E307-E318, 319-342
	vol 24	E1, E2, 3-15, E16, E17, 18-33, E34, 35-38, E39, 40-47, E48, 49-53, E58-E72, 73-124, 228, 259, 269,

7732-18-5	vol 24	277, 338, 353, 366, 386, 398, 406, 415, 421
	vol 26	1-427
	vol 27/28	E1-E6, 7-23, E24-E28, 29-44, E56-E85, 86-201, 580, 581
	vol 29	E1-E24, 25-44, E46-E51, 52-55, 57-96, 99, 170, E191, E228, 229-234
	vol 30	E1-E10, 11-22, E24-E33, 34-97, E105-E112, 113-156, E162-E165, 166-175, E184-E187, 188-193, E195-E197, 198-217, E220-E223, 224-253, E258, E259, 260-262, E263, E264, 265-267, E268-E275, 276-328, E330-E339, 340-372, E374-E382, 383-416, E428-E431, 432-446, E448-E450, 451-460, E461, E462, 463-466, E467-E473, 474-489
	vol 31	E1-E6, 7-10, E11-E15, 16-38, E39-E42, 43-46, E47-E52, 53-97, E98-E102, 103-126, E127-E130, 131-167, E168-E171, 172-205, E206-E212, 213-277, E278, E279, 280-296, E297, 298-306, E307-E311, 312-317, E318-E324, 325-333
	vol 32	E1-E3, 4-19, E20, 21-24, E25-E28, 29-50, E54, E55, 56-78, E79-E83, 84-165, 298, E330, E331, 336-339
	vol 35	1-6, 9, 11-13, E14, E15, 16-43, 58, 61-63, E64, 65-80, 94-102, 104-109, E110-E112, 113-185, 203-217, 219-229, E231, 232-237, E239, 240-247, 249-256, E257, 258-274, 277-288, E289, 290-300, 302-317
	vol 36	1-5, 8, 11-13, 15-17, E19, E20, 21-46, 54, 57-62, E63, 64-87, E88, 89, 90, E92, 93, 94, 96-98, 100, 102-112, 115-122, E124, 125-185, 206-225, 228, E229, 230-242, E243, 244-261, 268, E269, 270-281, 283-289, 294-302, E303, 304-341, 346-366, E367, 368-381, 402, E404, 405-411, 428-438, 441, 485
	vol 37	1, E2, E3, 4-6, E7, E8, 9-12, E13-E15, 16-21, E22, E23, 24-30, E31-E34, 35-41, E42-E46, 47-62, E63-E85, 86-183, E188, 189-192, E193, E194, 195, 196, E197-E199, 200-211, E212, E213, 214-220, E221-E229, 230-269, 272-274, E275, E276, 277-282, E283, E284, 285-288, E289-E291, 292-295, E296-E299, 300-306, E307, E308, 309-313, E314-E323, 324-365, 368, E369-E379, 380-422, 427, 428, 430, 432, E433, E434, 435-444, E445-E447, 448-454, E455, E456, 457-459, E460, 461-468, E469, E470, 471-480, E481, 482-484, E485-E491, 492-514
	vol 38	E1, E2, 3-8, E9-E12, 13-21, E23-E28, 29-40, E42-E46, 47-57, E59-E66, 67-90, 92-103, E104-E106, 107-113, E114-E117, 118-125, E126, E127, 128-134, E135-E141, 142-157, E161, 162, 163, E164, E165, 166, 168, E169, E170, 171-176, E177, E178, 179-185, E187-E189, 190-199, E201, E202, 203-215, E216, E217, 218-224, E225-E227, 228-234, E236-E238, 239-249, E250, 251, 252, 256-259, E260, 261-264, E265, E266, 267-270, E271, 272-275, E277, E278, 279-284, 286, E287, 288-293, E294-E296, 297-307, E311, E312, 313-318, E319, 320-325, E326, E327, 328-330, E332-E334, 335-339, E340-E343, 344-350, 353, 355, 356, E357, 358-362, E363, 364, 365, E366, E367, 368-375, E376, E377, 378-380, 384, E385, E386, 387-395, E396-E398, 399-404, 406, 407, E408-E410, 411-418, 421-425, E426, E427, 428-431, E433, 434-439, E440, 441-445, E446-E468, 449-455, 458, 459, 461, E464, E465, 466-473, 475, 476, E477, E478, 479-482, 485, E486, E487, 488-492, E493, 494-496, E497, 498-500, E501, E502, 503-506, 508-511, E512, 513, 514, 516, 519-525, 527, E528, 529-531, E534, 535, 536, 538, 540-544, 546, 547
7733-02-0	vol 23	212
	vol 26	278-282, 284, 285
7747-40-7	vol 31	80-85, 88, 89, 91, 226, 228, 229, 240-243, 251, 252, 255, 259, 262, 263, 267, 268, 271, 287, 288

7757-79-1	vol 23	E83, E84, E93, E94, 112, 113, 120, 142, 143, 245, 247, 338
	vol 26	E92, 96, 97, 373
	vol 30	122, 230, 388, 389, 436
	vol 31	92-95, 224, 232, 233, 253, 254, 260, 261, 286
	vol 32	E25, 39
	vol 33	E150, 151, 152, E175, 176-179, 182, E187, 188, E191, 192, E195, 196, E199, 200, 201
	vol 35	277-281
	vol 36	218-222, 352-356
7757-82-6	vol 23	64, 336
	vol 26	E3, E4, 16-33, 59-65, E74, 78-83, 284, 285, 289, 290
	vol 27/28	E76, E77, E79, 152, 153-156, 159
	vol 29	E49, 68
	vol 30	E30, 42-49, 204-207, 354-356
	vol 31	53, 122, 123, 136-138, 140, 145, 155, 161-164
	vol 32	E25, 39, 40, 43
7757-83-7	vol 26	290, 309
	vol 26	E1-E7, 8-71, 178, 179, E240, 247, 283-285, 289
7757-86-0	vol 31	115
7757-88-2	vol 26	E153-E159, 160-186
7757-93-9	vol 26	E191, E192, 231
7757-95-1	vol 26	E258, 259
7758-01-2	vol 30	142, 143, E220-E223, 224-257
7758-02-3	vol 24	E69, E70, 116, 117
	vol 30	141, 233-235, 392, 393, 425
	vol 33	136-138
	vol 36	318, 319
7758-05-6	vol 30	145, 146, 286, 287, 321, 322, 325-327, 366, E374-E382, 383-410, 421
7758-09-0	vol 26	E92, 98, 99
	vol 31	303
	vol 33	149, E172, 173, 174, 181, E185, 186, E189, 190, E193, 194, E197, 198, 202-205
7758-11-4	vol 31	108, 109, E168, 172, 173, E278, E279, 280-296
7758-19-2	vol 30	E29, E31, 59, 60, 83-89
7758-88-5	vol 22	97
7758-94-3	vol 32	E26, 34
7758-98-7	vol 23	45, 50
	vol 26	265
7761-88-8	vol 23	138
7772-04-7	vol 26	E5, 50-53
7773-08-4	vol 26	112
7775-09-9	vol 30	E30, 34-104
7775-11-3	vol 30	E30, 74, 75
7775-14-6	vol 26	85
7775-18-5	vol 30	E24-E33, 37-40
7775-19-1	vol 31	73, 74
7778-18-9	vol 26	E191, E192, 212-214, 220-223

7778-53-2	vol 31 vol 31	149, 165, E168-E171, 175-177, 184-187, E297, 298-306
7778-74-7	vol 30	137, 138
7778-77-0	vol 29 vol 31 vol 35	E87, 88-94 67, 68, 70-72, 75, 80-85, 88, 89, 91-95, E168, 174, 178-183, 203-205, E206-E212, 213-277, 328 4, 5, 9, 27-30, 32, 35, 62, 63, 72-75, 94, 96, 97, 101, 106, 148, 150-159, 162, 163, 224-229, 236, 237, 244, 245, 252-255, 266-269, 284-287, 294-299, 305-308
7778-80-5	vol 23 vol 26 vol 27/28 vol 30 vol 31 vol 32 vol 36	121 E92, 104, 105, 107 E81, E82, 155, 159, 168 123, 124, 231, 390 223, 237-239, 249, 250 39 235-242, 251-256, 258, 259, 274-278, 288, 289, 294, 329-331, 359-361, 363, 375-377, 403, 410, 411, 434, 435, 455, 458, 461, 463-465, 467, 472, 473, 474, 476, 477, 484, 485
7782-65-2	vol 21	E315, 321-324
7782-68-5	vol 30	279, 314-319, 321-324, 371, 372, 407, 408, 443, 457, 466, E467-E473, 474, 475, 477, 478, 481-489
7782-70-9	vol 26	E324
7782-82-3	vol 26	E305
7782-85-6	vol 31	E11
7782-99-2	vol 26	E92, 103
7783-00-8	vol 26	318, 319, 329, 330, 382, 383-387
7783-06-4	vol 27/28 vol 32	200, 201, 447-449, E750, 751-760 E1-E3, 4-19, E20, 21-24, E25-E28, 29-50, E51, 52-53, E54, E55, 56-78, E79-E83, 84-165, E166-E187, 188-326
7783-07-5	vol 32	E330, E331, 332-339
7783-11-1	vol 26	E113, E115
7783-18-8	vol 26	140
7783-19-9	vol 26	E331, 342, 343, 373
7783-20-2	vol 23 vol 26 vol 32	E10, E11, 64, 75, 295 E115, 132-137, 141-143, E144, E145, 147-151 39, 67
7783-28-0	vol 31	86, 87, 117, 119, 120, 281, 282, 293-296
7783-89-3	vol 30	214, 215, 242
7783-92-8	vol 30	79, 80
7783-97-3	vol 30	297
7784-05-6	vol 26	E350, 380-388
7784-41-0	vol 31	220
7784-42-1	vol 21	E302, 303-309
7785-87-7	vol 27/28	E65, E66, E77, E82, 105, 106, 155

7786-30-3	vol 23	E180, 256
	vol 26	E157, E158, 177, 186
	vol 27/28	E64-E66, E68, E74, E80, E81, 104, 106, 113, 146
	vol 35	76, 99, 273, 288, 300, 309
	vol 36	108, 121, 324, 478, 479, 481, 482
7787-39-5	vol 26	E240, E241, 242-247
7789-18-6	vol 30	454
	vol 33	E134, 135
7789-20-0	vol 24	E54, 55-57
	vol 27/28	E45, E46, 47-49
	vol 30	119, 120, 199, 227, 387
	vol 31	315, 332, 333
	vol 32	E327, 328, 329
	vol 37	184-187, 270, 271, 433, 434, 515
	vol 38	351
7789-23-3	vol 23	E10, E11, 63
7789-38-0	vol 30	E29, 62, 63, 192, E195-E197, 198-219
7790-37-6	vol 30	402, 439, 482
7790-69-4	vol 30	294, 295
	vol 33	208, E210, 211-213, 215, 217
7790-81-0	vol 30	481
7790-86-5	vol 22	98-107
7790-87-6	vol 22	112-113
7791-03-9	vol 32	298
7791-11-9	vol 30	93, 94, 139, 140, 171, 172, 174, 175
	vol 31	314
7803-51-2	vol 21	E281-E282, 283-301
7803-52-3	vol 21	E310, 311-314
7803-62-5	vol 21	E315, 316-320
7846-85-7	vol 22	187
8007-43-0	vol 35	201, 202
8032-32-4	vol 35	49
8062-15-5	vol 26	E191, E192, 235
8063-07-8	vol 29	E87, 88-94
8065-80-3	vol 36	182
9000-69-5	vol 35	179
	vol 36	3
9003-39-8	vol 35	90, 177, 206, 276
	vol 36	267, 340, 341
9004-64-2	vol 35	204
9004-65-3	vol 35	205
9005-64-5	vol 35	80, 182, 183
9005-65-6	vol 35	185
	vol 36	339
9005-66-7	vol 35	184

9049-37-0	vol 35 vol 36	164 4
10006-28-7	vol 31	165
10022-31-8	vol 23	122
10024-93-8	vol 22	151-187
10024-97-2	vol 27/28	761, 762
10025-74-8	vol 22	290-299
10025-76-0	vol 22	233-241
10025-87-3	vol 22	93, 108, 139, 182-187, 223, 241, 261, 280, 299, 320, 343, 355, 371, 380
10026-23-0	vol 26	E350, 370-373
10028-24-7	vol 29 vol 31	E87, 88-94 E11
10034-85-2	vol 32	E26, 31, E330, 339
10035-03-7	vol 26	E187, 220, 221
10039-32-4	vol 31	E11, 106, 107
10042-88-3	vol 22	272-280
10043-52-4	vol 29 vol 27/28 vol 31 vol 32 vol 35 vol 36	E87, 88-94 E66-E68, E75, E80, E81, 107-113, 146, 147 271 E25, E26, 40, 44 76, 99, 149, 273, 288, 300, 309 41, 108, 121, 161, 327, 478, 479, 481, 482
10049-21-5	vol 31	E11, E39
10051-44-2	vol 33	E128, 129, E253, 254, 255, E280, 281, E294, 295, E311, 312, E324, 325, E326, 327, E328, E329
10099-58-8	vol 22 vol 24	74-93 E64, E65, 95
10099-66-8	vol 22	378-380
10101-41-4	vol 26	222
10101-89-0	vol 31	E11, E98
10101-96-9	vol 26	E350, 374-376
10102-15-5	vol 26	E1-E7
10102-18-8	vol 26	E303-E305, 306-319, 398, 400
10102-20-2	vol 26	316, 317, E403, E404, 407-418, 425
10102-26-8	vol 31	E4
10103-46-5	vol 31	264-266, 271
10117-38-1	vol 26	E90-E92, 93-107, 327
10124-36-4	vol 26 vol 23	288 289
10124-37-5	vol 30	78
10137-74-3	vol 30	150
10138-41-7	vol 22	327, E328, 329-333, E334

10138-52-0	vol 22	246-261
10138-62-2	vol 22	311-320
10138-93-9	vol 31	273
10156-51-3	vol 31	E307
10192-30-0	vol 26	E115, 141-143, 150, 151
10196-04-0	vol 26	E5, 46-49, 66-70, E113-E116, 117-123, 126-143
10214-40-1	vol 26	E350, 377-379
10233-01-9	vol 31	302, 305, 306
10257-55-3	vol 26	E187-E193, 194-235
10294-49-2	vol 26	E260
10361-03-2	vol 31	46
10361-37-2	vol 24	E65, 96
	vol 29	E47, 57
	vol 36	325
10361-79-2	vol 22	123-139
10361-82-7	vol 22	208-223
10361-84-9	vol 22	1-18
10361-89-4	vol 31	E11
10361-91-8	vol 22	359-371
10361-92-9	vol 22	20, 21, E22, 23-31, E32, 33-34
10377-51-2	vol 30	284
10377-52-3	vol 30	296
	vol 31	E1-E6, 7-10
10431-47-7	vol 26	E322-E324, 325-330
10460-33-0	vol 20	E221, 222
10567-69-8	vol 30	397
11006-55-6	vol 35	163
	vol 36	170
11113-50-1	vol 31	118
12029-98-0	vol 30	479, 480
12057-24-8	vol 31	9, 10
12068-90-5	vol 29	E207, E208, 212
12124-97-9	vol 24	E60, 78, 79
	vol 27/28	E59, 90
12125-01-8	vol 30	464
12125-02-9	vol 20	13, 14
	vol 23	264
	vol 26	E5, 66-70, E74, 87, 88, E115, 138
	vol 27/28	E59, 89
	vol 30	424
	vol 31	90, 119, 120, 262, 263, 267, 268
	vol 32	39, 66
	vol 36	326

12136-45-7	vol 31	269, 270
12171-53-8	vol 36	212
12258-15-0	vol 23	106
12286-43-0	vol 35	215
12362-10-6	vol 31	E11
12529-99-6	vol 26	E331, 338, 339
13011-54-6	vol 31	E101
13126-12-0	vol 30	169, 170
	vol 33	E332, 333-335
13269-73-3	vol 35	317
13410-01-0	vol 26	311-313
13446-09-8	vol 30	280, E461, E462, 463-466
13446-29-2	vol 26	E153, 162, 165
13446-49-6	vol 30	403
13446-70-3	vol 30	237, E258, E259, 260-262
13446-71-4	vol 30	E29, 71, 93, 94, 139, 140, 148, E162-E165, 166-175
13446-76-9	vol 30	288-290, 367, 395, E428-E431, 432-446
13450-81-2	vol 26	E252
13450-90-3	vol 22	182, 183
13451-02-0	vol 26	E236, 237-239
13453-80-0	vol 31	E4
	vol 30	320
13453-86-6	vol 30	281
13453-87-7	vol 30	193
13454-75-6	vol 30	237, 262, E263, E264, 265-267
13454-81-4	vol 30	291, 292, 368, 396, 438, E448-E450, 451-460
13454-83-6	vol 33	133
13455-28-2	vol 30	400
13465-59-3	vol 22	19
13465-98-0	vol 26	E267, 268-270
13469-98-2	vol 22	45-47
13470-38-7	vol 22	48
13472-35-0	vol 31	E11, E39
13472-45-2	vol 31	151
13477-00-4	vol 30	E29, 73
13477-23-1	vol 26	E286, 287-290
13477-98-0	vol 30	401
13494-80-9	vol 25	E194, E195, 196-205
	vol 29	E213, 224

13536-53-3	vol 22	140-142
13536-73-7	vol 22	344-347
13536-79-3	vol 22	94, 95
13536-80-6	vol 22	188-195
13536-94-2	vol 32	E327, 328, 329
13537-18-3	vol 22	348-355
13543-71-9	vol 30	E1-E10, 11-23
13550-28-2	vol 30	E184-E187, 188-194
13568-71-3	vol 26	E248, E249, 250, 251
13569-80-7	vol 22	285-289
13572-98-0	vol 22	266-269
13597-44-9	vol 26	E271-E274, 275-285
13597-46-1	vol 26	E350, 389, 390
13597-78-9	vol 26	E331, 344, 345
13708-63-9	vol 22	270, 271
13709-38-1	vol 22	49-73
13709-42-7	vol 22	146-150
13709-46-1	vol 22	115-122
13709-94-9	vol 31	257, 258, 289, 290, 304
13718-26-8	vol 31	144, 159, 160, 163, 164
13718-59-7	vol 26	E349, 362-366
13759-87-0	vol 22	224-228
13759-88-1	vol 22	243, 244
13759-89-2	vol 22	372-375
13760-78-6	vol 22	307-310
13760-80-0	vol 22	357, 358
13760-83-3	vol 22	322-326
13761-91-8	vol 22	359-371
13763-67-2	vol 30	E29, 72, 95-97, 149, 173-175, E176-E178, 179-183
13765-03-2	vol 30	E268-E275, 276-329
13765-25-8	vol 22	230-232
13765-26-9	vol 22	203-207, 245
13768-48-4	vol 30	310
13769-43-2	vol 31	305, 306
13769-82-9	vol 31	E98
13769-83-0	vol 31	E98
13770-18-8	vol 23	81

13774-16-8	vol 31	E307-E311, 312-316
13774-25-9	vol 26	E155, E156, 168-172, 182-186
13780-18-2	vol 26	E349, 356, 357
13780-48-8	vol 22	242
13812-58-3	vol 26	E404, 423
13813-22-4	vol 22	96
13813-23-5	vol 22	143-145
13813-24-6	vol 22	196-202
13813-25-7	vol 22	229
13813-40-6	vol 22	284
13813-44-0	vol 22	376
13814-59-0	vol 26	E350, 391-393
13814-81-8	vol 26	E262, 263-266
13818-75-2	vol 22	262-265
13825-76-8	vol 22	321
13870-19-4	vol 30	486
13952-84-6	vol 21	E252, 276
	vol 22	16, 43, 142, 144, 193, 228, 244, 259, 265, 268, 278, 283, 297, 341, 347, 369, 375, 377
13967-90-3	vol 30	239
13982-78-0	vol 29	29, 32, 44, 97-99, 114, 115, 117, 121, 123, 125, 128, 129, 131, 138-140, 142-148, 154-162, 168, 169, 174-176, 181, 182, 183, 185, 186, 188, 189, 200, 201
14013-56-0	vol 26	E305, E331, 336, 337
14167-59-0	vol 24	E131-E134, 188
14187-32-7	vol 30	373, 418, 447
14335-33-2	vol 31	332, 333
14456-47-4	vol 22	281-283
14456-48-5	vol 22	300-303
14456-53-2	vol 22	381
14457-87-5	vol 22	109, 207
14459-36-0	vol 26	E351, 397-400
14460-28-7	vol 26	E271, 276, 277, 284, 285
14518-27-5	vol 31	E101
14590-38-6	vol 26	E349, 358-361
14732-16-2	vol 30	311, 404, 442, 487, 488
14767-09-0	vol 30	304
14802-36-9	vol 21	E234
14854-08-1	vol 38	524

14887-42-4	vol 31	E168
14887-48-0	vol 31	E11
14899-38-8	vol 26	E404, 420
14929-69-2	vol 26	E403, 406
14947-17-2	vol 22	72, 97, 122
14986-52-8	vol 22	108
15122-56-2	vol 26	E404, 424-426
15123-75-8	vol 30	298, 299, 323, 324, 369, 398, 440
15162-95-5	vol 30	240
15293-86-4	vol 26	E262, 263-266
15347-57-6	vol 33	72, 95
15457-71-3	vol 26	E324, E331, 340, 341
15457-72-4	vol 26	E324
15474-63-2	vol 22	304-306
15593-51-8	vol 26	E331, 332
15593-61-0	vol 26	E349, 354, 355
15702-34-8	vol 26	E349, 367
15819-50-8	vol 31	E11
15851-44-2	vol 26	E404, 421
15851-47-5	vol 26	E404, 427
15851-51-2	vol 26	E404, 422
15855-76-2	vol 26	E350, 394-396
15855-80-8	vol 26	E305
15857-44-0	vol 26	E350, 368, 369
15899-92-0	vol 26	E404, 419
16271-20-8	vol 31	E11
16731-55-8	vol 26	E108, E109, 110, 111
16761-12-9	vol 33	203
16788-57-1	vol 31	E168
16805-99-5	vol 36	76
16806-00-1	vol 36	299
16806-29-4	vol 35	256
16840-28-1	vol 36	79
17103-43-4	vol 36	84
17103-45-6	vol 36	86
17103-46-7	vol 35	1
17103-48-9	vol 36	207
17103-49-0	vol 36	298

17103-50-3	vol 35	13
17103-51-4	vol 35	104
17103-53-6	vol 35	109
17455-13-9	vol 35	36, 52, 53
17455-13-9	vol 36	283, 292, 293
18088-11-4	vol 31	316, 317
18607-98-2	vol 35	61, 62
18649-05-3	vol 31	E318-E324, 325-331
19077-98-6	vol 36	E63, 64-75, 462
19083-61-5	vol 30	252
19086-20-5	vol 26	E153, 162
19122-79-3	vol 29	E47, 56
19416-65-0	vol 22	50
19455-00-6	vol 33	E128, 129, E163, 164, 202
19455-20-0	vol 33	E119, 120, E142, 143, E157, 158, E189, 190, E191, 192
19455-21-1	vol 33	E121, 122, E159, 160, E193, 194, E195, 196
19578-81-5	vol 20	E254, 255
19630-06-9	vol 30	308, 309, 370, 441, 456
20281-00-9	vol 31	329-333
20427-58-1	vol 23	E156-E184, 187, 189, 191-193, 198-200, 202, 203, 208, 209, 213, 214, 217-219, 222-224, 226, 229, 237-241, 245, 247, 250, 251, 254-256, 260, 264-267
20427-59-2	vol 23	E10-E26, 31, 34, 37-39, 43-45, 49-52, 54, 55, 57, 61, 62, 64-79
20667-12-3	vol 23	83-97, 98-105, 107-143
20911-44-8	vol 26	E256
20911-45-9	vol 26	E256
21006-12-1	vol 26	E252
21041-95-2	vol 23	E270-E282, 284-292, 297, 300-303
21303-03-7	vol 33	E113, 114, 115, 216, 217
21312-10-7	vol 35	63
21908-53-2	vol 23 vol 29	E307-E317 E207
22024-52-3	vol 21	E234
22113-86-6	vol 27/28	723
22199-08-2	vol 36	215-222
22446-84-0	vol 30	307
22451-06-5	vol 26	E403, E404

22763-02-6	vol 31	E168
22763-03-7	vol 31	E168
23282-34-0	vol 33	205
23325-73-7	vol 35	207
24342-31-2	vol 35	277-281
24342-38-9	vol 36	59-62
24458-98-8	vol 26	E331, 333-335
24539-94-4	vol 20	E240, 241
25105-33-3	vol 31	248
25265-71-8	vol 32	E173, E175, 250
25265-71-8	vol 24	E331-E334, 372, E375-E378, 388, 395
	vol 27/28	676
25322-68-3	vol 32	E173-E175, 252, 253
	vol 35	180, 181, 197-199
	vol 36	111, 112, 183, 280, 281, 345
25340-17-4	vol 38	264
25425-97-2	vol 26	E331, 346, 347
25454-04-0	vol 26	E152
25512-65-0	vol 27/28	E643, 653
25659-29-4	vol 30	399
25837-80-3	vol 21	E252, 261-263, E281, E282, 290
26566-61-0	vol 35	40
26896-48-0	vol 32	259
26970-82-1	vol 26	E303
28090-46-2	vol 31	332, 333
28677-00-1	vol 29	E177, E178, 180, 181
28958-26-1	vol 30	241
29063-28-3	vol 21	E16, 48
29501-28-8	vol 26	E187
30981-48-7	vol 31	148
31632-63-0	vol 38	541
31692-85-0	vol 36	110
32702-66-2	vol 26	E256, 257
32736-64-4	vol 26	E5, E74, 87, E115, 127, 128, E144, E145, 146-149
32909-92-5	vol 35	252
33100-27-5	vol 32	E173-E176
34392-82-0	vol 36	83
35089-90-8	vol 23	264, 269
35788-00-2	vol 26	E260, 261

38333-62-9	vol 26	E349, 353
35943-12-5	vol 35	101
39588-36-8	vol 36	78
39719-87-4	vol 35	282-288
40016-88-4	vol 36	436
40854-09-9	vol 26	E153
41292-05-1	vol 26	E303
41292-06-2	vol 26	E322
42096-67-3	vol 30	484, 485
50292-03-0	vol 31	E318
51092-74-1	vol 26	E252, 253-255
51203-19-1	vol 35	216-218
51203-20-4	vol 35	246-248
51249-11-7	vol 36	443
51543-29-4	vol 36	E88, 89-91
51543-30-7	vol 36	E92, 93-95
51543-31-8	vol 35	59, 60
51543-32-9	vol 35	102, 103
51732-39-9	vol 35	12
53034-20-1	vol 30	252
53081-02-0	vol 35 vol 36	278-281 352-356
56444-82-7	vol 36	10
59224-69-0	vol 36	363
59375-51-8	vol 22	50
59826-67-4	vol 21	E7, 54
60365-38-0	vol 26	E248
60593-58-0	vol 31	E11
60593-59-1	vol 31	E11
60936-55-2	vol 26	E256
60943-67-1	vol 26	E286
61028-24-8	vol 31	E47, E98
61906-92-1	vol 30	252
62968-83-6	vol 30	252
63019-00-1	vol 38	542
63095-89-3	vol 32	E174-E176, 261
63233-31-8	vol 30	252
65177-07-3	vol 35	58

65177-18-6	vol 36	294, 295
65185-91-3	vol 31	E11
65410-83-5	vol 26	E248
65410-84-6	vol 26	E256
65597-32-2	vol 30	302, 303
66219-86-1	vol 36	228
66269-03-2	vol 36	209
66516-57-2	vol 26	E286
66905-89-3	vol 31	E11
66922-99-4	vol 31	E168
67417-37-2	vol 31	E11
70504-12-0	vol 30	300, 301
71119-13-6	vol 35	E239, 240-245
71119-14-7	vol 36	15-17
71119-15-8	vol 35	251
71119-16-9	vol 35	11
71119-18-1	vol 36	96
71119-19-2	vol 36	80
71119-20-5	vol 36	85
71119-21-6	vol 36	77
71119-22-7	vol 35	277
	vol 36	60, 61, 356
71119-23-8	vol 36	221, 222
71119-24-9	vol 36	301
71119-25-0	vol 35	316
71119-27-2	vol 35	249
	vol 36	81
71119-28-3	vol 36	82
71119-29-4	vol 35	315
71119-30-7	vol 35	314
71119-31-8	vol 35	312
71119-32-9	vol 35	310
71119-34-1	vol 36	448
71119-35-2	vol 36	364
71119-36-3	vol 36	447
71119-37-4	vol 36	284
71119-38-5	vol 36	300
71119-39-6	vol 36	437

71119-42-1	vol 35	214
71138-72-2	vol 36	296
71261-83-1	vol 36	357
71261-84-2	vol 36	210
71261-85-3	vol 36	211
71261-86-4	vol 36	214
71261-88-6	vol 36	402
71261-89-7	vol 36	58
71280-76-7	vol 36	208
71280-79-0	vol 36	213
71496-63-4	vol 36	268
71720-65-5	vol 36	87
71720-66-6	vol 36	449
72384-11-3	vol 36	115, 116
72860-77-6	vol 26	E153
72878-03-6	vol 26	E187
73621-77-9	vol 30	305, 306
73761-48-5	vol 30	252
74893-09-7	vol 31	E4
75042-11-4	vol 26	E248
75042-12-5	vol 26	E248
75042-13-6	vol 26	E271
76619-75-5	vol 36	224
76634-39-4	vol 36	223
77400-68-1	vol 36	8
77400-69-2	vol 36	7
77400-70-5	vol 36	55
77400-71-6	vol 36	56
77902-26-2	vol 26	E258
77902-27-3	vol 26	E258
77902-28-4	vol 26	E258
79832-54-5	vol 31	E307
79962-97-3	vol 35	254
81815-35-2	vol 36	9
82537-68-6	vol 36	347
83534-22-9	vol 26	E349
83753-29-1	vol 26	E153

84812-74-8	vol 36	351	
84812-75-9	vol 36	350	
84812-76-0	vol 35	212	
84812-77-1	vol 35	211	
84812-78-2	vol 35	210	
84812-79-3	vol 36	433	
84812-80-6	vol 36	432	
84812-81-7	vol 36	431	
84812-82-8	vol 36	118	
84812-83-9	vol 36	117	
84855-83-9	vol 36	119	
84930-17-6	vol 35	253	
84930-18-7	vol 35	255	
86729-19-3	vol 36	428, 430	
86729-20-6	vol 36	429	
86729-21-7	vol 35	209	
86729-22-8	vol 35	208	
86729-23-9	vol 36	349	
86729-24-0	vol 36	348	
89146-33-8	vol 26	E291, 292	
90004-53-9	vol 36	181	
91399-12-1	vol 20	E248, 249	
91399-13-2	vol 20	E252??	
91498-96-3	vol 26	E72	
91498-97-4	vol 26	E72	
96247-21-1	vol 26	E252	
101056-43-3	vol 31	E318	
101056-44-4	vol 31	E11	
101056-45-5	vol 31	E11, E39	
101056-46-6	vol 31	E11	
101056-47-7	vol 31	E168	
101056-48-8	vol 31	E168	
101056-49-9	vol 31	E168	
101056-50-2	vol 31	E168	
101056-52-4	vol 31	E307	
101917-67-3	vol 31	E11	
134732-17-3	vol 30	312	

AUTHOR INDEX

A

Abdel Hadi, I.
    vol 36   311
Abdullaev, G.B.
    vol 29   E213, E214, 224
Abduragimov, A.A.
    vol 31   E129, E130, 159-162
Abduragimova, R.A.
    vol 31   E129, E130, 144, 145, 159-164, E297, 302, 305, 306
Abegg, R.
    vol 23   E88, E96
Abrahams, S.C.
    vol 30   E269, E275
Abramov, I.I.
    vol 26   E155-E157, E159, 171, 172, 184, 185
Adeishvili, E.U.
    vol 23   E11, E13, E25, 61, 297
Adeyschvili, E.V.
    vol 23   E277, E278, E282
Adilova, A.A.
    vol 23   E180, E184, 256
Aditya, S.
    vol 23   E315, E317, 339
Adhart, O.
    vol 25   E301, E302
Affsprung, H. F.
    vol 37   E64, E67, E71-E73, E83, 144, 147, 148, E221, E225,
                  E227, E228, 240, 242, E369, E378, 400, 402
Agasyan, P.K.
    vol 25   E172
Ahmann, D.H.
    vol 25   E425-E427, 428
Ahrndts, T.
    vol 23   E156, E182
Aimonetto, S.
    vol 35   162, 163
    vol 36   169, 170
Akers, W.W.
    vol 27/28   E266, 272, 273
Akhmetov, K.M.
    vol 23   E13, E17, E18, E23, E25, 58
Akimov, V.P.
    vol 25   E230, E231, E234, 236
Akopyan, Z.A.
    vol 26   E404, E405, 414
Akselrud, N.V.
    vol 23   E16, E17, E25, 50, E171, E183, 212, E272, E281, 289
Al-Najjar, H.
    vol 24   E130-E134, 153, 156
    vol 32   E166-E169, E185, 198, 202, 207-209, 212
Al-Sahhaf, T.A.
    vol 27/28   748
Alberty, R.A.
    vol 31   E2, E6
Alcauskas, J.B.
    vol 27/28   E1-E6, 19, 20, E50, E51, 52-54
Aldrich, B.W.
    vol 27/28   E50, E51
Alekhin, Yu.V.
    vol 29   E1, E2, E5, E14, E16-E18, E21, 39
Aleksandrov, B.N.
    vol 29   E213, E214, 218
Aleksandrova, L. P.
    vol 37   E64, E67, E82, 137
Aleksandrova, M.V.
    vol 32   E26, E28, 32
Alekseeva, E.A.
    vol 31   E39-E42, E51, 69
Aleshkina, N.N.
    vol 33   E167, 168, 260

Aleshnick, J.J.
     vol 30   E195-E197, 214, 215
Alexander, D.W.
     vol 37   E64, E67, E68, E82, 126
Alexander, K.S.
     vol 35   E64, 81, 82, 84-88
     vol 36   187-191, 194, 196, 198, 200, 383, 385, 389, 391, 393,
                     395, 397, 413, 415, 417, 419, 421, 423, 425
Alfenaar, M.
     vol 27/28   719
Allemann, O.
     vol 36   330, 360
Allen, E.R.
     vol 23   E174-E176, E183, 235
Allibert, M.
     vol 25   E182, E183, 191
Allinne, M.
     vol 34   E14, 17, 181
     vol 35   115, 220, E231, 233, E239, 241, E257, 258,
                     282, 290, 302
     vol 36   E19, E20, 24, E63, 67
Allmand, A.
     vol 23   E1, E16, E25, E307-E309, E317
Allport, N.L.
     vol 34   16
Alner, D.J.
     vol 23   E90, E93, E96, 124-126
Alric, R.
     vol 34   E285, 291
     vol 35   272, 310-315
Althaus, H.
     vol 23   E16, E17, E25, 56, 57, E156, E170, E171, E182, E183,
                     191, 245, 248
Alwani, Z.
     vol 37   E64, E66, E76, E80, E83, E84, 148, E370, E377, E378
     vol 38   E9-E12, E59, E65, E66, E178, E201, E202
Amarell, G.
     vol 25   E93-E95
Ambrose, D.
     vol 29   E2, E6, E21
     vol 27/28   E24-E28
Amelin, A.G
     vol 21   E8, E9, E11, E12, 56-58, 65, 76-78
Amick, E.H.
     vol 27/28   319
Amirajafari, B.
     vol 27/28   E1-E5, E24-E28, 38, 194, 195
Amlie, R.F.
     vol 23   E86, E96, 140, E144, E145
Ammann, A.
     vol 30   E1, E4, 17, E24, E25, E32, 36, E105, E111, 118, E162,
                     E165, 167, E176, E178, 180.
Amron, I.
     vol 30   E268, E269, E273, 277, 314, 315, E467, E468, E470
Anderko, K.
     vol 25   E13-E18, E57, E58, E84, E85, E139-E141
Anderson, C.T.
     vol 23   E156, E182
Anderson, G.W.
     vol 35   1, 12, 13, E14, 16, 104, 107-108, 316
     vol 36   97, 299, 301, 443
Anderson, J.T.
     vol 25   E369-E371, 380
Anderson, R.H.
     vol 23   E174-E176, E183, 235
Anderson, W.R.
     vol 27/28   E248, E249, 262, E737, 743, 744
Andrew, L.M.
     vol 20   E62, 63, E103-E105, 106, E129, 131, E132-E136, 140,
                     E153-E157, 165, E202-E204, 206, E210-E213, 215
Andrews, J.C.
     vol 32   E25, E27, 29

Andrews, L.J.
    vol 37   E63, E68, E81, 107, E369, E372, E378, 389
    vol 38   E1, E2, 7, E9, E11, E12, 13, E23, E24, E28, 29, E42,
                    E43, E4547, E59, E60, E66, 68, E177, E178, 180, E187,
                    E189, 191, E201, E202, 205, E236-E238, 241, E265, E266,
                    267, E271, 272, E277, E278, 279, 323, E341, E343, 344,
                    391, 424
Andrianov, A.M.
    vol 26   E90-E92, 96-101
Anhorn, V.J.
    vol 30   421-426
Antikainen, P.J.
    vol 23   E90-E92, E96, 139
Antonetti, G.
    vol 31   E39-E42, E51, 96, 97, E101, E102, 124, 125, E129,
                    E130
Anvaer, B.I.
    vol 32   E1-E3, 8, E25, E26, E28, 30, 34, 35, 41
Aoki, M.
    vol 34   E14, E15, 35, 75, 159, E204, 209, 258, 262, 277,
                    279
    vol 35   2, 5, 8, E14, 17, 28, 55, 65, 74, 93,
                    E110-E112, 131, 158, 192
    vol 36   E19, E20, 30, 39, 47, 102, 107, 114, E124, 131, 159,
                    205, E243, 249, 255, 266, 285, 288, 291, 368, 376, 400,
                    405, 411
Apfel, O.
    vol 31   E39-E42, 43, E51, 53, E127, E130, 131, E206, E207,
                    E210, E212, 213, 221-224
Aquan-Yuen, M.
    vol 20   E103-E105, 113, E153-E157, 180
    vol 37   E315, E317, E322, 356, E366, 368
    vol 38   E419, 423
Ar, A.
    vol 27/28   731
Arai, Y.
    vol 26   E187, E193
    vol 27/28   E477, 480, 689, E737, 740-742
Arakawa, S.
    vol 37   E65, E67, E84, 103, E370, E372, E379, 422
    vol 38   E59, E60, E66, 87, E201, E202, 210
Araki, M.
    vol 37   E65, E67, E84, 179, 180, E370-E374, E379, 420-322
    vol 38   E9-E12, 21, E23, E24, E28, 40, E42-E44, E46, 57, E59-
                    E62, E66, 85-87, 103, E164, E165, 167, E169, E170, 175,
                    E177, E178, 184, 185, E187-E189, 199, E201, E202, 209,
Araki, M.
    vol 38   210
Aral, K.
    vol 27/28   222, 659
Aramian, K.T.
    vol 25   E369-E371
Archinard, P.
    vol 29   E1, E3, E14, E16-E18, E21, 28, E228, 229, 230,
    vol 29   233, 234
Aref"eva, R.P.
    vol 37   E42, E43, E46, 58, E65, E67, E84, 172, E222-E224,
                    E228, 263, 264, E315, E316, E322, 351, 354,
                    E370-E372, E379, 414, 415, E485, E486, E491, 510
    vol 38   E59, E60, E66, 82, 83
Arifov, A.A.
    vol 29   E213, E214, 219
Ariguib-Kbir, N.
    vol 23   E279, E282, 298, 299
Arii, K.
    vol 26   E1-E3, E5-E7, 44, 45, E71-E75
Arita, T.
    vol 35   E64, 83, 90, 275, 276
    vol 36   262, 267
Arkhipov, S.M.
    vol 30   E24, E25, E28, E29, E31, E32, 58, 71, 72, 93-97,
                    E105, E108-E111, 136, 139, 140, E162, E165, 169-172,
                    174, 175, E176-E178, 182, E220, E222, E223, 237, E258,
                    262, E263, E264, E268-E275, 279, 281-285, 287, 293

Arkhipov, S.M.
    vol 30   294, 296-310, 312, 317, 318, 320, 323, 324, 328,
                    E330-E332, E335, E338, E339, 369, 370, E374, E375, E378,
                    E381, 395, E428-E432, 438, 441, E448-E450, 455, E456,
                    E467, E468, E470-E472, 483
Arkhipova, M.S.
    vol 26   E153-E155, E158, 170
Arkipov, M.I.
    vol 23   E13, E18, E19, E25, E26, 49, 51, 52, 66-78, E163, E164,
                    E178, E183, 213, 214
Armbruster, M.H.
    vol 25   E33, E34, 41
Arnal, T.G.
    vol 26   E187, E192, E193, E236, E240, E241
Arnett, E.M.
    vol 27/28   723
Arnol´d, T.I.
    vol 23   E10, E24, E25, 54, 55
Arnold, D.S.
    vol 37   E64, E67-E69, E82, 122, 123
Arnold, M.R.
    vol 32   E79-E82, 88, 97, 112
Arora, B.R.
    vol 23   E156, E182
Arrhenius, S.
    vol 30   E105, E112
Arriaga, J.L.
    vol 24   E130-E134, 151, 175, 177
Artamanova, O.I.
    vol 31   E168, E171, 205, E206, E212, 219
Arutyunyan, M.G.
    vol 26   E404, E405, 414
Asano, K.
    vol 32   E169, E186, 213, 214
Ashmyan, K.D.
    vol 27/28   E57, E70-E72, E74, E85, 129
Askenasy, P.
    vol 31   E206, E208, E212, 226, 227, 262, 263
Astarita, G.
    vol 32   E79, E82
Ata, N.
    vol 27/28   E320
Atamanova, N.M.
    vol 25   E402-E404
Atkinson, L.P.
    vol 27/28   E50, E51
Atterberg, A.
    vol 26   E152
Attia, Y.A.
    vol 23   E10, E25
Atwood, K.
    vol 32   E79-E82, 88, 97, 112
Audrieth, L.
    vol 31   E14, E15
Audrieth, L.F.
    vol 22   214, 215
Auer-Welsbach, C.   -
    vol 22   35, 104, 135, 178, 219, 257, 295, 316, 339, 368
Aureggi, C.
    vol 25   E194, E195, 199
Autenrieth, W.
    vol 26   E236, 237, E240, E241, 242
Avdeeva, O.I.
    vol 27/28   E1-E5, E70, E71, E74, E77, E84, 124, 152
Aven, M.
    vol 25   E194, E195
Averbukh, M.A.
    vol 26   E348-E351, 361, 363
Averko-Antonovich, I.N.
    vol 30   E184-E187, 190, 191
Avico, U.
    vol 34   77, 308
    vol 35   145
    vol 36   33, 147

Aygaraeva, M.M.
    vol 25   E214, E215
Azarnoosh, A.
    vol 24   E39, 40-44
    vol 27/28   E281, 288, 289
Azorova, L.A.
    vol 30   E220, E222, E223, 239, E268, E270-E275, 286, 316,
                  321, 322, E374, E375, E378, E381, 397, E467, E468,
                  E470-E472

B

Babaeva, A.V.
    vol 30   E24, E25, E28, E30, E32, 42, 43
Babanly, M.B.
    vol 25   E194, E195
Babayan, G.G.
    vol 26   E404, E405, 414
    vol 31   E127, E129, E130, 143, 146, 147
Babenko, A.M.
    vol 26   E90-E92, 96-101
    vol 31   E39-E42, E47, E50, E51, 70-72, 86, 87, E206-E209, E212
Babinski, J.J.
    vol 25   E157-E159
Babkin, G.N.
    vol 25   E310, E317, E318
Babushkina, M.D.
    vol 26   E155-E157, E159, 180, 181, E187-E190, E193, 196, 197
Backus, H.S.
    vol 27/28   557, 558
Backx, P.
    vol 37   E65, E84, 187, E222, E228, 271
Badavamova, G.L.
    vol 25   E207-E208, 213, E214, E215, 220, E221, 223, 224, E225,
                  229, E230, E231, E234, 238, E242, 243
Badawi, A.A.
    vol 35   E110-E112, 138, 206
Badger, E.H.M.
    vol 32   E54, E55, 68
Badygina, L.I.
    vol 25   E194, E195
Baganz, E.
    vol 22   366
Baggio, R.F.
    vol 26   E248
Baggio, S.
    vol 26   E248
Bailey, J.A.
    vol 27/28   E750, 753, 757
Bailey, M.G.
    vol 31   E14, E15
Bakalov, V.
    vol 26   E153, E154, E157-E159, 173, 174, 177-179
Bakeev, M.I.
    vol 33   E25, E26, 30, 34, E57, E60, 76, E91, 94, 96
Bakeeva, S.S.
    vol 26   E348-E352
Baker, C.T.
    vol 23   E160, E162, E182, 249
Baker, E.G.
    vol 38   E114, E117, 119, 131, E135, E141, 144, E294, E295, 296,
                  297, 324, E501, E502, 503-505, 546, 547
Balabanovich, Ya.K.
    vol 31   E168, E171, 178, 179
Balashova, E.F.
    vol 31   E51, E52, 90, 91, E206-E209, 211, 212, 255, 256
Balashova, N.A.
    vol 23   E88, E96, E144, E145
Baldwin, R.R.
    vol 25   E425-E427, 428
Balej, J.
    vol 25   E13-E18, 32
Baletskaya, L.G.
    vol 25   E335-E337

Ball, T.R.
    vol 25   E33, E34
Ballintine, T.
    vol 29   E1, E5, E7, E14, E16-E18, E21, 41, E47, E48, 52, 63
Balog-Megyery, K. see Megyery-Balog, K.
Baltisberger, R.J.
    vol 29   E1, E5, E7, E14, E16-E18, E21, 41, E47, E48, 52, 63
Balyashko, V.M.
    vol 31   E127, E130, 151
Balyasnaya, A.M.
    vol 31   E127, E130, 139
Balyreva, V.A.
    vol 30   E220, E222, E223, 240, 241
Balzano, C.
    vol 32   E79, E82
Bancroft, W.D.
    vol 32   E180, E187, 290
Bandelin, F.J.
    vol 34   235, 238, 268, 269
    vol 35   72, 76, 79, 99, 268, 273, 286, 288, E289,
                  297, 300, 307, 309
    vol 36   105, 108, 120, 121, 474, 477-479, 481, 482
Banerjee, R.K.
    vol 23   E156, E182
Banerjee, S.
    vol 20   E3, 5, E19, 21, E73-E77, 87, E89-E91, 97, E103-E105,
                  115, E183, E184, 189
    vol 37   E65, E68, E84, 177, E370, E371, E379, 417
    vol 38   E1, E2, 8, E260, 263, E332, E334, 338, E341, E343, 350
Banick, C.J.
    vol 25   E103-E105, 108
Baranski, A.
    vol 25   E317, E318, 324
Barbaudy, J.
    vol 37   E63, E68, E81, 93
Barber, H.J.
    vol 34   162, 163
    vol 35   195, 196
    vol 36   52, 53
Barbieri, G.A.
    vol 23   E144, E145
Barclay, I.M.
    vol 21   E5, E12, 41
Barduhn, A.J.
    vol 24   E16, E17, 25, E65-E68, E70, E72, 107, 120
Barin, I.
    vol 25   E13-E18
Barker, T.V.
    vol 30   E428, E429, 431, 433, E448-450, 452
Barkova, L.V.
    vol 31   E308, E311
Barlow, M.
    vol 25   E330
Barone, G.
    vol 24   E1, E2, 6, E16, E17, 30
    vol 27/28   E1-E5, 179
    vol 37   E41, E42, E46, 50, E314, E316, E322, 335
Baroni, A.
    vol 23   E20, E26
Barrett, C.S.
    vol 25   E402-E404, 416
Barrett, W.H.
    vol 26   E1-E3, E5, 8, 9
Barsuk, S.D.
    vol 27/28   E297, 300-302
Barta, F.A.
    vol 30   E269, E274
Bartecki, A.
    vol 23   E94, E95, E97, 142, 143
Bartell, F.E.
    vol 20   E132-E136, 141
    vol 37   E63, E68, E72, E82, 114

```
Bartholomay, H.W.
     vol 25        E93-E95, 97
     vol 29        E213, E214, 216
Barton, J.R.
     vol 24        E375-E378, 379
Bashugyan, D.P.
     vol 31        E318, E320, E321, E324, 328
Baskov, A.
     vol 33        E57-E60, 61
Basova, E.P.
     vol 30        E195-E197, 201, 202, 212, 216, 217
Basset, J.Y.
     vol 33        E58-E60
Basu, U.P.
     vol 34        275, 276
     vol 35        E110, E111, 123
     vol 36        E124, 129, E243, 248, 466-468, 471, 483, 484
Batel, R.
     vol 25        E317, E318, 325
Battino, R.
     vol 27/28     E1-E5, 23, E56, E57, E84, 190, 191, E202-E203, E206,
                   E207, E215, 227, 231, E450-E454, 457-462, E493-E497,
                   503, 504, 506, 507, E582-E584, E592-E595, 625, 627,
                   683, E685, 687, E724, 725
Batyaev, I.M.
     vol 22        93, 108, 139, 182-187, 223, 241, 261, 280, 299, 320, 343,
                   355, 371, 380
Baubigny, H.
     vol 26        E267, 268
Bauer, G.C.
     vol 23        E170, E183
Baumann, K.
     vol 34        283, 284
     vol 36        258, 259, 277, 278
Bayne, A.H.
     vol 37        E197-E199, 201
Beaudoin, J.M.
     vol 27/28     E360, 366, 367
Becher, R.
     vol 34        E13, E14, 23, 50, 107, 108, 149, 297-301, E302,
                   305, 307, 309, 310, 314
     vol 35        E110, E111, 120, 144, 164, 178, 179
     vol 36        1-5
Beck, H.
     vol 25        E214, E215
Beck, R.P.
     vol 25        E57, E58, 62
Becke, A.
     vol 38        E294-E296, 307
Becker, B.
     vol 31        E212, 272
Becker, C.H.
     vol 34        258-261
     vol 35        200-202
Becker, Th.
     vol 32        E26, E28, 40
Beglov, B.M.
     vol 31        E206, E212, 276, 277
Bel´tschev, F.V.
     vol 31        E206, E207, E210, E212, 218, 237-239
Belden, B.C.
     vol 32        E180, E187, 290
Bell, H.C.
     vol 30        E24, -E26, E32, 35
Bell, J.M.
     vol 23        E21, E26
Bell, M.L.
     vol 27/28     673, 674
Bell, R.N.
     vol 31        E11, E15
Bell, R.P.
     vol 20        E132-E136, 138, E153-E157, 158
     vol 21        E1-E2, 4, E9-E11, 13, 28, 67, E81, E89
     vol 32        E166, E167, E170, E178, E179, E186, 190, 224,
```

Bell, R.P.
    vol 32   279, 280, 281
Bellon, P.
    vol 23   E144, E145
Belousov, A.I.
    vol 37   E66, E85
Belov, N.V.
    vol 23   E174, E176, E180, E183, E184, E280, E282
Ben-Naim, A.
    vol 24   E16, E17, 24, E54, 56
    vol 27/28   E1-E6, 16, 17, E45, E46, 47, E59, E68-E70, E74-E76,
               E80, E84, E85, 89, 116, 127, 149, 151, 163, 171,
               175, 176, 178, 180, 182, 190, 191, E450, E454, 456,
               E582-E592, 599, 604, 608, 612, 617, 621, E642-E644,
               650
    vol 37   E64, E65, E70, E83, 156, 186
    vol 38   E341, E343, 347, 351
Bender, S.Ya.
    vol 37   E42, E46, 49
Benditt, E.J.
    vol 35   152, 225
Benedict, M.
    vol 27/28   483
Benjamin, L.
    vol 25   E385-E387, 396
Benkovskii, V.G.
    vol 37   E314, E318, E322, 342, E369, E374, E378, 403
    vol 38   E225, E226, 230
Bennett, G.M.
    vol 37   E221, E227, E314, E315, E321
Bennett, H.C.
    vol 25   E1-E3, 5, E33, E34, 39, E44, E45, 48, E50, E51, 53, 54,
               E66, E67, E71, E72, 75, E76, E77, 79
Bennett, J.A.R.
    vol 25   E139-E141, 150, E385-E387, 394
Benrath, A.
    vol 30   E105, E108, E111, 117, 130, 131, E220, E223, 225, E374,
               E377, E380, 384
Bent, H.E.
    vol 25   E13-E18, 28, E33, E34, 40, E93-E95
Ben´yaminovic, O.A.
    vol 24   E196, E197, 200, 201
    vol 27/28   E248, E249, 256, 257, 276, E297, 300-302
Benz, R.
    vol 25   E263, E264, 265
Berchiesi, G.
    vol 33   3-8, E12, 13, E15, 18, 22-24, E145, 147, E150, 152,
               E229, 232, E237, 240
Berchiesi, M.A.
    vol 33   E237, 240
Berdyukova, V.A.
    vol 31   E127, E130, 150, E297, 303
Beremzhanov, B.A.
    vol 31   E11, E13, E14, E15, 23, E39-E42, E49, E51, 73, 74,
               E98, E102, 118, E168-E171, 180, 181, E206, E212,
               257, 258, E278, 289, 290, E297, 304,
Beerengarten, M.G.
    vol 21   E1-E3, E12, 14, 22, 30, E89
    vol 32   E170, E171, E182, E186, 226, 233286, 293, 295
Berg, A.G.
    vol 31   E168-E171, 190-194, 199-202, 216, E278, E297, E307,
               E308, E311, E318, E320
Berg, L.
    vol 30   E1-E4, 15, 16
Bergeim, F.H.
    vol 35   E110, E111, 113, 186, 207
Berger, J.E.
    vol 30   218, 256
Bergman, A.G.
    vol 31   E206-E208, E210, E212, 232, 233, 244, 245, 249,
               250, 274, 275, E278, E279, 282-285, 291, 292
    vol 33   E57-E60, 62, 64, 66, 67, 137, 165, E167, 168, E172,
               173, E175, 176, 219, 257, 260, 264, E267, 269

Berk, A.A.
    vol 31   E11, E15, 17, E127, E129, E130, 132, 133, 136-138,
                  155, 156
Berkengeim, T.I.
    vol 37   E63, E71, E72, E81, 101, E221, E225, E227, 231
    vol 38   E114-E117, 118
Berl, A.
    vol 32   E27, E28, 48
Berlin, M.A.
    vol 24   E257, E258, 283-298
    vol 27/28   694, 696-701
Bermzhanov, B.A.
    vol 30   E473, 489
Bernado, P.D.
    vol 35   E110-E112, 133, 170, 190
Berndt, A.F.
    vol 25   E214, E215
Bertazzoli, C.
    vol 35   6
    vol 36   104, 332, 365, 441, 442
Berthelot, M.
    vol 25   E13-E18
Berthier, P.
    vol 26   E262
Berton, A.
    vol 21   E302, 303, 308, 309
Bertoni, A.
    vol 30   E24, E25, E28, E29, E32, 73
Besserer, G.J.
    vol 32   E167, E186, 194, 195
Beuschlein, W.L.
    vol 26   E187-E189, E190, E195, 202, E240, E241, 245,
Beutier, D.
    vol 32   E54, E55, 65
Bevan, H.G.L.
    vol 35   142, 222
    vol 36   32, 70, 139, 232, 346, 358
Bevillard, P.
    vol 23   E1
Bezdel, L.S.
    vol 32   E172, E186, 238
Bezzubenko, A.A.
    vol 23   E148, 155
Bhattacharyya, R.
    vol 34   275, 276
    vol 35   E110, E111, 123
    vol 36   E124, 129, E243, 248, 466-468, 471, 483, 484
Biamonte, A.R.
    vol 35   79, 98
    vol 36   476, 480
Bianchini, A.
    vol 29   E213, E214, 221
Bichowsky, F.R.
    vol 26   E187-E189, E191, E192, 212
Biedermann, C.
    vol 23   E90, E91, E92, E96, E277, E281
Bijl, A.
    vol 27/28   E24-E28, 29, E57, E66-E69, E71, E72, E74-E76, E79, 80,
                  E84, 107, 114, 117-119, 148, 150, 162, 172, 181,
                  183
Bijl, H.C.
    vol 25   E402-E404, 408
Bilinski, H.
    vol 23   E311, E312, E316, E317, 342
Billett, F.
    vol 24   E1, E2, 7, E16, E17, 21, E59, E64-E71, 74, 95-99,
                  109, 115, 118, 119,
    vol 27/28   E5, 11, E68, E69, E72, E81, E84, 115, 121, 167
    vol 37   E63, E68, 116, E369, E372, E378, 390
    vol 38   E59, E60, E66, 72
Billiter, J.
    vol 30   E24-E26, E32, 53
Billman, G.W.
    vol 27/28   383, 284

Biltz, W.
    vol 25   E44, E45, 49
Binder, O.
    vol 23   E21, E22, E26, 43-45
Bircher, L.J.
    vol 24   E331-E334
    vol 27/28   E582-E586, E595, 597, 602, 606, 611, 616, 620, 622
                  624, 691
Biros, J.
    vol 25   E13-E18
Biryukov, V.P.
    vol 26   E404, E405, 420, 427
Bishimbaev, V.K.
    vol 26   E90-E92, 104, 105
Bishop, J.A.
    vol 30   E330-E332, E335, E336, E338, 364, 365
Bittner, H.
    vol 25   E326
Bittrich, H.J.
    vol 37   E65, E68, E71, E72, E84, 176, E485, E486, E488,
                  E489, E491, 512
Bjorklund, C.W.
    vol 25   E263, E264, 265
Black, C.
    vol 37   E31, E33, E34, 35, E42, E45, E46, 48, E63, E71,
                  E81, 104, 209, E221, E225, E227, 232, E314, E318
                  E321, 325, E455, E456, 457, E485, E486, E490, 494
    vol 38   E135, E136, E138, E141, 143
Blackman, G.E.
    vol 20   E6, E7, 10, E24, 25, E43, 44, E45, 46, E195, E196, 201,
                  E221, 222, E223, 224, E226, 227, E228, 229, E231, 232,
                  E233, 234, E235, 236, E238, 239, E240, 241, E242, 243,
                  E244, 245, E246, 247, E248, 249, E250, 251, E252, 253,
                  E254, 255, E256, 257
Blair, J.
    vol 25   E194, E195
Blais, C.
    vol 24   E130-E134, E218-E221, E257, E258, 270, 273, 278,
                  280, 356, 368
Blake, R.J.
    vol 32   E173, E187, 251
Blanc, M.Le.
    vol 30   E24-E26, E32, 34
Blanchard, K.C.
    vol 35   216, 249
    vol 36   81, 82, E88, 89, E92, 93
Blanco, R.E.
    vol 25   E425-E427, 429
Blanco-C, L.H.
    vol 27/28   E60, E66-E68, E85, 93, 94, 108, 109
Blank, H.
    vol 25   E434
Blarez, C.C.
    vol 30   E105, E112
Bleshinskii, S.V.
    vol 22   48, 114, 266, 284, 306
Bloomer, O.T.
    vol 27/28   E248, E249, 250-254
Blount, C.W.
    vol 27/28   E24-E28, E57, E72-E74, E85, E130-E142
Blount, H.N.
    vol 32   E182, E187, 298
Blumer, T.E.
    vol 27/28   539-542
Bobrova, V.V.
    vol 23   E11, E13, E14, E25, 60, 61, E277, E282, 296, 297
Bobrovnik, L.D.
    vol 26   E193, 218, 219
Bochkarev, N.F
    vol 31   E206-E208, E210, E212, 216, 232, 233
Bodmer, M.
    vol 31   E39-E42, E47, E49-E52, 65-68, 80-85, E206, E211,
                  E212, 240, 241

Bogoslovaskaya, T.T.
    vol 37   E314, E318, E322, 342, E369, E374, E378, 403
Bogoslovskaya, T.M.
    vol 38   E225, E226, 230
Bohon, R.L.
    vol 37   E63, E67, E68, E82, 111, E369, E371-E374, E378,
                 389
    vol 38   E23-E25, E28, 30, E42-E45, 48, E59, E60, E66, 70,
                 E236-E238, 243, E341-E343, 345
Bolander, B.
    vol 20   E153-E157, 171
    vol 37   E64, E72, E82, 138, E369, E378, 396
    vol 38   E9, E10, E12, 16, E23, E27, E28, 33, E42, E45, E46, 51
Bolden, P.L.
    vol 27/28   188
Bolton, W.
    vol 25   E274
Boncyk, L.
    vol 30   161, 258
Bondareva, O.S.
    vol 23   E18, E26, E174, E176, E180, E183, E184
Bondareva, T.I.
    vol 32   E176, E177, E187, 266
Bonhoeffer, K.F.
    vol 29   E1-E3, E14, E16-E18, E21, 26, E102-E105, E109, 113
                 E163-E165, 166
Bonnier, E.
    vol 25   E139-E141, E182, E183, 191
Boomer, E.H.
    vol 27/28   E303, 313, E320, 334, 335, E338, 347, 348
Booth, H.
    vol 31   E11, E15, E98, E99, E102, 105
Booth, H.S.
    vol 20   E73-E77, 79E103-E105, 114, E153-E157, 163
    vol 37   E63, E66, E81, 105, 108, E314, E316, E321, 326,
                 E369, E371, E378, 386, E485, E486, E490, 494
    vol 38   E177, E178, 179, E260, 261, E287, 288
Borovaya, F.E.
    vol 31   E206-E209, E212, 236
Bota, T.D.
    vol 21   E10-E12, 74
Bottger, W.
    vol 31   E2
Bottger, W.
    vol 23   E84, E96, 99
    vol 25   E1-E3, 4, E13-E18, 23, E33, E34, 38, E44, E45, 46, E57,
                 E58, 59, E71, E72, 73, E76, E77, 78, E385-E387, 389,
                 E402-E404, 407
Bottoms, R.R.
    vol 32   E80, E83
Bouaziz, R.
    vol 33   E58-E60
Bourgeois, J.
    vol 31   E168, E169, E171, 176, 177, E208, E212, 264-266
Bowers, W.G.
    vol 23   E171, E183
Bowersox, D.F.
    vol 25   E207, E208, 210, E214, E215, 216, E263, E264, 265, E271,
                 272, E274, 275, E282, 283, E301, E302, 305, E307, 308,
                 E434, 435
Boyer, F.L.
    vol 24   E331-E334
    vol 27/28   E582-E586, E595, 597, 602, 606, 611, 616, 620, 622,
                 624
Bozhovskaya, T.K.
    vol 27/28   E1-E5, E70, E71, E74, E76, E77, E84, 124, 152
Bradish, W.F.
    vol 27/28   E349, 351, 352
Bradley, R.S.
    vol 37   E64, E76, E83, 157, E369, E371, E372, E378, 406
Brady, A.P.
    vol 37   E64, E68, E82, 124
Brady, C.J.
    vol 38   E59-E63, E65, E66, 88, 89, E104-E106, 108, 109

Braghetti, M.
    vol 33   E15, 17, E58-E60, 136, 138, 144, E145, 148, E150,
                166, 171, 218, 220, 228, 234, E237, 258, 263
Braginskaya, R.I.
    vol 23   E20, E26
Braid, M.
    vol 29   E177, E179
Braley, S.A.
    vol 25   E369-E371
Brandon, J.
    vol 25   E263, E264, E271
Bratton, A.C.
    vol 35   216, 246, 249
    vol 36   81, 82, E88, 89, E92, 93
Brauer, G.
    vol 26   E262
Braun, A.
    vol 30   E108, E111, 130, 131
Braun, W.G.
    vol 37   E64, E76, E83, 152, E221, E226, E228, 246, E314
                E320-E322, 337
Braunstein, H.
    vol 25   E13-E18
Braunstein, J.
    vol 25   E13-E18
Brebrick, R.F.
vol 25   E194, E195, 200
Breckenridge, R.C.
    vol 31   E39-E42
Bregeault, J.M.
    vol 26   E271, E273
Breusov, O.N.
    vol 26   E403, E405, 406
    vol 30   E162, E165, 168, E176, E178, 181, E220, E222, E223,
                226, E258, 261, E263, E264, 267, E374, E375, E381,
                385, E428, E429, E431, 434, E448, E450, 453
Brewer, L.
    vol 25   E282
Brice, D.B.
    vol 26   E155, E156, E159, 165, 166
Bridgeman, O.C.
    vol 27/28   E50, E51
Britton, G.T.
    vol 25   E369-E371, 376, 377
Britton, H.T.S.
    vol 23   E83, E90, E96, 104, 138, E169, E183
Broadbank, R.W.C.
    vol 31   E307, E308, E311, E318, E324
Broadbent, C.P.
    vol 23   E178, E184, 269
Broadbent, D.
    vol 31   E11, E13-E15, 22
Broderson, H.J.
    vol 30   103, 219, 427
Brodeson, H.J.
    vol 22   107
Brodin, A.
    vol 35   264, 270, 271
Brodsky, A.E.
    vol 23   E307, E317
Brodsky, M.
    vol 31   E101, E102, 119, 120
Brollos, K.
    vol 37   E65, E83, 184, 185, E221, E226-E228, 247-249, E485,
                E490, E491
    vol 38   E341, E343
Bronsted, J.N.
    vol 21   E1, 13
    vol 30   E112, 147, E378, E381, 405
Bros, J.P.
    vol 25   E93-E95, 102
Brosche, A.
    vol 26   E1-E3, E5, 10-12, E72-E74, 76, 77, E90-E92, 93, 94, E108,
                E109, 110, 111

Brossmann, G.
     vol 25   E434
Brown, J.B.
     vol 25   E369-E371
Brown, O.L.
     vol 25   E103-E105, 112
Brown, R.L.
     vol 30   E29, E31-E33, 69, 90, E108, E109, E112, 133
     vol 37   E65, E67, E68, E83, 162, E188, 189, E370-E374,
                     E378, 408, E425, 426
     vol 38   E59-E62, E66, 79, 90
Brown, S.F.
     vol 30   E330, E333, E335, E339, 388, 389
Bruder, K.
     vol 31   E101, E102, 126
Brun, S.
     vol 31   E3, E6
Brune, U.
     vol 26   E187, E188, E193, 220
Brunisholz, G.
     vol 23   E92, E97
     vol 31   E39-E42, E47, E49-E52, 65-68, 80-85, E168, E169,
                     E171, 176, 177, E206, E208, E211, E212, 240, 241,
                     264-266, E308, E311, E320, E324
Brunner, E
     vol 27/28   E465, 474
Bruzzone, C.
     vol 25   E66, E67, 70, E71, E72, E73, E76, E77, 80, E207-E208,
                     212
Buchmeier, W.
     vol 26   E153, E158, E271, E273, E286
Buckley, W.D.
     vol 27/28   E202-E206, E208-E210, E214, 221, 226, 228, 232, 237,
                     E493, E494, E497, 499
Budantseva, L.S.
     vol 37   E65, E67, E71, E84, 167, E197-E199, 207, E221,
                     E223-E225, E228, 258, E275, E276, 281, E314, E317,
                     E318, E322, 347, E485, E486, E488-E490, 508
     vol 38   E114, E115, E117, 123, E135, E136, E141, 150
Budenholzer, R.A.
     vol 27/28   E281, E282, 286, 287
Buell, H.D.
     vol 30   E258, 260, E263, E264, 265, 266
Bugarzsky, S.
     vol 23   E307, E317
Bugli, G.
     vol 26   E252, E256
Bukacek, R.F.
     vol 27/28   382
Buketov, E.A.
     vol 23   E13, E17, E18, E23, E25, 58
     vol 26   E348-E352
Bukhman, S.P.
     vol 25   E13-E18, 31, E76, E77, 81, E173, E174, 175, 176, E285,
                     E286, 294, 296, E335-E337, 348, 352, E402-E404, 420
Bulina, V.A.
     vol 25   E207-E208, 213, E214, 219, E221, E225, 228,
                     E239, 241, E242, 244, E245, 246, E247, 248, E249,
                     250, E251, 252, E256, 257
Bulisceck, T.S.
     vol 31   E13, E14, E15, 37, 38
Bult, A.
     vol 36   217, 223, 224
Bunsen, R.
     vol 27/28   E1-E5, 7
Buogo, A.
     vol 35   6
     vol 36   104, 332, 365, 441, 442
Burd, S.D.
     vol 37   E64, E76, E83, 152, E221, E226, E228, 246, E314,
                     E320-E322, 337
Burger, A.
     vol 34   E14, E15, 41, 111, 150, 151, 167, 169-176, 293,
                     294

Burger, E.
        vol 25   E277
Burgess, W.M.
        vol 30   E1-E4, 12-14
Burk, W.
        vol 32   E184, 187, 314
Burkov, K.A.
        vol 23   E18, E26
Burlage, H.M.
        vol 34   150, 316, 322, 323
        vol 35   187, 230
        vol 36   49, 192, 226, 263, 439, 440
Burns, C.L.
        vol 25   E139-E141, 152
Burns, J.F.
        vol 27/28   E266, 272, 273
Bus´ko, E.H.
        vol 23   E18, E26
Busch, W.
        vol 23   E158, E182, 186
Busheina, I.S.
        vol 23   9, 82
Buslaev, Y.A.
        vol 30   E467, E468, E470, E472, 478
Busse, L.W.
        vol 35   146
Butler, J.A.V.
        vol 21   E5, E12, 41
Butler, J.N.
        vol 25   E1-E3, 9, E326, 328, E330, 333
        vol 31   E2, E6
Buttler, F.G.
        vol 26   E248, E249
Button, D.K.
        vol 38   E381, 384
Byeseda, J.J.
        vol 32   E1-E3, 19, E172, E173, E175, E184, E185, E187,
                        249, 304, 321
Bykhanov, I.M.
        vol 25   E194, E195
Bykharov, I.M.
        vol 29   E207, E208, 212
Bykova, I.N.
        vol 31   E307, E308, 312, 314, E318, E320, 325, 327
Bykovskaya, L.P.
        vol 29   E47, E51, 56
Byrne, J.E.
        vol 27/28   E495-E497, 504, 506, 507
Byrne, P.A.
        vol 27/28   E24-E28, 42, E50, E51, 55, E64-E68, E72, E74-E83,
                        E85, 104-105, 112, 113, 143, 146, 142, 153, 155-159,
                        165, 166, 168-170, 192, 193

C

Caddock, B.D.
        vol 37   E64, E72, E82, 178, E369, E375, E378, 392
Cady, G.H.
        vol 21   E281, E282, 301
Caillet, M.
        vol 25   E139-E141, 153
Calder, G.V.
        vol 20   E153-E157, 174
Calder, J.A.
        vol 37   E370, E372, E379, 412, E425, 429
        vol 38   E9, E11, E12, 18, 22, E23, E24, E28, 36, 41, E42, E43,
                        E46, 54, 58, E59, E60, E66, 80, 91, E164, E165, 166,
                        168, E169, E170, 172, 176, E177, E178, 183, 186, E187-
                        E189, 196, 200, E236-E238, 245, E250, 255, E265, E266,
                        269, 270, E271, 274, 276, E277, E278, 284, 285, E311,
                        E312, 314, E319, 320, E332-E334, 336, E341-E343, 348,
                        E352, 354, E357, 358, E360, 361, E363, 364, 372, E376,
                        E377, 380, E381, 383, E419, 420, E426, E427, 430, 432,
                        E501, E502, 506, 507, 525, 526, 544, 545

Caldwell, W.T.
    vol 36   302, 304, 364, 447-449
Calvert, H.T.
    vol 30   E105, E108, E112
Calzolari, F.
    vol 30   E105, E108, E111, 114, E162, E165, 166, E176, E178, 179
Cambi, L.
    vol 25   E57, E58, 60, E66, E67, 68
Cameron, F.K.
    vol 26   E187-E190, E193, 229, 230
Campbell, A.N.
    vol 23   E20, E26, E92, E97, 123
    vol 25   E139-E141, E182, E183, E285, E286, E402-E404
    vol 30   E1-E4, 18-22, E184-E187, 192, 193, E195-E197
Campbell, J.M.
    vol 25   E1-E3, 8
    vol 27/28   E1-E5, E24-E28, 38, 194, 195
Campenella, J.L.
    vol 25   E207, E208, 209
Campos Carles, A.
    vol 27/28   734, 735
Capell, R.G.
    vol 27/28   570
Carlson, R.G.
    vol 25   E194, E195
Carmichael, L.T.
    vol 32   E1-E3, 9, 10
Carritt, D.E.
    vol 23   E3, 4
Carswell, T.S
    vol 20   E6, E7, 8, 9
Carter, C.P.
    vol 22   E327
Carter, D.
    vol 24   E375-E378, 394, 410
Carter, H.D.
    vol 25   E139-E141
Carter, M.D.
    vol 25   E285, E286
Carter, R.T.
    vol 27/28   390, 391
Cassidy, R.M.
    vol 23   E272-E276, E281, 302
Casteneda, R.
    vol 24   E130-E133, 137, 155, 168, 171, 173, 178, 179,
                    E257, E258, 263, E331-E333, 341, 356, E375-E378,
                    381
Castrunuova, G.
    vol 27/28   E1-E5, 179
Cauquil, G.
    vol 21   E6, E12, 49, E281, 288, E315, 320
    vol 27/28   E637, 638
Cavazutti, G.
    vol 34   77, 308
    vol 35   145
    vol 36   33, 147
Cave, G.C.B.
    vol 20   E73-E77, 84, E153-E157, 178
    vol 37   E65, E71-E73, E84, 168, E221, E225, E228, 259
    vol 38   E42, E45, E46, 53
Chalova, E.P.
    vol 22   48, 114, 266, 284, 306
Chang, C.T.
    vol 25   E425-E427
Chang, F.T
    vol 21   E10, E12, 71-73
Chang, H.L.
    vol 27/28   E338, 345, 346, E520, 522, 524, 525
Chang, S.D.
    vol 27/28   E248, E249
Chang, T.L.
    vol 30   E105, E111, 120, E220, E223, 227, E374, E375, E381, 387
Chantooni, Jr., M.K.
    vol 30   373, E374, E381, 386, 417-419, 447

Chao, E.E.
    vol 26   E348-E352, 388, E404, E405, 426
Chao, F.
    vol 25   E335-E337, 346
Chao, K.-C.
    vol 27/28   E360, 368, 369, E370, 372, E522, 530, 531, E533,
                  534, 543-546, 548-550, 559-562, 680, 710, 711
Chapin, E.C.
    vol 38   E1, E2, 5
Chappelear, P.S.
    vol 27/28   E281, 292-294, E303, 309, 310, E320, 329-332, E737
Chappelow, C.C.
    vol 24   E130-E134, 157, 162, 165, 180, 183, 186, 195, 235,
                  237-239, 243, 245, 248-250
    vol 27/28   E202-E204, E210, E213-E215, 245, 247, E370, E493,
                  E497, 508, E724, 726
Chapuis-Goitreux, S.
    vol 31   E308, E311, E320, E324
Charykov, A.K.
    vol 37   E315, E318, E322, 353
Chateau, H.
    vol 26   E267, 269
Chatterji, H.G.
    vol 23   E10, E25
Chemodanov, D.I.
    vol 23   E179, E180, E184
Chen, R.J.J.
    vol 27/28   E281, 292-294, E303, 309, 310, E320, 329-332
Chen, S.-H.
    vol 27/28   723
Chen, Xialing.
    vol 23   334
Cheng, K.L.
    vol 26   E348-E352, 388, E404, E405, 426
Chernoglazova, F.S.
    vol 38   E23-E25, E27, E28, 37
Chernykh, L.V.
    vol 31   E11, E13-E15, 20, 21
Chernysheva, A.M.
    vol 23   E270, E279, E281, 291, 292
Cheung, H.
    vol 27/28   E266, 275
Chevalet, J.
    vol 25   E317, E318, 325
Chevreul,
    vol 26   E262
Chew, J.-N.
    vol 37   E63, E67, E71, E81, 109
Chey, W.
    vol 20   E153-E157, 174
Chikovani, N.S.
    vol 23   E179, E180, E184
Childs, A.F.
    vol 21   E281, E282, 283-287, 289-295, 300
Chimboulev, M.T.
    vol 26   E303-E305, 314-317, E403-E405, 412, 413
Chimbulev, M.T.   see Chimboulev, M.T.
Chiodelli, G.
    vol 33   E108, E109, E113, E116, E243, E289
Chiotti, P.
    vol 25   E425-E427, 433
Chiou, C.T.
    vol 20   E35, 39, E73-E77, 85, 88, E89-E91, 98
Chiranzelli, R.V.
    vol 25   E103-E105, 112
Chlopin, W.
    vol 30   E105, E111, 115
Choi, S.S.
    vol 29   E1-E3, E7, E14, E16-E18, E21, 30
Chrelashvili, M.V.
    vol 35   210-212
    vol 36   115-119, 350, 351, 431-433

Christian, S.D.
    vol 37   E64, E67, E71-E73, E83, 144, 147, 149, E221, E225,
                E227, E228, 240, 242, E369, E378, 400, 402
Christoff, A.
    vol 29   E1, E21, 25, E187
    vol 27/28   E1-E5, E58, E59, E84, 87, E640, E644, 645
Chu, T.C.
    vol 27/28   E303, 309, 310
Chuang, Y.D.
    vol 25   E425-E427
Chukhlantsev, V.G.
    vol 26   E348-E350, 355, 357, 359, 360, 367, 368, 371, 375, 378,
                380, 389, 392, 394, 395, 401
Chumakova, V.P.
    vol 33   E57-E60, 70, E77, E175, 80, 179, E267, 272
Chviruk, A.P.
    vol 29   E46, E48, 62
Ciavetta, L.
    vol 23   E277, E281
Ciceri, C.
    vol 35   6
    vol 36   104, 332, 365, 441, 442
Cingolani, A.
    vol 33   3-8, E12, 13, E15, 17, 18, 22-24, E35, E43, E57-
                E60, E105, E116, E123, E128, E134, 136, E142, 144,
                148, E150, 152, E175, 218, 228, E229, 232, 234,
                E237, 240, E253, E283, E297, E315, E326, E330, E332
Claes, P.
    vol 24   E130-E134, 139, 143, 169, 170, 190, 191
    vol 27/28   E202-E204, E214, 220, 223
Claire, Y.
    vol 25   E119-E121, 133
Clark, W.G.
    vol 34   E13, E14, 19
    vol 35   E110, E111, 117, 213
    vol 36   E19, E20, 26, 57, E124, 126, 225
Clarke, E.C.W.
    vol 32   E1-E3, 12, 13, E327, 328, 329
Claussen, W.F.
    vol 24   E1, E2, 8, E16, E17, 23
    vol 27/28   E1-E5, 12
    vol 37   E63, E67, E68, E82, 111, E369, E371-E374, E378, 389
    vol 38   E23-E25, E28, 30, E42-E45, 48, E59, E60, E66, 70, E236-
                E238, 243, E341-E343, 345
Claytor, E.E. Jr.
    vol 32   E79, E81-E83, 89, 122-125
Clements, H.E.
    vol 24   E218-E221, 230
Clever, H.L.
    vol 27/28   E56, E57, E84
    vol 29   E1, E7, E10, E21
    vol 30   E330, E333, E339, E428, E431, E462, E467, E473
Clifford, C.W.
    vol 37   E63, E71, E81, 88
Cloves, G.H.A.
    vol 38   E396, E398, 399, E440, 441, E446, E448, 449, E477,
                E478, 479, E486, E487, 488, E493, 494, 508-511, E512,
                513, 521, 522, 524, 527, E528, 529, E534, 535, 538,
                540-542
Coffin, R.L.
    vol 24   E1, E2, 10, E16, E17, 28, E34, 35, E61, E71,
                81-83, 122-124
    vol 27/28   E1-E6, 10, E64, E85, 103, 177
Coffinberry, A.S.
    vol 25   E434
Cogley, D.R.
    vol 25   E1-E3, 9
Cohen, A.
    vol 26   E187, E188, E192, 226-228, E236, 239, E240, E241, 246
Cohen, E.
    vol 23   E307, E317
    vol 25   E385-E387, 391
Cohen, J.
    vol 36   181

Cohen, Y.
    vol 27/28   E50, E51
Cohen-Adad, R.
    vol 31   E39-E42
Cola, M.
    vol 26   E248, E249
Coleman, C.F.
    vol 37   E64, E67, E71-E73, E83, 153, E221, E225, E228, 250
                   E314, E318, E322, 341
Coles, B.R.
    vol 25   E103-E105, 114
Coll, R.
    vol 24   E39
Collat, J.W.
    vol 23   E170, E171, E183, 189
Connolly, J.F.
    vol 37   E42, E46, 51, 52, E64, E76, E77, E83, 142, 143,
                   E296, E299, 300, 301, E369, E377, E378, 398, 399
                   E485, E490, 501, 502
Conrad, F.H.
    vol 26   E155, E156, E159, 165, 166, E187-E189, E190, E193, 202,
                   E240, E241, 245
Copeland, L.C.
    vol 23   E177, E183, 210
Copley, M.J.
    vol 21   E80, E82, E89, 102, 151, E252, 267, 274-278
Corby, T.C.
    vol 36   E124, 133, 182, 183, 203
    vol 37   E64, E68, E83, 154
Cordeiro, D.J.
    vol 27/28   E370, 377, 418, 419
Cornec, M.E.
    vol 30   E330-E333, E336-E338, 342, 349, 350
Corriez, P.
    vol 21   E302, 303, 308, 309
Corti, H.
    vol 31   E2, E6
Cosgrove, B.A.
    vol 27/28   E1-E5, 22, E45, E46, 48
Costa, M.
    vol 25   E335-E337, 346
Cote, P.A.
    vol 23   E17, E20, E26, 59
Counioux, J.J.
    vol 31   E39-E42
Counsell, J.F.
    vol 21   E80, E89
Cox, A.J.
    vol 23   E88, E96
Craig, D.N.
    vol 23   E83, E88, E96
Cramer, S.D.
    vol 27/28   E24-E28, E57, E72-E74, E85, 144, 145
Crenshaw, J.L.
    vol 25   E33, E34, 41, E385-E387, 392
Crescenzi, V.
    vol 24   E1, E2, 6, E16, E17, 30
    vol 37   E41, E42, E46, 50, E314, E316, E322, 335
Crossley, M.L.
    vol 36   9
Crovetto, R.
    vol 27/28   E1-E5, E24-E28, 41, 43, E45, E46, 49
    vol 31   E2, E6
Crowell, W.S.
    vol 23   E177, E183, 195, 196
Cucuel, F.
    vol 29   E1, E3, E14, E16-E18, E21, 27
Cukor, P.M.
    vol 27/28   E202-E204, E209, E210, E214, E215, 238, E453,
                    E454, 464, E493, E496, E495, 507
Culberson, O.L.
    vol 27/28   E24-E28, 30-32
Cunningham, G.I.
    vol 30   E24, E29, E31-33, 59, 60, 87-89

Currie, J.A.
    vol 26   E1, E2, E5, E6, 56, 57
Curtis, C.W.
    vol 27/28   547, 565, 566
Cusack, N.E.
    vol 25   E103-E105, 115
Cuta, F.
    vol 23   E84-E88, E91-E93, E96, 105, 117-122, E144, E145
Czetsch-Lindenwald, H.
    vol 34   E13-E15, 29, 30, 128-130, 132, 133, 135, 136,
    vol 34   138-141, 144-147, 164, 165, 250, 252, 254
    vol 35   80, 180-182

D

D´Abramo, G.
    vol 25   E93-E95, 101
D´Ans, J.
    vol 31   E11, E15, 26, 27, E98, E102, E168, E169, E171, 175
Daccomo, G.
    vol 20   E45, 47, 48
Dadze, T.P.
    vol 29   E1, E2, E5, E14, E16-E18, E21, 39
Dalakova, N.V.
    vol 29   E213, E214, 218
Dallas, H.S.
    vol 25   E425-E427
Dalu, G.
    vol 27/28   E50, E51
Damsma, J.E.
    vol 35   29, 63, 159, 229, 252, 254, 269, 287
    vol 36   40, 75, 160, 242, 334, 361
Danchenko, R.M.
    vol 23   E20, E26
Danckwerts, P.V.
    vol 32   E79, E82
Danforth, W.F.
    vol 27/28   E202, E203, E206, E207, E215, 227, 231, E582-E584,
                    E592-E595, 625, 627
Daniels, F.
    vol 25   E119-E121, 125
Danilchenko, P.T.
    vol 25   E57, E58, 64
Danilov, V.V.
    vol 23   E156, E182, E277, E279, E282, 294, 304
Danneil, A.
    vol 24   E48, 52
Darbinyan, G.M.
    vol 31   E127, E129, E130, 143, 146, 147
Dasent, W.E.
    vol 26   E260
Davalos, J.
    vol 27/28   E248, E249, 262, E737, 743, 744
Davies, C.W.
    vol 23   E171, E183
    vol 31   E2, E6
Davies, P.L.
    vol 37   E64, E74, E82, 128, E369, E375, E378, 392
Davis, W.W.
    vol 38   E396, E398, 399, E440, 441, E446, E448, 449, E477,
                    E478, 479, E486, E487, 488, E493, 494, E497, 498, 508-
                    511, E512, 513, 521, 522, 524, 527, E528, 529, E534,
                    535, 538, 540-542
Dawson, H.M.
    vol 23   E18, E19, E26, 64, 65
Dawson, L.R.
    vol 20   85, 168
    vol 30   218, 256
Day, H.M.
    vol 29   E213, E214, 222, 226
Dayananda, M.A.
    vol 25   E385-E387, 397

Dean, J.A.
    vol 30   E108, E109, E112, E196, E221, E223
    vol 31   E1
Dean, J.R.
    vol 23   E273-E276, E281, 302
Dean, M.R.
    vol 27/28   763
Dean, O.C.
    vol 25   E1-E3, E225, 226, E230, 232, E239, 240, E263,
                  E264, E282, E307, E326, E422, 423, E425-E427,
                  429, 430
Dean, R.B.
    vol 30   E112
Debruyn, L.
    vol 21   E4, E5, E12, 36, 39
Decanniere, L.
    vol 24   E130-E134, 139, 143, 169, 170, 190, 191
    vol 27/28   E202-E204, E214, 220, 223
Dede, L.
    vol 32   E26, E28, 40
Deetz, J.A.
    vol 32   E1-E3, 19, E172, E173, E175, E184, E185, E187,
                  249, 304, 321
Degruyter, C.J.
    vol 25   E84, E95, 87
Deguchi, T.
    vol 37   E65, E67, E84, 179, 180, E370-E374, E379, 420-422
    vol 38   E9-E12, 21, E23, E24, E28, 40, E42-E44, E46, 57, E59-
                  E62, E66, 85-87, 103, E165, E166, 167, E169, E170, 175,
                  E177, E178, 184, 185, E187-E189, 199, E201, E202, 209,
                  210
Dehedin-Favre, M.
    vol 23   E92, E97
Deineka, G.F.
    vol 26   E90-E92, 100, 101
Delepine,
    vol 21   E4, E5, E12, 35, 39
De Ligny, C.L.
    vol 27/28   719
    vol 37   E42-E44, E46, 54, E314, E316, E317, E322, 340,
                  E485-E487, E490, 504
    vol 38   E135, E136, E141, 147
Delimarskii, Yu.K.
    vol 29   E207, E208, 210
De Loos, Th.W.
    vol 24   E39
    vol 37   E315, E320, E321, E323, 357-362
Delury, R.H.
    vol 25   E402-E404, 411
Delves, R.T.
    vol 25   E194, E195, 198
Dem´yanets, L.N.
    vol 23   E280, E282
Demeester, W.A.T.
    vol 25   E402-E404, 415
Denessen, H.J.M.
    vol 27/28   719
Deniges, M.G.
    vol 26   E248
Denton, W.H.
    vol 24   E16, E17, 27, E65-E68, E71, 110
Dergacheva, M.B.
    vol 25   E57, E58, 65, E172, E182, E183, 192, E357-E359,
                  E369-E371, 384
Deright, R.
    vol 25   E357-E359, 365
Derrick, M.E.
    vol 29   E1, E7, E10, E21
Desai, P.D.
    vol 25   E1-E3, E33, E34, E44, E45, E50, E51, E119-E121,
                  E139-E141.E157-E159, E182, E183, E357-E359,
                  E369-E371, E385-E387
Deshmukh, R.D.
    vol 32   E79, E83

Deshpande, V.V.
    vol 23   E159, E162, E163, E165, E182, 215-224
Desre, P.
    vol 25   E139-E141, 153, E182, E183, 191
Desvignes, J.M.
    vol 30   E269, E274
DeVaney, W.E.
    vol 32   E167-E170, E186, 203, 221, 232
▸Devyatykh, G.G.
    vol 32   E169, E172, E179, E180, E182, E185, E186, 227,
                  228, E330, 332-335
Devyutykh, G.G.
    vol 21   E281, E282, 296-299, E302, 304-307, E310, 311-314,
                  E315, 316-319, 321-328
Dew, M.J.
    vol 37   E64, E76, E83, 157, E369, E371, E372, E378, 406
DeWet, W.J.
    vol 25   E277, 278, E285, E286, 289, E301, E302, E310,
                  E317, E318, 319
Dewijs, H.V.
    vol 23   E169, E178, E183, E272, E277, E281
Dey, A.K.
    vol 23   E10, E25
Dhabanandana, S.
    vol 31   E307, E308, E311, E318, E324
Dhar, N.R.
    vol 23   E10, E25, E83, E96, 102, 103, E157, E182
DiCapua, C.
    vol 30   E24, E25, E28, E29, E31, E32, 54, 73, E105, E108-E111, 126
Dickens, W.A.
    vol 26   E187, E193
Diepens, G.A.M.
    vol 24   E39, E197, 206, 207, 213-215
Dierksen, J.
    vol 26   E4-E6, 32, 53
Dietrich, H.G.
    vol 23   E156-E158, E163-E165, E167, E170, E172, E178, E182, 199,
                  200
DiFrancesco, R.
    vol 34   77, 308
    vol 35   145
    vol 36   33, 147
Dillard, E.F.
    vol 31   269, 270
Dillard, K.H.
    vol 27/28   553-555
Dim, A.
    vol 24   E331-E334, 344, 348
Dingman, J.C.
    vol 32   E79-E83, 155-159
Diogenov, G.G.
    vol 33   E35, 36-38, E40, 41, 42, E43, 44, E46, 47, 48, E50,
                  51, E57-E60, 63, 68, 70, E73, 74, 75, E77, 78,
                  80, E81, 82, 83, E88, 89, 133, E134, 135, 170, E175
                  177, 179, E210, 211, 212, 262, E267, 270-272, E332,
                  333, 335
Dirkse, T.P.
    vol 23   E84, E96, 129, E144, E145, 146, 147, E157, E158, E161,
                  E162, E171, E182, E183, 227, 228, 242, 243, E275, E281
Ditsent, V.E.
    vol 21   E9, E12, 66
Divers, E.
    vol 26   E291
Dixon, R.D.
    vol 27/28   E685, 687
Dixon, W.J.
    vol 30   E112
Djega-Mariadassou
    vol 26   E271, E273
Dmitrevskaya, O.I.
    vol 33   E15, 16, E99, 101, E109, 111, E121, 122, E125,
                  127, E150, 151, E161, 182, E187, 188, E189, E191,
                  192, E195, 196, E197, E199, 200, 201, 207, 209,
                  214, 216, E289, 290, 302, 323

Dobbins, J.T.
    vol 26   E187-E190, E193, 229, 230
Dodge, B.F.
    vol 27/28   319
Dodrokhotov, G.N.
    vol 23   E16, E17, E26
Doepping, O.
    vol 26   E262
Dolique, R.
    vol 26   E348-E351, 364, 402
    vol 34   E13, E14, 24, 114, 116, 127, 295, 296, 312, 313,
                   318, 319
    vol 35   171, 172
    vol 36   44, 172, 173, 260, 261
Domagala, R.F.
    vol 25   E422
Dombrovskaya, N.S.
    vol 31   E206, E208, E209, E212, 230, 231
Donahue, D.J.
    vol 20   E132-E136, 141
    vol 37   E63, E68, E72, E82, 114
Donald, M.B.
    vol 30   E108, E109, E112, 129
Donche, H.
    vol 25   E173, E174, 181
Donnan, F.G.
    vol 23   E1-E3, 7
Donnell, C.K.
    vol 36   302, 304, 364, 447-449
Donnelly, H.G.
    vol 27/28   E737, 738, 739
Donovan, J.E.
    vol 30   E330-E332, E336-E338, 341, 351-353, E382
Dortbudak, T.
    vol 25   E300, E307, E309, E314, 315, E316, E330
Douabul, A.A.
    vol 32   E1-E3, 33, E51, 52, 53
Dourson, R.H.
    vol 27/28   392, 393
Dowgird, A.
    vol 25   E285, E286, 297
Dragavtseva, N.A.
    vol 25   E173, E174, 176, E335-E337, 348
Dreisbach, R.R.
    vol 20   E73-E77, 80, E153-E157, 167, 225, 230
Dressler, H.
    vol 22   12, E32, 33, 50, 51, 61, 88, 102, 116, 117, 134,
                   147, 177, 189, 204, 225, 230, 236, 242, 270, 276,
                   286, 308, 314, 323, 366
Druz, N.A.
    vol 26   E403, E405, 406
Druzhinin, I.G.
    vol 31   E98, E99, E101, E102, 112, E127, E130, 148
Druzinina, E.P.
    vol 25   E93-E95, 99
Dryden, I.G.C.
    vol 32   E26-E28, 46, 49, E54, E55, 69-71
Dubeau, C.
    vol 32   E330, E331, 337, 338
Dubois, S.
    vol 35   E110-E112, 136, 166, 167
Dubova, N.M.
    vol 25   E173, E174
Dubovaya, V.K.
    vol 26   E187, E191, E193, 215
Dudakov, V.G.
    vol 31   E98, E99, E101, E102, 115, 116
Dudereov, N.G.
    vol 23   E280, E282
Duffy, J.R.
    vol 27/28   E24-E28, 33, E66-E68, E70, E74, E75, E84, 110,
                  111, 122, 123, 147
DuPont de Nemours and Co.Inc
    vol 32   E184, E187, 302

Duque-Estrada, E.
    vol 37   E197-E199, 201
Durand, R.
    vol 37   E63, E66, E81, 106, E221, E223, E224, E227, 233,
                  E314, E318, E321, 327, E485, E486, E490, 495
Durante, M.
    vol 26   E267, 269
Durdyev, M.
    vol 26   E348-E351, 372, 373
Dure, G.
    vol 20   E1, 2
Durel, M.P.
    vol 34   E14, 17, 181
    vol 35   E110, E111, 115, 220, E231, 233, E239, 241,
                  E257, 258, 282, 290, 302
    vol 36   E19, E20, 24, E63, 67
Duruz, J.J.
    vol 33   E119, E121, E123, E125, E161, E197, E226, E248,
                  E250, E278, E292, E303, E308, E311, E313, E315,
                  E317, E319, E321
Durymanova, M.A.
    vol 26   E1-E6, 16, 17, 38, 59-62
Dymond, J.H.
    vol 27/28   E702, 705
Dyrssen, D.
    vol 23   E91, E92, E96, 139, E275, E277, E281, E309, E317, 331
Dyson, W.H.
    vol 23   E160, E162, E182, 253

Dzholdasova, R.M.
    vol 25   E207-E208, E214, E215, 217, 218, E230, E231
Dzhuraev, Kh.Sh.
    vol 22   101, 175, 217, 254, 292, 350, 364
Dziuba, E.Z.
    vol 25   E194, E195, 201
Dzuev, A.D.
    vol 31   E278, E279, 282, 285

E

Eakin, B.E.
    vol 27/28   E248, E249
    vol 32   E167-E170, E186, 203, 221, 232
Earl, Ch.B.
    vol 26   E152
Eastman, E.D.
    vol 25   E369-E371, 374
Eberley, N.E.
    vol 23   E177, E183, 195, 196
Ebert, M.
    vol 26   E348-E351, 376
Ebian, A.R.
    vol 36   280, 281, 371
Eckers, W.
    vol 26   E153, E158, E252, E256, E258, E271, E273, E286
Eckstrom, H.C.
    vol 30   218, 256
Economy, J.
    vol 31   E13-E15, 37, 38
Edelman, B.A.
    vol 25   E194, E195
Edush, T.V.
    vol 31   E127, E130, 150, E297, 303
Edwards, T.J.
    vol 25   E135
    vol 32   E54, E55
Efanov, L.N.
    vol 26   E1, E6, 15
Efimova, N.M.
    vol 32   E183, E184, E187, 324
Efremov, A.A.
    vol 37   E221, E225, E227, 239, E314, E318, E332, 334, E485,
                  E488, E490, 500

Efremova, G.D.
    vol 37   E221, E226, E228, 251
Eganhouse, R.P.
    vol 38   E236-E238, 245, E250, 255, E311, E312, 314, E319, 320,
                  E332-E334, 337, E341-E343, 348, E352, 354, E357, 358,
                  E360, 361, E363, 364, 372, E419, 420
Egger, K.
    vol 23   E170, E183, 247
Eggers, H.
    vol 25   E44, E45, 49
Eggert, G.L.
    vol 25   E103-E105, 111
Egorov-Tismenko, Yu, K.
    vol 23   E180, E184
Egorova, V.I.
    vol 32   E181, E182, E187, 301
Eichner, M.
    vol 26   E262
Eidinoff, M.L.
    vol 20   E35, 36, E132-E136
Eilert, A.
    vol 25   E66, E67, 69
Eisen, H.
    vol 35   46, 89, 189
    vol 36   201, 264, 282, 342, 387
Eisenberg, B.
    vol 32   E79, E82, E83
Ejima, A.
    vol 34   E285, 286, 292
    vol 35   3, 22, 68, 263
Ejima, A.
    vol 36   103, 287, E404, 408
El-Sayed, A.A.
    vol 35   E110-E112, 138, 206
El-Suradi, S.M.
    vol 26   E153, E158, E187, E192, E248, E271, E273, E286
Elbishlawi, M.
    vol 27/28   E511, 515, E522, 523
Elghorche-Choubani, C.
    vol 23   E174, E183, 262, 263
Elia, V.
    vol 27/28   E1-E5, E85
Ellender, J.H.
    vol 21   E80, E89
Ellington, R.T.
    vol 27/28   E248, E249
Elliott, D.G.
    vol 27/28   E281, 292-294
Elliott, R.P.
    vol 25   E422
ElSabbagh, H.
    vol 35   30, 31
Elten, M.
    vol 26   E291
Eltsov, N.M.
    vol 23   E13, E14, E25, 60, E277, E282, 296
Elworthy, P.H.
    vol 36   E124, 132, 133, 176-178, 182, 183, 202, 203
    vol 37   E64, E68, E83, 154
Endovitskaya, M.R.
    vol 31   E278, 286
Engelen, B.
    vol 26   E153, E158, E248, E252, E256, E258, E271, E273, E286
Engelhardt, G.
    vol 26   E187-E190, E193, 211
Englin, B.A.
    vol 37   E13-E15, 17, E22, E23, 24, E31, E33, E34, 36, E64
                  E71, E72, E82, 141, E197, E199, 202, E212, E213,
                  215, E221, E224, E225, E227, 238, 272, E275, E276,
                  277, E289-E291, 292, E314, E318, E319, E322, 333,
                  E369, E375, E378, 397, 432, E433, E434, 435, 444,
                  E445, E447, 449, E455, E456, 458, 464, 478, E485,
                  E488-E490, 499

Englin, B.A.
    vol 38   E23, E27, E28, 34, E59, E62, E63, E66, 76, 93, 97,
                     E114-E117, 120, 132, E135, E138, E139, E141, 145, E177,
                     E178, 181, E187, E189, 193, 194, 220-222, E260, 262,
                     264, E265, E266, 268, E271, 273, E277, E278, 282, E287,
                     289, 293, E311, E312, 313, 322, 325, 373, 374, 393, 394,
                     425, 462, 463, E464, E465, 467
English, J.P.
    vol 34   E302, 303
    vol 35   11, E110, E111, 114, 219, E231, 232, E239,
                     240, 250, 251, 256, 317
    vol 36   E63, 66, E124, 125, 206, 207, E228, 229, E243, 244
                     E269, 270, 279, 284, 296-298, 305, 366, 437, 438
Enikeev, R.Sh.
    vol 25   E207-E208, 213, E214, E215, 218, 219, E221, 223,
                     E225, 227, 228, E230, E234, E239, 241, E242, 246,
                     E247, 248, E249, 250, E251, 252, E253, 255, E256,
                     257
Epstein, L.F.
    vol 25   E301, E302, 303, E317, E318
Erahler, B.V.
    vol 23   E274, E281, 300
Eraskov, R.Sh.
    vol 30   E473, 489
Erdenbaeva, M.I.
    vol 26   E348-E352
Erdmann, H.
    vol 31   E308, E311, E318, E324
Erdos, E.
    vol 26   E240, E241, 247
Erickson, E.E.
    vol 37   E64, E67-E69, E82, 122, 123
Eriksrud, E.
    vol 25   E335-E337, 354
Erlykov, A.M.
    vol 33   E57-E60, 63, E81, 83, 170, E210, 262, E332, 335
Ermola, L.P.
    vol 23   E90, E96
Ermolaev, M.I.
    vol 26   E293, 301, 302
Espenbetov, A.V.
    vol 25   E13-E18, E33, E34, E402-E404
Esterson, G.L.
    vol 24   E375-E378, 379
Etard, A.
    vol 26   E260
    vol 30   E105, E111
Eto, Y.
    vol 32   E171, E172, E186, 239, 240
Ettisch, G.
    vol 23   E18, E26
Ettmayer, P.
    vol 25   E285, E286, 298, E425-E427, 432
Eucken, A.
    vol 27/28   E1-E5, E72, E74, E84, 120
Euler, H.
    vol 23   E177, E184, E277, E282
Evans, E.J.
    vol 25   E369-E371
Evans, J.F.
    vol 32   E182, E187, 298
Evans, D.F.
    vol 27/28   683, 723
Evdokimova, K.A.
    vol 33   E57-E60, 62, E172, 173, E175, 176, 264, E267, 269
Everson, H.E.
    vol 20   E73-E77, 79E103-E105, 114, E153-E157, 163
    vol 37   E63, E66, E81, 105, 108, E314, E316, E321, 326,
                     E369, E371, E378, 386, E485, E486, E490, 494
    vol 38   E177, E178, 179, E260, 261, E287, 288
Evokimov, A.N.
    vol 26   E5, E6, 71
Exner, J.
    vol 37   E485, E488, E490, 505

Eyring, L.
      vol 25   E225, E230, E231, E234, E256
Ezerskii, M.L.
      vol 36   286, 312, 436
Ezheleva, A.E.
      vol 21   E281, E282, 296-299, E302, 304-307, E310, 311-314,
                    E315, 316-319, 322-328
      vol 24   E257, E258, 262, 279, E299-E302, 308, 330,
                    E331-E333, 341, 370, E375-E378, 387, 407
      vol 32   E169, E172, E179, E180, E182, E185, E186, 227, 228,
                    E330, 332-335
Ezhova, T.A.
      vol 22   25, 80, 99, 126, 160, 211, 234, 250, 290, 311,
                    331, 348, 361, 378

F

Fainberg, A.
      vol 29   E177, E179
Fairchild, W.R.
      vol 27/28   E266, 272, 273
Faith, H.E.
      vol 35   1, 12, 13, E14, 16, 104, 107, 109, 316
      vol 36   97, 299, 301, 443
Fanelli, R.
      vol 32   E185, 317
Farhi, L.E.
      vol 27/28   731
Farkas, E.J.
      vol 37   E197-E199, 203
Farnell, R.G.W.
      vol 26   E187, E188, E192, 194
Fauser, G.
      vol 32   E172, 175, E187
Fauvel, A.
      vol 23   E159, E161, E163, E182, 201, 202
Fay, H.
      vol 25   E157-E159
Fedorciv, V.E.
      vol 30   E461, E462, 464
Fedotova, T.D.
      vol 30   E461, E462, 464
Feillolay, A.
      vol 27/28   E1-E5, E61, E62, E85, 98
Feitknecht, W.
      vol 23   E1, E13, E16, E17, E25, E156, E157, E169, E170, E171,
                    E182, E183, 191, E271, E272, E281
Feix, G.
      vol 37   E65, E68, E71, E72, E84, 176, E485, E486, E488,
                    E489, E491, 512
Felsing, W.A.
      vol 32   E330, E331, 336, 339
Feree, J.
      vol 25   E13-E18, 22, E33, E34, 37, E266, E282, E285, 295
Ferloni, P.
      vol 33   E103,
Fernandez-Prini, R.
      vol 27/28   E1-E5, E24-E28, 41, E45, E46, 49
      vol 31   E2, E6
Ferrein, A.
      vol 31   E98, E102
Ferroni, G.
      vol 31   E39-E42, E51, 96, 97, E101, E102, 124, 125, E129,
                    E130
Ferry, J.D.
      vol 23   E3, 4, E21, E26, E315, E317
Fialkov, YA.A.
      vol 23   E16, E17, E25, 50, 212, E272, E281, 289
Fichtner, D.A.
      vol 24   E196, E197, 206, 207, 213-215
Field, J.P.
      vol 20   237
Field, L.R.
      vol 27/28   E450, E451, E454, 457, E493, E494, E497, 503

Fielder, M.
    vol 25   E103-E105, 115
Filatova, T.N.
    vol 23   E171, E177, E183
Filhol, E.
    vol 31   E11, E15
Filipescu, L.
    vol 31   E51, E52, E206, E207, E212, 242, 243, 273
Filippov, T.S.
    vol 20   E153-E157, 166
Filippov, V.V.
    vol 38   E59, E62, E63, E66, 71
Filippova, L.M.
    vol 25   E1-E3, 11, E13-E18, 30, E33, E34, 43, E76, E77,
                  E82, E119-E121, 132, E139-E141, 156, E157-E159,
                  170, E182, E183, 193, E335-E337, E385-E387, 401,
                  E402-E404, 419
Filyas, Yu.I.
    vol 37   E64, E76, E83, 158
Fink, S.
    vol 25   E258, E259, E263, E271, E274, E277, E301,
                  E302, 305, E310, E311, E317, E318, 323
Finke, A.
    vol 22   13, 14, 17, 18
Fischer, F.
    vol 27/28   580, 581
Fisk, Z.
    vol 25   E103-E105, 114
Fitzer, E.
    vol 25   E301, E302
Flad, D.
    vol 25   E207-E208, E214, E221, E225
Flatt, R.
    vol 23   E92, E97
    vol 31   E168, E169, E171, 176, 177, E208, E212, 264-266,
                  E308, E311, E320, E324
Fleck, J.
    vol 30   E105, E108, E109, E111, 127, 128
Fleischer, A.
    vol 23   E144, E145
Fleitman, A.H.
    vol 25   E258, E259, E263, E264, E271, E274, E301,
                  E302, E330
Fleming, K.A.
    vol 20   237
Fleury, D.
    vol 24   E130-E134, 140-142, 144, 145, E218-E221, 232
Flottman, F.
    vol 30   E105, E111, 116, E374, E375, E379, E380, 383
Foda, A.
    vol 35   30, 31
Foerster, F.
    vol 26   E1-E3, E5, 10-12, E72-E74, 76, 77, E90-E92, 93, 94, E108,
                  E109, 110, 111
Fogg, P.G.T.
    vol 32   E171, E186
Fogh, I.
    vol 25   E84, E85, 86
Fontaine, R.
    vol 24   E130-E134, 139, 143, 169, 170, 190, 191,
                  E218-E221, 222, 223, 227, 232, 233, E257, E258,
                  261, 265, E299-E302
    vol 27/28   E202-E204, E214, 220, 223
Foote, H.W.
    vol 30   E330-333, E335-338, 340, 344-348, 354-358, E380, E382
Fordyce, R.G.
    vol 38   E1, E2, 5
Forgan, R.
    vol 24   E16, E17, 27, E65-E68, E71, 110
Forker, G.M.
    vol 31   E1
Fornasini, M.L.
    vol 25   E214, E215, E221, E225, E230, E231, E239,
                  E245, E253, E425-E427

Forsberg, H.C.
    vol 25   E425-E427
Fortunatov, N.S.
    vol 29   E207, E208, 209
Forziati, A.E.
    vol 25   E13-E18
Foucault, J.
    vol 34   E13, E14, 24, 114, 116, 127, 295, 296, 312, 313,
                 318, 319
    vol 35   171, 172
    vol 36   44, 172, 173, 260, 261
Fox, Ch.L., Jr.
    vol 34   180
    vol 35   215
    vol 36   58, 208-216, 227, 268, 357, 402
Fox, P.L.
    vol 34   180
    vol 35   215
    vol 36   58, 208-214, 216, 227, 268, 357, 402
Francesconi, A.Z.
    vol 27/28   E629, 632-634
Franck, E.U.
    vol 24   E48, 52
    vol 27/28   E629, 632-634
Franck, G.
    vol 25   E103-E105, 118
Frank, H.S.
    vol 37   E70, E84
Frankland, E.
    vol 24   E331-E334, 355
Franks, F.
    vol 37   E64, E67, E76, E77, E82, 132
    vol 38   E294-E296, 300, E376, E377, 379, E426, E427, 429, E464,
    vol 38   E465, 468
Franzosini, P.
    vol 33   E15, E17, E35, E43, E57-E60, E103, E105, E116,
                 E119, E123, E128, E134, 136, 138, E142, 144, E145,
                 148, E150, E154, 166, 171, E175, 218, 220, 228,
                 234, E237, E253, 258, 263, E283, E297, E315, E326,
                 E330, E332
Frasunyak, V.M.
    vol 25   E194, E195
Frazier, A.W.
    vol 31   269, 270
Freed, V.H.
    vol 20   E73-E77, 85
Freedman, A.
    vol 30   E467, E468, E471, E472
Freegarde, M.
    vol 37   E366, E516
    vol 38   E158, 159, E308, 309
Freeman, D.H.
    vol 37   E65, E68, E84, 174, E188, 192
    vol 38   E236-E238, 249, E250, 257, E386, 389, 390, E396-E398,
                 404, E405, 407, E419, 422, E433, 435, 436, 438, 439,
                 E441, 444, 445, E446-E448, 454, E456, 459, E477, E478,
                 482, E483, 485, E486, E487, 491, 492
Freeman, J.H.
    vol 22   E328
Freiburg, C.
    vol 26   E248
Freidson, G.S.
    vol 21   E8, E9, E11, E12, 56-58, 65, 76-78
Fricke, R.
    vol 23   E156, E163, E183, 203, E307, E317
Fridman, G.B.
    vol 23   E179, E184
Fridman, Ya.D.
    vol 23   E277, E282
Friedel, C.
    vol 24   E331-E334, 343, 359
    vol 27/28   E582, E591, E595, 619

Friesen, W.T.
    vol 35   29, 63, 159, 229, 252, 254, 269, 287
    vol 36   40, 75, 160, 242, 334, 361
Frilette, V.J.
    vol 38   E1, E2, 6
Frisk, A.R.
    vol 35   153, 154, 228
    vol 36   156, 241, 252, 275, 455, 458, 461, 463-465, 472, 473,
                  485
Frolich, P.K.
    vol 27/28   E266, 267, E281, 282, E303, 304, E320, E349, 350,
                  E465, 466, E511, 512, 559, 560, 635, 636
Frolkov, A.Z.
    vol 25   E207-E208, E214, E221, E225, E234, E230, 235,
                  E242, E253, 254
Fronberg, M.G.
    vol 25   E182, E183, 187
Froning, H.R.
    vol 27/28   199, 200
    vol 32   E79, E81-E83, 89, 122-125
Frost, B.R.T.
    vol 25   E425-E427
Fruhwirth, O.
    vol 23   E170, E171, E183
Frydman, M.
    vol 26   E187, E190, E191, E193, 209, 210
Fuhner, H.
    vol 37   E42, E43, E46, 47, E314, E316, E321, 324, E369,
                  E371, E377, 381, E485, E486, E490, 492
    vol 38   E59, E66, 67, E135, E136, E141, 142, E201, E202, 203
Fujicka, G.S.
    vol 21   E281, E282, 301
Fulton, J.W.
    vol 23   E158, E163, E165, E168, E170, E171, E182, 229
Funk, R.
    vol 30   E1, E4, 11, E184-E187, 188, E268, E273, 276
Furman, A.A.
    vol 20   E153-E157, 166
    vol 38   E59, E62, E63, E66, 71
Furmanov, A.S.
    vol 32   E182, E182, E187, 301
Furmer, I.E.
    vol 21   E1-E3, E8, E9, E11, E12, 14, 22, 30, 56-58, 65, 76-78, E89
    vol 32   E170, E171, E182-E184, E186, E187, 226, 233, 286,
                  293, 295, 324
Fursova, A.F.
    vol 30   E109, E110, E112, 137, 138
Furtig, H.B.
    vol 23   E161, E182, 244
Fuseya, L.G.
    vol 23   E308, E310, E317, 321

G

Gaarenstroom, S.W.
    vol 23   E144, E145
Gabler, C.
    vol 31   E11, E15, E98, E99, E102, 106, 107, E206, E207,
                  E212, 214, 225
Gabriel, A.
    vol 31   E11, E15, 17, E127, E129, E130, 132, 133, 136-138,
                  155, 156
Galaktionova, O.V.
    vol 22   136, 179, 220, 238, 258, 277, 296, 317, 340, 352
Galalenko, N.P.
    vol 25   E194, E195
Galasyn, V.
    vol 22   96, 145, 202, 229, 269
Gale, R.P.
    vol 24   E16, E17, 25, E65-E68, E70, E72, 107, 120
Galea, J.
    vol 31   E39-E42, E51, 96, 97, E101, E102, 124, 125, E129,
                  E130

Galinker, I.S.
     vol 23   E94, E97, 107
     vol 26   E1-E3, E5, 14
Galkin, N.P.
     vol 22   72, 97, 122, 207
Gallardo, M.A.
     vol 27/28   E640, E642-E644, 651, 652, 660, 695
Galstyan, V.D.
     vol 31   E98, E102, 114, E129, E130, 149, 165
Galus, Z.
     vol 25   E285, E286, 290, 297, E317, E318, 324, E326, 334
Galushkina, R.A.
     vol 31   E208, E212, 244, 245, 274, 275
Gami.D.C.
     vol 27/28   E248, E249, 250-254
Gamol´skii, A.M.
     vol 32   E54, 61, 62
Gamsjager, H.
     vol 32   E1-E3, E25, E26, E28, 36-38, E51
Ganelina, E.Sh.
     vol 26   E404, E405, 420-424, 427
Ganz, S.N.
     vol 23   E20, E26
     vol 31   E47, E49, E51, E52, 77, 78, E206, E207, E210, E212,
                     260, 261
Garb, S.
     vol 36   154, 164, 234, 238, 239
Garcia Onandia, A.
     vol 35   143
Garcia Onandia, T.
     vol 36   140, 250, 453, 454, 456, 457, 459, 460, 469, 470
Gardner, G.C.
     vol 27/28   E248, E249, 264, 265
Gardner, G.R.
     vol 24   E331-E333, 344, 348
Garrett, A.B.
     vol 23   E84-E88, E91-E93, E96, 105, 117-122, E144, E145,
                     E308, E310, E311, E313, E317, 322-329, 336-338
Garrod-Thomas, R.N.
     vol 25   E1-E3, E157-E159, E301, E302, E335-E337, 340
Garst, A.W.
     vol 27/28   183, 184
     vol 32   E79, E81-E83, 92, 93, 104-106, 129, 130, 149-151
Garton, G.
     vol 20   E6, E7, 10, E24, 25, E43, 44, E45, 46, E195, E196, 201,
                     E221, 222, E223, 224, E228, 227, E228, 229, E231, 232,
                     E233, 234, E235, 236, E238, 239, E240, 241, E242, 243,
                     E244, 245, E246, 247, E248, 249, E250, 251, E252, 253,
                     E254, 255, E256, 257
Gasanova, Kh.D.
     vol 31   E297, 305, 306
Gasco, M.R.
     vol 35   162, 163
     vol 36   169, 170
Gattow, G.
     vol 32   E185, 318
Gaune-Escard, M.
     vol 25   E93-E95, 102
Gavrish, M.L.
     vol 23   E94, E97, 107
Gavze, M.N.
     vol 25   E301, E302
Gayer, K.H.
     vol 23   E270, E273-E276, E281
Gayfullin, A.Sh.
     vol 25   E13-E18, 30, E119-E121, 132, E157-E159,
                     E335-E337, E385-E387, E402-E404
Gayler, M.L.V.
     vol 25   E139-E141, 148
Gay Lussac, J.L.
     vol 30   E105, E111
Gedan, H.
     vol 37   E65, E68, E71, E72, E84, 176, E485, E486, E488,
     vol 37   E489, E491, 512

Geese, J.W.
    vol 26   E187, E192, E193, E240, E241
Geffcken, G.
    vol 30   E220, E221, E223, 228, 229
Geiderovich, O.I.
    vol 23   E90, E96
Geilmann, W.
    vol 26   · E348-E351, 379
Geller, E.B.
    vol 27/28   E451, E452, E454, 458-461
Genco, J.
    vol 24   E196, E197, 206, 207, 213-215
Gendrano, M.C.
    vol 37   E64, E71-E73, E83, 145
Gent, M.
    vol 37   E64, E68, E82, 132
Gerardin, C.A.
    vol 30   E105, E112
Gerasimov, Ya.I.
    vol 25   E230, E234
    vol 30   E220, E222, E223, 233-234
Gerencser-Nemeth, M.
    vol 34   E302, 306, 315
    vol 36   E303, 309, 339
Gerrard, W.
    vol 21   E4, E6-E9, E12, 16, 33, 46, 48, 54, 59-63, 69, E80-E91,
                  98-101, 103-114, 117-128, 131-134, 148-150, 152-170,
                  173-177, 183, 185, 189, 191, 193-195, 198, 202, 203,
                  205, 206, 209, 211, 213-215, 217-222, 224, 227-230,
                  232, 233, E281, E282, E302
    vol 24   E130-E134, 150, 176, 193, 234, 241, 242, 247,
                  E257, E258, 264, 266, 268, 272, 274-276, 281,
                  282, E299-E301, 305, 312, 313, 315-320, 322,
                  324-327, 329, 347, 349, 354, 362, 373, 374,
                  E375-E377, 384, 389, 390, 391, 393, 396, 397
    vol 32   E20, E166-E168, E170-E173, E175, E176, E178-E182,
                  E184, E186, 206, 225, 227, 230, 242, 244, 247, 255,
                  256, 264, 271-273, 282-285, 288, 289, 292, 294,
                  296, 297, 300, 303
Gerraughty, R.
    vol 36   186, 193, 195, 197, 199, E367, 382, 384, 388, 390, 392
                  394, 396, 412, 414, 416, 418, 420, 422, 424
Gerstner, F.
    vol 29   E1, E3, E14, E16-E18, E21, 27
Gerver, J.
    vol 27/28   E24-E28, 29, E57, E66-E69, E71, E72, E74-E76,
                  E79, E80, E84, 107, 114, 117-119, 148, 150,
                  162, 172, 181, 183
Gessner, M.
    vol 23   E311, E312, E316, E317, 342
Gester, G.C.Jr.
    vol 37   E314, E315, E321
Gestrich, W.
    vol 32   E173, E175, E187, 252
Getzen, F.W.
    vol 20   E62, E99, E100, 102, E103-E105, 109, 110
    vol 38   E236-E238, 244, E332-E443, 335, E341-E343, 346, E386,
                  387, E396-E398, 401
Ghanem, A.
    vol 35   E14, 20, 30, 31, 37-42
Ghanem, N.A.
    vol 37   E485, E488, E490, 505
Ghione, M.
    vol 35   6
    vol 36   104, 332, 365, 441, 442
Ghosh, S.
    vol 23   E10, E25
Gibanel, F.
    vol 27/28   E643, E644, 654
Gileadi, E.
    vol 25   E84, E85, 92
Gilfillan, E.S.
    vol 25   E33, E34, 40, E93-E95

Gill, S.J.
    vol 20   E73-E77, E89-E91, E153-E157
    vol 37   E14, E15, E42, E43, E46, E65, E70, E84, E222, E229, E315, E316, E323, E370, E371, E374, E379, E486, E491
    vol 38   E24, E28
Gillespie, P.C.
    vol 32   E54, E55, 73-77
Gilligan, D.R.
    vol 36   154, 155, 164, 165, 233, 234, 238, 239, 251, 257, 273, 274, 329, 335, 359, 362, 412, 414, 416, 418, 420, 422, 424
Gilliland, E.R.
    vol 32   E166, E186, 191
Gilman, T.S.
    vol 29   E213, E214, 215
Gilyazova, G.N.
    vol 31   E212, 271
Gimblett, F.G.R.
    vol 31   E2, E6
Gimel´shtein, V.G.
    vol 33   E35, E43, E46, E50, 54, 55, E58-E60, E81, 83, E88, 89, 90, E134, 134, E175, 177, 178, E210, 212, E267, 271, E332, 333-335
Gin, G.
    vol 26   E262
Ginnings, P.M.
    vol 26   E90-E92
Ginsberg, E.
    vol 21   E80, E82, E89, 102, 151, E252, 267, 274-278
Ginzberg, D.M.
    vol 32   E54, 63, 64
Gioia, F.
    vol 32   E79, E82
Giorgi, A.L.
    vol 25   E421
Girich, T.E.
    vol 31   E39-E42, E47, E49, E51, E52, 77, 78, 92-95, E206, E207, E210, E212, 260, 261
Gishler, P.E.
    vol 26   E187, E190, E193, 203, 204
Gjaldbaek, J.C.
    vol 24   E130-E134, 138, 148, 225, E257, E258, 267, E375-E377, 384
    vol 27/28 E1-E5, 13, E202-E205, E214, 218, E450, E454, 455, E493, E494, E497, 500, E582, E590, E595, 598, 603, 637, 639, 655, 657
    vol 29   E102-E109, 126, 130, E133, E137, 141, E177-E179, 184
Gjedebo, F.
    vol 30   E105, E111, 117, E220, E223, 225, E374, E377, E380, 384
Gjessing, E.T.
    vol 29   E1, E5, E21, 43, E51, 95, 96
Gladkikh, I.P.
    vol 25   E13-E18
Gladkovskaya, A.A.
    vol 31   E208, E212, 244, 245, 249, 250, 274, 275
Gladyshev, V.P.
    vol 25   E1-E3, 10, E33, E34, E135, 136, 138, E172, E194, E195, 204, E214, E215
Glanville, J.W.
    vol 24   E251, 253, 254
Glaser, M.
    vol 27/28 E370, 373-375
Gleiser, M.
    vol 25   E1-E3, E33, E34, E44, E45, E50, E51, E119-E121, E139-E141, E157-E159, E182, E183, E357-E359, E369-E371, E385-E387
Glew, D.N.
    vol 29   E1-E4, E7, E14, E16-E18, E21, 33-35, E46, E48, 59-61
    vol 32   E1-E3, 12, 13, E327, 328, 329
    vol 38   E187-E189, 192
Gniewosz, S.
    vol 27/28 566

Goadby, H.K.
    vol 31   E11, E15, E98, E99, E102, 105
Gocken, N.A.
    vol 21   E10, E12, 71-73
Goddard, J.B.
    vol 25   E1-E3, 8
Gogoleva, T.V.
    vol 21   E2, E4-E8, E10-E12, 21, 38, 45, 53, 75
Gogorishvili, P.V.
    vol 36   115, 116
Goldman, S.
    vol 37   E225, E228, 255, 271
    vol 37   E65, E71-E73, E84, 163, 187, E221, E222, E224,
Goldman, S.
    vol 20   E73-E77, 83, E153-E157, 173
Goldschmidt, J.M.E.
    vol26   E187, E188, E192, 226-228, E236, 239, E240, E241
Golonka, L.
    vol 25   E194, E195
Golub, A.M.
    vol 22   20, 37, 75, 76, 90, 91, 95
Golubeva, M.S.
    vol 33   E57-E60, 64, E167, 168, 260
Golubkov, A.V.
    vol 25   E194, E195
Golubkova, V.V.
    vol 33   E26, 29, E31, 33, 49, 84, E91, 93
Golutvin, Yu.M.
    vol 32   E20, 21, E54, 58, 78
Gorbachev, S.V.
    vol 37   E66, E85
Gordin, V.L.
    vol 25   E173.E174
Gordley, N.P.
    vol 25   E194, E195
Gordon, J.E.
    vol 38   E250, 252-254
Gordon, L.I.
    vol 27/28   E50, E51
Gorgeu, A.
    vol 24   E331-E334, 343, 359
    vol 26   E248, 251
    vol 27/28   E582, E591, E595, 619
Gorman, M.A.
    vol 20   E202-E204, 205, E210-E213, 214
Gorowara, B.
    vol 27/28   E685, 687
Gorozhankin, E.V.
    vol 31   E206, E207, E212, 253, 254
Goto, S.
    vol 34   E14, E15, 45
    vol 35   262
    vol 36   E367, 370
Gotze, G.
    vol 20   E202-E204
Gouy, M.
    vol 25   E139-E141, E157-E159, E182, E183, E301, E302, E357-E359,
                  E369-E371, E385-E387, E402-E404
Grace, R.E.
    vol 25   E385-E387, 395, 397
Gracheva, N.V.
    vol 32   E183, E184, E187, 324
Gradkikh, I.P.
    vol 25   E1-E3, E33, E34, E44, E45, E50, E51
Grady, L.T.
    vol 35   91, 100
Graham, L.
    vol 27/28   382
Gramkee, B.E.
    vol 25   E369-E371, 378
Grassi.L.
    vol 27/28   E1-E5, 179
Green, W.J.
    vol 37   E70, E84

Gregory, M.D.
    vol 37   E64, E71-E73, E83, 249, E221, E225, E228, 242, E378,
                  402
Grieble, D.
    vol 29   E1, E5, E7, E14, E16-E18, E21, 41, E47, E48, 52, 63
Griffin, R.B.
    vol 25   E221, 222, E225, E274
Griffiths, J.E.
    vol 30   E1-E4, 18-22
Grigor, T.I.
    vol 22   112, 113, 196, 304, 305
Grigor´eva, E.A.
    vol 33   64
Grigorevoa, M.J.
    vol 25   E402-E404
Grigorvich, Z.I.
    vol 22   124, 125, E127, 128-131, E152, 153, E156, 157, E162,
                  164, 170, 172
Grigoryan, K.G.
    vol 31   E129, E130, E318, E320, E321, E324, 328
Grigsby, R.D.
    vol 37   E64, E67, E73, E83, 147
Grishko, N.I.
    vol 32   E181, E182, E187, 301
Griswold, J.
    vol 37   E63, E67, E71, E81, 109
Grobe, G.
    vol 25   E1-E3, 7
Groll, W.
    vol 25   E326, 329
Gromakov, S.D.
    vol 25   E194, E195
Gronlund, F.
    vol 25   E335-E337, 356
Groschuff, E.
    vol 37   63, E71-E73, E81, 87
Groscuff, E.
    vol 30   E467, E468, E470, E472, 474, 475
Gross, P.
    vol 30   254
Gross, P.M.
    vol 20   E103-E105, 112, E132-E136, 137, E152-E157, 158,
                  E202-E204, 205, E210-E213, 214
    vol 37   E63, E68, E81, 96, 100, E369, E372, E377, 384
Grosse, A.V.
    vol 25   E93-E95, 100
Grove, N.H.
    vol 27/28   565
Gschneider, K.A.
    vol 25   E221, 222, E225, E230, E231, E234, E256, E274
Gubbins, K.E.
    vol 27/28   E1-E6, E79, E85, 160, 161
Gubeli, A.O.
    vol 26   E342-E352, 382-387
Gubergrits, M.Ya.
    vol 37   E65, E67, E84, 164, E314, E317, E322, 343, E366,
                  367, E370, E371, E379, 409, E485-E487, E490, 506,
                  E516, 517
    vol 38   E135, E136, E138, E141, 148, E158, 160, E201, E202, 208,
                  E225, E226, 231, E294-E296, 302, E308, 310, E326, E327,
                  330, 331, 375, E381, 382, E456, 457, E483, 484, 515,
                  E517, 518, 532, 533, 537, 539
Gubieva, D.N.
    vol26   E342-E352, 382-387
Gudkova, V.I.
    vol 22   74, 98, 123, 151, 208
Gudtsov, N.T.
    vol 25   E301, E302
Guerry, D.Jr.
    vol 27/28   E202-E205, E214, 219, E450, E451, 463, E499,
                  E640-E644, 647, 653, 661, 691, 707
Guichard, M.
    vol 30   E467, E468, E470, E472, 476

Guinasso, H.L.Jr.
        vol 27/28   E50, E51
Guln, J.A.
        vol 27/28   547, 565, 566
Gulyamov, Yu.M.
        Vol 26   E5, E6, 58
        vol 31   E39-E42, E47, E49, E51, E52, 77, 78, 92-95, E206,
                     E207, E210, E212, 260, 261
Guminichenko, L.V.
        vol 25   E221, 223, E225, 228, E239, 241
Guminski, c.
        vol 25   E330, 334, E335-E337, 355
Gunn, R.D.
        vol 27/28   E320
Guntz, A.
        vol 25   E13-E18, 22, E33, E34, 37
Gupta, S.C.
        vol 26   E187, E188, E192, E193, 217
Gupta, M.
        vol 35   146
Gupta, M.K.
        vol 27/28   E248, E249, 264, 265
Guryanova, O.N.
        vol 25   E84, E85, E335-E337
Guseva, A. N.
        vol 37   E13-E15, 16, E64, E76, E78, E79, E82, 133, E212,
                     E213, 214, E221-E224, E226, E227, 237, 270, E369,
                     E370, E374, E376-E378, 393, 423, 443, E445, E447,
                     448, E485, E486, E489, E490, 497, 515
        vol 38   E9, E12, 14, E23, E25, E26, E28, 32, E42, E44-E46, 50,
                     E59, E65, E66, 75, 101, E104-E106, 107, E177, E178,
                     181, E201, E202, 207
Gusyakov, V.P.
        vol 34   E13-E15, 26-28, 37, 38, 49, 53, 59-66, 68-71, 85-90,
                     92-95, 97, 98, 101-104, 106, 131, 134, 137, 142,
                     143, 148, E204, 207, 210, 211, 215, 217, 219, 226,
                     228-231, 251, 253, 255-257, 264
        vol 35   E110-E112, 129, 183-185, 198
        vol 36   E303, 306, 313-327, 345
Guter, M.
        vol 27/28   E248, E249, E477
Guthrie, F.
        vol 31   E98, E102
Gutierrez, F.H.
        vol 34   161, 177, 179, 192-199, 263, 272, 273, 281, 317
        vol 35   194, 238
        vol 36   6-8, 10, 14, 18, 50, 51, 55, 56, 343
Guttierrez Losz, C.
        vol 27/28   E640, E642-E644, 651, 652, 654, 660, 695
Gutmanaite, H.
        vol 26   E303-E305, 308, E322-E324, 326, E331, 335, 337, 339,
                     341, 343, 345, 347
Guy, S.
        vol 23   E178, E184, 269
Gyani, P.P.
        vol 23   E92, E97
Gyunashyan, A.P.
        vol 31   E127, E129, E130, 152, 153, 165

H

Haar, L.
        vol 29   E190-E192, 193, 194, 196-199
Haas, J.L.Jr.
        vol 27/28   E74, E85
Haberli, E.
        vol 23   E169, E171, E183
Haccuria, M.
        vol 24   E218-E221, 231, 240, E414, 416, 417
        vol 27/28   575
        vol 32   E170, E186, 215
Hagar, H.
        vol 26   E153, E154, E158

Hagara, A.
    vol 27/28   658
Hagerman, G.
    vol 35   153, 154, 228
    vol 36   156, 241, 252, 275, 455, 458, 461, 463-465, 472, 473,
                 485
Haggin, J.H.S.
    vol 27/28   336, 337
Hagisawa, H.
    vol 23   E169, E183, 187
    vol 26   E153-E156, E158, 161, 168, 169
Hahn, E.
    vol 26   E113-E116, 121-123, 129, 130
Haidegger, E.
    vol 27/28   720
Haiduc, C.
    vol 31   E212, 267, 268
Hajicek, O.
    vol 25   E385-E387
Hala, J.
    vol 21   E8, E12, 64
Halban, H.
    vol 21   E83-E87, E89, 178, 184, 186-188, 190, 192, 194, 196,
                 199-201, 204, 207, 208, 212, 216, 223, 231, 279, 280
Halford, J.O.
    vol 23   E93, E97, 116

Hall, C.E.
    vol 31   E171, 172, 173
Halmos, Z.
    vol 33   E58-E60
Hamer, W.J.
    vol 23   E83, E88, E96
Hames, D.A.
    vol 29   E1-E4, E7, E14, E16-E18, E21, 33-35, E46, E48, 59-61
Hamlin, W.E.
    vol 35   105
Hammick, D.L.
    vol 26   E1, E2, E5, E6, 56, 57
    vol 31   E11, E15, E98, E102, 105
Hammond, J.S.
    vol 23   E144, E145
Hammond, L.
    vol 24   E414, 418, 419, 420
    vol 27/28   730
Hanano, M.
    vol 35   E64, 67, E110-E112, 139, E257, 261
    vol 36   E124, 136, E404, 406, 407, 409
Handa, Y.P.
    vol 27/28   E1-E5, 23
Hanks, J.H.
    vol 29   E87
Hannaert, H.
    vol 24   E218-E221, 231, 240, E414, 416, 417
    vol 27/28   575
    vol 32   E170, E186, 215
Hansen, M.
    vol 25   E13-E18, E57, E58, E84, E85, E139-E141
Hantzsch, A.
    vol 23   E156, E182
Hanus, J.
    vol 26   E293, 295
Haraguchi, K.
    vol 29   E46-E49, 54, 55, 57, 58, 64, 66-71
Harding, R.D.
    vol 31   E307, E308, E311, E318, E324
Haring, M.M.
    vol 25   E139-E141, 149, E157-E159, 167
    vol 29   E213, E214, 223
Harkins, W. D.
    vol 37   E63, E67, E81, 103
    vol 38   E187, E189, 190, E201, E202, 204
Harman, T.C.
    vol 25   E194, E195

Harned, H.
    vol 23   E309, E317
Harris, H.G.
    vol 24   E130-E134, 161, 163, 164
    vol 27/28   E202, E210, E211, E214, 243, 244, 246
Harris, N.L.
    vol 25   E1-E3, E13-E18, E33, E34, E44, E45, E55, E57, E58,
                   E66, E67, E76, E77, E84, E85, E103-E105, E119-E121,
                   E134, E135, E139-E141, E157-E159, E172-E174, E182,
                   E183, E258, E259, E263, E268, E271, E274, E277,
                   E282, E284-E286, E301, E302, E307, E310, E314,
                   E316-E318, E326, 327, E330, 332, E335-E337,
                   E357-E359, E369-E371, E385-E387, E402-E404
Hartel, G.H.
    vol 32   E174, E175, E183, E184, E187, 253, 258, 259, 261, 325
Hartley, G.S.
    vol 30   E105, E108, E111, 156
Hartley, H.
    vol 26   E1-E2, E5, 8, 9
Hartmann, G.
    vol 23   E145
Harvey, W. W.
    vol 26   E348-E352, 393, 396, 397
Hasegawa, S.
    vol 36   289, 294
Hasegawa, T.
    vol 30   E374, E375, E382, 414-416, E428, E429, E431, 444-446,
                   E448-E450, 458-460
Hatchard, C.G.
    vol 37   E366, E516
    vol 38   E158, 159, E308, E309
Hatem, S.
    vol 21   E4, E5, E12, 27, 37, 42-44
Hatfield, M.R.
    vol 25   E157-E159, 167
Hathes, W.
    vol 20   E153-E157
Hatibarua, J.R.
    vol 33   E58-E60
Hatzl, H.
    vol 23   E174, E183, 230
Haul, R.A.W.
    vol 25   E277, 278, E285, E286, 289, E301, E302, E310,
                   E317, E318, 319
Hauptshein, M.
    vol 29   E177, E179
Hawking, F.
    vol 35   149
    vol 36   41, 161
Hawkins, D.T.
    vol 25   E1-E3, E33, E34, E44, E45, E50, E51, E119-E121,
                   E139-E141, E157-E159, E182, E183, E357-E359,
                   E369-E371, E385-E387
Haworth, H.W.
    vol 20   E59, 60, E62, 64, E132-E136, 143
Hayano, I.
    vol 37   E64, E76, E79, E81, 119, 120, E188
Hayashi, M.
    vol 37   E63, E67, E68, E82, 118
Hayduk, W.
    vol 21   E5-E7, E9, E10, E12, 47, 50, 55, 70
    vol 24   E130-E134, 136, 137, 140-142, 144-146, 152, 154,
                   155, 168, 171, 173, 178, 179, E218-E221, 222,
                   223, 227, 232, 233, E257, E258, 261, 263, 265,
                   270, 273, 278, 280, E299-E302, E331-E333, 342,
                   357, 358, 369, E375-E378, 381, E411
    vol 27/28   E202-E206, E208-E210, E214, 221, 226, 228, 232,
                   237, E493, E494, E497, 501, E582, E595
    vol 32   E166-E168, E171-E173, E175, E176, E178, E179, E182,
                   E184, E186, 199, 241, 243, 248, 263, 270, 274, 287, 299
Hayek, E.
    vol 23   E20, E26, E174, E183, 230
Hays, S.E.
    vol 35   91, 100

Hayward, A.
    vol 23   E21, E26, 32, 33
Hayworth, C.B.
    vol 20   E153-E157, 164
Hayworth, K. E.
    vol 37   E221, E224, E226-E228, 243-245, E315, E320, E323,
                  338, 339
Hazlewood, F.J.
    vol 33   E58-E60
Headridge, J.B.
    vol 23   9, 82
Hearn, B.
    vol 23   E21, E26, 32, 33
Hebert, J.
    vol 23   E17, E20, E26, 59
Hegarty, M.J.
    vol 27/28   E248, E249, 264, 265
Heidman, J.L.
    vol 38   E59-E63, E65, E66, 88, 89, E104-E106, 108, 109, E136,
                  E137, E139-E141, 156, 157

Heinsen, A.
    vol 26   E113-E116, 134, 135, E144, E145, 150, 151
Hekster, Ch.A.
    vol 35   62, 75, 97, 101, 253, 255, 299, 308
Hekster, Y.A.
    vol 35   29, 63, 159, 229, 252, 254, 269, 287
    vol 36   40, 75, 160, 242, 256, 276, 334, 361, 363, 377, 403,
                  434, 435
Helander, S.
    vol 35   153, 154, 228
    vol 36   156, 241, 252, 275, 455, 458, 461, 463-465, 472, 473,
                  485
Hellriegel, E.
    vol 23   E18, E26
Hellwig, K.C.
    vol 26   E1-E5, 13, 34, 35
Hendrich, G.
    vol 23   E159, E163-E166, E182, 207-209
Henning, J.
    vol 26   E153, E158, E271, E273
Henry, T.H.
    vol 25   E369-E371
Hensel, W.E.Jr.
    vol 27/28   760
Henson, B.J.
    vol 27/28   547, 565, 566
Hentz, R.R.
    vol 21   E2, E8, E12, 20
Hepel, T.
    vol 23   E15, E25
Hepler, L.G.
    vol 23   E165, E168-E173, 183, 266, 267
Herbert, G.M.
    vol 23   E23, E26
Herning, P.E.
    vol 25   E194, E195, 205
Herscovici, J.
    vol 21   E10-E12, 74
Hertzberg, G.
    vol 27/28   E1-E5, E72, E74, E84, 120
Hervier, B.
    vol 26   E267, 269
Herz, W.
    vol 23   E158, E165, E167, E182, E183, 264, E308, 310, E317, 335
    vol 37   E63, E66, E81, E85, E314, E315, E321
Herzog, G.W.
    vol 23   E170, E171, E183
Hess, C.B.
    vol 25   E357-E359, 363
Hess, L.G.
    vol 24   E411, 412, 413
    vol 27/28   648

Heuston, F.T.
    vol 26   E271-E273, 275
Heycock, C.T.
    vol 25   E13-E18, 20, E139-E141, 143, E157-E159, 161,
                E182, E183, 185, E402-E404, 406
Heyne, R.
    vol 25   E300
Hickling, A.
    vol 25   E285, E286, E335-E337
Hicks, B.L.
    vol 27/28   E281, 283-285
Hicks, C. P.
    vol 21   E80, 89
    vol 37   E73, E85, E222, E229, E319, E323
Hideshima, K.
    vol 30   E31, E33, 83, 84
Hiebenthal, F.Z.
    vol 23   E308, E310, E317, 335
Hietanen, S.
    vol 23   E90-E92, E96
Higuchi, T.
    vol 35   E110-E112, 124, 146, 175
    vol 36   E124, 130, 184
Higuchi, W.I.
    vol 35   133, 170, 190
Hikari, S.
    vol 29   E1, E5, E7, E14, E16-E18, E21, 40, E47, E49, E50
                E51, 72-86, E87, 88-94, 100, 101
Hildebrand, D.A.
    vol 29   E1, E5, E7, E14, E16-E18, E21, 41, E47, E48, 52, 63
Hildebrand, J.H.
    vol 23   E171, E183
    vol 24   E218-E221, E299-E302
    vol 25   E13-E18, E369-E371, 374
    vol 27/28   E213, E214, 229, 682, 684, E702, 704
    vol 29   E213, E214, 215
    vol 31   E1, E6
Hill, A.E.
    vol 30   E330-E332, E335-E339, 341, 351-353, 364-366, E374,
                E375, E379, E380, E382, 388-391
    vol 37   E63, E67, E71-E73, E81, 89, 90
Hill, E.S.
    vol 24   E422, 424, 425
Hill, L.M.
    vol 26   E115, E116, 124, 125
Hilpert, K.
    vol 25   E66, E67, E103-E105
Hilpert, S.
    vol 38   E236-E238, 239
Hinds, J.I.D.
    vol 26   E236
Hine, J.
    vol 20   E59, 60, E62, 64, E132-E136, 143
Hinnuber, J.
    vol 25   E172-E174, E268, E277, E282, E284-E286, E425-E427
Hinzner, F.W.
    vol 25   E357-E359
Hirai, T.
    vol 30   E31, E33, 83, 84
Hirano, H.
    vol 35   26, 32, 35, 61
Hiraoka, H.
    vol 27/28   E213, E214, 229, 684
Hiroshi, H.
    vol 26   E115, E116, 126, 131
Hirschler, A.E.
    vol 23   E308, E310, E317, 322-325, 336-338
Hiura, M.
    vol 34   56, 57, 82, 96
    vol 35   69, 77, 78
    vol 36   143, 144, 151, 166, 168, 372-374, 378, 379
Hiza, M.J.
    vol 27/28   E248, E249, 263, E477, 481, 482

Hoegfeldt, E.
    vol 37   E64, E72, E82, 138, E369, E378, 396
    vol 38   E9, E10, E12, 16, E23, E27, E28, 33, E42, E43, E46, 51
Hofer, F.
    vol 23   E16, E17, E25, 56, 57, E156, E171, E182, 248
Hoffmann, A.E.
    vol 27/28   E320, 322, 323, E465, 468, E511, 514
Hoftijzer, P.J.
    vol 32   E54, 56, 57, 66, 67, 72
Hogan, J.J.
    vol 27/28   E266, 267, E281, 282, E303, 304, E320, E349, 350,
                  E465, 466, E511, 512, 559, 560, 635, 636
Hogfeldt, E.
    vol 20   E153-E157, 171
Hohenstein, W.P.
    vol 38   E1, E2, 6
Hohn, H.
    vol 25   E301, E302
Holden, R.B.
    vol 25   E55
Holland, H.C.
    vol 23   E179, E184
Hollerer, I.
    vol 23   E170, E171, E183
Hollifield, H.C.
    vol 20   E1
Holm, K.
    vol 30   E112, 153
Holz, E.
    vol 35   143
    vol 36   140, 250, 453, 454, 456, 457, 459, 460, 469, 470
Holz, S.
    vol 35   143
    vol 36   140, 250, 453, 454, 456, 457, 459, 460, 469, 470
Holzl, F.
    vol 26   E90-E92, 102, 103
Honda, K.
    vol 25   E139-E141, 147, E157-E159, 164, E402-E404, 414,
Hopfner, A.
    vol 21   E80, E89, 92-97, E234, 235-243, E252, 253-260
Hopfner, H.-M.
    vol 21   E80, E89, 92, 93, 96, 97, E234, E252, 253-256, 259, 260
Hopkins, B.S.
    vol 22   E152, E156, 214, 215
Hoppel, H.-E.
    vol 21   E81-E83, E89, 135, 171, 172, E234
Hopson, W. H.
    vol 37   E63, E71, E82, 117, E314, E316, E318, E321, 329,
                  E485, E486, E490, E496
Hori, S.
    vol 26   E153, E158, E271, E273
Horiba, S.
    vol 37   E66, E81
Horiuchi, A.
    vol 24   E1, E2, 9, E59, E65-E71, 73, 105, 111, 112, 114,
                  116
    vol 27/28   E1-E6, E70, E74, E79, E80, E85, 128, 164,
Horiuti, J.
    vol 27/28   E493, E494, E497, 498, E641, E644, 646, 655, 662,
                  E685, 686, 693
Horn, A.B.
    vol 27/28   E24-E28, 30
Horner, L.
    vol 25   E1-E3, 12
Horsley, G.W.
    vol 25   E301, E302
Horsman-van den Dool, L.E.W.
    vol 21   E1-E3, E11, E12, 15, 23, 25, 26, 79
Horvath, M.
    vol 34   E302, 306, 315
    vol 36   E303, 309, 339
Hoskins, J.C.
    vol 27/28   188

Hottovy, J.D.
    vol 27/28   397-399, 407-409, 444-446
Ho-Tun, E.
    vol 26   E153, E158
Hovorka, V.
    vol 26   E293, 295
Howard, W.B.
    vol 24   E299-E301, 304
Howell, W.W.
    vol 23   E308, E311, E317, 326, 327
Hoyt, C.S.
    vol 25   E157-E159, 165
Hronec, M.
    vol 27/28   658
Hsi, C.
    vol 27/28   E248, E249, 258, E477, 478, 479, 483, 484
Hsieh, Y.Y.
    vol 30   E105, E111, 120, E220, E223, 227, E374, E375, E381, 387
Hsu, C.C.
    vol 24   E375-E378, 379
Huang, S.S.-S
    vol 27/28   E533, 536, 538
    vol 32   E169-E171, E186, 218-220, 231, 234
Hudson, D.
    vol 26   E108, E109, 112
Hudson, D.R.
    vol 25   E357-E359, 367
Huff, H.
    vol 37   E64, E68, E82, 124
Hug, E.
    vol 36   E19, E20, 21, E63, 64
Hughes, F.
    vol 26   E152
Hughes, R.E.
    vol 32   E166, E168, E186, 192, 193
Hulbert, R.W.
    vol 30   E24-E26, E30-E32, 37-40
Hulett, G.A.
    vol 23   E307, E317
    vol 25   E402-E404, 411
Hultgren, R.
    vol 25   E1-E3, E33, E34, E44, E45, E50, E51, E119-E121,
                 E139-E141, E157-E159, E182, E183, E357-E359,
                 E369-E371, E385-E387
Hultquist, M.E.
    vol 36   9
Hume-Rothery, W.
    vol 25   E369-E371, 383
Humm, W.
    vol 26   E187-E190, E193, 201
Humme, H.
    vol 23   E163, E183, 203
Humphrey, K.C.
    vol 31   E98, E102
Humphreys, R.E.
    vol 23   31, E158, E180, E182, 190, E270, E281, 285
Humphreys, W.J.
    vol 25   E335-E337, 338, E357-E359, 360
Hung, H.H.
    vol 25   E425-E427
Hung, J.H.
    vol 24   E1, E2, 3, E16, E17, 18, E60-E64, E71, 78, 79,
                 84-87, 89-94, 121
    vol 27/28   E1-E6, 15, E59-E63, E85, 90-92, 95-97, 99
Hunt, H.
    vol 30   161, 257, 420-426
Hunt, L.J.
    vol 27/28   E338, 345, 346
Hunt, M.R.
    vol 23   E21, E26, 32, 33
Huntjens, F.J.
    vol 32   E54, 56, 57, 66, 67, 72
Hurlen, T.
    vol 25   E285, E286, 299, E335-E337, 354

Hursh, J.B.
    vol 29   E1, E5, E7, E21, 44, E48, E51, 65, 97-99, 176
Hussey, C.L.
    vol 24   E130-E134, 184, 185, 187, 188
Hutchison, C.A.
    vol 20   40, E132-E136, 139, 149, E153-E157, 162, 182
Hutting, G.F.
    vol 23   E156, E159, E162, E182, 205, 206
Hwang, S.-C.
    vol 27/28   E522, 528, 529, E737, 745, 746
I

Iandelli, A.
    vol 25   E214, E215, E221, E225, E230, E231, E234, E256
Ibraheem, Y.
    vol 35   4, 20, 37-42
Ichigaki, T.
    vol 25   E402-404, 414
Ichihashi, T.
    vol 35   26, 32, 61
Ichikawa, F.
    vol 23   E272, E276
Iggena, H.
    vol 25   E57, E58, 59, E71, E72, 73, E76, E77, 78,
                    E385-E387, 389, E402-E404, 407
Ignateva, L.A.
    vol 25   E173, E174, E335-E337, 350, 351
Igolinskava, I.M.
    vol 25   E84, E85, E335-E337
Igolinskii, V.A.
    vol 25   E84, E85, 91, E139-E141, E335-E337
Iijima, H.
    vol 34   E14, E15, 44, E204, 214
    vol 36   E367, 369
Iijima, T.
    vol 26   E74, E75, 85
Il'Inskii, Vl.P.
    vol 30   E24-E26, E29-E32
Il'yasov, I.I.
    vol 33   E26, 28, E31, 32, E57-E60, 67, 137, 165, 219, 257
Ilavsky, J.
    vol 27/28   658
Ilyukhin, V.V.
    vol 23   E280, E282
Imadsu, A.
    vol 31   E11, E15, E39-E42, 44, 45
Inoue, T.
    vol 35   3, 22, 68, 263
    vol 36   103, 287, 371, E404, 408
Inoue, Y.
    vol 25   E13-E18, 29
Inouye, K.
    vol 25   E385-E387, 391
Iofa, Z.A.
    vol 23   E162, E183, 211
Iovi, A.
    vol 31   E212, 267, 268
Ipatieff, V.N.
    vol 24   E251, 252
    vol 27/28   E511, 515, 516
Irvin, N.M.
    vol 25   E258, E259, E268, E277, E282, E284-E286, E288, E301
Isaacs, E.E.
    vol 32   E79-E83, 94, 98-104, 138, 139, 160-165, E176, E177,
                    E187, 268
Isbin, H.S.
    vol 30   98, 99, 101, 157-159
Ishida, K.
    vol 21   E3, E10, E12, 27, 29, 31, 32, E89
    vol 24   E251, 255, 256
Ishigaki, T.
    vol 25   E139-E141, 147, E157-E159, 164
Ishikawa, F.
    vol 26   E113-E116, 118, 119, 126, 131, 136, 137

Ishimori, T.
    vol 25   E425-E427
Ishizaki, N.
    vol 29   E1, E5, E7, E14, E16-E18, E21, 40, E47, E49
                E50, E51, 72-86, E87, 88-94
Islamov, D.U.
    vol 29   E213, E214, 219
Ito, H.
    vol 25   E103-E105, 108
    vol 26   E262
Ito, K.
    vol 34   E13-E15, 32, 36, 178
    vol 35   E110-E112, 128, 130
    vol 37   E65, E66, E84, 166, E370, E371, E379, 411, E425, 428
Ivanov, P.T.
    vol 26   E1-E3, E5, E6, 48, 49, 66, 67, E74, E75, 87, 88,
                E115, E116
Ivanova, I.N.
    vcl 23   E18, E26
Ivanovsky, F.P.
    vol 27/28   E629, 630
Ivashchenko, A.N.
    vol 23   E174, E176, E183
Ivleva, T.I.
    vol 31   E127, E130, 150, E297, 303
Izmailova, A.A.
    vol 22   80, 99, 126, 160, 211, 234, 250

J

Jaacobi, M.   see Yaacobi
Jablonski, Z.
    vol 23   E179, E184, 257
Jaccard, P.
    vol 22   21, E22, 24, 26, 77, 79, 84, E152, 155, E156, 159,
                167, 209, 210, 212, 246, E247, 248, 251, 327, E328,
                330, 332, E336, 359, 360, 362
Jackson, D.J.D.
    vol 23   E178, E184, 269
Jackson, J.L.
    vol 32   E79-E83, 155-159
Jacobs, J.
    vol 27/28   E281, 291, 387-389
Jacoby, R.M.
    vol 27/28   200, 201
Jadot, R.
    vol 24   E130-E134, 135, 147, 149, 167, 172, 174, 189, 192,
                194, 226, 229, 246, E257, E258, 260, 271
Jaeger, A.
    vol 23   E11, E25, 63, E313, E317, 319
    vol 37   E63, E66, E81, 91, E369, E371, E377, 380
Jaeger, W.
    vol 25   E402-E404
Jager, L.(Jaeger)
    vol 23   E177, E184, 236
    vol 26   E1-E3, E5, E6, 42, 43, E72-E74, 82, 83, E271, E273
Janecke, E.
    vol 31   E168, E169, E171, 186, 187, E297, 298, 300, 301
Janecke, E.
    vol 25   E13-E18, E157-E159, 163, E402-E404, 410
Jangg, G.
    vol 25   E66, E67, E84, E85, E103-E105, 113, E157-E159, E173,
                E174, 177, 179, E206, E258, E259, 260, E263, E264, E267,
                E268, E277, 279, E285, E286, 291, 292, 298, E300-E302,
                E307, E309, E310, 312, E314, 315, E316-E318, 321, E326,
                329, E330, E335-E337, 344, 345, 349, E357-359, 368, 399,
                E422, 424, E425-E427, 431, 432
    vol 29   E213, E214, 217, 220, 225
Janghorbani, M.
    vol 30   102
Janickis, J. (see also Janitzki, J.)
    vol 26   E303-E305, 308, E322-E324, 326, E331, 335, 337, 339
                341-347

Janitzki, J.
    vol 26   E303-E305, 306, 307, E322-E324, 325, E331, 333, 334,
                    336, 338, 340
Japas, M.L.
    vol 27/28   E1-E5, E24-E28, 41, E45, E46, 49
Jarowski, C.I.
    vol 35   E14, E15, 21, 23, 25, 34, 44
Jaulmes, P.
    vol 31   E3, E6
Jayaraman, A.
    vol 25   E194, E195
Jeffes, J. H. E
    vol 37   E63, E71-E73, E81, 102
Jelinek, H.
    vol 23   E148, E150, 151, 152
Jellinek, K.
    vol 23   E16, E25, 39, E89, E96, 108
Jenkins, S.H.
    vol 23   E10, E25, 31, E158, E180, E182, 190, E270, E281, 285
Jentoft, R.E.
    vol 38   E23-E26, E28, 31, E42-E45, 494
Jepson, W.B.
    vol 29   E190, E192, 200, 201
Jeszenovics, A.
    vol 26   E187, E193
Jirsa, F.
    vol 23   E148, E150, 151, 152
Jockers, R.
    vol 20   E122, 123, E202-E204, 208
Johansson, L.G.
    vol 26   E252
Johns, R. G. S.
    vol 37   E63, E71-E74, E82, 113
Johnson, C.A.
    vol 27/28   E303, 313, E320, 334, 335, E338, 347, 348
Johnson, H. H.
    vol 37   E64, E68, 87, 132
Johnson, J.R.
    vol 37   E64, E67, E71-E73, E83, 144, 147, E221, E225, E227,
    vol 37   240, E369, E378, 400
Johnson, K.W.R.
    vol 25   E263, E264, 265
Johnson, S.A.
    vol 29   E1, E7, E10, E21
Johnson, W.B.
    vol 27/28   319
Johnson, W.H.
    vol 20   E73-E77, 81, E132-E136, 142, E153-E157, 168, E202-E204,
                    207, E210-E213, 216
Johnston, H.L.
    vol 23   E10, E13-E15, E17, E18, E25, 46-48, E84-E88, E91-E93, E96,
                    105, 117-122, E144, E145, E148-E150, 153, 154, E161,
                    E182, E310, E317
Johnston, J.
    vol 23   E156-E158, E163-E165, E167, E170, E172, E182, 199, 200
Johnston, W.H.
    vol 37   E64, E71, E82, 121, E369, E371, E378, 391
Johnstone, H.F.
    vol 26   E271, E273
Johnstone, N.B.B.
    vol 24   E62, E69-E71, 88, 117
Jones, E.V.
    vol 23   E21, E23, E26
Jones, J.H.
    vol 32   E79, E81-E83, 89, 122-125
    vol 37   E63, E71, E82, 117, E314, E316, E318, E321, 329,
                    330, E485, E486, E490, 496
Jones, J.R.
    vol 20   E73-E77, 82, E132-E136, 144, E153-E157, 172, E210-E213,
                    217
    vol 37   E64, E66, E82, 134, E369, E378, 394
    vol 38   E59, E62, E66, 73, E277, E278, 281

Jonsson, J.A.
    vol 37   E42-E44, E46, 61, E315, E317, E323, 363, E485,
                    E487, E491, 514
    vol 38   E135-E137, E141, 155, E225-E227, 234
Jordan, M.
    vol 22   69
Joris, G.G.
    vol 37   E31, E33, E34, 35, E42, E45, E46, 48, E63, E71,
                    E81, 104, 209, E221, E225, E227, 232, E314, E318,
                    E321, 325, E455, E456, 457, E485, E488, E490, 493
    vol 38   E135, E138, E141, 143
Jou, F.-Y.
    vol 32   E79-E81, E83, 109-111, 113-115
Joyner, R.A.
    vol 25   E139-E141, 146, E357-E359, 362
Jung, M.
    vol 26   E153, E158, E271, E273

K

Kabadi, M.B.
    vol 23   E159, E162, E163, , E165, E182, 215-224
Kabanova, B.N.
    vol 23   E91, E96
Kachkimbaeva, S.A.
    vol 22   48, 114, 266, 284, 306
Kadyrova, Z.O.
    vol 23   E10, E24, E25
Kagan, I.K.
    vol 25   E402-E404
Kageyama, T.
    vol 26   E74, E75, 85
Kahn, S.M.
    vol 26   E404, E405, 425
Kahre, L.C.
    vol 27/28   E281, 295, 296, E303, 311, 312
Kaidalova, T.A.
    vol 23   E313, E317, 333
Kairbaeva, A.A.
    vol 25   E335-E337, 352, E402-E404, 420
Kakimoto, Y.
    vol 27/28   688, 690
Kakovskii, I.A.
    vol 25   E369-E371
Kalenberg, L.
    vol 25   E285, E286, 287
Kalra, H.
    vol 27/28   E266, 280, 749
    vol 32   E166-E168, E170, E186, 200, 201, 228, 229
Kaltwasser, K.
    vol 25   E33, E34, 42
Kamada, A.
    vol 34   E14, E15, 35, 42, 75, 159, 160, 200-203, E204, 209,
                    238, 262, 277-279
    vol 35   2, 5, 8, E14, 17, 18, 28, 55, 56, 59, 60, 65,
                    93, 102, 103, E110-E112, 131, 135, 158, 192,
                    193, 217, 218, 247, 248
    vol 36   E19, E20, 30, 31, 39, 47, 48, E88, 90, 91, E92, 94,
                    95, 102, 107, 114, E124, 131, 159, 205, E243, 249,
                    253, 285, 288, 291, 376, 400, 405, 411, 427
Kameda, T.
    vol 23   E169, E170, E183
Kamenev, A.I.
    vol 25   E172
Kamenskaya, A.N.
    vol 22   92, 106, 138, 181, 222, 240, 260, 279, 298, 319, 342,
                    354, 370, 379
Kamgar-Parsi, B.
    vol 29   E6, E7, E10, E21
Kaminishi, G.
    vol 27/28   736, E737, 740-742
Kamisawa, K.
    vol 26   E271, E273

Kanda, F.A.
      vol 25   E119-E121
Kaneniwa, N.
      vol 34   E14, E15, 44, E204, 214
      vol 35   E64, 66, 67, E110-E112, 137, 139, E257,
                    259-261, 274
      vol 36   E124, 134-136, 171, E367, 369, E404, 406, 407, 409
Kanke, M.
      vol 34   E13, E14, 43, 168
      vol 35   E110-E112, 134
Kapantsyan, E.E.
      vol 26   E404, E405, 414
Kapustinsky, A.F.
      vol 32   E1-E3, 8, E25, E26, E28, 30, 34, 35, 41
Karapetyan, T.I.
      vol 31   E318, E320, E321, E324, 328
Karataeva, I.M.
      vol 30   E268-E270, E272, E273, 288, 289, 292, 295, E330, E335,
                    E339, 367, 368, E374, E375, E378, E379, E381, 399-402,
                    E428-431, 439, E448-E450, 454, E467, E472, 481, 482
Karlsson, R.
      vol 37   E64, E71-E73, E83, 159
Karmazina, V.D.
      vol 26   E271, E272, E274, 281, 282
Karnakhov, A.S.
      vol 30   E109, E110, E112, 137, 138
Karnaukhov, A.S.
      vol 31   E98, E101, E102, 110
Karpinski, Z.J.
      vol 25   E135, 137
Kartzmark, A.A.
      vol 25   E402-E404
Kartzmark, E.M.
      vol 25   E182, E183
      vol 30   E184-E185, 192, 193, E195-E197
Kasai, J.
      vol 26   E187, E192
Kashcheev, G.N.
      vol 33   E40, 42, E73, 75
Kashina, N.I.
      vol 30   E24, E25, E28, E29, E31, E32, 58, 71, 72, 93-97, E105,
                    E108-E111, 136, 139, 140, 148, 149, E162, E164, E165,
                    168, 171, 172-175, E176-E178, 181, 182, E220, E222,
                    E223, 226, E258, 261, E263, E264, 267, E268-E274,
                    283, 287, 290, 293, E374, E375, E378, E381, 385,
                    E428-431, 434, E448-E450, 453
Kasperik, A.S.
      vol 27/28   E511, 514
Katayama, T.
      vol 37   E315, E318, E322, 352
Katz, D.L.
      vol 24   E39, 46, 47
      vol 27/28   E24-E28, 196-199, E737, 738, 739
Katzaev, G.A.
      vol 25   E84, E85, E139-E141, E335-E337, 350
Kawai, M.
      vol 27/28   E1-E6, 173, 174
Kawakara, H.
      vol 29   E1, E5, E7, E14, E16-E18, E21, 40, E47, E49, E50
                    E51, 72-86, E87, 88-94, 100, 101
Kawamoto, T.
      vol 29   E1, E5, E7, E14, E16-E18, E21, 40, E47, E49, E50
                    51, 72-86, E87, 88-94
Kawashiro, T.
      vol 27/28   743, 735
Kawata, M.
      vol 35   262
      vol 36   E367, 370
Kawata, Y.
      vol 29   E51, 100, 101
Kay, W.B.
      vol 24   E196, E197, 198, 199, 204-207, 210-215
      vol 29   E190, E192, 203-206
      vol 37   E64, E76, E80, E82, 127

Kazantsev, A.A.
     vol 31   E206, E207, E212, 217
Kazantsev, M.
     vol 25   E369-E371, 372,
Kazarnmovski, I.A.
     vol 23   E144, E145
Kbir-Ariguib, N.
     vol 23   E174, E183, 262, 263
Kedvessy, G.
     vol 35   E110-E112, 141, 203-205
     vol 36   311
Keefer, R. M.
     vol 20   E62, 63, E103-E105, 106, E129, 131, E132-E136, 140,
                    E153-E157, 165, E202-E204, 206, E210-E213, 215
     vol 37   E63, E68, E81, 107, E369, E372, E378, 387
     vol 38   E1, E2, 7, E9, E11, E12, 13, E23, E24, E28, 29, E42,
                    E43, E45, 47, E59, E60, E66, 68, E177, E178, 180, E187,
                    E189, 191, E201, E202, 205, E236-E238, 241, E265, E266,
                    267, E271, 272, E277, E278, 279, 323, E341, E343, 344,
                    391, 424
Keevil, T.A.
     vol 24   E299-E302, 310
     vol 27/28   708
Keight, D.G.
     vol 23   E10, E25, 31, E158, E180, E182, 190, E270, E281, 285
Kelley, K.K.
     vol 25   E1-E3, E33, E34, E44, E45, E50, E51, E119-E121, E139-E141,
                    E157-E159, E182, E183, E357-E359, E369-E371, E385-E387
Kells, M.C.
     vol 25   E55
Kelman, L.R.
     vol 25   E55, E263
Kemmerer, A.L.
     vol 23   E179, E184
Kemmerick, M.
     vol 25   E434
Kemula, W.
     vol 25   E285, E286, 290
Kendall, J.
     vol 32   E25, E27, 29
Kendall, P.
     vol 25   E103-E105, 115
Kennedy, G.C.
     vol 25   E194, E195
Kennedy, S.W.
     vol 33   E332
Kent, R.L.
     vol 32   E79, E82, E83
Kenton, F.H.
     vol 27/28   E77, E84, 154
Kerp, W.
     vol 25   E1-E3, 4, E13-E18, 23, E33, E34, 38, E44, E45, 46, E57,
                    E58, 59, E71, E72, 73, E76, E77, 78, E385-E387, 389,
                    E402-E404, 407
Kesans, A.
     vol 23   E11, E25
Kester, D.
     vol 27/28   E50, E51
Kestovnikov, N.N.
     vol 25   E230, E234
Ketelle, B.H.
     vol 23   E92, E96
Ketkovich, V.I.
     vol 33   E145
Khachikyan, A.V.
     vol 31   E318, E320, E321, E324, 328
Khadorchenko, V.V.
     vol 23   E11, E25
Khaitina, M.V.
     vol 33   E237, 266, E303, 305, E308, 310
Khalil, S.A.H.
     vol 36   310, 337, 340, 341

Khallieva, Sh.D.
    vol 31   E39-E42, E47, E49-E51, 75, 76, 88, 89, E206, E207,
                    E212, 259
Khanapina, K.
    vol 25   E230, E231, E234, 237
Khawam, M.N.
    vol 34   E13-E15, 29, 30, 128-130, 132, 133, 135, 136,
                    138-141, 144-147, 164, 165, 250, 252, 254
    vol 35   80, 180-182
Khikharev, M.I.
    vol 31   79
Khmeliova, M.G.
    vol 23   E313, E317, 333
Khodakovskii, I.L.
    vol 23   E83, E86, E89-E91, E96, 135-137, E180, E181, E184, 261
    vol 29   E1, E21
Khokhlov, S.F.
    vol 32   E183, E184, E187, 323
Khomyak, V.V.
    vol 25   E194, E195
Khomyakov, V.I.
    vol 22   72, 97, 122, 207
Khvala, M.A.
    vol 25   E194, E195
Kidnay, A.J.
    vol 27/28   E248, E249, 262-265, E477, 481, 482, E737,
                    743, 744, 747, 748
Kidyarov, E.I.
    vol 30   E268-E275, 279, 281-285, 287, 290, 291, 293, 294,
                    296-310, 312, 317, 318, 320, 323, 324, 328, E330,
                    E332, E335, E338, E339, 369, 370, E374, E378, E381,
                    E428-E431, 441, E448-E450, 455, 456, E467E468,
                    E471, E472, 483
Kiers, C.T.
    vol 26   E286
Kienle, R.H.
    vol 34   1-12, E14, 20, 73, 78-80, 83, 91
Kikuth, W.
    vol 36   E124, 127, E228, 230, E243, 245, E269, 270
Kindrick, R.C.
    vol 32   E79-E82, 88, 97, 112
King, A.D.
    vol 27/28   E1-E5, 100, 101, 188, 189
King, C.J.
    vol 38   E277, E278, 283, 286
King, F.E.
    vol 22   78, E156, 157, E162
King, M.B.
    vol 24   E130-E134, 153, 156
    vol 32   E166-E169, E185, 198, 202, 207-209, 212
King, R.H.
    vol 35   91, 100
Kirchmayr, H.R.
    vol 25   E157-E159, 168, E206, E215, E221, E225, E230,
                    E231, E239, E242, E245, E247, E249, E253, E274,
                    E285, E286, 292, E314, E335-E337, 345, E385-E387, 399
Kirchnerova, J.
    vol 20   E73-E77, 84, E153-E157, 178
    vol 37   E65, E71-E73, E84, 168, E221, E225, E228, 259
    vol 38   E42, E45, E46, 53
Kirgintsev, A.N.
    vol 30   E105, E108, E111, 148-150, E162, E164, E165,
                    173, E176-E178, 183, E220, E222, E223, 237, 238,
                    E258, 262, E263, E264, E374, E375, E378, E381, 395,
                    396, E428-E431, 438, E448-E450
    vol 31   E47, E48, E51, E212
Kirmse, E.M.
    vol 22   1-19, E22, 23, 27-29, 31, E32, 33, 34, 36, 38-45, 47,
                    49, 52-60, 62-71, 73, 88, 89, 94, 100, 102, 103, 115,
                    119-121, E127, 128, 132-134, 137, 140-144, 146, 148-
                    150, E162, 165, 176, 177, 180, 188, 189, 191, 193, 197,
                    199, 203, 205, 206, 213, 216, 218, 221, 224, 225, 227,
                    228, 231-233, 235-237, 239, 242, 244, 245, E247, 249,
                    255, 256, 259, 262, 264, 265, 267, 268, 271, 273-276,

Kirmse, E.M.
    vol 22   278, 282, 283, 285, 287-289, 291, 293, 294, 297, 300, 302, 303, 307, 309, 312-315, 318, 322, 324-326, E328, 329, 333, E334-E336, 337, 338, 341, 344, 346, 347, 349, 351, 357, 358, 363, 365-367, 369, 372, 374-377

Kirsanova, L.D.
    vol 31   E278, E279, 293-296

Kirshenbaum, I.
    vol 20   E35, 36, E132-E136

Kisan, W.
    vol 32   E184, E187, 314

Kisarov, V.M.
    vol 20   E153-E157, 170

Kiss, A.V.
    vol 32   E1-E3, E20, 22-24

Kitao, K.
    vol 34   E14, E15, 42, 160, 200-203
    vol 35   E14, 18, 56, 59, 60, 102, 103, E110-E112, 135, 193, 217, 218, 247, 248
    vol 36   E19, E20, 31, 48, E88, 90, 91, E92, 94, 95

Kito, S.
    vol 37   E65, E66, E84, 166, E370, E371, E379, 411, E425, 428

Klamut, C.J.
    vol 25   E258, E259, E263, E264, E271, E274, E301, E302

Klaschka, J.T.
    vol 24   E16, E17, 27, E65-E68, E71, 110

Klasens, H. A.
    vol 26   E256, E258

Klassen, H.B.
    vol 36   217, 223, 224

Klebanov, G.S.
    vol 26   E1-E3, E5, E6, 54, E90-E92, 95, E303-E305, 309, 310, E322-E324, 327, 328
    vol 30   E195-E197, 201, 202, 212, 216, 217

Klecka, M. E.
    vol 37   E63, E67, E71, E81, 109

Klehr, E.H.
    vol 29   E102-E104, E106, E109, 128, 131, E149, E150, E153 155, 158, E177-E179, 180, 183, 185, 186, E187, 189

Klein, H.R.
    vol 35   91, 100

Klein, O.
    vol 23   E157, E182

Klemenc, A.
    vol 20   E73-E77, 78, E89-E91, 92, E99, E100, 101, E103-E105, 107, 108

Klement, W.
    vol 25   E194, E195

Klemm, W.
    vol 23   E144, E145
    vol 25   E84, E85, 88

Klevens, H. B.
    vol 37   E63, E68, E82, 110, E369, E372, E375, E378, 388
    vol 38   E59, E60, E66, 69, E201, E202, 206, E236, E238, 242, E277, E278, 280, E396-E398, 400, E440, 442, E446, E448, 450, E477, E478, 480, E486, E487, 488, E493, 495, E497, 499, E534, 536

Klimecek, R.
    vol 26   E1-E3, E5, E6, 42, 43, E72-E74, 82, 83

Klimenko, N.N.
    vol 32   E54, 61, 62

Klosky, S.
    vol 23   E84, E94, E96, 111

Klushina, T. V.
    vol 26   E348-E352, 381

Klymer, S.D.
    vol 32   E79, E83

Klyukas, Yu.E.
    vol 25   E76, E77, 82

Knacke, O.
    vol 25   E13-E18

Knapp, H.
    vol 27/28   97, E712, 718

Knazko, L.
    vol 34   E13, E15, 34, 99, 100, 105
Knoche, H.
    vol 23   E20, E26
Knudsen, M.
    vol 27/28   E50, E51
Knutter, R.
    vol 25   E103-E105
Kobatake, Y.
    vol 27/28   682, E702, 704
Kobayashi, H.
    vol 29   E51, 72-86, E87, 88-94, 100, 101
Kobayashi, R.
    vol 24   E39, 46, 47
    vol 32   E169, E171, E178, E186, 213, 214, 237, 276,
    vol 27/28   E248, E249, 255, 259-261, E266, 274, 277, 278,
                 E281, 292, 294, E303, 309, 310, E320, 329-322,
                 E338, 344-346, E360, 365, 378-381, 385, 386,
                 394-396, 400-403, E522, 524, 525, 528, 529,
                 539-542, 556, E737, 745, 746.
Kobayashi, Y.
    vol 25   E425-E427
Kobe, K.A.
    vol 26   E1-E5, 13, 34, 35
    vol 27/28   E77, E84, 154
    vol 30   98, 99, 101, 157-159
    vol 31   E11-E15, 30-36, E98, E102, E127, E130, 157, 158
Kobelev, V.A.
    vol 20   E153-E157, 160
Koene
    vol 26   E252
Koenigsbuch, M.
    vol 26   E187, E188, E192, 226-228
Kofman, L. S.
    vol 37   E7, E8, 10, E22, E23, 25, E31-E34, 38
Kogan, I.M.
    vol 26   E%, E6, 71
Kogan, L.M.
    vol 22   74, 98, 123, 151
Kohle, H.
    vol 29   E1, E3, E14, E16-E18, E21, 27
Kohn, J.P.
    vol 27/28   E320, 324-326, 333, 336, 337, E338, 342, 343, E349,
                 351, 352, 355-357, E360, 366, 367, E370, 376, 377,
                 397-399, 407-409, 412-426, 432-435,
                 441-446, E750, 754, 755
Kohr, D.A.
    vol 23   E84, E89, E96, 98
Koishi, M.
    vol 26   E187, E192
Kol´ba, V.I.
    vol 31   E39-E42, E47, E48, E51, 79
Kolb, E.D.
    vol 23   E181, E184, 246
Kolesnikova, T.L.
    vol 32   E79, E83
Koliushklo, I.I.
    vol 27/28   44, E629, 631
Kollmann, K.
    vol 25   E301, E302, E310, E317, E318, E335-E337, 341
Kolotilova, V.Ya.
    vol 31   E307, E308, 312, 314, E318, E320, 325, 327
Koltai, L.
    vol 26   E187, E193
Kolthoff, I.M.
    vol 23   E169, E170, E183
    vol 30   373, E374, E381, 386, 417-419, 447
Komar, H.P.
    vol 23   E90, E96
Komarenko, V.G.
    vol 27/28   E582, E583, E586-E588, E595, 607
Komatsu, M.
    vol 34   E14, E15, 45
    vol 35   262

```
Komatsu, M.
    vol 36       E367, 370
Kondratev, V. P.
    vol 37       E66, E85
Koneva, N.V.
    vol 29       E46, E48, 62
Kononov, A.V.
    vol 23       E20, E26, E178, E184
Kononova, I.V.
    vol 26       E1-E6, 29, E72-E74, 81
Konovalova, N.A.
    vol 22       92, 106, 138, 181, 222, 240, 260, 279, 298, 319, 342,
                 354, 370, 379
Koonce, K.T.
    vol 27/28    E338, 344, E360, 365, 396, 400-403
Koopman, D.E.
    vol 30       E24, E28, E31-E33, 56, 57, 87-89
Kopylov, V. V.
    vol 37       E66, E85
Korenman, I. M.
    vol 23       E90, E96
    vol 37       E42, E43, E46, 58, E65, E67, E84, 172, E222-E224,
                 E228, 263, 264, E315, E316, E322, 351, 354, E370-
                 E372, E379, 414, 415, E485, E486, E491, 510
    vol 38       E59, E60, E66, 82, 83
Korf, D.M.
    vol 31       E127, E130, 139
Kornfeld, E.C.
    vol 36       302, 304, 364, 447-449
Kornyev, A.N.
    vol 23       E280, E282
Korobkina, N.P.
    vol 25       E33, E34, 43, E157-E159, 170, E385-E387, 401
Korotkevich, B.I.
    vol 30       E105, E108, E111, 148, 149, E162, E164, E165, 173
Korovin, S.S.
    vol 22       136, 179, 220, 238, 258, 277, 296, 317, 340, 352
Korshunov, B.G.
    vol 22       74, 98, 123, 151, 208
Korshunov, V.I.
    vol 25       E230, E231
Korshunov, V.N.
    vol 25       E1-E3, E13-E18, E44, E45, E50, E51
Koryakina, N.S.
    vol 23       E280, E282
Korzhenyak, N.G.
    vol 30       E28, E31-E33, 82, 85, 86
Kosyakov, N.E.
    vol 32       E183, E184, E187, 323
Kotel´nikova, L.P.
    vol 26       218, 219
Kothner, P.
    vol 31       E308, E311, E318, E324
Kotova, N.A.
    vol 25       E84, E85, E335-E337
Kovachev, Ts.
    vol 26       E153, E154, E157-E159, 173, 174, 177-179
Kovalenko, P.N.
    vol 23       E16, E25, E90, E96, E170, E183
Kovaleva, S.V.
    vol 25       E135, 138, E194, E195, 204
Kovba, L.D.
    vol 23       E88, E96, E144, E145
Kozhukhovskij, A.A.
    vol 31       E206-E209, E212, 236
Kozicki, H.
    vol 23       E20, E26
Kozin, L.F.
    vol 25       E1-E3, E33, E34, E34, E44, E45, E50, E51, E55, E57,
                 E58, 65, E66, E67, E71, E72, E76, E77, E83, E84, E85,
                 E93-E95, E103-E105, 106, 110, 117, E119-E121, 129,
                 E134, E135, E139-E141, E157-E159, E172-E174, E182,
                 E183, 193, E194, E195, 206, E214, E215, E2221, E225,
                 E230, E234, E239, E242, E245, E247, E249, E251, E253,
```

Kozin, L.F.
    vol 25   E256, E258, E259, E263, e267, E268, E271, E274, E277,
                    E282, E284-E286, E300-E302, E307, E309, E310,
                    E316-E318, E326, E330, E335-E337, E357-E359,
                    E369-E371, 384, E385-E387, 398, E402-E404, E421,
                    E422, E425-E427
Kozina, T.M.
    vol 23   E177, E184, 258
Kozintseva, T.N.
    vol 32   E1-E3, 11, E25, E28, 42-44
Kozitskii, V.P.
    vol 30   E105, E111, 150, E176-E178, 183
Kozlov, K.
    vol 23   E83, E86, E89-E91, E96, 135-137
Kozlovskii, A.J.
    vol 25   E214, E215
Kozlovskii, M.T.
    vol 25   E33, E34, E135, E173, E174, 178, E285, E286, 293,
                    296, E335-E337, 342
Kozlowska-Kolodziej, B.
    vol 23   E94, E95, E97, 142, 143
Kragas, T.K.
    vol 32   E171, E178, E186, 237, 276
Krahl, M.E.
    vol 38   E395, 398, 399, E440, 441, E446, E448, 449, E477, E478,
                    479, E486, E487, 488, E493, 494, E497, 498, /509-511,
                    E512, 513, 421, 522, 524, 527, E528, 529, E534, 535,
                    538, 540-542
Krasil´shtschikov, A.I.
    vol 31   E51, E52, 188, 189, E206, E207, E210,
                    E212, 215, 228, 229
Krasnoschchekova, P.Ya.
    vol 38   E135, E136, E138, E141, 148, E158, 160, E201,
                    E202, 208, E225, E226, 231, E294-E296, 302, E308, 310,
                    E326, E327, 330, 331, 375, E381, 382, E456, 457, E483,
                    484, 515, E517, 518, 532, 533, 537, 539
Krasnoshchekova, R.Ya.
    vol 37   E65, E67, E84, 164, E314, E317, E323, 343,
                    E366, 367, E370, E371, E379, 409, E485, E486, E487,
                    E490, 506, E516, 517
Krasnova, I.E.
    vol 25   E285, E286, E317, E318
Kraus, C.A.
    vol 30   E1-E4, 12-14
Krause, L.
    vol 26   E331, 332
Kraut, K.
    vol 25   E13-E18
Krebaeva, Sh.D.
    vol 25   E214, E215, 218, E221, E225, 227, E230, E234,
                    E253, 255
Krebs, B.
    vol 32   E185, 318
Krebs, H.A.
    vol 36   138, 162, 163
Krejci, J.
    vol 31   E206, E207, E212, 248
Kremann, R.
    vol 25   E57, E58, E84, E85
Kremers
    vol 30   E195-E197
Krepelka, H.
    vol 29   E228, 231, 232
Kresheck, G.C.
    vol 24   E1, E2, 4, E16, E17, 20, E54, 55, 57
Kresrovnikov, A.N.
    vol 25   E194, E195
Krestovikov, A.N.
    vol 29   E207, E208, 212
Kretschmer, C.B.
    vol 24   E331-E333, 335-337, 340, 350-352, 356, 363-365,
                    368
Krickova, J.
    vol 26   E153, E154, E157, E159, 175, 176

Kristensen, B.
    vol 25   E335-E337, 356
Kriven, W.M.
    vol 33   E332
Krucz, G.
    vol 26   E152
Kruger, B.
    vol 23   E18, E26
Kruger-Thiemer, E.
    vol 34   58, 67, 72, 183-185, 218, 227, 233, 265-267
    vol 35   E110-E112, 147, 148, 150, 223, 224, 226,
                    235-237, 243, 244, 245, 265-267, 283-285,
    vol 35   E289, 293-295, 304-306
    vol 36   11-13, 15-17, E19, E20, 34-36, E63, 71-73, 148, 152,
                    E228, 235-237
Kryukova, T.A.
    vol 26   E291, 292
Krzyzanowska,   T.
    vol 37   E13, E15, 21, E31-E34, 41, E42, E43, E46, 59, E65,
                    E68, E84, 173, E212, E213, 218, E222, E228, 265,
                    E283, E288, E289-E291, 295, E296, E299, 306, 312,
                    E315, E316, E322, 355, E370, E371, E379, 416, E445,
                    E447, 451, 466, 468, E469, E470, 475, 480, E481,
                    484, E485, E486, E491, 511
    vol 38   E9, E12, 20, E23, E28, 39, E42, E46, 56, E59, E60, E66,
                    84, 96, 99, E114, E117, 125, E126, E127, 130, 134, E135,
                    E141, E154, E169, E170, 174, E187, E189, 198, 215, 224,
                    E225-E227, 233, 235, 259, E271, 275
Kublik, Z.
    vol 25   E135, 137, E335-E337, 353, E357-E359
Kubo, K.
    vol 34   E13, E15, 42, 160, 200-203
    vol 35   E14, 18, 56, 59, 60, 102, 103, E110-E112,
                    135, 193, 217, 218, 247, 248
    vol 36   E19, E20, 31, 48, E88, 90, 91, E92, 94, 95
Kubota, H.
    vol 32   E166-E168, E170, E186, 200, 201, 228, 229
Kucharov, Kh.
    vol 30   78, 81
Kudchadker, A. P.
    vol 37   E64, E67, E76, E77, 129-131, E221, E223, E224,
                    E226, E227, 235, 236, E314, E316, E320, E321, 331,
                    332
Kudinov, I.B.
    vol 30   E374, E375, E379, E381, 404, E428-E431, 442, E467,
                    E468, E471, E472, 487, 488
Kudrina, L.
    vol 26   E293, 301, 302
Kudryavtsev, N.T.
    vol 23   E270, E272, E279, 295
Kudyrov, B.I.
    vol 30   E448-E450, 455, 456, E467, E468, E471, E472, 483
Kuhlmann, A.Z.
    vol 23   E10, E25, 27, E84, E96, 101, E158, E182, 185, E270, E281,
                    283
Kuhnert-Brandstatter, M.
    vol 34   293, 294
    vol 35   106, 107, E110-E112, 125, 156, 160, 188
Kuleshov, V.A.
    vol 25   E1-E3, 10
Kuliev, A.A.
    vol 25   E194, E195
    vol 29   E213, E214, 224
Kulikow, E.
    vol 23   E178, E184, 257
Kumar, K.
    vol 26   E187, E188, E192, E193, 217
Kunanbaev, G.S.
    vol 31   E39-E42, 73, 74
Kunanbaeva, G.S.
    vol 31   E11, E13-E15, 23, E49, E51, E98, E102, 118
Kunde, V.G
    vol 27/28   E50, E51

Kunev, D.K.
     vol 26   E403-E405, 411-413
Kuntz, R.R.
     vol 29   E1, E3, E14, E16-E18, E21, 31, E102-E109, 110
                   112, 116, 119, 120, 122, 132, E163-E165, 167
                   E177, E179, 180
Kuntze, O.
     vol 31   E206, E208, E212, 220
Kuptsova, N.V.
     vol 23   E313, E317, 333
Kurata, F.
     vol 27/28   E750, 754, 755
Kurbanbekov, A.
     vol 22   101, 175, 217, 254, 292, 350, 364
Kurbanov, A.A.
     vol 25   E194, E195
Kurnakov, N.S.
     vol 25   E13-E18, 24, E33, E34, 36, E44, E45, 47, E50,
                   E52, E119-E121
Kuroda, T.
     vol 20   E183, E184, E190, E195, E196
Kurota, O.
     vol 26   E187, E191, E193, 223
Kushrov, Kh.
     vol 30   E24, E33
Kusma, I.B.
     vol 25   E206
Kutna, I.M.
     vol 34   E13, E15, 37, 38, 134, 142, 143, 148, E204, 210,
                   211, 251
     vol 35   E110-E112, 129, 183-185, 198
     vol 36   345
Kutnetsova, Z.M.
     vol 30   E461, E462, 463, 464
Kuwahara, M.
     vol 20   E6, E&, 11-14
Kuz´menko, A.L.
     vol 30   E380, E381, 409, 410
Kuz´minykh, I.N.
     vol 26   E155-E157, E159, 180, 181, E187-E190, E193, 196,
                   197, E271-E274, 278-280
Kuzina, V.A.
     vol 30   E24, E25, E28, E29, E31, E32, 58, 71, 93, E105, E108-E111,
                   136, 139, 140, E162, E164, E165, 169, 170, 174, 175,
                   E268, E270-E274, 282-284, 287, 290, 298-306, E374,
                   E378, E428-E431, E467, E468, E471, E472, 483
Kuzmany, P.
     vol 30   254
Kuzmina, G.A.
     vol 25   E194, E195
Kuznetsov, A.I.
     vol 21   E2, E4-E8, E10-E12, 21, 38, 45, 53, 75
Kuznetsov, D.I.
     vol 31   E206, E209, E212, 236
Kuznetsov, F.A.
     vol 25   E230, E231, E234
Kuznetsov, V.G.
     vol 31   E206-E209, E212, 230, 231
Kuznetsov, V.N.
     vol 23   E83, E86, E89-E91, E96, 135-1737
Kuznetsova, A.G.
     vol 26   E1, E3-E7, 18, 29-31, 33, 39-41, 63-65, E72-E75
                   78-81, E90-E92, 104-107, E271-E274, 278-280
Kuznetsova, G.P.
     vol 22   112, 113, 196, 304, 305
     vol 31   E307, E308, 312, 314, E314, E320, 325, 327
Kuznetsova, N.K.
     vol 25   E1-E3, E13-E18, E33, E34, E44, E45, E50, E51
Kvyatkovskii, A.N.
     vol 23   E10, E24, E25
Kwiecinska, A.
     vol 23   E24, E26, E179, E184, 259

L

Labash, J.A.
    vol 26   E1-E6, 36, 37, 46, 47, 68-70, E113-E116, 138, 139
Labendzinski, St.
    vol 23   E308, E317
Labrocca, P.J.
    vol 18   27-29
Lacey, W.N.
    vol 24   E48, 53, 216, 217, E251, 253, 254, E422, 424-429
    vol 27/28   E266, 268-271, E281, 283-287, E297, 298, 299,
                  E303, 305-307, 318, E320, 321, E338, 339-341,
                  E360, 361-366, 383, 384, 390-393, E465, 467,
                  469-471, E511, 513, 557, 558, 571, E750, 751,
    vol 32   E166-E168, E186, 196, 197, 204, 205,
Lach, J.L.
    vol 35   E110-E112, 124, 175
    vol 36   E124, 130, 181, 184
Ladeinova, L.V.
    vol 23   E158, E180, E182, E184, 237-241, E270, E279, E281, 291,
                  292
Laffitte, M.
    vol 33   E145, 147, E150
Lajtai, I.
    vol 32   E1-E3, E20, 22-24
Lal, D.
    vol 32   E79-E83, 107, 152-154
Lamb, A.B.
    vol 30   E470, E472, 479, 480
Lamoreaux, R.H.
    vol 25   E282
Lamure, J.
    vol 23   E23, E26
Landsberg, R.
    vol 23   E161, E182, 244
Lane, W.H.
    vol 38   E1, E2, 3, 4
Lange, A.A.
    vol 25   E13-E18, 31, E76, E77, 81, E173, E174, 175,
                  E285, E286, 294, 296, E335-E337, 352, E402-E404, 420
Lange, N.A.
    vol 31   E1, E6
Langecker, H.
    vol 34   51, 74, E204, 206, 216, 234
    vol 35   E110-E112, 121, 155, E289, 291, 292, 296, 303
    vol 36   E19, E20, 28, 38, E63, 69, E124, 128, 145, 157, 231,
    vol 36   E243, 247, 253, 450-452, 462, 475
Lannung, A.
    vol 27/28   E1-E5, 13, E202-E205, E214, 218, E450, E454, 455,
                  E493, E494, E497, 500, E511, 513, E582-E586,
                  E595, 598, 603, 637, 639, 655, 657
Lapan, A.A.
    vol 23   E20, E26
Larin, G. M.
    vol 37   E221, E225, E227, 239, E314, E318, E322, 334, E485,
                  E488, E490, 500
Larionova, I.S.
    vol 30   E220, E222, E223, 240, 241
Larsen, E.R.
    vol 27/28   E24-E28
Larson, W.D.
    vol 30   E428, E429, E431, 435, 436
Latimer, W.M.
    vol 31   E1, E6,
Latypov, Z.M.
    vol 25   E194, E195
Laube, E.
    vol 25   E206
Laudise, R.A.
    vol 23   E181, E184, 246
Laue, E.
    vol 23   E84, E86-E91, E93, E96, 112-115
Lauffenburger, R.
    vol 31   E1, E3, E4, E6, 9, 10, E101, E102, E307, E308,

Lavrentev, V.I.
    vol 25   E230.E234
Lavut, E.A.
    vol 26   E403-E405, 407-410, 417-419
Lawrence, G.J.
    vol 23   E178, E184
Lawrenson, I.J.
    vol 27/28   E24-E28
Lawson, G.J.
    vol 23   269
Lawson, J.D.
    vol 32   E79, E81-E83, 92, 93, 104-106, 129, 130, 149-151
Lawson, J.D.
    vol 27/28   184, 185
Lawson, W.P.
    vol 25   E194, E195
Lazarus, J.H.
    vol 35   E14, E15, 21, 23, 25, 34, 44
Leach, J.G.
    vol 23   E3
Leary, J.A.
    vol 25   E207, E208, 210, E214, E215, 216, E263, E264, 265, E271,
                 272, E274, 275, E282, 283, E301, E302, 305, E307, 308,
                 E434, 435
Lebeau, P.
    vol 24   E1, E2, 15, E16, E17, 33, E34, 38, E218-E221, 228,
                 E257, E258, 259, 269, 277, E331-E333, 338, 353,
                 366, E375-E378, 386, 398, 406, E414, 415
Lebedeva, E.N.
    vol 22   136, 179, 220, 238, 258, 277, 296, 317, 340, 352
Lebel, H.
    vol 36   E19, E20, 22, E63, 65
LeBreton, J.G.
    vol 24   E48, 49-51
Lee, H.C.
    vol 25   E425-E427
Lee, J.I.
    vol 32   E1-E3, 14-18, E79-E83, 90, 91, 98-103, 116, 126-
                 128, 131-135, 141-148
Lee, T.S.
    vol 25   E425-E427, 433
Lee, Y.S.
    vol 25   E425-E427
Legler, E.
    vol 25   E173, E174, 177
Legret, D.
    vol 27/28   E465, E511, 521, E522, 532, E533, 535, 537
Lehoczky, S.L.
    vol 25   E194, E195, 203
Lehrman, A.
    vol 33   E26, 27, 72, E91, 92
Leibush, A.G.
    vol 32   E79, E81, E83, 86, 87, 95, 96, 118, 140
Leifer, E.
    vol 33   72
Leinonen, P. J.
    vol 37   E64, E68, E83, 160, E221, E223, E224, E228, 254,
                 E275, E276, 280, E296-E299, 303, E314, E322, 344
Leipper, A.
    vol 31   E11-E15, 30, E127, E130, 157, 158
Leipziger, S.
    vol 27/28   382, 578, 579
Lel´kin, K.P.
    vol 26   E1-E6, 15
Leland, H.L.
    vol 23   E148-E150, 153, 154
Lemaire, H.P.
    vol 23   E20, E26, E92, E97, 123
Lemley, J.
    vol 23   E313, E317, 328, 329
Lenoir, J-Y.
    vol 24   E130-E134, 158, 236, 244, E299-E301, 306, 307,
                 309, 311, 314, 320, 321, 323, 328, E331-E333, 339,
                 345, 346, 360, 361, 367, 371, 372, E375-E377, 380,

Lenoir, J-Y.
      vol 24   345, 346, 360, 361, 367, 371, 372, E375-E377, 380,
                      385, 388, 392, 395, 402, 405
      vol 27/28   E202, E203, E214, 236, E493, E497, 510, E629,
                      E641, E642, E644, 672, 676-678, E702, 706, 709,
                      715, 722, 727, 729
      vol 32   E167-E169, E172, E175, E176, E178, E180, E182-
                      E186, 210, 222, 245, 246, 250, 267, 275, 291,
                      305, 315, 316, 319, 322, 326
Lentz, H.
      vol 27/28   E629, 632-634
Leonesi, D.
      vol 33   3-8, E12, 13, E15, 17, 18, 22-24, E58-E60, 136,
                      138, 144, 148, E150, 152, 166, 171, 218, 220, 228,
                      234, E237, 258, 263
Lepeshkov, I.
      vol 30   E105, E108, E109-E112, 134, 135, 137, 138, E268,
                      E270-E273, E275, 280, 313, 321, 322, 325-327, E374,
                      E375, E378-E381, 399-402, 406, E428-E431, 437,
                      E461, E462, E467, E468, E471, E472, 481, 482, 484, 485
Lepeshkov, I.N.
      vol 31   E98, E101, E102, 122, 123, E308, E311
Lepilina, R.G.
      vol 23   E177, E184, 258
Leschinskaya, Z.I.
      vol 26   E348-E352, 361, 363, 381
Lessels, R.W.M.
      vol 21   E281, E282, 283-287, 289-295, 300
Lesteva, T. M.
      vol 37   E65, E67, E71, E84, 167, E197-E199, 207, E221,
                      E223-E225, E228, 258, E275, E276, 281, E314, E317,
                      E318, E322, 347, E485, E486, E488-E490, 508
      vol 38   E114, E115, E117, 123, E135, E136, E141, 150
Levelt Sengers, J.M.H.
      vol 29   E190-E192, 193, 194, 196-199
Levina, M.G.
      vol 23   E277, E282
Levinton, L.Ya
      vol 27/28   44, E629, 631
Levitan, N.I.
      vol 34   E13, E14, 19
      vol 35   E110, E111, 117, 213
      vol 36   E19, E20, 26, 57, E124, 126, 225
Levitskaya, E.P.
      vol 27/28   E248, E249
Levitskaya, S.A.
      vol 25   E173, E174, 180, E335-E337, 347
Levitskaya, T.D.
      vol 25   E194, E195
      vol 29   E207, E208, 212
Levy, E.M.
      vol 21   E83, E84, E89, 179, 180
Lewis, B.
      vol 25   E194, E195, 198
Lewis, G.G.
      vol 31   E11, E13-E15, 22
Lewis, J.B.
      vol 25   E139-E141, 150, E385-E387, 394
      vol 26   E293, E294
Lewis, N.B.
      vol 26   E1-E4, E6, 19-26
Leya, S.
      vol 34   E13, E14, 23, 50, 107, 108, 149, 297-301, E302,
      vol 34   305, 307, 309, 310, 314
      vol 35   E110, E111, 120, 144, 164, 178, 179
      vol 36   1-5
Leyko, J.
      vol 32   E54, E55, 59, 60
Li, Y.-H.
      vol 27/28   553, 555
Liabastre, A. A.
      vol 37   E2, E3, 6, E13-E15, 19, E42, E43, E46, 55, E64, E67, E83,
                      155, E193, E194, 196, E197-E199, 206, E221, E223, E224,
                      E228, 253, E369, E371-E374, E378, 405, E433, E434, 437

Liao, P.-K.
    vol 25   E194, E195
Lichtenthaler, R.N.
    vol 27/28   E370, 373-375
    vol 37   E315, E320, E321, E323, 357-362
Liebhafsky, H.A.
    vol 25   E84, E85, 89
Liebl, G.
    vol 25   E103-E105, E173, E174, E317, E318, E335-E337
Lightfoot, J.H.
    vol 27/28   E511, 514
Lihl, F.
    vol 25   E173, E174, 177, E285, E286, E302, E310, E311,
                  E317-E318, E335-E337
Likhol´ot, N.M.
    vol 34   E13-E15, 27, 33, 37, 38, 49, 53, 59-66, 68-71, 81,
                  85-90, 92-95, 97, 131, 134, 137, 142, 143, 148,
                  228-232, 237, 251, 253, 255-257, 264
    vol 35   E110-E112, 127, 129, 161, 183-185, 198
    vol 36   E303, 306, 307, 313-327, 336, 345
Lilich, L.S.
    vol 23   E18, E26
    vol 31   E11, E13-E15, 20, 21, E39-E42, E51, 69
Liminga, R.
    vol 30   E269, E275
Lin, H.M.
    vol 27/28   E360, 368, 369, E370, 372, E511, 520, E522, 530,
                  531, E533, 534, 543-546, 548-550, 561-563, 680,
                  710, 711, E737
Lin, I.J.
    vol 24   E61, E71, 80
Lin, J.-L.
    vol 26   E348-E351
Lin, L.J.
    vol 24   E16, E17, 19
Lin, O.H.
    vol 34   152, 158, 166
Lin, P.J.
    vol 24   E130-E134, 166
    vol 27/28   E202, E203, E209-E213, E215, 241, 673, 674
Lindauer, G.C.
    vol 25   E93-E95
Lindqvist, O.
    vol 26   E252
Lindsay, Jr.W.T.
    vol 31   E13-E15, 37, 38
Linford, R.G.
    vol 32   175, E187, 265
Linford, R.G.
    vol 24   E299-E302
Lingafelter, E.C.
    vol 32   E79, E83, 84, 85
Linke, W.F.
    vol 26   E90-E92
    vol 29   E228
    vol 30   E24, E25, E30-E32, 76, 77, E195-E197, 200, E330-E332,
                  E337, E338, 343
Lippold, B.H.
    vol 34   287-290
Lipshits, B.M.
    vol 23   E161, E182, 231-234
Lissant, K. J.
    vol 37   E63, E67, E82, 112, E197-E199, 200, E221, E227,
                  234, E314, E316, E321, 328
Litvin, B.N.
    vol 23   E180, E184
Litvin, V.P.
    vol 32   E54, 63, 64
Litvinenko, M.S.
    vol 32   E26-E28, 45, 47, 50
Ljungstroem, E.
    vol 26   E252
Lobachev, A.N.
    vol 23   E280, E282

Lombardi, R.B.
    vol 36   109-112
Lomov, A.L.
    vol 25   E230, E234
Long, F. A.
    vol 37   E63, E82, 115
Loomis, A.G.
    vol 25   E57, E58, 61, E369-E371
Lopez, M.C.
    vol 27/28   660, 695
Loprest, F.J.
    vol 30   E374, E375, E381, 403
Lorenzo, A.
    vol 27/28   E750, 756, 758, 759
Lott, W.A.
    vol 35   E110, E111, 113, 186, 207
Lotter, Yu. G.
    vol 37   E314, E320-E322, 348
    vol 38   E294-E296, 304
Lovchikov, V.S.
    vol 23   E161, E182, 231-234
Low, M.
    vol 20   E73-E77, 78, E89-E91, 92, E99, E100, 101, E103-E105,
                  107, 108
Lozhkina, J.G.
    vol 23   E270, E279, E281, 291, 292
Lu, B. C. Y.
    vol 27/28   E248, E249, 258, E266, 279, E477, 478, 481, 483,
                  484
    vol 37   E31-E34, 39, E42-E46, 56, E65, E67, E68, E71-E73,
                  E83, 161, E283, E284, 286, E289-E291, 293, E296-
                  E299, 304, E307, E308, 310, E314, E317-E319, E322,
                  345, E369, E372, E373, E375, E379, 407, 469, E470,
                  473, E481, 482, E485-E490, 507
    vol 38   E9-E12, 17, E23, E24, E27, E28, 35, E42, E43, E45, E46,
                  52, E59-E63, E66, 78, E114-E117, 122, E126, E127, 128,
                  E135, E136, E138, E139, E141, 149, E216, E217, 219
Luborsky, E.
    vol 25   E301, E302
Lucas, M.
    vol 27/28   E1-E5, E61, E62, E85, 98
Lucasiewicz, T.
    vol 30   E270, E272, E275, 319, E467, E472
Lugscheider, E.
    vol 25   E258, E259, E263, E264, E335-E337, 349
    vol 29   E213, E214, 217, 220, 225
Lugscheider, W.
    vol 25   E206, E214, E215, E221, E230, E231, E239, E242,
                  E245, E247, E249, E253, E267
Lui, Chin-sun
    vol 31   E13-E15, 37, 38
Luks, K.D.
    vol 27/28   333, E370, 377, 397-399, 412-426, 432-435,
                  441-446, 539-542
Lumme, P.
    vol 23   E275, E277, E281
Lusby, G.R.
    vol 26   E1-E6, 36, 37, 46, 47, 68-70, E113-116, 138, 139
Lutsik, V.P.
    vol 23   E156, E182
Lutsyk, A. I.
    vol 24   E1, E2, 14, E16, E17, 32, E34, 37, E59, E72, 75-77
    vol 27/28   E1-E5, E58, E59, E85, 88
    vol 37   E42, E43, E46, 60, E212, E213, 219, E222, E223,
                  E228, 266, E307, E308, 313, E445-E447, 453, E485,
                  E487, E491, 513
Lutz, H.D.
    vol 26   E153, E158, E187, E192, E248, E252, E256, E258,
                  E271, E273, E286
Lux, H.
    vol 29   E1, E3, E14, E16-E18, E21, 27
Lyalina, R.B.
    vol 30   E471, E472, 486

Lyon, A.M.
    vol 20   40, E132-E136, 139, 149, E153-E157, 162, 182
Lyubimov, E.I.
    vol 22   93, 261, 280, 299, 320, 343, 355, 371, 380

M

Maass, D.
    vol 32   E1-E3, 6, 7
Maass, O.
    vol 26   E187, E190, E193, 203, 204
MacDonald, D.D.
    vol 23   E88, E96
Machala, J.
    vol 26   E1-E3, E5, E6, 42, 43, E72-E74, 82, 83
Mackay, D.
    vol 20   E103-E105, 113, E153-E157, 180
    vol 37   E64, E65, E68, E83, E84, 160, 165, E188, 190, E221,
                  E223, E224, E228, 254, 256, 257, E275, E276, 280,
                  E296-E299, 303, E314-E317, E322, 344, 346, 356,
                  E366, 368, E370, E371, E379, 410, E425, 427
    vol 38   E161, 163, E236-E238, 246, E294, E296, 303, E311, E312,
                  315, E319, 321, E332-E334, 337, E341-E343, 348, E352,
                  355, 356, E357, 359, E360, 362, E363, 364, E366, E367,
                  368, E386, 388, 392, E396-E398, 402, E419, 423, E426,
                  E427, 4341, E433, 434, 437, E441, 443, E446-E448, 452,
                  461, 475, 476, E477, E478, 481, E486, E487, 490, E493,
                  496, E497, 500, E512, 514, 520, 522, E528, 530, 531,
                  543
Macrygeorgos, C.A.
    vol 27/28   E750, 756
Mader, W.J.
    vol 35   91, 100
Maeda, A.
    vol 27/28   E737, 740
Maehara, S.
    vol 29   E1, E5, E7, E14, E16-E18, E21, 40, E47, E49-E51
                  72-86, E87, 88-94
Maey, E.
    vol 25   E1-E3, E13-E18, 19, E33, E34, E357-E359
Magel, T.T.
    vol 25   E425-E427
Mahapatra, P.
    vol 23   E315, E317, 339
Maier, C.G.
    vol 23   E152, E182
Maier, S.
    vol 27/28   E465, 474
Maijs, L.
    vol 23   E16, E26, E90, E96, E272, E281, E309, E317
Mains, G.J.
    vol 29   E1, E3, E14, E16-E18, E21, 31, E102-E109, 110
                  112, 116, 119, 120, 122, 132, E163-E165, 167, E177
                  E179, 180
Makarov, S.Z.
    vol 23   237-241
Makarova, I.A.
    vol 25   E76, E77, 81
Makarova, J.A.
    vol 25   E13-E18, 31
Makin, A.V.
    vol 31   E98, E99, E101, E102, 110-112, 121-123
Makovskaya, G.V.
    vol 23   E13, E16, E18, E19, E25, 79
Makranczy, J.
    vol 21   E1-E3, E12, 14, 22, 30, 33, E89
    vol 32   E166-E168, E170, E172, E182, E185, E186, 188, 189,
    vol 27/28   E202-E209, E215, 216, 225, 230, 233-235, E582-E588,
                  E590-E595, 600, 605, 609, 613, 618, 623, 626, 628,
                  667-671
    vol 32   226, 233, 286, 293, 295
Makrides, A.C.
    vol 25   E326, 328, E330, 333

Maksimov, B.A.
    vol 23   E280, E282
Maladkar, V.K.
    vol 21   E5-E9, E12, 46, 48, 54, 59-63, 69
Malan, H.P.
    vol 25   E425-E427
Maleczkine, S.M.
    vol 27/28   667-671
Maleschenko, K.F.
    vol 20   E35, 37
Malesh, W.
    vol 34   235, 238, 268, 269
    vol 35   72, 76, 96, 99, 268, 273, 286, 288, E289,
                 297, 300, 309
    vol 36   105, 108, 120, 121, 474, 477-479, 481, 482
Malik, S.K.
    vol 24   E1, E2, 10, E16, E17, 28, E34, 35, E61, E71,
                 81-83, 122-124
    vol 27/28   E1-E6, 10, E64, E85, 103, 177
Malinovskii, Ya.A.
    vol 23   E174, E176, E180, E183, E184
Maltsev, Yu.T.
    vol 25   E385-E387
Malyshev, A.A.
    vol 30   E380, E381, 409, 410
Malysheva, T.V.
    vol 32   E20, 21, E54, 58, 78
Mamonova, G.
    vol 26   E348-E352
Mamutova, Z.A.
    vol 25   E214, E215
Manakan, D. A.
    vol 37   E197-E199, 201
Manes, M.
    vol 20   E35, 39, E73-E77, 88, E89-E91, 98
Mangan, J.L.
    vol 26   E293, 299, 300
Manning, W.P.
    vol 32   E1-E3, 19, E172, E173, E175, E184, E185, E187, 249, 304,
                 321
Manvelyan, M.G.
    vol 31   E98, E102, 114, E129, E130, 149, 165
Manzhelii, V.G.
    vol 27/28   E582, E584, E586-E588, E595, 607
Maraman, W.J.
    vol 25   E263, E264, 265
Marchenkov, V.F.
    vol 21   E7, E12, 51, 52
Marchenkova, R.G.
    vol 21   E7, E12, 51, 52
Marcus, Y.
    vol 23   E272, E281
Mardakhaev, B.N.
    vol 26   E305, 320, 321
Marecek, J.
    vol 26   E240, E241, 247
Margulis, E.V.
    vol 26   E153, E158, 164, E187, E188, E193, 195, E236, 238
                 E240, E241, 243, E248, E249, 250, E252, 253,
                 E256, 257, E258, 259, E260, 261, E262, 266, E267
                 270, E271, E272, E274, 284, 285, E286, 287-290, E293, 296
Mark, C.B.
    vol 20   E73-E77, 82, E132-E136, 144, E153-E157, 172
Markant, H.P.
    vol 26   E153-E155, E158, 162, 163, 167
Markarov, S.Z.
    vol 23   E10, E24, E25, 54, 55, E158, E180, E182, E184
Markovic, M.
    vol 23   E311, E312, E316, E317
Marsh, K.N.
    vol 31   E2, E6
Marshall, A.L.
    vol 25   E301, E302, 303

Marshall, W.L.
    vol23   E21, E23, E26
    vol 31   E11, E13-E15, 24, 25, E171, 172, 173, 182, 183
Marson, H.W.
    vol 35   1, 12, 13, E14, 16, 104, 107, 109, 316
    vol 36   97, 299, 301, 443
Martin, J.L.
    vol 32   E80, E82, E83, 108
Marten, E.A.
    vol 23   E3
Martin, A.R.
    vol 24   E218-E221
    vol 35   142, 222
    vol 36   32, 70, 139, 232, 346, 358, E367, 380, 381, 386, 398,
                   401
Martinek, A.
    vol 35   106, 107, E110-E112, 125, 156, 160, 188
Martinson, I.G.
    vol 23   E277, E279, E282, 294, 304
Martynova, O.I.
    vol 23   E23, E26
Marusawa, T.
    vol 26   E153-E155, E158, 160, E187-E193, 198, 213
Marvel, C.S.
    vol 21   E80, E82, E89, 102, 151, E252, 267, 274-278
Maslennikova, V.Ya.
    vol 24   E48
Mason, J.T.
    vol 25   E425-E427, 433
Massaldi, H.A.
    vol 38   E277, E278, 283, 286
Masson, D.R.
    vol 25   E194, E195
Masson, J.
    vol 23   E92, E97
Masson, M.R.
    vol 26   E404, E405, 415, 416
Massoth, F.E.
    vol 27/28   760
Masterton, W. L.
    vol 37   E64, E71-E73, E83, 145
Masuoka, H.
    vol 27/28   222, 659, 688, 690
    vol 32   E171, E172, E186, 239, 240
Matheson, I.B.C.
    vol 27/28   E1-E5, 189
Mathewson, C.H.
    vol 29   E213, E214, 222, 226
Mather, A.E.
    vol 32   E1-E3, 14-18, E79-E83, 90, 91, 94, 98-103, 107-111,
                   113-116, 126-128, 131-139, 141-148, 152-154, 160-165,
                   E176, E177, E187, 268
Mathieu, M.P.
    vol 24   E218-E221, 231, 240, E414, 416, 417
    vol 27/28   575
    vol 32   E170, E186, 215
Mathis, H.B.
    vol 25   E314
Mathur, R.P.P.
    vol 23   E84, E85, E94, E96, 102, 103
Matous, J.
    vol 21   E82, E89, 142-147
Matroff, G.
    vol26   E248, E249
Matsumara, H.
    vol 34   E13, E14, 25, 52, 57, 76, 82, 96
    vol 35   69-71, 77, 78
    vol 36   143, 144, 151, 166, 168, 372-374, 378, 379
Matsuno, T.
    vol 26   E187, E192
Matsyk, L.V.
    vol 26   E74, E75, 86, 89
Matthes, F.
    vol 25   E207-E208, E214, E221, E225

Matthews, F.W.
    vol 26   E187, E192
Matvienko, V.G.
    vol 32   E180, E187, 302-311
Maude, A.H.
    vol 27/28   569
Mauger, J.W.
    vol 35   E64, 81, 82, 84-88
    vol 36   186-191, 193-200, E367, 382-385, 388-397, 412-425
Maurer, R.J.
    vol 25   E357-E359, 366
Maxwell, J.
    vol 25   E285, E286, E335-E337
May, L.
    vol 23   E170, E183
May, W. E.
    vol 37   E65, E68, E84, 174, E188, 192
    vol 38   E236-E238, 249, E250, 257, E386, 389, 390, E396-E398,
                  404, E405, 407, E419, 422, E433, 435, 436, 438, E441,
                  444, 445, E446-E448, 454, E456, 459, E477, E478, 482,
                  E483, 485, E486, E487, 491, 492
Mayer, S.W.
    vol 29   E207, E208, 211
Mayer, W.
    vol 31   E1, E3, E6, 8
Mayfield, F.D.
    vol 27/28   E320, 322, 323, E465, 468, E511, 514
Mayforth, F.R.
    vol 24   E299-E301, 304
Mazukami, S.
    vol 35   94, 95
Mazzetti, C.
    vol 30   E105, E111
McAmis, A.J.
    vol 32   E330, E331, 336, 339
McAuliffe, C.
    vol 24   E1, E2, 11, E16, E17, 29, E34, 36
    vol 27/28   E1-E5
    vol 37   E2, E3, 4, E7, E8, 9, 11, 12, E13-E15, 18, 27-30,
                  E31, E32, E34, 37, E42, E43, E46, 53, E64, E68,
                  E82, E83, 135, E193, E194, 195, E197-E199, 204,
                  210, 211, E212, E213, 216, E221, E223, E224, E227,
                  241, 273, 274, E275, E276, 278, E283, E284, 285,
                  E296-E299, 302, E307, E308, 309, E314, E322, 336,
                  E369, E372, E378, 395, 401, E433, E434, 436,
                  438-442, E445-E447, 450, 454, E460, 461, E469,
                  E470, 471, 472, E485-E487, E490, 503
    vol 38   E9, E11, E12, 15, E59, E60, E66, 74, 77, 92, 94, 100,
                  110, 111, E114, E115, E117, 121, E135, E136, E141, 146,
                  E169, E170, 171, E187-E189, 193, 195, 211, 212, E216,
                  E217, 218, E225, E226, 229, E294, E296, 301, E326, E327,
                  329
McBain, J.W.
    vol 25   E369-E371, 376, 377
McCall, M.T.
    vol 26   E187, E188, E192
McCants, J. F.
    vol 37   E63, E71, E82, 117, E314, E316, E318, E321, 329,
                  330, E485, E486, E490, 496
McCarty, C.N.
    vol 22   86, 87, 173, 214, 215, 252, 253
McCoubrey, J.C.
    vol 21   E281, E282, 283-287, 289-295, 300
McCoy, H.N.
    vol 25   E234
McCrosky, C.R.
    vol 26   E348-E351, 365, 366
    vol 30   E258, 260, E263, E264, 265, 266
McCurdy, K.G.
    vol 23   E165, E168-E173, E183, 266, 267
McDaniel, A.S.
    vol 27/28   E202-E205, E214, 217, 224, E491, 492, E493, E494,
                  E497, 498, 502, 505, E582-E586, E588-E591, E595, 596

McDevit, W. F.
    vol 37   E62, E82, 115
McDonald, A.
    vol 34   E13, E14, 22, 126
    vol 35   E110, E111, 119, 174
McDowell, L.A.
    vol 23   E10, E13-E15, E17, E18, E25, 46-48, E161, E182, E310, E317
McGee, K.A.
    vol 27/28   E74, E85, 601, 610, 615
McHale, J.L.
    vol 27/28   E724, 725
McIlroy, R.A.
    vol 26   E158, E159, 186
McIntosh, A.O.
    vol 26   E187, E192
McKetta, J. J.
    vol 24   E39, 40-45, E48, 49-51
    vol27/28   E24-E28, 30-32, 196-199, E281, 288-290, E320,
                    327, 328, 427-431, 436-438
    vol 37   E64, E67, E76, E77, 129-131, E221, E223, E224,
                    E226, E227, 235, 236, E314, E316, E320, E321,
                    331, 332
McLauchlan, W.H.
    vol 32   E1-E3, 4, E25, E28, 39
McMillan, J.A.
    vol 23   E144, E145
McNeil, K.M.
    vol 32   E79, E82
Mebane, W.M.
    vol 26   E187-E190, E193, 229, 230
Meerburg, P.A.
    vol 30   E330, E332, E335, E338, 371, E374, E378, E380, 407
                    , E461, E462, 465, E467, E468, E470
Mees, G.
    vol 25   E369-E371, 382
Megyery-Balog, K.
    vol 27/28   E202-E209, E215, 216, 225, 230, 233-235, E582-E588,
                    E590-E595, 600, 605, 609, 613, 618, 623, 626, 628
    vol 32   E166-E168, E185, 188, 189
Mehl, R.F.
    vol 25   E402-E404, 416
Mehra, M.C.
    vol 26   E348-E352, 382-387, E404, E405, 425
Mehrotra, A.H.
    vol 27/28   576, 577
Mehta, S.C.
    vol 35   E110-E112, 133, 170, 190
Meier, R.
    vol 36   330, 360
Meisel, T.
    vol 33   E58-E60
Mel´nikov, O.K.
    vol 23   E174, E176, E183
Melbye, G.S.
    vol 23   E11, E25, 34
Mellor, J.W.
    vol 26   E90-E92, E113, E116
    vol 30   E105, E111
Menzel, H.
    vol 31   E11, E12, E14, E15, 28, 29, E98, E99, E102, 106,
                    107, E206, E207, E212, 214, 225
Menzies, A.W.
    vol 31   E98, E102
Menzinger, F.
    vol 25   E93, E95, 101
Merbach, A.
    vol 22   21, E22, 24, 26, 77, 79, 84, E152, 155, E156, 159,
                    167, 209, 210, 212, 246, E247, 248, 251, 327, E328,
                    330, 332, E336, 359, 360, 362
Merilainen, P.
    vol 23   E89, E96, 127, 128
Merlo, F.
    vol 25   E66, E67, 70.E71, E72, E76, E77, 80, E207, E208,
                    212, E214, E215, E221, E225, E230, E231, E239, E245,

Merlo, F.
    vol 25   E253, E425-E427
Merriam, M.F.
    vol 25   E103-E105, 114
Merrill, R.C.
    vol 27/28   333, 412-417, 420-426, 432-435, 441-443
Mertins, CH.
    vol 26   E248, E271, E273
Mertke, I.
    vol 25   E230
Merz, V.
    vol 25   E13-E18
Merzon, V.V.
    vol 26   E404, E405, 420, 427
Meshali, M.
    vol 35   E14, 20, 30, 31, 37-42
Mesmer, R.E.
    vol 31   E171, 172, 173
Mesorama, J.M.P.
    vol 26   E187, E192, E193, E236, E240, E241
Messing, A.F.
    vol 25   E225, 226, E230, 232, E239, 240, E263, E264, 282,
                  E307, E326, E422, 423, E425-E427, 430,
Messow, U.
    vol 21   E2, E12, 18, 19
Metcalf, A.D.
    vol 20   E183, E184, 188, E190, 194, E195, E196, 200
Meteva, K.B.
    vol 31   E308, E311, E318, E320, E323, E324, 329-333
Metzer, A.
    vol 24   E61, E69, E71, E72, 113
von Meyenburg, H.
    vol 36   330, 360
Meyer, E.J.E.
    vol 34   84
Meyer, M.
    vol27/28   732
Mezosi, J.
    vol 36   311
Michels, A.
    vol27/28   E24-E28, 29, E57, E66-E69, E71, E72, E74-E76, E79,
                  E80, E84, 107, 114, 117-119, 148, 150, 162, 172,
                  181, 183, E248, E249
Michels, H.J.
    vol 33   E119, E121, E123, E125, E226, E248, E250, E278,
                  E292, E303, E308, E311, E313, E315, E317, E319,
                  E321
Micka, Z.
    vol26   E348-E351, 376
Mieli, A.
    vol 31   E11, E14, E15, 16, E168, E171, 174
Mikadze, I.I.
    vol 35   208-212
    vol 36   117-119, 347, 349-351, 428-433
Mikhailova, E.M.
    vol 30   E268, E270, E273, 286, 316, E374, E378, E381,
                  E467, E468, E472
Mikhailova, M.N.
    vol 31   E127, E130, 134, 135
Mikhalev, V.A.
    vol 36   328, 333
Mikheev, N.B.
    vol 22   92, 106, 138, 181, 222, 240, 260, 279, 298, 319,
                  342, 354, 370, 379
Mikuloski, T.
    vol 23   E24, E26, E179, E184, 259
Milazzo, C.
    vol 31   E2, E6
Miller, J.
    vol 20   E73-E77, 86, E89-E91, 96, E153-E157, 181
Miller, K.W.
    vol 24   E218-E221, 224, E414, 418, 419, 420
    vol 27/28   730

Miller, L.A.
      vol26   E152
Miller, R.C.
      vol27/28   E248, E249, 263, E477, 481, 482
Milligan, L. H.
      vol 37   E63, E66, E81, 92, E314, E315, E321, E485, 490
Milosovich, G.
      vol 35   168
Milyutina, N.A.
      vol 23   E181, E184, 252, E280, E282, 293, E313, E317, 332
Minardi, A.
      vol 25   E301, E302, 304, E310
Minchenko, S.P.
      vol 33   140, E141, E142, 143, 222
Minder, W.
      vol 23   E16, E17, E25, 56, 57, E157, E171, E182
Miner, W.N.
      vol 25   E434
Mingoia, Q.
      vol 34   112, 113, 115, 153, 154, 157
Minich, M.A.
      vol 33   149, E172, 174, 181, E185, 186, E189, 190, E193,
                   194, E198, 198, 202-205
Minitskii, P.L.
      vol 30   E268-E270, E273, E279, 285, 291, 317, 318, E330,
                   E332, E335, E338, E448-E450, E467, E468, E472
Mioduski, T.
      vol 30   E163, E165
Miralles, M.J.
      vol 36   E367, 380, 381, 386, 398, 401
Miravitlles, M.L.
      vol 30   23, 100, 194, 329
Mirenkova, T.F.
      vol 31   E168, E171, 205, E206, E212, 219
Mirlina, S.Ya
      vol 23   E162, E183, 211
Mirnaya, T.A.
      vol 33   E108, E109, 112, E113, 115, 156, 246, E289, 291
Mirnyi, V.N.
      vol 33   E108, E109, 112, E113, 115, 156, 246, E289, 291
Miroshina, O.S.
      vol 31   E47, E49, E51, E52, 77, 78
Mirsaidov, U.
      vol 22   101, 175, 217, 254, 292, 350, 364
Miseta, M.
      vol 35   E110-E112, 141, 203-205
Mishnina, T.A.
      vol 27/28   E1-E5, E70, E71, E74, E76, E77, E84, 124, 152
Mishra, K.K.
      vol 23   E156, E182
Mitchell, A.J.
      vol 26   E248, E249
Mitchell, S.
      vol 38   E236, E238, 240
Mitniskii, P.L
      vol 30   E268-E270, E273, 279, 285, 291, 317, 318, E330, E332,
                   E335, E338, E448-E450, 458-460, E462
Miyamoto, H.
      vol 30   E330, E333, E339, E374, E375, E382, 414-416,
                   E428, E429, E431, 444-446, E448-E450, 458-460,
                   E462, E467, E473
Miyano, Y.
      vol 27/28   688, 690
Mizetskaya, I.P.
      vol 25   E194, E195
Mochalov, K.I.
      vol 26   E1, E5, E6, 52
Modak, S.
      vol 34   180
      vol 35   215
      vol 36   58, 208-216, 227, 268, 357, 402
Moeller, T.
      vol 22   96, 145, 192, 194, 195, 197, 200-202, 229, 269
      vol 23   E271, E272, E281, 289

```
Moeller, T.
      vol 31        E3, E6
Moers, K.
      vol 25        E300
Moesveld, A.L.T.
      vol 25        E402-E404, 415
Moffatt, W.G.
      vol 25        E83
Mohanlal, K.
      vol 23        E10, E25, E83, E96, E157, E182
Moir, J.
      vol 23        E163-E165, E167, E183, 265
Moisa, L.P.
      vol 23        E271, E281
Moisiejeva, N.B.
      vol 23        E162, E183, 211
Moles, E.
      vol 30        E472, 477
Momtaz, R.Sh.
      vol 31        E307, E308, E310, 315-317, E320, E324
Monaghan, P.R.
      vol 32        E79, E81-E83, 119-121
Monfort, J.P.
      vol 24        E130-E134, 151, 175, 177
Monina, S.S.
      vol 26        E1, E5, E6.52
Monk, C.B.
      vol 20        E210-E213, 217
      vol 31        E2, E6
      vol 37        E64, E66, E82, 134, E369, E378, 394
      vol 38        E59, E62, E66, 73, E277, E278, 281
Monro, D. C.
      vol 37        E64, E76, E83, 157, E369, E371, E372, E378, 406
Monroe, G.S.
      vol 27/28     E511, 515, 516
Moore, B.C.
      vol 37        E63, E71-E74, E82, 113
Moore, T.F.
      vol 32        E79-E83, 155-159
Moraht, H.
      vol 26        E152
Moran, W.G.
      vol 25        E103-E105
Morawietz, W.
      vol 25        E33, E34, E103-E105, 116
Morey, G.W.
      vol 31        46
Morgen, L.T.
      vol 33        E35, 38, E46, 48, 133
Morishita, T.
      vol 34        E14, E15, 42, 160, 200-203
      vol 35        E14, 18, 56, 59, 60, 102, 103, E110-E112,
                    135, 193, 217, 218, 247, 248
      vol 36        E19, E20, 31, 48, E88, 90, 91, E92, 94
Morisue, T.
      vol 21        E3, E10, E12, 26, 29, 31, 32, E89
Morozova, V.E.
      vol 25        E335-E337, 350
Morrison, B.H.
      vol 25        E425-E427, 429
Morrison, D
      vol 26        E260
Morrison, T. J.
      vol 24        E1, E2, 7, E16, E17, 21, E59, E64-E71, 74, 88,
                    95-99, 109, 115, 117-119
      vol 27/28     E1-E5, 11, E68, E69, E72, E81, E84, 115, 121, 167,
      vol 37        E63, E68, E82, 116, E369, E372, E378, 390
      vol 38        E59, E60, E66, 72
Mortimer, G.A.
      vol 24        E218-E221, 230
Morvay, J.
      vol 36        311
Moser, H.
      vol 25        E189-E121
```

Moser, H.C.
    vol 29   E1-E3, E7, E21, 29, E102-E105, E109, 114, 115
                    E133-E134, E137, 138, 139, E149, E150, E153, 154
                    157, E163-E165, E177-E179, 182, E187, 188
Moshkevich, A.S.
    vol 25   E157-E159, 169
Mosso, U.
    vol 20   E116, 117
Moudgil, B.M.
    vol 24   E16, E17, 19
    vol 27/28   E1-E5, 18
Moule, D. C.
    vol 37   E64, E71, E72, E83, 146
Movlanov, Sh.
    vol 29   E213, E214, 224
Moy, J. A. E.
    vol 37   E63, E71-E73, E81
Mraw, S.C.
    vol 27/28   745, 746
Mraz, R.
    vol 31   E51, E52, E206, E207, E211, E212, 251, 252, E278,
                    287, 288
Muccitelli, J.A.
    vol 27/28   E1-E5, 21, E57, E58, E63, E64, E85, 86, 102, 186, 187
Mueller, J.
    vol 23   E11, E25
Mueller, R.
    vol 22   105, 110, 111
Muhlbauer, H.G.
    vol 32   E79, E81-E83, 119-121
Mukhamedieva, Sh.M.
    vol 25   E214, E215, 217
Mukimova, D.S.
    vol 29   E47, E51, 56
Mulder, G.J.
    vol 30   E105, E111
    vol 31   E11, E15, E98, E102, E127, E130
Muller, A.
    vol 31   E98, E102
Muller, E.
    vol 23   E11-E13, E18, E25, 35-38, E159, E161, E163, E182, 201, 202
Muller, J.
    vol 23   E159, E161, E163, E182, 201, 202
Muller, L.
    vol 23   E161, E182, 244
Muller, R.
    vol 25   E57, E58, E84, E85
Muller-Clemm, H.
    vol 26   E187-E190, E193, 199, 200
Mulley, R.A.
    vol 20   E183, E184, 188, E190, 194, E195, E196, 200
Mullins, L.J.
    vol 25   E263, E264, 265
Munter, P.A.
    vol 30   E29, E31-E33, 69, 90, E108, E109, E112, 133
Muromtsev, B.A.
    vol 31   E308, E311
Murooka, T.
    vol 26   E113-E116, 118, 119, 136, 137, E271-E273, 283
Murphy, A.J.
    vol 25   E357-E359, 364
    vol 29   E214
Murphy, G.M.
    vol 20   E35, 36, E132-E136
Murrieta-Guevara, F.
    vol 27/28   E712, 716, 717
    vol 32   E180, E187, 312, 313
Musaev, N.Yu
    vol 30   E24, E33, 78, 81
Musbally, G.M.
    vol 30   E184-E187, 192, 193, E195-E197
Muspratt, J.S.
    vol 26   E248, E249, E258, E271, E274

Mustafa, I.
    vol 25   E172
Muthman, W.
    vol 25   E214, E215
Muthmann, F.
    vol 31   E206, E208, E212, 220
Myl, J.
    vol 31   E168-E171, 203, 204
Mylius, F.
    vol 30   E1, E4, 11, E184-E187, 188, E268, E273, 276
    vol 25   E84, E85

N

Nabiev, M.N.
    vol 26   E187, E191, E193, 215
    vol 30   E24, E33, 81
Nadirov, E.G.
    vol 33   25, E26, 30, 34, 76, E91, 94, 96
Nagai, H.
    vol 29   E46-E49, 54, 55, 57, 58, 64, 66-71
    vol 37   E65, E67, E84, 179, 180, E370-E374, E379, 420, 421
    vol 38   E9-E12, 21, E23, E24, E28, 40, E42-E44, E46, 57, E59-
                  E62, E66, 85, 86, 103, E164, E165, 167, E169, E170,
                  175, E177, E178, 184, 185, E187-E189, 199, E201, E202,
                  209
Nagai, T.
    vol 35   36, 51-53, 58
    vol 36   137, 141, 142, 146, 149, 167, 174, 175, 180, 185, 283,
                  289, 290, 292-295
Nagaoka, H.
    vol 25   E301, E303, E310
Nagata, K.
    vol 35   94, 95
Nagino, Y.
    vol 30   E31, E33, 83, 84
Nagy, B.
    vol 27/28   E24-E28, 33, E66-E68, E70, E74, E84, 110, 111, 122,
                  123, 147
Naito, D.-I.
    vol 26   E187-E193, 213
Nakahara, T.
    vol 32   E169, E186, 213, 214
Nakamori, I.
    vol 30   E31, E33, 83, 84
Nakamura, M.
    vol 29   E1, E5, E7, E14, E16-E18, E21, 40, E47, E49, E50
                  E51, 72-86, E87, 88-94, 100, 101
Nakano, M.
    vol 35   E64, 83, 90, 275, 276
    vol 36   262, 267
Nakano, Y.
    vol 24   E1, E2, 5, E16, E17, 22, E65-E68, E71, 100-104,
                  106, 108
Nakaoka, A.
    vol 26   E187, E191, E193, 223
Nallet, A.
    vol 30   E29, E31-E33, 55, 70, 91, 92, E108, E109, E112, 121, 132
Nambu, B.
    vol 36   283, 289, 290, 292-295
Nambu, N.
    vol 35   36, 51-53, 58
Namiot, A. Yu.
    vol 24   E1, E2, 13, E16, E17, 26
    vol 27/28   E1-E5, 14, E24-E28, 36, 37, E57, E70-E72, E74, E85,
                  129, E370, 371
    vol 37   E41, E46, 49
    vol 38   E294-E296, 304, E464, E465, 470
Nasipuri, R.N.
    vol 36   310, 335, 340, 341
Nasir, P.
    vol 32   E80, E81, E83, 136, 137
Nason, H, K.
    vol 20   E6, E7, 8, 9

Natarajan, G. S.
    vol 37   E2, E3, 5, E22, E23, 26, E197-E199, 205, E275,
                    E276, 279, 282, E455, E456, 459, E460, 463
    vol 38   112, 113, 292
Nauruzov, M.H.
    vol 38   E225, E226, 230
Navratil, J.
    vol 26   E1, E3, E6, 55, E72-E75, 84
Nazarov, B.F.
    vol 25   E317, E318
Nazarova, L.A.
    vol 31   E308, E311
Neal, J.L. Jr.
    vol 26   E348-E351, 365, 366
Nebenzahl, L.L.
    vol 37   1
Nederbragt, G.W.
    vol 27/28   E248, E249, E281
Neese, N.
    vol 31   E98, E102
Neis, N.D.
    vol 30   E24-E26, E30-E32, 37-40
Neish, W.J.P.
    vol 35   E110, E111, 122, 176,
    vol 36   E19, E20, 29, 46
Nejedlik, J.F.
    vol 25   E134, E258, E259, E263, E264, E271, E274, E282, E284
Nelson, H.D.
    vol 20   E132-E136, 147, E153-E157, 179, E202-E204, 209,
                    E210-E213, 220
    vol 37   E42-E44, E46, 54
    vol 38   E135, E136, E141, 147
Nemstov, M. S.
    vol 37   E65, E67, E71, E84, 167, E197-E199, 207
    vol 38   E114, E115, E117, 123, E135, E136, E141, 150
Neokladnova, L.N.
    vol 32   E181, E182, E187, 301
Nerad, A.J.
    vol 25   E55, E263, E301, E302
Nesbitt, R.U.Jr.
    vol 35   277-281
    vol 36   58-62, 218-222, 352-356
Nesse, G.J.
    vol 30   E374, E375, E378, E381, 412
Nessler, F.
    vol 31   E206, E208, E212, 226, 227, 262, 263
Nesterova, A.K.
    vol 33   E57-E60, 66
Neumann, A.
    vol 27/28   E737
Neville, F.H.
    vol 25   E13-E18, 20, E139-E141, 143, E157-E159, 161,
                    E182, E183, E185, E402-E404, 406
Newitt, D.M.
    vol 27/28   E248, E249, E477
Newman, J.
    vol 32   E54, E55
Newman, M.
    vol 20   E153-E157, 164
Newnham, R.
    vol 25   E194, E195
Ng, H-J.
    vol 32   E166-E168, E170, E186, 200, 201, 217, 228, 229
Ng, S.
    vol 24   E130-E134, 161, 163, 164
    vol 27/28   E202, E210, E211, E214, 243, 244, 246, E533, 536
Nichols, N. F.
    vol 37   E14, E15, E42, E43, E46, E65, E70, E84
Nichols, N.F.
    vol 20   E73-E77, E89-E91, E153-E157
    vol 38   E24, E28
Nicklasson, M.
    vol 35   264, 270, 271

Niedergesaess, U.
    vol 22   31, 33, 36, 47, 89, 94, 103, 132, 141, 143, 176,
                    191, 197, 218, 227, 237, 256, 264, 267, 275, 282,
                    294, 302, 314, E334-E336, 338, 346, 367, 374, 376
Nielson, S.
    vol 25   E194, E195
Niepel, W.
    vol 21   E82, E89, 142-147
Nigmatulina, A.A.
    vol 25   E157-E159, E181-E183, 188
Nigmetova, R.Sh.
    vol 25   E172, E182, E183, E230, E357-E359
Niini, A.
    vol 37   E63, E67, E72, E81, 99
Nikishin, G.D.
    vol 25   E207-E208, E214, E221, E225, E234, E235, E242, E253,
                    254
Nikitina, E.A.
    vol 31   E307, E308, E311, E318, E324
Nikitina, I.E.
    vol 24   E196, E197, 200, 201
    vol 27/28   E248, E249, 256, 257, 276
Nikolaev, N.S.
    vol 30   E467, E468, E470, E72, 478
Nikushina, N.L.
    vol 25   E369-E371, 384
Nilsson, G.
    vol 26   E187, E190, E191, E193, 207-210
Ninomiya, Y.
    vol 30   E268, E269, E274
Nirenberg, Z.
    vol 31   E129, E130, 166, 167
Nizhnik, A.T.
    vol 25   E93-E95, 96
Noda, K.
    vol 21   E3, E10, E12, 26, 29, 31, 32, E89
    vol 24   E251, 255, 256
Noer, S.
    vol 25   E334-E337
Nohka, J.
    vol 27/28   689
Nogami, H.
    vol 36   137, 141, 142, 146, 149, 167, 174, 175, 180, 185
Noonan, E.
    vol 30   E105, E111, 119, 199
Norberg-Schutz, Chr.
    vol 26   E1-E3, E5, 10-12, E72-E74, 76, 77, E90-E92, 93, 94,
                    E108, E109, 110, 111
Nordenskjold, N.
    vol 30   E105, E112
North, E.
    vol 25   E157-E159
Northam, J.J.
    vol 35   105
Northey, E.H.
    vol 36   9
Norton, F.J.
    vol 25   E301, E302, 303
Nosek, M.V.
    vol 25   E139-E141, 154, E157-E159, 171, E182, E183,
                    189, E402-E404, 418
Novak, J.
    vol 37   E42-E44, E46, 61
    vol 38   E135-E137, E141, 155, E225-E227, 234
Novak, J.P.
    vol 21   E82, E89, 142-147
Novikov, G.I.
    vol 30   E380, E381, 409, 410
Novoselova, A.V.
    vol 25   E194, E195
Novotny, H.
    vol 25   E206, E326
Nurakhmetov, N.N.
    vol 30   E473, 489

Nurminskii, N.N.
    vol 33   E35, 36, E40, E43, E58-E60, E134, 135, E175, 177,
                  E210, 212
Nutting, E.G.
    vol 20   E103-E105
Nyqvist, H.
    vol 35   264, 270, 271
Nyvlt, J.
    vol 26   E1, E3, E6, 55, E72-E75, 84, E153, E154, E157, E159,
                  175, 176

O

O´Grady, T. M.
    vol 37   E64, E76, E83, 150
O´Kane, D.F.
    vol 25   E194, E195
O´Reilly, W.F.
    vol 27/28   539-542
O´Sullivan, T.D.
    vol 27/28   E24-E28, 35, E57, E71, E72, E74, E85, 125, 126,
Obukhov, A.P.
    vol 31   E127, E130, 134, 135
Odarich, V.F.
    vol 26   E5, E6, 58
Oelert, H.
    vol 34   283, 284
    vol 36   258, 259, 277, 278
Oelsen, W.
    vol 25   E301, E302, E310, E317-E318
Oelsner, L.
    vol 22   31, 33, 36, 47, 89, 94, 103, 132, 141, 143, 176, 191,
                  197, 218, 227, 237, 256, 264, 267, 275, 282, 294, 302,
                  314, E334-E336, 338, 346, 367, 374, 376
Oey, T.S.
    vol 30   E24, E28, E29, E31-E33, 56, 57, 59, 60, 87-89
Offenbach, J.A.
    vol 30   E24, E25, E33, 79, 80, E220, E222, E223, 242
Oganesyan, E.B.
    vol 31   114, E129, E130, 165
Ogata, H.
    vol 34   E285, 286, 292
    vol 35   3, 22, 68, 263
    vol 36   103, 287, 371, E404, 408
Ogawa, F.
    vol 25   E103-E105, 108
Ogg, A.
    vol 25   E357-E359
Ogston, A.G.
    vol 20   E30, 32
Ohta, Y.
    vol 27/28   731
Okabe, T.
    vol 26   E153, E158, E262, E271, E273
Okouchi, S.
    vol 29   E1, E5, E7, E14, E16-E18, E21, 42, E102-E105
                  E109, 111, 118, 124, 127, E165, 170, 171
Okuwaki, A.
    vol 26   E262
Olcese, G.L.
    vol 25   E214
Olds, R.H.
    vol 27/28   E297, 298, 299, E303, 306, 307, E360, 361-364
Olson, J.D.
    vol 27/28   649
Omarova, A.F.
    vol 25   E310, E317, E318
Omarova, N.D.
    vol 25   E33, E34, 43, E119-E121, 132, E335-E337,
                  E385-E387, 401, E402-E404, 419
Omelchenko, A.V.
    vol 25   E194, E195, 197
Omori, K.
    vol 26   E262

Onat, E.
    vol 29   El, E2, E5, E7, El4, El6-El8, E21, 37
Onstott, E.I.
    vol 25   El-E3, 8
Opalovski, A.A.
    vol 30   E461, E462, 463
Oppenheimer, H.
    vol 37   E63, E67, E81, 103
    vol 38   E187, E189, 190, E201, E202, 204
Opredelnikova, L.V.
    vol 31   E278, E279, 282
Oratovskii, V.I.
    vol 32   E54, 61, 62
Orekhov, I.I.
    vol 31   E168, E171, 178, 179
Organesyan, E.B.
    vol 31   E98, E102
Orr, R.J.
    vol 20   E3, 4, E15, 16, E17, 18, E19, 20, E22, 23, E26, 27,
                  E28, 29, E33, 34, E35, 38, E41, 42, E49, 50, E51, 52,
                  E53, 54, E55, 56, E57, 58, E59, 61, E67, 68, E69, 70,
                  E71, 72, E73-E77, E118, 119, E120, 121, E122, 124,
                  E125, 126, E127, 128, E129, 130, E132-E136, 148
Osipova, G.A.
    vol 25   E214, E215, 218, E221, E225, 227, E230, E234, E253, 255
Ostapczuk, P.
    vol 25   E335-E337, 353, E357-E359
Ostapkevich, N.A.
    vol 26   El-E3, E5, E6, 54, E90-E92, 95, E303-E305, 309, 310,
                  E322-E324, 327, 328
Osterheld, R.
    vol 31   E14, E15
Osugi, A.
    vol 25   E13-E18, 29
Oteeva, G.Z.
    vol 25   E221, 223, E225
Othmer, D.F.
    vol 20   E153-E157, 161
Otto, F.D.
    vol 32   E79-E83, 90, 91, 94, 98-103, 107-111, 113-116, 126-
                  128, 138, 139, 141-148, 152-154, E176, E177, E187, 268
Otuka, Y.
    vol 26   E187, E189, E190, E193, 205, 206
Overdick, F.
    vol 26   E113-E116, 140
Owen, E.A.
    vol 29   E213, E214, 227
Owens, J.L.
    vol 32   E54, E55, 73-77
Ozanne, N.
    vol 32   E330, E331, 337, 338
Ozerova, N.A.
    vol 29   El, E21

P

Paal, T.
    vol 34   46-48
    vol 36   444-446
Pahlevanzadeh, H.
    vol 32   E166-E168, E182, E184, E186, 199, 299
Pajaczkowska, A.
    vol 25   E194, E195, 201
Pakhapill, Yu.A.
    vol 38   E456, 457, E483, 484, 515, E517, 518, 532, 533, 537,
                  539
Pakhomov, V.I.
    vol 30   E268-E270, E271, E273, E274, 286, 316, E374, E378,
                  E381, E467, E471, E472
Pakshver, A.B.
    vol 23   E13, E18, E25, 51, E163, E164, E178, E183, 213, 214
Palenzona, A.
    vol 25   E214, E215, E221, E225, E230, E231, E234, E256

Palkina, N.A.
 vol 26   E1, E3, E4, E6, 27, 28, 50, 51
Palmaer, E.
 vol 25   E301, E302, E317, E318
Palman, H.
 vol 25   E84, E85, E173, E174, 179, E258, E259, 260, E277,
        279, E285, E286, 291, E310, 312, E317, E318, 321,
        E335-E337, 344, E357-E359, 368, E422, 424, E425-E427,
        431
Palmer, M.G.
 vol 21   E281, E282, 283-287, 289-295, 300
Pan, K.
 vol 26   E348-E351
Panchenkov, G.M.
 vol 21   E2, E4, E5, E6, E7, E8, E10-E12, 21, 38, 45, 53, 75
Panin, O.B.
 vol 33   E58-E60
Panina, L.S.
 vol 25   E310
Pannetier, G.
 vol 26   E252, E256, E271, E273
Panson, A.J.
 vol 31   E13-E15, 37, 38
Pape, D.
 vol 21   E2, E12, 18, 19
Paranchich, S.Yu.
 vol 25   E194, E195
Paravano, N.
 vol 31   E11, E14, E15, 16, E168, E171, 174
Parcher, J.F.
 vol 24   E130-E134, 166, 184, 185, 187, 188
 vol 27/28 E202, E203, E209-E213, E215, 241, E370, 673, 674
Parent, J.D.
 vol 27/28 E248, E249, 250-254
Pariaud, J.
 vol 29   E1, E3, E14, E16-E18, E21, 28, E228, 229, 230, 233, 234
Parikh, J.S.
 vol 27/28 382
Paris, R.A.
 vol 30   E28, E29, E31-E33, 55, 70, 91, 92, E108, E109,
        E112, 221, 132
Parke, M.H.
 vol 20   E6, E7, 10, E24, 25, E43, 44, E45, 46, E195, E196, 201,
        E221, 222, E223, 224, E226, 227, E228, 229, E231,
        232, E233, 234, E235, 236, E238, 239, E240, 241, E242,
        243, E244, 245, E246, 247, E248, 249, E250, 251, E252,
        253, E254, 255, E256, 257
Parker, C.A.
 vol 38   E158, 159, E308, 309
Parker, E.G.
 vol 31   E168, E169, E171, 184, 185
Parkman, M.F.
 vol 25   E268, 269, E277, 280, E301, E302, 306, E310,
        E311, E317, E318, 322
Parks, G.S.
 vol 23   E156, E182
Parks, R.D.
 vol 25   E93-E95, 101
Parks, W.G.
 vol 25   E103-E105, E207, E208, 209, E422
Parnov, E. I.
 vol 37   E13-E15, 16, E64, E76-E78, E79, E82, 133, E212,
        E213, 214
 vol 38   E9, E12, 14, E23, E25, E26, E28, 32, E42, E44-E46, 50,
        E59, E65, E66, 75, 101, E104-E106, 107, E177, E178,
        181, E201, E202, 207
Parravano, N.
 vol 25   E369-E371, 375
Parry, G.S.
 vol 33   E58-E60
Parsons, G.H.
 vol 20   E150
Parsons, L.B.
 vol 32   E173, E175, E187, 254

Parsons, R.
    vol 26   E240, E241
Paruta, A.N.
    vol 34   E13-E15, 31
    vol 35   E64, 81, 82, 84-88
    vol 36   186-191, 193-200, E367, 382-385, 388-397, 412-425
Pashinkin, A.S.
    vol 26   E348-E352
Patkausas, R.M.
    vol 26   E303-E305, 311-313
Patyi, L.
    vol 21   E1-E3, E12, 14, 22, 30, E89
    vol 27/28   E202-E209, E215, 216, 225, 230, 233-235
    vol 32   E166-E168, E170, E171, E182, E185, E186, 188, 189,
                    226, 233, 286, 293, 295
Paul, J.
    vol 29   E87
Paul, M.A.
    vol 38   E250, 251, E352, 353
Paul, R.C.
    vol 30   102, 160
Paulsen, O.
    vol 20   E129
Pavia, R. A.
    vol 37   E64, E72, E73, E82, 125
Pavlov, S. Yu.
    vol 37   E7, E8, 10, E22, E23, 25, E31-E34, 38
Pavlov, V.L.
    vol 33   E26, 29, E31, 33, 49, 84, E91, 93
Pavlova, S.P.
    vol 37   E7, E8, 10, E22, E23, 25, E31-E34, 38
Pavlovich, P.
    vol 25   E119-E121, 123
Pawlewsky, B.
    vol 30   E105, E111, 113
Pean de St.Cilles, L.
    vol 26   E262
Peer, A.A.
    vol 27/28   E266, 267, E281, 282, E303, 304, E320, E349, 350,
                    E465, 466, E511, 512, 559, 560, 635, 636
Peisakhov, I.L.
    vol 26   E271, E272, E274, 281, 282
Pekova, I.
    vol 26   E348-E351, 376
Peled, E.
    vol 25   E84, E85, 92
Pellini, G.
    vol 25   E194, E195, 199
Pepper, D.S.
    vol 35   152, 225
Per´kova, N.N.
    vol 36   286, 312, 436
Perdok, W.G.
    vol 26   E256, E258
Perekupko, T.V.
    vol 26   E74, E75, 86, 89
Perel´man, F.N.
    vol 30   E28, E31-E33, 82, 85, 86
Perinet, G.
    vol 26   E305, 318, 319, E324, 329, 330
Perov, E.I.
    vol 25   E230, E231, E234, 236
Peshchevitskii, B.I.
    vol 23   E148, 155
Peshkov, W.
    vol 25   E385-E387, 393
Pesin, Ya.M.
    vol 26   E262, 265
Peter, K.
    vol 37   E65, E83, 184, 185
    vol 38   E342, E344
Peters, C.J.
    vol 27/28   E370, 373-375

Petersen, H.Jr.
    vol 35   E64, 81, 82, 84-88
    vol 36   187-191, 194, 196, 198, 200, 383, 385, 389, 391, 393,
                 395, 397, 413, 415, 417, 419, 421, 423, 425
Petersen, J.
    vol 33   95
Petot, C.
    vol 25   E182, E183, 191
Petot-Ervas, G.
    vol 25   E139-E141, 153, E182, E183, 191
Petrov, M.R.
    vol 31   E127, E130, 154
Phelphs, R.E.
    vol 27/28   E248, E249, 262, E737, 743, 744
Phillips, A.W.
    vol 30   E470, E472, 479, 480
Phillips, N.D.
    vol 26   E153-E155, E158, 162, 163, 167
Piantoni, G.
    vol 33   3-8, E12, 13, E15, 18, 22-24, 138, E150, 166, 171,
                 220, E229, 232, E237, 258, 263
Piatekiewicz, J.
    vol 32   E54, E55, 60
Piater, J.Z.
    vol 23   E270, E272, E274-E276, E281, 286, 287
Pichugina, A.
    vol 29   E207, E208
Piercey, A.G.A.
    vol 27/28   E303, 313, E338, 347, 348
Pierotti, R. A.
    vol 37   E2, E3, 6, E13-E15, 19, E42, E43, E46, 55,
                 E64, E67, E83, 155, E193, E194, 196,
                 E197-E199, 206
Pietaszewska, J.
    vol 30   E270, E272, E275, 319, E467, E472
Piiper, J.
    vol 27/28   732-735
Pike, F. P.
    vol 37   E64, E67-E69, E82, 122, 123
Pikulina, N.S.
    vol 32   E54, 63, 64
Pinaev, G.F.
    vol 26   E348-E351, 369
Pinaev, V.A.
    vol 26   E153-E158, 182, 183
Pinto, L.
    vol 23   E170, E183, 247
Pispisa, B.
    vol 24   E1, E2, 6, E16, E17, 30
    vol 37   E41, E42, E46, 50
Pistorius, C.W.F.
    vol 25   E119-E121, 130, 131
Pitteloud, M.N.
    vol 22   21, E22, 24, 26, 77, 79, 84, E152, 155, E156, 159, 167,
                 209, 210, 212, 246, E247, 248, 251, 327, E328, 330, 332,
                 E336, 359, 360, 362
Piuze, J.
    vol 23   E94, E97
Plaksin, I.N.
    vol 25   E330, 331, E369-E371, 379
    vol 29   E214
Plank, C. A.
    vol 37   E64, E67-E69, E82, 122, 123
Planovskii, A.N.
    vol 26   E5, E6, 71
Planting, P.J.
    vol 25   E330
Plate, A. F.
    vol 37   E13-E15, 17, E22, E23, 24, E31, E33, 36,
                 E64, E71, E72, E82, 141, E197-E199, 202,
                 E212, E213, 215
    vol 38   E23, E27, E28, 34, E59, E62, E63, E66, 76, 93, 97,
                 E114-E117, 120, 132, E135, E138, E139, E141, 145, E177,
                 E178, 182, E187, E189, 194, 213, 220-222, E260, 262,

Plate, A. F.
     vol 38   264, E265, E266, 268, E271, 273, E277, E278, 282, E287,
                     289, 293, E311, E312, 313, 322, 325, 373, 374, 393,
                     394, 425, 462, 463, E464, E465, 467
Plate, W.
     vol 25   E284
Platford, R.F.
     vol 31   E39-E42, E98, E99, E102, 117, E278, E279
Platt, J.H.
     vol 26   E108, E109, 122
Pleskov, Yu.V.
     vol 23   E91, E96
Plotnikov, V.A.
     vol 29   E207, E208, 209
Plummer, A.W.
     vol 26   E187, E193
Plummer, M.N.
     vol 36   154, 155, 164, 165, 233, 234, 238, 239, 251, 257, 273,
                     274, 329, 335, 359, 362
Pluzhnikova, M.F.
     vol 24   E257, E258, 283-298
     vol 27/28   694, 696-701
Pochtakova, E.I.
     vol 33   E9, 11, E12, 14, 20, 21, E50, 52, 56, E57-E60,
                     65, 69, E77, 79, E85, 86, 87, 98, E99, 100, 104
                     E105, 106, 107, E108, E109, 110, E116, 117, 118,
                     E119, 120, E123, 124, E125, 126, E128, 129, E145,
                     146, 153, E154, 155, E157, 158, E159, 160, E161,
                     162, E163, 164, E183, 184, E185, E189, E199, 206,
                     E229, 231, 242, E243, 245, E248, 249, E250, 252,
                     E253, 255
Pocock, F.J.
     vol 23   E1, 6, E21, E26, 29, 30
Podbornova, N.I.
     vol 23   E13, E18, E25, 51, 52, E163, E164, E178, E183, 213, 214
Podkorytova, N.V.
     vol 25   E317, E318
Pokric, B.
     vol 23   E170, E183
Polak, J.
     vol 37   E31-E34, 39, E42-E46, 56, E65, E67, E68,
                     E71-E73, E83, 161
     vol 38   E9-E12, 17, E23, E24, E27, E28, 35, E42, E43, E45, E46,
                     52, E59-E63, E66, 78, E114-E117, 122, E126, E127, 128,
                     E135, E136, E138, E139, E141, 149, E216, E217, 219
Poleva, G.A.
     vol 30   E268, E271-E275, 297-299, 302, 303, E330, E332,
                     E335, E339, 370, 441, E448-E450, 455, 456
Polglase, M.F.
     vol 24   E1, E2, 8, E16, E17, 23
     vol 27/28   E1-E5, 12
Pollitzer, F.
     vol 32   E1-E3, E26, E28, 31
Poloshinzewa, E. N.
     vol 37   E63, E71-E73, E81, 97
Polosin, V.A.
     vol 31   E206-E208, E212, 234, 235
Polyakova, T.P.
     vol 25   E173, E174
Polyshchuk, S.A.
     vol 23   E313, E317, 333
Polyvyanni, I.R.
     vol 23   E181, E184, 252, E280, E282, 293, E313, E317, 332
Pomeroy, R.D.
     vol 27/28   318, 569
Pomianowski, A.
     vol 23   E15, E25
Ponomareva, E.I.
     vol 23   E11, E13, E25, E156, E171, E182, 260
Ponter, A.B.
     vol 24   E331-E333, 344, 348
Poon, D.P.L.
     vol 27/28   E266, 279

```
Popova, M.Ya.
     vol 29        El, E21
Popova, Z.V.
     vol 31        E98, E99, E102, 108, 109, E278, E279
Popp, O.
     vol 25        E13-E18
Poppl, L.
     vol 33        E58-E60
Poroshkova, M.A.
     vol 26        El-E6, 31, 41, E72-E75, 79
Porter, E.G.
     vol 24        E414, 418, 419, 420
     vol 27/28     730
Portnov, M.A.
     vol 36        328, 383
Porubszky, I.
     vol 26        E187, E193
Posnjak, E.
     vol 23        E21, E22, E26, 40-42
Postmus, C.
     vol 23        E158, E161, E162, E171, E182, 227, 228, E275, E281
Poston.R.S.
     vol 27/28     E320, 327, 328, 427-431
Poston.T.M.
     vol 21        E10, E12, 71-73
Postovskii, I.Ya.
     vol 34        E13, E14, 21, 109, 182, E204, 205, 270, E302,
                   304, 320, 321
     vol 35        E110, E111, 118, 214, 221, E231, 234, 242
     vol 36        E19, E20, 27, 42, E63, 68, E243, 246, E269, 272
Potapov, V.F.
     vol 24        E257, E258, 283-298
     vol 27/28     684, 696-701
Potemin, S.S.
     vol 33        E58-E60, E229, 233, E237, 241, 261, E267, 272, 274
Pound, B.G.
     vol 23        E88, E96
Povalyaeva, N.V.
     vol 32        E180, E187, 309-311
Powell, R.J.
     vol 27/28     E702, 703, 721
Pozhidaeva, T.N.
     vol 26        E404, E405, 424
Pozzoli, S.
     vol 29        E213, E214, 221
Prabhakara, C.
     vol 27/28     E50, E51
Prahl, W.
     vol 20        E153-E157
Prapaitrakul, W.
     vol 27/28     100, 101
Prasad, B.
     vol 23        E10, E17, E25, 53, E315, E317, 339
Pratt, J.N.
     vol 25        E425-E427
Prausnitz, J.M.
     vol 21        E2, E3, E12, 17, 24, 34
     vol 24        E130-E134, 157, 161-165, 180, 183, 186, 195, 235,
                   237-239, 243, 245, 248-250
     vol 27/28     E24-E28, E202-E204, E209-E211, E213-E215, 238,
                   239, 243-247, E314, 315, E370, E453-E456, 464,
                   E493, E496, E497, 508, 509, E724, 726
     vol 32        E54, E55, E166-E168, E171, E176, E177, E180,
                   E185-E187, 211, 223, 235, 236, 306-308, 320
Predel, B.
     vol 25        E93-E95, 97, E139-E141, 155, E182, E183, 190
Prener, J.S.
     vol 25        E194, E195
Preston, D.G.
     vol 29        E214
Price, A.R.
     vol 27/28     E248, E249, 255, E266, 274, 378-381
Price, E.G.
     vol 25        E307, E316
```

Price, L.C.
    vol 27/28   E24-E28, 39, 40, E57, E72-E74, E85, 130-142
    vol 37   E13-E15, 20, E31-E34, 40, E42-E44, E46, 57,
                 62, E65, E68, E84, 169, E188, 191, E212,
                 E213, 217, 220
    vol 38   E9, E11, E12, 19, E23, E24, E28, 38, E42, E43, E46, 55,
                 E59, E60, E66, 81, 95, 98, 102, E114, E115, E117, 124,
                 E126, E127, 129, 133, E135-E137, E140, E141, 151, E161,
                 162, E169, E170, 173, E187-E189, 197, 214, 223, E225,
                 E226, 232, 258, E287, 290, 291
Prideaux, E.B.R.
    vol 31   E2, E6
Prikhod´ko, M.U.
    vol 29   E207, E208, 210
Prime, G.E.
    vol 25   E422
Prisyazhnyi, V.D.
    vol 33   E108, E109, 112, E113, 115, 156, 246, E289, 291
Pritze, M.
    vol 26   E348-352, 398
Procenko, P.I.
    vol 31   150
Prodany, N.W.
    vol 27/28   303, 308, E314, 316, 317, 410, 411
Prosyanov, N.N.
    vol 20   E153-E157, 177
Protsenko, P.I.
    vol 31   E127, E130, E297, 303
Pruntsev, A.E.
    vol 30   E270, E275, 328
Pryanishnikova, M. A.
    vol 37   E13, E15, 17, E22, E23, 24, E31, E33, E34,
                 36, E64, E71, , E72, E82, 141, E197, E199,
                 202, E212, E213, 215
    vol 38   E23, E27, E28, 34, E59, E62, E63, E66, 76, 93,
                 97, E114-E117, 120, 132, E135, E138, E139, E141, 145,
                 E177, E178, 182, E187, E189, 194, 213, 220-222, E260,
                 262-264, E265, E266, 268, E271, 273, E277, E278, 282,
                 E287, 289, 293, E311, E312, 313, 322, 325, 373, 374,
                 393, 394, 425, 462, 463, E464, E465, 467
Pryor, W.A.
    vol 38   E23-E26, E28, 31, E42-E45, 49
Prytherch, W.E.
    vol 25   E139-E141
Prytz, M.
    vol 23   E170, E183
Pucar, Z.
    vol 23   E170, E183
Puech, R.
    vol 34   E285, 291
Puech, R.
    vol 35   272, 310-315
Pulver, R.
    vol 34   271, 274, 280, 282
Pulver, R.
    vol 35   E110-E112, 151, 227
Pulver, R.
    vol 36   E19, E20, 37, E63, 74, 153, E228, 240
Punin, Yu.O.
    vol 31   E168, E171, 205, E206, E212, 219
Puri, P.S.
    vol 27/28   E370, 376
Pushin, N.A.
    vol 25   E119-E121, 127, E139-E141, 144, E157-E159, 162,
                 E182, E183, 186, E385-E387, 390, E402-E404, 409
Pushkin, A.G.
    vol 32   E183, E184, E187, 323
Putley, E.H.
    vol 25   E294, E195
P´yankov, V.A.
    vol 23   E181, E184
Pyatnitskii, I.V.
    vol 26   E348-E351, 372, 373

Q

Qi, S.
        vol 29   E48
Quadrafoglio, F.
        vol 24   E1, E2, 6, E16, E17, 30
        vol 37   E41, E42, E46, 50
Quentin, K-E.
        vol 20   E1, 2
Quill, L.L.
        vol 22   E152, E156, E162
Quist, A.S.
        vol 31   E2, E6

R

Racster, L.V.
        vol 22   81, 161
Raffofort, Z.
        vol 20   E129
Ragg, M.
        vol 23   E3, 5, E21, E26, 28, E158, E182, 188, E315, E317, 318
Rai, M.N.
        vol 23   E156, E182
Rainer, W.
        vol 32   E1-E3, E26, E28, 38, E51
Rakhimbaev, D.
        vol 29   E213, E214, 219
Rakhimova, A.
        vol 22   101, 175, 217, 254, 292, 350, 364
Rakova, N.N.
        vol 31   E127, E130, 151
Ramaiah, N.A.
        vol 26   E187, E188, E192, E193, 217, 231
Ramanujam, S.
        vol 27/28   578, 579
Rammelsberg, C.
        vol 26   E248, E256, E258
Rammelsberg, F.C.
        vol 31   E1, E3, E6
Randall, M.
        vol 23   E93, E97, 116
Rashkovich, L.N.
        vol 31   E307, E308, E310, E311, 315-317, E318, E320, E323,
                      E324, 329-333
Rassonskaya, I.S.
        vol 33   E145
Ratnikova, V.D.
        vol 31   E318-E321, E324, 326
Raub, E.
        vol 25   E284
Ravdel, A.A.
        vol 23   E156, E182, E277, E279, E282, 294, 304
        vol 25   E157-E159, 169
Ravich, M.I.
        vol 31   E11, E13, E15, 18, 19, E98, E99, E102, 108, 109,
                      E127-E130, 140, E168, E169, E171, 195-198, E278,
                      E279, 280, E297, 299
        vol 33   E145
Ray, B.
        vol 25   E194, E195, 196
Reamer, H.H.
        vol 24   E48, 53, E196, E197, 208, 209, 216, 217
        vol 27/28   E266, 270, 271, E303, 306, 307, E338, 339-341,
                      E360, 361-364, E365, 404-406, E465, 469-471,
                      E750, 751, 752
        vol 32   E166-E168, E186, 196, 197, 204, 205
Reavis, J.G.
        vol 25   E263, E264, 265
Rebert, C. J.
        vol 37   E64, E76, E80, E82, 127
Rebiere, G.
        vol 23   E84, E96, 100

Rebrova, T.I.
    vol 23   E10, E24, E25
Redman, M.J.
    vol 26   E348-E352, 393, 396, 397
Reglin, W.
    vol 31   E1, E3, E4, E6, 7
Regosz, P.
    vol 34   46-48
    vol 36   444-446
Reichardt, H.
    vol 29   E1-E3, E14, E16-E18, E21, 26, E102-E105, E109, 113
                  E163-E165, 166
Reichle, R.A.
    vol 23   E165, E168-E173, E183, 266, 267
Reid, R.C.
    vol 25   E258, E259
Reinders, W.
    vol 25   E357-E359, 361
Reinke, L.
    vol 32   E173, E175, E187, 252
Reinmann, R.
    vol 23   E271, E272, E281
Reitsamer, G.
    vol 23   E170, E171, E183
Rejlek, M.
    vol 26   E1-E3, E5, E6.42, 43, E72-E74, 82, 83
Remoissent, M.
    vol 30   E269, E274
Remy, H.
    vol 23   E10, E25, 27, E84, E96, 101, E158, E182, 185, E270,
                  E281, 283
Renault, P.
    vol 24   E130-E134, 158, 236, 244, E299-E301, 306, 307,
                  309, 311, 314, 320, 321, 328, E331-E333, 339, 345,
                  346, 360, 361, 367, 371, 372, E375-E377, 380, 385,
                  392, 402, 405
    vol 27/28   E202, E203, E214, 236, E493, E497, 510, E641,
                  E642, E644, 672, 676-678, E702, 706, 709, E712,
                  715, 722
    vol 32   E167, E168, E172, E175, E176, E178, E180, E182-
                  E186, 210, 222, 245, 246, 250, 267, 275, 291, 305,
                  315, 316, 319, 322, 326
Rengemo, T.
    vol 26   E187, E188, E190, E191, E193, 207-210, 222
Renier, J.J.
    vol 30   E428, E429, E431, 435, 436
Renon, H,
    vol 24   E130-E134, 158-160, 181, 182, 236, 244,
                  E299-E301, 306, 307, 309, 311, 314, 320, 321, 328,
                  E331-E333, 339, 345, 346, 360, 361, 367, 371, 372,
                  E375-E377, 380, 385, 388, 392, 395, 402, 405
    vol 27/28   E202, E203, E209, E210, E213-E215, 236, 240, 242,
                  E465, E493, E497, 510, E511, 521, E522, 532, E533,
                  535, 537, E641, E642, E644, 672, 676-678, E702,
                  706, 709, E712, 715, 722, 727, 729
    vol 32   E54, E55, 65, E167, E168, E172, E175, E176, E178,
                  E180, E182-E186, 222, 245, 246, 250, 267, 275,
                  291, 305, 315, 316, 319, 322, 326
Rettich, T.R.
    vol 27/28   E1-E5, 23
Revzina, T.V.
    vol 26   E403, E405, 406
    vol 30   E162, E164, E165, 168, 171, 172, E176, E178, 181,
                  E220, E222, E223, 226, E258, 261, E263, E264, 267,
                  E374, E375, E381, 385, 3980-394, E428, E429, E431,
                  434, E448-E450, 453
Rey, J.
    vol 25   E119-E121, 133
Reznikov, M.I.
    vol 23   E23, E26
Rhodes, E.
    vol 33   E58-E60
Rhymer, P.W.
    vol 23   E271, E272, E281, 289

Rhys, D.W.
    vol 25   E307, E316
Ricci, F.P.
    vol 25   E93-E95, 101
Ricci, J.E.
    vol 30   E24-E26, E28-E33, 41, 44-48, 61-66, 74-77, 79, 80,
                   E105, E108, E109, E111, 123, 124, 141-146, E195-E197,
                   198, 200, 203-211, 213-214, E220-E223, 224, 230-232,
                   235, 236, 242, E268, E269, E271, 277, 314, 315, E330-E332,
                   E335-E338, 343, 359-363, E374, E375, E378-E381, 403,
                   412, E467, E458, E471, E472
Rice, D.A.
    vol 24   E16, E17, 25
Rice, P.A.
    vol 24   E65-E68, E70, E72, 107, 120
Richards, F.A.
    vol 27/28   E50, E51
Richards, T.W.
    vol 25   E1-E3, E119-E121, 125, 126, E157-E159,
                   E301, E302, E335-E337, 340
Richards, W.L.
    vol 27/28   200, 201
Richardson, M.J.
    vol 29   E190-E192, 200-202
Richon, D.
    vol 24   E130-E134, 159, 160, 181, 182
    vol 27/28   E202, E203, E209, E210, E213, E215, 240, 242,
                   E511, 521, E522, 532, E533, 535, 537
Richter, P.W.
    vol 25   E119-E121, 130, 131
Ridenour, W.P.
    vol 27/28   570
Ridgion, J.M.
    vol 23   E23, E26
Riegger, E.
    vol 32   E79, E83, 84, 85
Riess, W.
    vol 35   4, 7, 9, 10, 27, 54, 73, 92, 157, 191, 298,
                   301
    vol 36   96-101, 106, 113, 122, 123, 158, 204, 254, 265, 331,
                   344, 375, 399, 410, 426
Riley, G.A.
    vol 23   E21, E26, E315, E317
Riley, H.L.
    vol 23   E23, E26
Riley, J.P.
    vol 27/28   E50, E51
    vol 32   E1-E3, 33, E51, 52, 53
Ripan, R.
    vol 26   E348-E351, 353, 354, 356, 358, 362, 370, 374,
                   377, 390, 391
Rittener, A.
    vol 32   E27, E28, 48
Rivas, O.R.
    vol 27/28   E202, E203, E209, E210, E215, 239
    vol 32   E176, E177, E180, E185, E187, 269, 306-308, 320
Rivett, A.C.D.
    vol 26   E1-E4, E6, 19-26
Robart, J.
    vol 26   E187, E188, E192
Robbins, D.
    vol 26   E90-E92
Robert, J.
    vol 25   E103-E105
Roberts, E.A.O.
    vol 29   E213, E214, 227
Roberts, L.R.
    vol 27/28   E281, 288, 289
Robertson, J.B.
    vol 37   E63, E71, E81, 98
Robertson, R.E.
    vol 38   E187-E189, 192

Robinson, D.B.
    vol 27/28   E266, 280, 439, 440, 447-449, E533, 536, 538,
                 553-555, 749, E750, 753, 756-759
    vol 32   E166-E168, E170, E171, E186, 192-195, 200, 201,
                 217-220, 228, 229, 231, 234
Robinson, R.A.
    vol 20   E195, E196
    vol 23   E90, E96, 104, E169, 183
    vol 27/28   E50, E51
    vol 31   E2, E6
Roblin, R.O.
    vol 34   E302, 303
    vol 35   1, 11-13, E14, 16, 104, 107, 109, E110, E111,
                 114, 219, E231, 232, E239, 240, 250, 251,
                 256, 316, 317
    vol 36   E19, E20, 23, 54, E63, 66, 76-80, 83-87, 96, 97, E124,
                 125, 206, 207, E228, 229, E243, 244, E269, 270, 279,
                 284, 296-301, 305, 366, 437, 438, 443
Rochester, C.H.
    vol 20   E150, 152
Roddy, J W.
    vol 37   E64, E67, E71-E73, E83, 153
Rodgers, J.C.
    vol 26   E187, E191-E193, 224, 225
Rodiger, W.
    vol 23   E157, E182
Rodin, I.V.
    vol 26   E153, E158, 164, E187, E188, E193, 195, E236,
                 238, E240, E241, 243, E248, E249, 250, E252, 253,
                 E256, 257, E258, 259, E260, 261, E262, E266,
                 E267, 270, E271, E272, E274, 284, 285, E286-E290,
                 E293, 296
Rodin, Yu.A.
    vol 32   E79, E83
Rodionova, N.A.
    vol 31   E208, E212, 246, 247, E278, E279, 293-296
Rodman, C.J.
    vol 27/28   569
Rodnyanskii, I.M.
    vol 26   E1-E3, E5, 14
Rodot, H.
    vol 25   E194, E195
Rodot, M.
    vol 25   E194, E195
Rodriguez, A.T.
    vol 27/28   E712, 716, 717
    vol 32   E180, E187, 213, 313
Roeder, A.
    vol 25   E33, E34
Rogers, B.L.
    vol 27/28   E314, 315
Rogne, A.K.G.
    vol 29   E1, E5, E21, 43, E51, 95, 96
Rogojski, J.-B.
    vol 26   E260, E262
Rogowicz, J.
    vol 26   E240, E241, 244
Rohdewald, P.
    vol 34   E13, E15, 40, 117-125, 186-191, E204, 213, 240-249
Rohrig, A.
    vol 26   E248, E256, E293, E294
Rolfe, C.
    vol 25   E369-E371, 383
Rollet, A.P.
    vol 31   E1, E3, E4, E6, 9, 10
Romano, A.J.
    vol 25   E258, E259, E263, E264, E271, E274, E301, E302
Romashchenko, I.M.
    vol 26   E1, E6, 15
Romsay, O.B.
    vol 20   E59, 60, E62, 64, E132-E136, 143
Roof, J.G.
    vol 37   E42, E46, E66, E80, E83
    vol 38   E59, E65, E66, E135, E140, E141, E225, E226, E294, E296

Roos, G.D.
    vol 25   E119-E121, 124
Roozeboom, B.
    vol 31   E308, E311, E320, E324
Rordam, H.N.K.
    vol 23   E83, E96
Rose, F.
    vol 25   E84, E85
Rose, F.L.
    vol 36   32, 70, 139, 232, 346, 358
Rosenbaum, C. J.
    vol 37   E63, E71, E72, E81, 95
Rosenberg, H.S.
    vol 29   E190, E192, 203-206
Rosenheim, A.
    vol 26   E331, 332, E348-E352, 398
    vol 31   E1, E3, E5, E6, 7
Rosenthal, W.
    vol 24   E125, 126-129, E299-E301, 303, E374-E377, 382
Rosin, J.
    vol 30   413
Roslonek.H.
    vol 25   E330, 334
Roslyakova, O.N.
    vol 31   E127, E130, 154
Rossi, S.S.
    vol 38   E332, E334, 339, 340, E446-E448, 455, E456, 460
Rossmanith, K.
    vol 22   35, 46, 104, 109, 118, 135, 178, 190, 219, 226, 243,
                  254, 257, 263, 281, 295, 301, 316, 321, 339, 345,
                  356, 368, 373, 381
Rostocker, W.
    vol 25   E422
Rotatiu, G.J.
    vol 29   E213, E214, 215
Roth, J.
    vol 33   E58-E60
Roth, W.L.
    vol 23   E308, E317
Rothacker, D.
    vol 25   E139-E141, 155, E182, E183, 190
Rothmund, V.
    vol 30   E220, E221, E223, 243-253
Rotner, Yu, M.
    vol 25   E194, E195
Rowlinson, J.S.
    vol 29   E190-E192, 195, 200-202
Royce, H.D.
    vol 25   E285, E286, 287
Rozentaveig, S.A.
    vol 23   E275, E281, 300
Ruban, L.M.
    vol 25   E1-E3, 10, E33, E34
Rubenbauer, J.
    vol 23   E163, E164, E183, 192
Rubin, L.C.
    vol 27/28   485
Rubleva, V.V.
    vol 31   E127, E130, 150, E297, 303
Ruchnova, S.A.
    vol 31   E208, E212, 246, 247
Rudakov, E.S.
    vol 24   E1, E2, 14, E16, E17, 32, E34, 37, E59, E72,
                  75-77
    vol 27/28   E1-E5, E58, E59, E85, 88
    vol 37   E42, E43, E46, 60, E212, E213, 219
Rudy, B.C.
    vol 35   E14, E15, 19, 24, 33, 43, 45, 47-50, 57
Ruetschi, P.
    vol 23   E86, E96, 140, E144, E145
Ruhemann,
    vol 27/28   E248, E249, E477

Ruhl, G.
    vol 23   E21, E26
    vol 26   E252, 254, 255, E262, 263, 264, E271-E273, 276,
                277, E293, 297, 298
Rupprecht, H.
    vol 35   140, 177
Rushing, S.S.
    vol 26   E187, E191-E193, 224, 225
Russell, A.S.
    vol 25   E258, E259, E268, E277, E282, E284-E286, 288,
                E301, E302, E310, E317, E318, E335-E337, 339, E425-E427
Rustamov, K.A.
    vol 31   E297, 302
Rustamov, P.G.
    vol 26   E305, 320, 321
Rusz, L.
    vol 27/28   E202-E209, E215, 216, 225, 230, 233-235, E582-E588,
                E590-E595, 600, 605, 609, 613, 618, 623, 628, 663-671
    vol 32   E166-E168, E185, 188, 189
Rutenber, C.B.
    vol 23   E92, E96
Ryan, D.E.
    vol 23   E273-E276, E281, 302
Ryazantseva, A.P.
    vol 23   E270, E275, E281
Rychly, R.
    vol 26   E153, E154, E157, E159, 175, 176
Ryzhova, N.G.
    vol 23   E23, E26
Rza-Zade, P.F.
    vol 31   E129, E130, 144, 145, 159-164, E297, 302

S

Sabbah, R.
    vol 26   E305, 318, 319, E324, 329, 330
Sada, E.
    vol 37   E65, E66, E84, 166
Sadamoto, S.
    vol 32   E171, E172, E186, 239, 240
Sadeghi, N.
    vol 23   E92, E97, 130-133
Sadilenko, A.S.
    vol 21   E1-E3, E12, 14, 22, 30, E89
    vol 32   E170, E171, E182, E186, 226, 233, 286, 293, 295
Safonov, V.V.
    vol 22   74, 98
Safranova, T.P.
    vol 24   E422, 423
    vol 27/28   574
Sagadieva, K.Zh.
    vol 25   E207-E208, 211, 213, E214, E215, 217, 218,
                220, E221, 223, 224, E225, 229, E230, E231,
                233, E234, 237, 238, E242, 243, E285, E286,
                293, E335-E337, 342
Sagara, H.
    vol 27/28   E477, 480
Sage, B.H.
    vol 24   E48, 53, E196, E197, 208, 209, 216, 217, E251,
                253, 254, E422, 426-429
    vol 27/28   E266, 268-271, E281, 283-287, 291, E297, 298, 299,
                E303, 305-307, E320, 321, E338, 339-341, E360,
                361-364, E365, 383, 384, 387-393, 404-406, E465,
                467, 469-471, E511, 513, 557, 558, E750, 751, 752
    vol 32   E1-E3, 9, 10, E166-E168, E186, 196, 197, 204, 205
Sah, R.
    vol 37   E65, E71-E73, E84, 175
Sahgal, A.
    vol 21   E5-E7, E9, E10, E12, 47, 50, 55, 70
    vol 24   E331-E334
    vol 32   E171-E173, E175, E176, E178, E179, E186, 241, 243
                248, 263, 270, 274, 287
Sahr, E.
    vol 31   E11, E12, E14, E15, 28, 29

Saijo, A.
    vol 29   E51, 100, 101
Saillard, E.
    vol 26   E158, E159
Saito, A.
    vol 25   E425-E427
    vol 27/28   222, E477, 480, 559, 689, E737, 740-742
Sakai, M.
    vol 24   E251, 255, 256
Sakharov, Yu.G.
    vol 22   25, 80, 99, 126, 160, 211, 234, 250, 272, 290, 311,
    vol 22   331, 348, 361
Sakharova, N.N.
    vol 22   80, 99, 126, 160, 211, 234, 250, 378
Salem, T.M.
    vol 23   E308, E317, 330
Salib, N.N.
    vol 36   280, 281
Salkind, A.
    vol 23   E144, E145, E160, E162, E182
Salomon, M.
    vol 30   E163, E165, E330, E333, 339, E428, E431, E462, E467,
                  E473
Samoilov, P.P.
    vol 30   E461, E162, 464
Sanchez, F.M.E.
    vol 35   E110-E112, 126
Sanchez, M.
    vol 24   E39
Sandercock, J.A.W.
    vol 32   E166, E168, E186, 192, 193
Sandmann, B.J.
    vol 35   277-281
    vol 36   59-62, 218-223, 352-356
Sanemasa, I.
    vol 23   E307, E317
    vol 29   E1, E5, E12, E14, E16-E18, E21, 38, 45, E46-E49, 54
                  55, 57, 58, 64, 66-71
    vol 37   E65, E67, E84, 179, 180
Sanemasa, J.
    vol 38   E9-E12, 21, E23, E24, E28, 40, E42-E44, E46, 57, E59-
                  E62, E66, 85-87, 103, E164, E165-167, E169, E170, 175,
                  E177, E178, 184, 185, E187-E189, 199, E201, E202, 209,
                  210
Sanesi, M.
    vol 33   E35, E43, E57-E60, E103, E105, E116, E119, E123,
                  E128, E134, E142, E175, E253, E283, E297, E315,
                  E326, E330, E332
Sanfourche, A.A.
    vol 31   E4, E6
Sano, F.
    vol 30   E374, E375, E382, 414-416, E428, E429, E431,
                  444-446, E448-E450, 458-460
Sano H.
    vol 26   E187, E188, E191-E193, 232
Sapozhnikova, N.V.
    vol 34   E13, E14, 21, 109, 182, E204, 205, 270, E302, 304,
                  311, 320, 321
    vol 35   E110, E111, 118, 214, 221, E231, 234, 242
    vol 36   E19, E20, 27, 42, E63, 68, E243, 246, E269, 272
Sarapulova, I.F.
    vol 33   E40, 41, 42, E43, 44, E46, 47, E57-E60, 68, E73,
                  74, 75, E81, 82
Sarashina.E.
    vol 27/28   689
Sarbu, L.
    vol 21   E10-E12, 74
Sarieva, L.S.
    vol 25   E135, 138, E194, E195, 204
Sarkisov, Yu.S.
    vol 23   E179, E180, E184
Sasagawa, S.
    vol 36   289, 294

Sasaki, K.
    vol 23   E174, E183
Sasaki, S.
    vol 29   E1, E5, E7, E14, E16-E18, E21, 42, E102-E105
                  109, 111, 118, 124, 127, E165, 170, 171
Sasaki, T.
    vol 37   E63, E67, E68, E82, 118
Sasim, D.
    vol 25   E335-E337, 355
Sato, H.
    vol 26   E271-E273, 283
Sato, T.
    vol 23   E272, E276, E281
Saugier-Cohen Adad, M.
    vol 23   E174, E183, 262, 263
Savage, D.W.
    vol 32   E79, E82
Savich, R.F.
    vol 31   E11, E13-E15, 23, E39-E42, 73, 74, E98, E102, 118,
                  E168, E171, 180, 181, E206, E212, 257, 258, E278,
                  289, 290, E297, 304
Savvina, Ya.D.
    vol 27/28   E349, 353, 354, 358, 359, E465, 472, 473, E511,
                  518, 519, E522, 526, 527
Saxena, A.C.
    vol 27/28   439, 440, 447-449
Sayamyan, E.A.
    vol 31   E98, E102, 114, E129, E130, 143, 165, E318, E320,
                  E321, E324, 328
Saylor, J. H.
    vol 20   E103-E105, 112, E132-E136, 137, E153-E157, 158,
                  E202-E204, 205, E210-E213, 214
    vol 37   E63, E68, E81, 96, 100
Sayun, M.G.
    vol 25   E239
Sayward, J.M.
    vol 34   1-12, E14, 20, 73, 78-80, 83, 91
Scaletti, U.
    vol 30   E24, E25, E28, E31, E32, 54, 67, E105, E108-E111, 126
Scatchard, G.
    vol 31   E39-E42
Scatturin, V.
    vol 23   E144, E145
Schaafsma, J.G.
    vol 24   E422, 426, 427
    vol 27/28   E266, 268, 269
Schadler, H.W.
    vol 25   E385-E387, 395
Schatzberg, P.
    vol 37   E64, E72, E82, 136
    vol 38   E225, E226, 228, E294, E295, 299, E326, E327, 328,
                  E376, E377, 378, 395, E426, E427, 428, E464, E465, 466
Scheeline, H.W.
    vol 32   E166, E186, 191
Scheffer, F. E. C.
    vol 37   E42, E46, E66, E80, E85
Scheraga, H.A.
    vol 24   E1, E2, 4, E16, E17, 20, E54, 55, 57
Schevchik, J.E.
    vol 31   E308, E311
Schick, K.
    vol 23   E308, E317, 320
Schiff, H.
    vol 31   E98, E102,
Schiffer, B.
    vol 30   E105, E111, 117, E220, E223, 225, E374, E377, 380, 384
Schindler, P.
    vol 23   E16, E17, E25, 56, 57, E152, E170, E171, E182, E183, 191,
                  245, 247, 248, E270, E272, E275, E281, 290
    vol 32   E1-E3, E25, E26, E28, 36-38, E51
Schiraldi, A.
    vol 33   E108, E109, E113, E116, E134, E243, E289
Schlosing, T.
    vol 30   E105, E112

Schlosser, W.
    vol 25   E369-E371
Schmalenberg, U.
    vol 22   30
Schmandt, W.
    vol 30   E24-E26, E32, 34
Schmedding, D.W.
    vol 20   E35, 39, E73-E77, 88, E89-E91, 98
Schmid, H.
    vol 23   E174, E183, 230
Schmidt, W.
    vol 25   E84, E85, 90, E335-E337
Schmitt, R.E.
    vol 25   E1-E3, 12
Schneck, H.
    vol 25   E182, E183, 187
Schneider, G. M.
    vol 20   E202-E204
    vol 37   E64-E66, E76, E80, E83, E84, 148, 184, 185
    vol 38   E9-E12, E59, E65, E66, E178, E201, E202, E341, E343
Schneider, H.
    vol 24   E1, E2, 4, E16, E17, 20, E54, 55, 57
Schneider, R.F.
    vol 25   E369-E371
Schneider, R.Th.
    vol 26   E152
Schnellbach, W.
    vol 30   E108, E112, 413
Schneller, G.H.
    vol 35   78, 79
    vol 36   476, 480
Schnyders, H.
    vol 23   E86, E96, E144, E145, 147
Schoch, E.P.
    vol 23   E307, E308, E317
    vol 24   E299-E301, 304
    vol 27/28   320, 322, 323, E465, 468, E511, 513
Scholander, P.F.
    vol 27/28   E50, E51
Scholder, R.
    vol 23   E159, E163-E166, E182, E277, E278, E281, E315, E317
Schonfeld, F.W.
    vol 25   E434
Schramke, E.
    vol 29   E213, E214, 215
Schreiner, O.
    vol 31   E11-E15, 26, 27, E98, E102, E168, E169, E171, 175,
Schrier, L.A.
    vol 23   E160, E162, E182, 253
Schriver, G.W.
    vol 27/28   723
Schroder, W.
    vol 27/28   E1-E5, 34
Schroeder, E.
    vol 36   E19, E20, 22, E63, 65
Schroeder, W.C.
    vol 31   E11, E15, 17, E127, E129, E130, 132, 133, 136-138,
    vol 31   155, 156
Schuberth, H.
    vol 31   E101, E102, 126
Schuhmann, H.
    vol 25   E33, E34, 42
Schuller, A.
    vol 25   E13-E18, 25
Schulten, M.A.
    vol 31   E3, E6
Schulz, L.G.
    vol 25   E119-E121, 128
Schulze, A.
    vol 25   E402-E404, 413
Schulze, R.C.
    vol 25   E277
Schupp, O.E.
    vol 25   E335-E337, 343

Schurch, A.
    vol 23   E170, E183, 245
Schurmann, H.K.
    vol 25   E93-E95, 101
Schwab, G.M.
    vol 23   E145
Schwartz, E.
    vol 32   E174, E175, E187, 262
Schwarz, F. P.
    vol 20   E73-E77, 86, E89-E91, 95, 96, E153-E157, 181
    vol 37   E65, E67, E84, 178, E197-E199, 208
    vol 38   E236-E238, 247, 248, E250, 256, E311, E312, 316-318,
                  E366, E367, 369-371, E396-E398, 403, E405, 406, E419,
                  421, E446-E448, 453, E456, 458, 516, E517, 519
Schwarz, R.
    vol 26   E137-E190, E193, 199, 200
Schweitzer, D.
    vol 33   E26, 27
Schweitzer, D.G.
    vol 25   E425-E427
Sciamanna, S.F.
    vol 32   E174, E175, E187, 256
Scott, R. L.
    vol 37   E73, E85
    vol 38   E64, E66, E139
Scudder, N.P.
    vol 27/28   318, 569
Sebastian, H.M.
    vol 27/28   E360, 368, 369, E370, E511, 520, E522, 530, 531,
                  E533, 534, 543-546, 548-550, 561-564, 680, 710, 711
Sechenov,
    vol 32   E25, E27
Sedova, V.A.
    vol 26   E4, E5, E7, 33, 65, E74, E75
Segre, G.J.
    vol 25   E425-E427
Seidell, A.
    vol 29   E228
    vol 30   E195-E197
    vol 31   E51, E52
Sekiguchi, K.
    vol 34   E13-E15, 25, 32, 36, 43, 52, 54, 55, 168, 178
    vol 35   E110-E112, 128, 130, 134
Sekikawa, H.
    vol 35   E64, 83, 90, 275, 276
    vol 36   262, 267
Sekiya, M.
    vol 26   E187, E192
Selivanova, G.A.
    vol 23   E270, E272, E279, 295
Selivanova, N.M.
    vol 26   E348-E352, 361, 363, 381
Selleck, F.T.
    vol 32   E1-E3, 9, 10, E166-E168, E186, 204, 205,
Selmeczi, B.
    vol 35   E110-E112, 141, 203-205
Semibratova, N.M.
    vol 25   E139-E141, 154, E157-E159, 171, E402-E404, 418
Semishin, V.I.
    vol 26   E155-E157, E159, 171, 172, 184, 185
Senderens, J.B.
    vol 31   E14, E15
Sengers, J.V.
    vol 29   E6, E7, E10, E1
Senkowski, B.Z.
    vol 35   E14, E15, 19, 24, 33, 43, 45, 47-50, 57
Senturk, N.H.
    vol 27/28   749
Serafimov, L. A.
    vol 37   E7, E8, 10, E22, E23, 25, E31-E34, 38
Serbrennikov, V.V.
    vol 25   E230, E231, E234, 236
    vol 30   E220, E222, E223, 240, 241

```
Sergeeva, G.S.
     vol 33      E35, 37
Sergienko, I.D.
     vol 32      E183, E184, E187, 323
Setoyama, K.
     vol 26      E187, E192, E193
Seubert, K.
     vol 26      E291
Seward, R.P.
     vol 21      E9, E12, 68
Seybold, K.
     vol 33      E58-E60
Sgoll, G.B.
     vol 34      287-290
Shabashova, M.L.
     vol 26      E262, 265
Shadiakhy, A.
     vol 21      261-266, 268-273
Shah, I.S.
     vol 26      E153-E155, E158, 162, 163, 167
Shah, N.H.
     vol 35      E14, E15, 21, 23, 25, 34, 44
Shakhova, S.F.
     vol 27/28   675, 679, 681, E712, 713, 714, 728
     vol 32      E176, E177, E187, 266
Shakhparonov, M.I.
     vol 31      E206-E208, E212, 234, 235
Shakhtakhtinskii, M.G.
     vol 29      E213, E214, 224
Shalaevskaya, V.N.
     vol 25      E84, E85, 91, E139-E141, E335-E337
Shalamov, A.E.
     vol 25      E157-E159, 171
Shalygin, V.A.
     vol 20      E153-E157, 177
Shammasov, R.E.
     vol 30      E24, E33, 78, 81
Sharma, J.K.
     vol 26      E187, E188, E192, E193, 231
Sharov, V.G.
     vol 20      E1
Shaw, C.S.
     vol 23      E10, E25
Shcherbakova, L.G.
     vol 31      E11, E13, E15, 18, 19, E127, E130
Shelton, F.K.
     vol 26      E256
Shenderi, E.P.
     vol 27/28   E629, 630
Shenkin, Ya.S.
     vol 31      E206-E208, 246, 247, 253, 254
Shepherd, R.G.
     vol 35      216, 246, 249
     vol 36      81, 82, E88, 89, E92, 93
Sherman, W.V.
     vol 21      E2, E8, E12, 20
Sheth, P.R.
     vol 35      E14, E15, 21, 23, 25, 34, 44
Shevchik, E.
     vol 31      E318, E320, E323, E324, 329-333
Shevtsova, Z.N.
     vol 22      74, 98, 123, 151, 208
Shibazaki, T.
     vol 35      3, 22, 68, 263
     vol 36      103, 287, 371, E404, 408
Shibuya, M.
     vol 30      E330, E331, E338, 372, E467, E470, , E472
Shiino, H.
     vol 26      E187, E193
Shim, J.
     vol 27/28   E320, 324-326
Shimidzu, T.
     vol 26      E291
```

Shineider, H.
    vol 30   E112
Shiomi, Ts.
    vol 31   E11, E15, E98, E99, E102, 103, 104
Shipman, L.M.
    vol 27/28 E349, 355-357
Shirinskikh, A.V.
    vol 25   E402-E404
Shishkina, L.A.
    vol 30   E24, E25, E28, E32, 50
Shishkov, Yu.D.
    vol 22   72, 97, 122, 207
Shiu, W. Y.
    vol 20   E103-E105, 113, E153-E157, 180
    vol 37   E65, E68, E84, 165, E188, 190
    vol 38   E161, 163, E236-E238, 246, E294, E296, 303, E311, E312,
                    315, E319, 321, E332-E334, 337, E341-E343, 349, 355,
                     356, E357, 359, E360, 363, E363, 365, E366, E367, 368,
                     E386, 388, 392, E396-E398, 402, E419, 423, E426, E427,
                     431, E433, 434, 437, 461, 475, 476, E477, E478, 481,
                     E486, E487, 490, E493, 496, E497, 500, E512, 514, 520,
                     522, E528, 530, 531, 543
Shkadova, A.I.
    vol 34   E13-E15, 39, 110, E204, 212, 239
    vol 35   E110-E112, 132, 169
    vol 36   E303, 308, 338
Shklar, R.Sh.
    vol 25   E369-E371
Shklovskaya, R.M.
    vol 30   E162, E164, E165, 169, 170, E220, E222,
                     E223, 237, E258, 262, E263, E264, E268-E275, 279,
                     281, 282, 284, 285, 287, 290, 291, 294, 296-303, 312,
                     317, 318, 320, 323, 324, E330, E332, E335, E338,
                     E339, 369, 370, E374, E375, E378, E381, 395,
                     E428-E431, 438, 441, E448-E450, 455, 456, E467,
                     E468, E471, E472, 483
Shlyapnikov, D.S.
    vol 23   E3, 8, E23, E26, 80, E180, E181, E184, 268, E280, E282,
                     305, 306, E316, E317, 340, 341
Shneerson, A.L.
    vol 32   E79, E81, E83, 86, 87, 95, 96, 118, 140
Shokin, I.N.
    vol 26   E90-E92, 104, 105
Sholette, W.P.
    vol 23   E160, E162, E182, 253
Shoor, S.K.
    vol 27/28 E1-E6, E79, E85, 160, 161
Shorina, E.V.
    vol 23   E10, E24, E25, 54, 55
Short, I.
    vol 21   E5-E7, E9, E10, E12, 47, 50, 55, 70
    vol 24   E331-E334
    vol 32   E171-E173, E175, E176, E178, E179, E186, 241, 243,
                     248, 263, 270, 274, 287
Short, O.A.
    vol 23   E177, E184, 210
Shpunt, S.J.
    vol 31   E39-E42, E47, E48, E51, 54-64
Shtern, E.K.
    vol 23   E3, 8, E23, E26, 80, E180, E181, E184, 268, E280, E282,
                     305, 306, E316, E317, 340, 341
Shternina, E.B.
    vol 31   E98, E99, E101, E102, 115, 116
Shtrum, E.L.
    vol 23   E275, E281, 300
Shulgina, I.M.
    vol 26   E403-E405, 409, 410
Shunk, F.
    vol 25   E173, E174
Shvedov, V.P.
    vol 25   E207-E208, E214, E221, E225, E234, 235, E242,
                     E253, 254

Shvelashvili, A.E.
    vol 35   210-212
    vol 36   117-119, 350, 351, 431-433
Sidgwick, N.V.
    vol 20   E183, E184, 185-187, E190, 191-193, E195, E196,
                  197-199
Siede, B.
    vol 25   E119-E121
Siekierski, S.
    vol 30   E163, E165,
Signoetti, Ciranni, E.
    vol 34   77, 308
    vol 35   145
    vol 36   33, 147
Sillen, L.C.
    vol 23   E90-E92, E96
    vol 26   E187, E188, E190, 191, 201-210, 222
Silver, L.
    vol 32   E54, E55, 68
Simensen, M.
    vol 36   E19, E20, 22, E63, 65
Simmons, J.P.
    vol 30   E184-E187, 189
Simnick.J.J.
    vol 27/28   E360, 368, 369, E511, 520, E522, 530, 531, E533,
                  534, 543-546, 548, 549, 561-564, 680, 710, 711
Simon, A.
    vol 26   E155, E159, E187, E189, E193, E236, E240, E241
    vol 27/28   E360, 368, 369, E511, 520, E522, 530, 531, E533,
                  534, 543-546, 548, 549, 561-564, 680, 710, 711
Simon, E.
    vol 37   E63, E67, E81, 103
    vol 38   E187, E189, 190, E201, E202, 204
Simonelli, A.P.
    vol 35   E110-E112, 133, 170, 190
Simonov, M.A.
    vol 23   E180, E184
Simpson, P.
    vol 24   E130-E133, 136, 146, 152, 154
Simulin, Y.N.
    vol 38   E23-E25, E27, E28, 37
Singh, A.D.
    vol 26   E271, E273
Singh, B.N.
    vol 23   E156, E182
Singh, R.P.
    vol 37   E65, E71-E73, 84, 175
Sinistri, C.
    vol 33   E237
Sinyavskaya, R.P.
    vol 27/28   44, E629, 631
Sircar, S.C.
    vol 23   E10, E17, E25, 53
Sirotin, A.G.
    vol 24   E196, E197, 200, 201
    vol 27/28   E248, E249, 256, 257, 276
Sisi, J.C.
    vol 32   E330, E331, 337, 338
Sjogren, B.
    vol 35   153, 154, 228
    vol 36   241, 252, 275, 455, 458, 461, 463-465, 472, 473, 485
Skell, P.
    vol 33   E91, 92
Skirrow, G.
    vol 27/28   E50, E51
Skorobogatova, V.I.
    vol 32   E20, 21, E54, 58, 78
Skripka, V. G.
    vol 24   E196, E197, 200, 201
    vol 27/28   E24-E28, 36, 37, E57, E70-E72, E74, E85, 129,
                  E248, E249, 256, 257, 276, E297, 300-302, E370,
                  371
    vol 37   E65, E76, E77, E83, E84, 170, 171

Skripka, V. G.
    vol 38   E135, E138, E141, 152, 153, E294-E296, 304-306, E464,
                    E465, 470, 472, 473
Sloan, E.D.
    vol 27/28   748
Smaaberg, R.
    vol 25   E285, E286, 299
Smeeth, A.G.
    vol 23   E90, E93, E96, 124-126
Smija, J.
    vol 30   E270, E272, E275, 319, E467, E472
Smit, J.H.
    vol 20   E132-E136, 147, E153-E157, 179, E202-E204, 209,
                    E210-E213, 220
Smith, A.R.
    vol 25   E301, E302
Smith, D.L.
    vol 25   E182, E183
Smith, F.E.
    vol 25   E402-E404, 412
Smith, F.J.
    vol 23   E23, E26
Smith, G.McP.
    vol 25   E1-E3, 5, E33, E34, 39, E44, E45, 48, E50, E51,
                    53, 54, E66, E67, E71, E72, 75, E76, E77, 79
Smith, M.J.S.
    vol 24   E16, E17, 27, E65-E68, E71, 110
Smith, M.L.
    vol 22   E328
Smith, N.O.
    vol 27/28   E24-E28, 33, 35, E57, E60, E66-E68, E70-E72, E74,
                    E75, E84, 93, 94, 108-111, 122, 123, 125, 126, 147
Smith, R.M.
    vol 31   E2, E6
Smith, S.B.
    vol 30   E374, E375, E378, E380, 408, E467, E468, E472
Smits, A.
    vol 25   E57, E58, E84, E85
Smurigina, T.V.
    vol 25   E76, E77
Smyth, C.P.
    vol 25   E119-E121, 126
Snyder, J. R.
    vol 37   E64, E76, E77, E82, 139, 140
Sobin, S.S.
    vol 35   173
    vol 36   43, 179
Sobol, S.I.
    vol 23   E20, E26, E178, E184
Soboleva, L.V.
    vol 30   E471, E472, 486
Sobolevskaya, N.G.
    vol 30   E162, E165, 168, E176, E178, 181, E220, E222,
                    E223, 226, 261, E263, E264, 267, E374, E375, E381,
                    385, E428-E431, 434, E448-E450, 453
Sobr, J
    vol 21   E82, E89, 142-147
Sochevanov, V.G.
    vol 23   E160, E162, E165, E167, E183, 225, 226
Soffar, F.
    vol 34   E285, 286, 292
Sogawa, K.
    vol 29   E1, E5, E7, E14, E16-E18, E21, 40, E47, E49, E50
                    E51, 72-86, E87, 88-94
Sokalova, O.N.
    vol 31   E307, E308, E311, E318, E324
Sokolov, N.M.
    vol 33   E9, 10, E12, E15, E50, 53, E57-E60, 65, 69, E77,
                    E85, 97, E99, 100-102, E103, E105, E108, E109, 110,
                    E113, 114, E116, E119, 120, E121, 122, E123, E125,
                    127, E128, 139, 140, E141, E142, 143, E145, 146, 149,
                    E167, 169, E172, 174, 180-182, E183, 184, 186,
                    E187, E189, 190, E191, 192, E193, 194, E195, 196,
                    E197, 198, E199, 200-205, 207, 209, 213-217, 221,

Sokolov, N.M.
    vol 33   222, 224, 225, E226, 227, E229, 230, 231, 235,
                  E237, 238, 242, E243, 244, 245, 247, E248, E250,
                  251, E253, 254, 256, 259, 265, 266, E267, 268,
                  273, 275-277, E278, 279, E280, 281, 282, E283, 284,
                  E285, 286, E287, 288, E289, E292, 293, E294, 295,
                  296, E297, 298-302, E303, 304, 305, E306, 307,
                  E308, 309, 310, 312, E313, 314, E315, 316, E317,
                  318, E319, 320, E321, 322, 323, E324, 325, E326,
                  327, E328, 328, 329, E330, 331
Sokolov, S.J.
    vol 31   E278, 281
Sokolova, E.I.
    vol 26   E1-E6, 30, 40, E72-E74, 78
Solc, Z.
    vol 31   E168-E171, 203, 204
Solenchyk, B.
    vol 31   E129, E130, 166, 167
Soloman, E.
    vol 27/28   485-490
Solov´ev, A.P.
    vol 31   E51, E52, 90, 91, E206-E209, E211, E212, 255, 256
Solov´ev, M.A.
    vol 22   182, 183, 185-187
Solov´eva, V.D.
    vol 23   E11, E13, E14, E25, 60, 61, 260, E277, E282, 296, 297
Solovyeva, V.D.
    vol 23   260, E277, E282
Somait, F.A.
    vol 27/28   E737, 747
Somasundaran, P.
    vol 24   E16, E17, 19
    vol 27/28   E1-E5, 18
Sonnenberg, H.
    vol 34   283, 284
    vol 36   258, 259, 277, 278
Sorokin, V.I.
    vol 29   E1, E2, E4, E5, E14, E16-E18, E21, 36, 39
Soshmikov, V.I.
    vol 25   E194, E195, 197
Sotova, N.N.
    vol 26   E1-E6, 29-31, 39-41, 44, E72-E75, 78-81
Spack, A.
    vol 30   E330-E333, E336-E338, 342, 349, 350
Speakman, J.C.
    vol 36   138, 162, 163
Spencer, J.F.
    vol 25   E119-E121, E157-E159, E194, E195, E335-E337,
                  E402-E404
Spencer, J.N.
    vol 29   E1, E3, E7, E14, E16-E18, E21, 32, E102-E109
                  117, 121, 123, 125, 129, E133-E137, 140, 142-146
                  E147, 148, E149-E153, 156, 159-162, E163-E165,
                  168, 169, E172, E173, 174, 175, E177, E179, 181
Spencer, J.R.
    vol 27/28   E511, 517, E522, 523
Spengler, H.
    vol 25   E103-E105, E173, E174, 196, E317, E318, E335-E337
Speranskaya, E.F.
    vol 25   E310
Spereni, G.
    vol 25   E57, E58, 60, E66, E67, 68
Spicer, W.M.
    vol 25   E93-E95, 97, E103-E105, 107
    vol 29   E213, E214, 216
Spivakovskii, V.B.
    vol 23   E13, E16, E18, E19, E25, 79, E271, E281
Sprake, C.H.S.
    vol 29   E2, E6, E21
Srb, V.
    vol 31   E51, E52, E206, E207, E211, E212, 251, 252, E278,
                  287, 288
Sreenathan, B.R.
    vol 30   101, 160

Stanford, J.W.
     vol 34   180
     vol 35   215
     vol 36   58, 208-214, 216, 227, 268, 357, 402
Stanley, D.R.
     vol 27/28   E50, E51
Stapp, F.P.
     vol 27/28   318, 571
Starkovich, J.A.
     vol 30   102
Stasevich, N.H.
     vol 23   E10, E24, E25, 54, 55
Staudenmayer, L.
     vol 31   E15, E168, E171
Staufenbiel, E.
     vol 23   E277, E278, E281, 288, E315, E317
Staveley, L. A. K.
     vol 27/28   E248, E249
     vol 37   E63, E71-E74, E81, E82, 102, 113
Ste.-Marie, J.
     vol 23   E89, E90, E96, 141, E170, E172, E173, E178, E183, 250,
                   251, 254, 255
Stearns, R. S.
     vol 37   E63, E67, E81, 103
     vol 38   E187, E189, 190, E201, E202, 204
Steiner, B.
     vol 23   E156, E159, E162, E182, 205, 206
Stenbeck, S.
     vol 29   E214
Stepanova, G.S.
     vol 27/28   E303, E465, 475, 476, E511
Stepanova, I.N.
     vol 24   E257, E258, 283-298
     vol 27/28   694, 696-701
Stepanova, Z.G.
     vol 21   E1-E3, E12, 14, 22, 30, E89
     vol 32   E170, E171, E182-E184, E186, E187, 226, 233,
                   286, 293, 295, 324
Stepin, B.D.
     vol 22   112, 113, 196, 304, 305
     vol 31   E307, E308, 312, 314, E318, E320, 325, 327
Sterner, C.
     vol 27/28   E737
Stewart, J.F.
     vol 23   E1, 6, E21, E26, 29, 30
Stock, A.
     vol 29   E1, E3, E14, E16-E18, E21, 27
Stockl, F.
     vol 30   E24, E29, E32, 49, 67
Stoessell, R.K.
     vol 27/28   E24-E28, 42, E50, E51, 55, E64-E68, E72, E74-E83,
                   E85, 104-106, 112, 113, 143, 146, 153, 155-159, 165,
                   166, 168-170, 192, 193
Stoica, T.
     vol 21   E10-E12, 74
Stokes, R.H.
     vol 20   E195, E196
     vol 31   E2, E6
Stoller, L.
     vol 24   E1, E2, 10, E16, E17, 28, E34, 35, E61, E71,
                   81-83, 122-124
     vol 27/28   E1-E6, 10, E64, E85, 103, 177
Storonkin, A.V.
     vol 33   E35, 39, E43, 45, E57-E60, 71, E229, 233, E237,
                   241, 261, E267, 272, 274
Strachan, A.N.
     vol 20   237
Strakosch, E.A.
     vol 34   E13, E14, 19
     vol 35   E110, E111, 117, 213
     vol 36   E19, E20, 26, 53, E124, 126, 225
Streitweiser, A.
     vol 24   E299-E302, 310
     vol 27/28   708

Streitweiser, A.
     vol 37   1
Stricker, H.
     vol 36   150
Stubley, D.
     vol 29   E190, E192, 195
Stuckley, J. M.
     vol 37   E63, E68, E81, 100
Suetaka, T.
     vol 24   E1, E2, 9, E59, E65-E71, 73, 105, 111, 112, 114,
                    116
     vol 27/28   E1-E6, E70, E74, E79, E80, E85, 128, 164
Sugakevich, I.P.
     vol 23   E20, E26
Sukhanov, L.P.
     vol 31   E39-E42, E47, E48, E51, 79
Sukmans´ka, I.V.
     vol 34   E13-E15, 26, 28, 98, 101-104, 106
Sultanov, R. G.
     vol 27/28   E24-E28, 36, 37, E370, 371
     vol 37   E65, E76, E77, E83
     vol 38   E464, E465, 472, 473
Sundermann, F.W.
     vol 35   152, 225
Sunwoo, C.
     vol 35   46, 89, 189
     vol 36   201, 264, 282, 342, 387
Susak, N.J.
     vol 27/28   E74, E85
Suter, R.
     vol 34   271, 274, 280, 282
     vol 35   E110-E112, 151, 227
     vol 36   E19, E20, 37, E63, 74, 153, E228, 241
Sutton, C.
     vol 38   E9, E11, E12, 18, 22, E23, E24, E28, 35, E42, E43, E46,
                    54, 58, E59, E60, E66, 80, 91, E164, E165, 166, 168,
                    E169, E170, 172, 176, E177, E178, 183, 186, E187-E189,
                    196, 200, 259, E265, E266, 269, 270, E271, 274, 276,
                    E427, 430, 432, E464, E465, 471, 474, E501, E502, 506,
                    E277, E278, 284, 285, E376, E377, 380, E381, 383, E426,
                    507, 525, 526, 544, 545
Suvarova, N.V.
     vol 23   E280, E282
Suzuki, A.
     vol 36   137, 141, 142, 146, 149, 167, 174, 175, 180, 185
Suzuki, T.
     vol 26   E262
Svetlova, G.M.
     vol 32   E169, E186, 216
Svirchevskaya, E.G.
     vol 23   E11, E13, E14, E25, 60, 260, E277, E282, 296
Svrcek, W.Y.
     vol 27/28   576, 577
Sweeney, C.W.
     vol 32   E174, E176, E183, E187, 260
Swenson, T.
     vol 30   E24-E26, E29, E32, 62, 63, E105, E109, E111, 142,
                    143, E195-E197, E220, E221, E223
Swinehart, D.F.
     vol 23   E158, E163, E165, E168, E170, E171, E182, 229
Swoboda, O.
     vol 26   E187, E188, E191, E192, 214, 233-235
Sysoyev, L.N.
     vol 23   E313, E317, 332
Szebenyi, I.
     vol 27/28   720
Szekely, A.
     vol 27/28   720
Szeliga, J.
     vol 37   E13-E15, 21, E31-E34, 41, E42, E43, E46,
                    59, E65, E68, E84, 173, E212, E213, 218

Szeliga, J.
    vol 38   E9, E12, 20, E23, E28, 39, E42, E46, 56, E59, E60, E66,
                  84, 96, 99, E114, E117, 123, E126, E127, 130, 134, E135,
                  E141, 154, E169, E170, 174, E187, E189, 198, 215, 224,
                  E225-E227, 233, 235, E271, 275
Szendrei, T.
    vol 26   E153, E158

T

Taber, W.C.
    vol 23   E21, E26
Tadros, M.E.
    vol 26   E187, E188, E192
Tagawa, K.
    vol 34   E14, E15, 45
    vol 35   262
    vol 36   E367, 370
Taillon, R.
    vol 23   E17, E20, E26, 59, E272-E277, E281, 303
Takahashi, S.
    vol 23   E174, E183
    vol 26   E187, E191, E192, E193, 223
Takayama, K.
    vol 35   36, 51-53, 58
    vol 36   283, 289, 290, 292-295
Takeda, S.
    vol 23   E16, E25
Takubo, T.
    vol 34   56, 57, 76, 82, 96
    vol 35   69-71, 77
    vol 36   143, 144, 151, 166, 168, 372-374, 378, 379
Tama, A. A.
    vol 37   E64, E67, E73, E83, 147
Tammann, G.
    vol 25   E13-E18, 21, E33, E34, 35, E119-E121, E139-E141, 142,
                  E157-E159, 160, E172, E182, E183, 184, E268, E277, E282,
                  E284-E286, E301, E302, E317, E318, E335-E337, 341, E357-
                  E359, E369-E371, 373, E385-E387, 388, E402-E404, 405,
                  E425-E427
    vol 29   E47, 53
Tamminen, V.
    vol 23   E17, E26, 81
Tannanaeva, N.N.
    vol 25   E103-E105, 106
Tarantino, S.
    vol 26   E248, E249
Tarasenkova, O.S.
    vol 23   E180, E184, 256
Tarasov, A.A.
    vol 33   E35, 39, E43, 45, E58-E60, 71
Tarasova, G.N.
    vol 30   E268, E270-E274, 280, 311, 313, 325-327, E374,
                  E375, E379-E381, 398, 404, 406, E428-E431,
                  437, 440, 442, E461, E462, E467, E468, E471, E472,
                  484-486, 487, 488
Tarassenkaw, D. N.

    vol 37   E63, E71-E73, E81, 97
Tarrer, A.R.
    vol 27/28   547, 565, 566
Tartar, H.V.
    vol 32   E79, E83, 84, 85
Tarullo, M.
    vol 27/28   E24-E28, E57, E72-E74, E85, 130-142
Tatarinova, V.A.
    vol 30   E428-E431, 443, E448-E450, 457, E461, E462,
                  466, E467, E468, E472
Tatuoka, S.
    vol 30   E268, E269, E274, 278
Tauch, E.J.
    vol 27/28   E266, 267, E281, 282, E303, 304, E320, E349, 350,
                  E465, 466, E511, 512, 559, 560, 635, 636

Tawashi, R.
    vol 34   E13-E15, 29, 30, 128-130, 132, 133, 135, 136,
                  138-141, 144-147, 164, 165
    vol 35   E110-E112, 136, 165-167
Taylor, A.E.
    vol 30   E105, E108, E111, 151, 152, 154, 155
Taylor, D.F.
    vol 25   E139-E141, 152
Taylor, D.R.
    vol 24   E299-E302, 310
    vol 27/28   708
Taylor, H.S.
    vol 37   E31, E33, E34, 35, E42, E45, E46, 48, E63,
                  E71, E81, 104, E209
    vol 38   E135, E138, E141, 143
Taylor, J.A.
    vol 26   E152
Taylor, M.D.
    vol 22   E328
Taylor, N.W.
    vol 21   E1, E11
Taylor, P.
    vol 31   E14, E15
Teeter, G.A.
    vol 25   E135, 136
Tebbe, U.
    vol 27/28   732
Tel´nyuk, E.N.
    vol 23   E181, E184
Telepneva, A.E.
    vol 26   E1-E4, E6, 16, 17, 38, 59-62
Tember, G.A.
    vol 25   E135, 136
Temmerman, E.
    vol 25   E173, E174, 181
Templeton, C.C.
    vol 26   E187, E191-E193, 224, 225
Tenu, R.
    vol 31   E39-E42
Teodorovich, V.P.
    vol 32   E172, E186, 238
Tereshchenko, L.Ya.
    vol 31   E168, E171, 178, 179
Terpstra, P.
    vol 26   E256, E258
Terrana, J.D.
    vol 26   E152
Terres, E.
    vol 23   E21, E26, E177, E184
    vol 26   E113-E116, 121-123, 129, 130, 134, 135, 140,
                  E144, E145, 150, 151, E252, 254, 255, E262, 263,
                  264, E271-E273, 276, 277, E293, 297, 298
Teslo, S.P.
    vol 26   E5, E6, 58
Teusdea, G.M.
    vol 21   E10-E12, 74
Thibault, M.
    vol 25   E103-E105
Thomas, W.H.
    vol 38   E332, E334, 339, 340, E446-E448, 455, E456, 460
Thomas, J.S.
    vol 23   E1, 7
Thompson, E.S.
    vol 25   E301, E302
Thompson, H.E.
    vol 25   E157-E159, 166
Thompson, W. H.
    vol 37   E64, E66, E76-E79, E82, 139, 140
Thomsen, E.S.
    vol 24   E130-E134, 138, 148, 225, E257, E258, 267,
                  E375-E377, 384
Thorne, R.L.
    vol 38   E250, 252-254

Thornhill, D.G.T.
    vol 32   E176, E187, 265
Thrasher, R.D.
    vol 31   269, 270
Thurston, W. M.
    vol 37   E64, E71, E72, E83, 146
Thury, G.
    vol 32   E1-E3, E20, 22-24
Tichborne, C.R.
    vol 26   E271-E273, 275
Tichy, D.
    vol 31   E51, E52, E206, E207, E211, E212, 251, 252, E278,
               287, 288
Tilden, W.A.
    vol 31   E98, E102
Tilquin, B.
    vol 24   E130-E134, 139, 143, 169, 170, 190, 191
    vol 27/28 E202-E204, E214, 220, 223
Tilton, V.V.
    vol 24   E411, 412, 413
    vol 27/28 648
Timofeev, S.I.
    vol 30   E330, E332, E335, E339, 370
Timofeeva, E.G.
    vol 21   E7, E12, 51, 52
Timofeyev, G.
    vol 25   E157-E159
Timonin, V.E.
    vol 21   E7, E12, 51, 52
Timoshenko, Yu.M.
    vol 31   E212, 271
Tirschmann, J.
    vol 22   31, E32, 33, 36, 47, 89, 94, 103, 132, 141, 143,
               176, 191, 197, 218, 227, 237, 256, 264, 267, 275,
               282, 294, 302, 314, E334-E336, 337, 338, 346, 367,
               374, 376
Toda, S.
    vol 30   E105, E111, 122
Todheide, K.
    vol 24   E48, 52
Toibaev, B.K.
    vol 25   E173, E174
Toischer, K.
    vol 23   E156, E182
Tokareva, A.G.
    vol 30   E268, E271, E274, 300, 301, 308, 309, E330,
               E332, E335, E338, 369
Tokunaga, J.
    vol 27/28 E1-E6, 173, 174
Tomashik, V.N.
    vol 25   E194, E195
Tomashevsky, G.P.
    vol 26   E348-E351, 367, 368, 371, 394, 395, 401
Tomii, R.
    vol 23   E16, E25
Tominaga, T.
    vol 27/28 E685, 687
Tomlinson, J.W.
    vol 23   E88, E96
Tomoda, T.
    vol 29   E51, 72-86, E87, 88-94, 100, 101
Tonelli, P.L.
    vol 33   E35, E43, E58-E60, E105, E116, E119, E123, E128,
               E134, E142, E175, E253, E283, E297, E315, E326
Toner, D.F.
    vol 25   E317, E318, 320
Torochesnikova, N.S.
    vol 26   E1-E6, 29-31, 39-41, 64, E72-E75, 78-81
Torochestnikov, N.S.
    vol 31   E278, E279, 293-296
Toropova, V.F.
    vol 26   E248-E352, 399, 400
Toul, F.
    vol 29   E228, 231, 232

Toyota, H.
    vol 20   E6, E7, 11-14
Trachenberg, I.
    vol 23   E160, E162, E182, 249
Traube, W.
    vol 23   E15, E25
Traul´ko, I.V.
    vol 30   E380, E381, 409, 410
Treadwell, W.D.
    vol 30   E1-E4, 17, E24, E25, E32, 36, E105, E111, 118,
                  E162, E165, 167, E176, E178, 180
Trebukhov, A.A.
    vol 25   E357-E359
Trefouel, M.
    vol 34   E14, 18
    vol 35   E110, E111, 116
    vol 36   E19, E20, 25
Tremaine, P.R.
    vol 31   E14, E15
Tremper, K.K.
    vol 21   E2, E3, E12, 17, 24, 34
    vol 32   E166-E168, E171, E186, 211, 223, 235, 236
Trendafelov, D.
    vol 26   E153, E154, E157-E159, 173, 174, 177-179
Treshchina, N.I.
    vol 27/28   572, 573
Treybal, R.E.
    vol 20   E153-E157, 164
Trimble, F.
    vol 30   254
Troitskaya, N.I.
    vol 30   E268, E271, E272, E274, 304
Trueger, E.
    vol 20   E153-E157, 161
Trukhanova, E.A.
    vol 26   E90-E92, 106, 107
Tschungaeff, L.
    vol 30   E105, E111, 115
Tseff, A.L.
    vol 23   E20, E26
Tsibulevskaya, T.A.
    vol 30   E162, E164, E165, 169, 170, E268, E272,
                  E273, E275, 281, 323, 324, E467, E468,
                  E471, E472, 483
Tsiklis, D.S.
    vol 24   E48
    vol 32   E169, E186, 216
Tsindrik, N.M.
    vol 33   E9, 10, 19, E50, 53, 97, 102, E113, 114, 208, E210,
                  213, 215, 217, E237, 239, 266
Tskitishvili, M.G.
    vol 35   208-212
    vol 36   115-119, 347, 349-351, 428-433
Tsonopoulos, C.
    vol 37   E65, E68, E72, E73, E76, E78, E79, E84,
                  181, 182
    vol 38   E59-E63, E65, E66, 88, 89, E104-E106, 108, 109, E136,
                  E137, E139-E141, 156, 157
Tsuchiya, S.
    vol 34   56, 57, 76, 82, 96
    vol 35   69-71, 77, 78
    vol 36   143, 144, 151, 166, 168, 372-374, 378, 379
Tsuda, Y.
    vol 34   E13, E14, 43, 168
Tsuyuki, N.
    vol 26   E187, E192
Tsybnlevskii, A.M.
    vol 24   E257, E258, 283-298
    vol 27/28   694, 696-701
Tuck, D.G.
    vol 21   E8, E12, 64
    vol 29   E1-E3, E7, E14, E16-E18, E21, 30

Tugolukov, V. M.
    vol 37   E13, E15, 17, E22, E23, 24, E31, E33, E34,
                  36, E64, E71, E72, E82, 141, E197-E199,
                  202, E212, E213, 215
    vol 38   E23, E27, E28, 34, E59, E62, E63, E66, 73, 93, 97,
                  E114-E117, 120, 132, E135, E138, E139, E141, 145, E177,
                  E178, 182, E187, E189, 194, 213, 220-222, E260, 262,
                  264, E265, E266, 268, E271, 273, E277, E278, 282, E287,
                  289, 293, E311, E312, 313, 322, 325, 373, 374, 393,
                  394, 425, 462, 463, E464, E465, 467
Tukhtaev, S.
    vol 30   E24, E33, 78, 81
    vol 31   E206, E212, 276, 277
Tumanova, N.Kh.
    vol 29   E207, E208, 210
Tunell, G.
    vol 23   E21, E22, E26, 40-42
Tung, T.
    vol 25   E194, E195
Turner, S.L.
    vol 20   E183, E184, 185-187, E190, 191-193, E195, E196, 197-199
Turnetskaya, A.F.
    vol 30   E105, E106, E109, E111, 134, 135
Turolla, E.
    vol 35   6
    vol 36   104, 332, 365, 441, 442
Tusheva, L.A.
    vol 31   E127, E130, 148
Tyrell, V.
    vol 23   E275, E281, E309, E317, 331

U

Ubbelohde, A.R.
    vol 33   E58-E60, E119, E121, E123, E125, E161, E197, E199,
                  E226, E248, E250, E278, E292, E303, E308, E311,
                  E313, E315, E317, E319, E321, E330, E332
Uchida, J.I.
    vol 26   E187-E193, 213
Udovenko, V. V.
    vol 37   E64, E67, E82, 137
Uehara, K.
    vol 27/28   E248
Ugai, Ya.A.
    vol 25   E173, E174
Ugorets, M.I.
    vol 23   E13, E17, E18, E23, E25, 58
Ukraintseva, E.A.
    vol 31   113, E127, E130, 141, 142
Ul´yanova, T.P.
    vol 31   E168, E171, 205, E206, E212, 219
Umano, S.
    vol 24   E1, E2, 5, E16, E17, 22, E65-E68, E71, 100-104, 106,
                  108
    vol 37   E64, E76, E79, E82, 119, 120, E188
Umehara, T.
    vol 24   E1, E2, 9, E59, E65-E71, 73, 105, 111, 112, 114, 116
    vol 27/28   E1-E6, E70, E74, E79, E80, E85, 128, 164
Umezawa, T.
    vol 30   E268, E269, E274, 278
Upthegrove, W.R.
    vol 25   E385-E387, 400, E402-E404, 417
Urazov, G.G.
    vol 23   E161, E182, 231-234
Urey, H.C.
    vol 20   E35, 36, E132-E136
Urieta, J.S.
    vol 27/28   E640, E642-E644, 651, 652, 654, 660, 695
Urosova, M.A.
    vol 31   E127, E130, 151
Usachev, P.v.
    vol 25   E194, E195

Usenove, K.A.
    vol 25   E214, E215, 218, E221, E225, 227, E230,
                    E234, E245, E253, 255
Uspenskii, S. P.
    vol 37   E63, E67, E72, E81, 94

V

Valvani, S. C.
    vol 20   E3, 4, 5, E15, 16, E17, 18, E19, 20, 21, E22, 23, E26,
                    27, E28, 29, E33, 34, E35, 38, E41, 42, E49, 50, E51, 52,
                    E53, 54, E55, 56, E57, 58, E59, 61, E67, 68, E69, 70, E71,
                    72, E73-E77, 87, E89-E91, E103-E105, 115, E118, 119, E120,
                    121, E122, 124, E125, 126, E127, 128, E129, 130, E132-
                    E136, 148, E183, E184, 189
    vol 37   E65, E68, E84, 177
    vol 38   E1, E2, 8, E260, 263, E332, E334, 338, E341, E343, 350
van Konynenburg, P. H.
    vol 37   E73, E85, E222, E229, E319, E323
Vance, J.E.
    vol 30   E330-E333, E335-E338, 340, 344-348, 354-358,
                    E380, E382
Vandenbosch, R.
    vol 23   E158, E161, E162, E171, E182, 227, 228, E275, E281
Vanderhart, D.
    vol 23   E84, E96
Van der Kooi, H.J.
    vol 27/28   E370, 393-375
Van der Linden, T.
    vol 26   E370, 373-375
Vanderlugt, L.A.
    vol 23   E84, E96, 129, E144, E145, 147
Vanderzee, C.E.
    vol 31   E2, E6
Van Heteran
    vol 25   E139-E141, 145
Van Horn, L.D.
    vol 27/28   385, 386, 394, 395, 556
Vanjusheva, L.N.
    vol 31   E11, E13-E15, 20, 21
van Konynenberg, P.H.
    vol 38   E64, E66, E141
van Krevelen, D.W.
    vol 32   E54, 56, 57, 66, 67, 72
Van Lent, P.H.
    vol 25   E139-E141, 151
Vanstone, E.
    vol 25   E13-E18, 26, 27
Vanyarko, V.G.
    vol 25   E194, E195
Vanyukov, A.V.
    vol 25   E194, E195
    vol 29   E207, E208, 212
Vargo, E.J.
    vol 25   E134, E217, E174, E277, E282, E284

Vasil´eva, N.A.
    vol 27/28   694, 696-701
Vasil´kova, I.V.
    vol 33   E35, 39, E43, 45, E57-E60, 71, E229, 233, E237,
                    241, 261, E267, 272, 274
Vasilenko, N.A.
    vol 26   E113-E116, 120, 127, 128, 132, 133, 141-143, E144
                    E145, 146-149
Vasilev, K.H.
    vol 26   E303-E305, 316, 317, E403-E405, 412, 413
Vasileva, N.A.
    vol 24   E257, E258, 283-298
Vassilev, H.
    vol 26   E303-E305, 314, 315, E403-E405, 411
Vaughn, J.W.
    vol 30   218, 256
Vaugn, T.H.
    vol 20   E103-E105

Vdovkina, T.W.
        vol 30   E268, E271-E275, 284, 296, 297, 305, 306, 320, 441
Vei, D.
        vol 32   E183, E184, E187, 324
Vejrosta, J.
        vol 24   E196, E197, 202, 203
        vol 37   E42-E44, E46, 61
        vol 38   E135, E136, E137, E141, 155, E225-E227, 234
Velikanova, L.V.
        vol 31   E278, 283, 284, 291, 292
Velikovskii, A.S.
        vol 27/28   E303, E349, E465, 475, 476, E522, 551, 552
Vemulapalli, G.K.
        vol 29   E165
Vengel, P.F.
        vol 25   E194, E195
Venkatachalam, K. A.
        vol 37   5, E22, E23, 26, E197-E199, 205
        vol 38   112, 113, 292
Verbeek, F.
        vol 25   E173, E174, 181
Verendyakina, N.A.
        vol 31   E51, 90, 91, E206-E209, E211, E212, 255, 256
Vereshchagina, V.I.
        vol 31   E278, 286
Vericeanu, G.
        vol 26   E248-E251, 353, 354, 356, 358, 362, 370, 374, 374
                       377, 390, 391
Vermillion, H. E.
        vol 37   E63, E66, E81
Verpiaetse, H.
        vol 25   E173, E174, 181
Vesala, A.
        vol 20   E89-E91, 93, 94, E103-E105, 111, E132-E136, 146,
                       E153-E157, 175, 176, E210-E213, 218, 219
Veselitskaya, T.A.
        vol 36   328, 333
Vetkina, L.S.
        vol 31   E307-E309, E311, 313
Victoria, A.P.
        vol 30   E472
Vier, D.T.
        vol 25   E421
Vinogradov, E.E.
        vol 30   E220, E222, E223, 239, E268-E275, 280, 286,
                       288, 289, 292, 295, 311, 313, 316, 321, 322,
                       325-327, E330, E335, E339, 367, 368, E374, E376,
                       E378-E381, 397-402, 404, 406, E428-E431,
                       437, 439, 440, 442, E448-E450, 454, E461, E462,
                       E467, E468, E471, E472, 481, 482, 484, 485, 487, 488
Vlasov, G.A.
        vol 30   E24, E25, E28, E32, 50
Vlasova, T.L.
        vol 31   E168, E171, 178, 179
Vodrazka, W.
        vol 32   E174, E175, E187, 262
Vogel, A.
        vol 29   E102-E109, 126, 130, E133, E137, 141, E177-E179, 184
                       E207, E208
Vohland, P.
        vol 31   E101, E102, 126
Voigt, A.F.
        vol 29   E1-E3, E7, E14, E16-E18, E21, 29, 32, E102-E109,
                       114, 115, 117, 121, 123, 125, 129, E133-E137,
                       138-140, 142-146, E147, 148, E149-E153, 154-162
                       E163-E165, 168, 169, E172, E173, 174, 175, E177-
                       E179, 181-183, 185, 186, E187, 188, 189
Vokhrysheva, L.E.
        vol 25   E239
Vol, A.E.
        vol 25   E402-E404
Volkamer, K.
        vol 32   E174, E175, 262

Volkov, A.G.
    vol 25   E1-E3, E33, E34, E44, E45, E50, E51
Volkova, V.N.
    vol 25   E173, E174
Volkova, V.P.
    vol 26   E348-E351, 369
Volosatova, N.S.
    vol 25   E194, E195
Volova, L.M.
    vol 27/28   E477
Volpe, D.
    vol 27/28   E1-E5, 179
Volquartz.
    vol 21   E1, 13
von Meyenburg, H.
    vol 36   330, 360
Vorob´eva, O.I.
    vol 26   E403-E405, 407-410, 417, 418
Vorob´eva, T.A.
    vol 31   E39-E42, E47, E50, E51, 70-72, 86, 87, E206-E209,
                    E212
Vorona, Yu.V.
    vol 25   E194, E195
Voronina, L.V.
    vol 31   E168-E171, 180, 181, E206, E212, 257, 258, E278,
                    289, 290, E297, 304
Voronskaya, G.N.
    vol 22   136, 179, 220, 238, 258, 277, 296, 317, 340, 352
Vorotnitskaya, L.T.
    vol 26   E155-E157, E159, 171, 172, 184, 185
Vorozhtsov, N.N.
    vol 20   E153-E157, 160
Vos, A.
    vol 26   E286
Voskanyan, S.S.
    vol 31   E129, E130, 149
Vosolsobe, J.
    vol 31   E51, E52, E206, E207, E211, E212, 251, 252, E278,
                    287, 288
Vree, T.B.
    vol 35   29, 62, 63, 75, 97, 101, 159, 229, 252,
                    253-255, 269, 287, 298, 308
    vol 36   40, 75, 160, 242, 256, 276, 334, 361, 363, 377, 403,
                    434, 435
Vriesenga, J.
    vol 23   E83, E96
Vulikh, A.I.
    vol 30   E105, E108, E111, 148, 149, E162, E164, E165, 173
Vybornova, Ya, I.
    vol 27/28   E303, E465, 475, 476, E511

W

Wadso, I.
    vol 20   E73-E77, E89-E91, E153-E157
    vol 37   E14, E15, E42, E43, E46, E65, E70, E84
    vol 38   E24, E28
Waerstad, K.R.
    vol 31   269, 270
Wagner, J.G.
    vol 35   105
Wahlgren, S.
    vol 34   155, 156
    vol 35   197, 199
Walch, W.
    vol 27/28   E737
Wald, F.
    vol 25   E83
Wald, M.
    vol 30   254
Waldeck, W.F.
    vol 30   E184-E187, 189
Waldmann, K.
    vol 26   E155, E159, E187, E189, E193, E236, E240, E241

Walfisz, A.
      vol 27/28   586
Walker, R.A.
      vol 25   E425-E427
Walker, R.D.
      vol 27/28   E1-E6, E79, E85, 160, 161
Wallace, R.E.
      vol 29   E87
Walls, H.A.
      vol 25   E385-E387, 400, E402-E404, 417
Walls, W.S.
      vol 27/28   763
Walter, E.B.
      vol 24   E130-E133, 136, 146, 152, 154
Walter-Levy, L.
      vol 23   E23, E26
Walton, J. H.
      vol 37   E63, E71, E72, E81, 95
Wang, J.Y.N.
      vol 25   E55, 56, E258, E259, 261, E263, E264, E301, E302,
                    E335-E337
Wang, L.C.
      vol 25   E425-E427
Wang, R.H.
      vol 27/28   E281, 288-290, 436-438
Wang, D.I.-J.
      vol 27/28   E266, 275
Warman, J.W.
      vol 21   E1-E3, E11, E12, 15, 23, 25, 26, 79
Wasik, S.P.
      vol 37   E65, E67, E68, E83, E84, 162, 174, E188, 189, 192
      vol 38   E59-E62, E66, 79, 90, E236-E238, 247, 249, E250, 257,
                    E311, E312, 317, E366, E367, 369-371, E386, 389, 390,
                    E396-E398, 404, E405, 407, E419, 422, E433, 435, 436,
                    438, 439, E441, 444, 445, E446-E448, 454, E456, 459,
                    E477, E478, 482, E483, 485, E486, E487, 491, 492
Watanabe.T.
      vol 30   E330, E331, E338, 372, E467, E470, E472
Watari, N.
      vol 34   E14, E15, 44, E204, 214
      vol 35   E64, 66, 67, E110-E112, 137, 139, E257,
                    259-261, 274
      vol 36   E124, 134-136, 171, E367, 369, E404, 406, 407, 409
Waters, J.A.
      vol 24   E218-E221, 230
Watters, J.I.
      vol 25   E335-E337, 343
Wauchope, R.D.
      vol 20   E62, E99, E100, 102, E103-E105, 109, 110
      vol 38   E236-E238, 244, E332-E334, 335, E3412-E343, 346, E386,
                    387, E396-E398, 401, E446-E448, 451
Weast, R.C.
      vol 20   E45
Weatherford, W.D.
      vol 27/28   570
Webster, D.C.
      vol 27/28   E303, 305, E320, 321, E465, 467, E511, 513
Weeks, J.R.
      vol 25   E258, E259, 262, E263, 265, E268, 270, E271, 273,
                    E274, 276, E277, 281, E301, E302, 303, E310, E311,
                    313, E317, E318, 323, E330, E425-E427
Wehe, A.H.
      vol 24   E39, 45
Weibke, F.
      vol 25   E44, E45, 49
Weidmann, H.
      vol 23   E157, E182
Weihs, G.
      vol 25   E66, E67
Weil, L.
      vol 20   E1, 2
Weil, S.A.
      vol 27/28   578, 579

Weiner, L.G.
    vol 25   E369-E371, 381
Weiner, P.H.
    vol 24   E130-E134, 184, 185, 187, 188
Weinstein, L.
    vol 34   E13, E14, 22, 126
    vol 35   E110, E111, 119, 174
Weisberg, J.
    vol 26   E187, E188, E192, 216
Weiss, P.
    vol 25   E84, E85, 88
Weiss, R.F.
    vol 27/28   E50, E51
Weith, W.
    vol 25   E13-E18
Welsh, T.W.B.
    vol 22   103
    vol 30   104, 219, 427
Weltman, C.
    vol 30   E24-E26, E30-E32, 74, 75
Wen, W.-Y.
    vol 24   E1, E23, E16, E17, 18, E60-E64, E71, 78, 79, 84-87,
                  89-94, 121
    vol 27/28   E1-E5, 15, 21, E57-E64, E85, 86, 90-92, 95-97, 99,
Wendrow, B.
    vol 31   E11-E15, 31-36, E98, E102, E127, E130
Wenger, L.M.
    vol 27/28   E24-E28, E57, E72-E74, E85, 130, 142
                  102, 186, 187
Werner, E.
    vol 20   E30, 31, E65, 66, E150, 151
West, D.H.
    vol 22   82, 83, E152, 153, E162, 163, 166, 169, 171
Westlake, T.N.
    vol 24   E130-E134, 184, 185, 187-188
Wetlaufer, D.B.
    vol 24   E1, E2, 10, E16, E17, 28, E34, 35, E61, E71, 81-83,
                  122-124
    vol 27/28   E1-E6, 10, E64, E85, 103, 177
Wetton, E.A.M.
    vol 31   E11, E13-E15, 22,
Whaley, D.K.
    vol 25   E268, E277, E310, E311, E317, E318
Wheeler, F.I.
    vol 30   E428, E429, E431, 432, E448-E450, 451
Wheeler, O.H.
    vol 21   E83, E84, E89, 179, 180
Whitby, C.S.
    vol 23   E86, E96
White, A.G.
    vol 25   E434
White, C.M.
    vol 25   E369-E371, 378
White, J.C.
    vol 25   E139-E141, 149
    vol 29   E213, E214, 223
White, R.C.
    vol 20   E153-E157, 161
Whitley, F.J.
    vol 27/28   567
Whitman, C.I.
    vol 25   E55
Whitworth, C.W.
    vol 34   258-261
    vol 35   200-202
Wichterle, I.
    vol 24   E196, E197, 202, 203
    vol 27/28   E248, E249, 259-261, E266, 277, 278
Wiebe, R.
    vol 24   E331-E333, 335-337, 340, 350-352, 363-365, 368
Wiers, B.
    vol 23   E144, E145, 146
Wiese, H.C.
    vol 27/28   E281, 291, 385-389, 404-406

Wiesenberg, D.A.
     vol 27/28   E50, E51
Wijen, A.J.M.
     vol 24   E39
Wilcock, R.J.
     vol 27/28   E202, E203, E206, E207, E215, 227, 231, E452-E454,
                  462, E582-E584, E592-E595, 625, 627, E724, 725
Wilf, J.
     vol 24   E16, E17, 24, E54, 56
     vol 27/28   E1-E5, 16, E45, E46, 47
     vol 37   E64, E65, E70, E83, 156, 186
     vol 38   E341, E343, 347, 351
Wilhelm, E.
     vol 27/28   E1-E5, 23, E202, E203, E206, E207, E215, 227, 231,
                  E450-E454, 457-462, E493-E497, 503, 504, 506, 507,
                  E582-E584, E590-E595, 625, 627, E685, 687, E724,
                  725
Wilkinson, J.H.
     vol 34   162, 163
     vol 35   195, 196
     vol 36   52, 53
Wilkinson, W.D.
     vol 25   E55, E263
Williams, B.
     vol 27/28   E303, 308, E314, 316, 317, 410, 411
Williams, D.J.
     vol 25   E194, E195
Williams, E.J.
     vol 25   E57, E58, 63
Williams, J.H.
     vol 34   E302, 303
     vol 35   11, E110, E111, 114, 219, E231, 232, E239,
                  240, 250, 251, 256, 317
     vol 36   E63, 66, E124, 125, 206, 207, E228, 229, E243, 244,
                  E269, 270
Williams, T.C.
     vol 25   E369-E371
Willis, W.D.
     vol 26   E152
Willson, H.S.
     vol 30   E330-E332, E335, E336, E338, 364, 365
Wilson, A.S.
     vol 25   E425-E427, 428
Wilson, G. M.
     vol 27/28   E248, E249
     vol 37   E65, E68, E72, E73, E76, E78, E79, E84,
                  181, 182
     vol 38   E59-E63, E65, E66, 88, 89, E104-E106, 108, 109
Windaus, A.
     vol 26   E236, 237, E240, E241, 242
Windhaber, K.
     vol 27/28   E465, 474
Windmaisser, F.
     vol 30   E24, E25, E30-E32, 49, 67
Wing, J.
     vol 20   E73-E77, 81, E132-E136, 142, E153-E157, 168, E202-E204,
     vol 20   207, E210-E213, 216
     vol 37   E64, E71, E82, 121
Winkler, J.
     vol 25   E134
Winkler, L.W.
     vol 27/28   E1-E5, 8, 9, E50, E51, E582, E590, E595, 614
Winnek, P.S.
     vol 34   E302, 303
     vol 35   114, 219, E231, 232, E239, 240, 250, 251,
Winnek, P.S.
     vol 35   1, 11-13, E14, 16, 104, 107, 109, E110, E111,
                  256, 316, 317
     vol 36   E19, E20, 23, 54, E63, 66, 76-80, 83-87, 96, 97, E124,
                  125, 206, 207, E228, 229, E243, 244, E269, 270, 279,
                  284, 296-301, 305, 366, 437, 438, 443
Winograd, N.
     vol 23   E144, E145

Winteler, F.
     vol 30   E28, E33, 51, 52, E105, E108, E109, E111, 125
Winter, H.
     vol 25   E1-E3, 4, E13, -E18, 23, E33, E34, 38, E44, E45, 46
Winterhagen, H.
     vol 25   E369-E371
Wishnia A.
     vol 24   E1, E2, 12, E16, E17, 31, E414, 421
     vol 37   E42, E46
Witteman, G.W.
     vol 25   E421
Wohler, L.
     vol 26   E4-E6, 32, 53
Wolf, W.
     vol 25   E1-E3, 7
Wolff, H.
     vol 21   E80, E84, E89, 92-97, 115, 116, 129, 130, 135-138, 171,
                   172, 181, 182, 225, 226, E234, 235-251, E252, 253-266,
                   268-273
Wolkoff, A.W.
     vol 38   E294, E296, 303, E426, E427, 431
Woo, L.
     vol 23   E84, E94, E96, 111
Wood, C.E.C.
     vol 20   E150, 152
Wood, J.
     vol 23   E158, E163, E182, 193
Wood, T.
     vol 24   E331-E333, 344, 348
Woodward, A.E.
     vol 23   E174-E176, E183, 235
Woolley, J.C.
     vol 25   E194, E195
Woolmer, R.N.
     vol 27/28   565
Woontner, L.
     vol 23   E270, E273-E276, E281, 301
Worley J. D.
     vol 37   E64, E83, 151
Worthinton, H.E.C.
     vol 36   E124, 132, 176-178, 202
Wrigge, W.
     vol 26   E348-E351, 379
Wright, R.
     vol 30   E105, E108, E111, E374, E375, 382, 411
Wu, D.
     vol 29   E48
Wu, Z.
     vol 27/28   97, E712, 718
Wunderlich, R.
     vol 30   E105, E111, 117, E220, E223, 225, E374,
                   E377, E380, 384
Wurtz, R.
     vol 21   E80-E84, E89, 115, 116, 129, 130, 137, 138, 181,
                   182, 225, 226, E234, 244-251, E252
Wurz, O.
     vol 26   E187, E188, E191, E192, 214, 233-235
Wyatt, D.K.
     vol 35   91, 100
Wymer, R.G.
     vol 25   E425-E427

Y

Yaacobi M.
     vol 24   E16, E17, 24, E54, 56
     vol 27/28   E1-E6, 16, 17, E45, E46, 47, E59, E68-E70, E74-E76,
                   E80, E84, E85, 89, 116, 127, 149, 151, 163, 171,
                   175, 176, 178, 180, 182, E450, E454, 456, E582-592
                   E595, 599, 604, 608, 612, 617, 621, E642-E644, 650
     vol 37   E64, E65, E70, E83, 156, 186
     vol 38   E341, E343, 347, 351
Yaggee, F.L.
     vol 25   E55-E263

Yakimets, E.M.
     vol 26   E153-E155, E158, 170
Yakobi, N.Y.
     vol 30   238, E263, E264, E374, E375, E378, E381,
                   396, E448-E450
Yalkowsky S. H.
     vol 20   E3, 4, 5, E15, 16, E17, 18, E19, 20, 21, E22, 23, E26,
                   27, E28, 29, E33, 34, E35, 38, E41, 42, E49, 50, E51, 52,
                   E53, 54, E55, 56, E57, 58, E59, 61, E67, 68, E69, 70, E71,
                   72, E73-E77, 87, E89-E91, E103-E105, 115, E118, 119, E120,
                   121, E122, 124, E125, 126, E127, 128, E129, 130, E132-
                   E136, 148, E184, 189
     vol 37   E65, E68, E84, 177
     vol 38   E1, E2, 8, E260, 263, E332, E334, 338, E341, E343, 350
Yamada, H.
     vol 35   26, 32, 35, 61
Yamada, T.
     vol 29   E1, E5, E7, E14, E16-E18, E21, 40, E47, E49, E50
                   E51, 72-86, E87, 88-94, 100, 101
Yamamoto, S.
     vol 27/28   E1-E6, 19, 20, E50, E51, 52-54
Yamane, T.
     vol 38   E464, E465, 469
Yamazaki, M.
     vol 34   E14, E15, 35, 75, 159, E204, 236, 262, 277-279
     vol 35   2, 5, 8, E14, 17, 28, 55, 65, 74, 93, 131,
                   158, 192
     vol 36   E19, E20, 30, 39, 47, 102, 107, 114, E124, 131, 159,
                   205, E243, 249, 255, 266, 285, 288, 291, 368, 376,
                   400, 405, 411, 427
Yan-Sho-Syan, G.V.
     vol 25   E139-E141, 154, E157-E159, 171, E182, E183,
                   189, E402-E404, 418
Yanagaze, R.
     vol 25   E103-E105, 108
Yanick, N.S.
     vol 30   E24-E26, E30, E32, 44-48, E105, E111, 123, 124
Yanitskii, I.V.
     vol 26   E303-E305, 311-313
Yankovich, V.N.
     vol 22   20, 37, 75, 76, 90, 91, 95
Yano, T.
     vol 24   E1, E2, 9, E59, E65-E71, 105, 111, 112, 114, 116
     vol 27/28   E1-E6, E70, E74, E79, E80, E85, 128, 164
Yaron, I.
     vol 31   E129, E130, 166, 167
Yaroshenko, L.B.
     vol 26   E1, E3-E6, 18, 63, E74-E75
Yarym-Agaev, N.L.
     vol 27/28   44, E629, 631
Yasterbova, L.F.
     vol 22   112, 113, 196, 304, 305
     vol 31   E129, E130, 140
Yasuda, M.
     vol 26   E113, E114, E116, 117
Yasue, T.
     vol 26   E187, E193
Yata, N.
     vol 34   E14, E15, 35, 42, 75, 159, 160, 200-203, E204,
                   209, 236, 262, 277-279
     vol 35   2, 5, 8, E14, 17, 18, 28, 55, 56, 59, 60, 65,
                   74, 93, 102, 103, E110-E112, 131, 135, 158,
                   192, 193, 217, 218, 247, 248
     vol 36   E19, E20, 30, 31, 47, 48, E88, 90, 91, E92, 94, 95,
                   102, 107, 114, E124, 131, 205, E243, 249, 255, 266,
                   285, 288, 291, 368, 376, 400, 405, 411, 427
Yatsenko, S.P.
     vol 25   E93-E95, 99
Yavorskii, V.T.
     vol 26   E74, E75, 86, 89
Yelkin, A.E.
     vol 23   E180, E181, E184, 261
Yevdokimova, G.V.
     vol 31   E127, E130, 151

Yokoyama, C.
    vol 27/28   222, 659
Yorizane, M.
    vol 27/28   688, 690
Yoshida, F.
    vol 38   E464, E465, 469
Yoshida, Z.
    vol 25   E330
Yoshimura, S.
    vol 27/28   688, 690
Yosim, S.J.
    vol 29   E207, E208, 211
Youness, T.
    vol 25   E335-E337, 343
Young, A.S.
    vol 25   E194, E195
Young, C.L.
    vol 37   E73, E85
    vol 38   E64, E66
Yousef, R.T.
    vol 34   250, 252, 254
    vol 35   80, 180-182
Yu, T.L.
    vol 25   E425-E427
Yugai, M.R.
    vol 31   E206, E212, 276, 277
Yusapova, E.N.
    vol 23   E11, E13, E25, 260

Z

Zabransky, Z.
    vol 23   E272, E281
Zachariasen, W.H.
    vol 30   E269, E274
Zadneprovsky, G.M.
    vol 23   E313, E317, 333
Zagrebina, L.A.
    vol 26   E1, E4-E6, 38, 59, 60
Zaichko, L.F.
    vol 25   E173, E174
Zakharov, M.S.
    vol 25   E93-E95, E135, E173, E174, E335-E337
Zakharov, Z.A.
    vol 25   E335-E337
Zakharova, E.A.
    vol 25   E173, E174, E335-E337, 351
Zakharova, M.I.
    vol 29   E214
Zanetti, R.
    vol 23   E144, E145
Zangen, M.
    vol 26   E187, E188, E192, 226-228, E236, 239, E240, E241, 246
Zapponi, P.P.
    vol 25   E157-E159, 167
Zaripova, L.G.
    vol 25   E194, E195
Zasosov, V.A.
    vol 36   328, 333
Zavaglio, V.
    vol 35   6
    vol 36   104, 332, 365, 441, 442
Zborilova, L.
    vol 31   E206, E207, E212, 248
Zebrava, A.I.
    vol 25   E335-E337, 347, E385-E387, 401, E402-E404, 419
Zebrava, A.I.
    vol 25   E1-E3, 11, E13-E18, 30, E33, E34, 43, E76, E77,
                 82, E119-E121, 132, E135, E139-E141, E156,
                 E157-E159, 170, E173, E174, 178, 180, E181, E183,
                 188, 193, E207, E208, 211, 213, E214, E215, 217-220,
                 E221, 223, 224, E225, 228, 229, E230, E231, 233, E234,
                 237, 238, E239, 241, E242, 243, E245, 246, E247, 248,
                 E249, 250, E251, 252, E256, 257, E285, E286, 295,

Zeck, S.
    vol 27/28   97, E712, 718
Zeininger, H.
    vol 27/28   761, 762
Zel´venski, Ya.D.
    vol 20   E153-E157
Zelikman, A.N.
    vol 31   E39-E42, E47, E48, E51, E127, E130, 151
Zellhoefer, G.F.
    vol 21   E80-E82, E89, 102, 151, E252, 267, 274-278
Zelvinsky, Y.D.
    vol 27/28   E629, 630
Zerbe, C.
    vol 27/28   580, 581
Zhadanovich, L.A.
    vol 24   E196, E197, 200, 201
    vol 27/28   E248, E249, 256, 257, 276
Zheldybaeva, b.
    vol 25   E207-E208, 211
Zherdienko, L.P.
    vol 30   E268, E271, E273, 312
Zhigur´yanova, S.A.
    vol 26   E267, 270
Zhikharev, M.I.
    vol 31   E127, E130, 154
Zhilina, T.A.
    vol 22   92, 106, 138, 181, 222, 240, 260, 279, 298, 319, 342,
                   354, 370, 379
Zholovskaya, R.M.
    vol 30   E268, E271, E273
Zhorzholiana, N.B.
    vol 35   210-212
    vol 36   117-119, 350, 351, 431-433
Zhou, Weifang
    vol 23   E314, E315, E317, 334
Zhumakanov, V.Z.
    vol 25   E1-E3, 11, E33, E34, E76, E77, 82, E139-E141,
                   156, E182, E183, 193
Zhuze, T.P.
    vol 24   E422, 423
    vol 27/28   574
Ziegel, G.
    vol 25   E84, E85, 92
Zielinski, A.Z.
    vol 20   E153-E157, 169
Zil´berman, Ya.I.
    vol 26   E1-E3, E5, E6, 48, 49, 66, 67, E74, E75, 87, 88,
                   E115, E116
Ziller, K.H.
    vol 35   140, 177
Zimmels, Y.
    vol 24   E69, E72, 113
Zimmerer, R.O., Jr
    vol 35   91, 100
Zimmerman, P.A.
    vol 22   192, 193, 194, 198, 200, 201
Zlomov, V.P.
    vol 25   E194, E195
Zmija   see Smija, J.
Zolotareva, M.N.
    vol 21   E9, E12, 66
Zolotukhin, V.D.
    vol 23   E181, E184
Zorin, A.D.
    vol 21   E281, E282, 296-299, E302, 304-307, E310, 311-314
                   E315, 316-319, 321-328
    vol 24   E257, E258, 262, 279, E299-E302, 308, 330, E331-E333,
                   341, 370, E375-E378, 387, 407
Zoroatskaya, I.V.
    vol 25   E194, E195
Zubchenko, Yu.P.
    vol 27/28   675, 679, 681, E712, 713, 714, 728

Zuccaro, P.
    vol 34   77, 308
    vol 35   145
    vol 36   33, 147
Zucker, D.
    vol 25   E55
Zueva, M.V.
    vol 21   E281, E282, 296-299, E302, 304-307, E310, 311-314,
                   E315, 316-319, 321-328
Zukovsky, G.J.
    vol 25   E1-E3, 6, E44, E45, 47, E50, E51, 52
Zutic, V.
    vol 25   E317, E318, E325
Zvagolskya, E.V.
    vol 25   E93-E95, 96
Zvjagincev, O.E.
    vol 25   E13-E18
Zvorykin, A.J.
    vol 31   E206-E209, E212, 230, 231
Zvorykin, A.Y.
    vol 31   E307-E309, E311, 313, E318-E321, 324, 326
Zwietasch, K.J.
    vol 22   11, 29, 31, E32, 33, 36, 47, 89, 94, 103, 132, 141,
                   143, 176, 191, 197, 218, 227, 235, 237, 255, 256,
                   264, 267, 274, 275, 282, 293, 294, 302, 313, 314,
                   333, E334-E336, 337, 338, 346, 351, 365, 367, 374, 376
Zyuzina, L.F.
    vol 31   E51, E52, 90, 91, E206-E209, E211, E212, 255, 256

SOLUBILITY DATA SERIES

Volume 1	H. L. Clever, *Helium and Neon*
Volume 2	H. L. Clever, *Krypton, Xenon and Radon*
Volume 3	M. Salomon, *Silver Azide, Cyanide, Cyanamides, Cyanate, Selenocyanate and Thiocyanate*
Volume 4	H. L. Clever, *Argon*
Volume 5/6	C. L. Young, *Hydrogen and Deuterium*
Volume 7	R. Battino, *Oxygen and Ozone*
Volume 8	C. L. Young, *Oxides of Nitrogen*
Volume 9	W. Hayduk, *Ethane*
Volume 10	R. Battino, *Nitrogen and Air*
Volume 11	B. Scrosati and C. A. Vincent, *Alkali Metal, Alkaline Earth Metal and Ammonium Halides. Amide Solvents*
Volume 12	C. L. Young, *Sulfur Dioxide, Chlorine, Fluorine and Chlorine Oxides*
Volume 13	S. Siekierski, T. Mioduski and M. Salomon, *Scandium, Yttrium, Lanthanum and Lanthanide Nitrates*
Volume 14	H. Miyamoto, M. Salomon and H. L. Clever, *Alkaline Earth Metal Halates*
Volume 15	A. F. M. Barton, *Alcohols with Water*
Volume 16/17	E. Tomlinson and A. Regosz, *Antibiotics: I. β-Lactam Antibiotics*
Volume 18	O. Popovych, *Tetraphenylborates*
Volume 19	C. L. Young, *Cumulative Index: Volumes 1–18*
Volume 20	A. L. Horvath and F. W. Getzen, *Halogenated Benzenes, Toluenes and Phenols with Water*
Volume 21	C. L. Young and P. G. T. Fogg, *Ammonia, Amines, Phosphine, Arsine, Stibine, Silane, Germane and Stannane in Organic Solvents*
Volume 22	T. Mioduski and M. Salomon, *Scandium, Yttrium, Lanthanum and Lanthanide Halides in Nonaqueous Solvents*
Volume 23	T. P. Dirkse, *Copper, Silver, Gold, and Zinc, Cadmium, Mercury Oxides and Hydroxides*
Volume 24	W. Hayduk, *Propane, Butane and 2-Methylpropane*
Volume 25	C. Hirayama, Z. Galus and C. Guminski, *Metals in Mercury*
Volume 26	M. R. Masson, H. D. Lutz and B. Engelen, *Sulfites, Selenites and Tellurites*
Volume 27/28	H. L. Clever and C. L. Young, *Methane*
Volume 29	H. L. Clever, *Mercury in Liquids, Compressed Gases, Molten Salts and Other Elements*
Volume 30	H. Miyamoto and M. Salomon, *Alkali Metal Halates, Ammonium Iodate and Iodic Acid*
Volume 31	J. Eysseltová and T. P. Dirkse, *Alkali Metal Orthophosphates*
Volume 32	P. G. T. Fogg and C. L. Young, *Hydrogen Sulfide, Deuterium Sulfide and Hydrogen Selenide*
Volume 33	P. Franzosini, *Molten Alkali Metal Alkanoates*
Volume 34	A. N. Paruta and R. Piekos, *4-Aminobenzenesulfonamides. Part I: Non-cyclic Substituents*
Volume 35	A. N. Paruta and R. Piekos, *4-Aminobenzenesulfonamides. Part II: 5-Membered Heterocyclic Substituents*
Volume 36	A. N. Paruta and R. Piekos, *4-Aminobenzenesulfonamides. Part III: 6-Membered Heterocyclic Substituents and Miscellaneous Systems*
Volume 37	D. G. Shaw, *Hydrocarbons with Water and Seawater. Part I: Hydrocarbons C_5 to C_7*
Volume 38	D. G. Shaw, *Hydrocarbons with Water and Seawater. Part II: Hydrocarbons C_8 to C_{36}*
Volume 39	C. L. Young, *Cumulative Index: Volumes 20–38*
Volume 40	J. Hala, *Halides, Oxyhalides and Salts of Halogen Complexes of Titanium, Zirconium, Hafnium, Vanadium, Niobium and Tantalum*
Volume 41	C.-Y. Chan, I. N. Lepeshkov and K. H. Khoo, *Alkaline Earth Metal Perchlorates*